Acute Respiratory Distress Syndrome

Cellular and Molecular Mechanisms
and Clinical Management

NATO ASI Series

Advanced Science Institutes Series

A series presenting the results of activities sponsored by the NATO Science Committee, which aims at the dissemination of advanced scientific and technological knowledge, with a view to strengthening links between scientific communities.

The series is published by an international board of publishers in conjunction with the NATO Scientific Affairs Division

A	**Life Sciences**	Plenum Publishing Corporation
B	**Physics**	New York and London
C	**Mathematical**	Kluwer Academic Publishers
	and Physical Sciences	Dordrecht, Boston, and London
D	**Behavioral and Social Sciences**	
E	**Applied Sciences**	
F	**Computer and Systems Sciences**	Springer-Verlag
G	**Ecological Sciences**	Berlin, Heidelberg, New York, London,
H	**Cell Biology**	Paris, Tokyo, Hong Kong, and Barcelona
I	**Global Environmental Change**	

PARTNERSHIP SUB-SERIES

1. Disarmament Technologies	Kluwer Academic Publishers
2. Environment	Springer-Verlag
3. High Technology	Kluwer Academic Publishers
4. Science and Technology Policy	Kluwer Academic Publishers
5. Computer Networking	Kluwer Academic Publishers

The Partnership Sub-Series incorporates activities undertaken in collaboration with NATO's Cooperation Partners, the countries of the CIS and Central and Eastern Europe, in Priority Areas of concern to those countries.

Recent Volumes in this Series:

Volume 295 — Prions and Brain Diseases in Animals and Humans
    edited by Douglas R. O. Morrison

Volume 296 — Free Radicals, Oxidative Stress, and Antioxidants: Pathological and
    Physiological Significance
    edited by Tomris Özben

Volume 297 — Acute Respiratory Distress Syndrome: Cellular and Molecular
    Mechanisms and Clinical Management
    edited by Sadis Matalon and Jacob Iasha Sznajder

Volume 298 — Angiogenesis: Models, Modulators, and Clinical Applications
    edited by Michael E. Maragoudakis

Series A: Life Sciences

Acute Respiratory Distress Syndrome

Cellular and Molecular Mechanisms and Clinical Management

Edited by

Sadis Matalon

The University of Alabama at Birmingham
Birmingham, Alabama

and

Jacob Iasha Sznajder

University of Illinois at Chicago
Chicago, Illinois

Springer Science+Business Media, LLC

Proceedings of a NATO Advanced Study Institute on
Acute Respiratory Distress Syndrome: Cellular and Molecular Mechanisms and
Clinical Management,
held June 15 – 25, 1997,
in Corfu, Greece

NATO-PCO-DATA BASE

The electronic index to the NATO ASI Series provides full bibliographical references (with keywords and/or abstracts) to about 50,000 contributions from international scientists published in all sections of the NATO ASI Series. Access to the NATO-PCO-DATA BASE is possible via a CD-ROM "NATO Science and Technology Disk" with user-friendly retrieval software in English, French, and German (©WTV GmbH and DATAWARE Technologies, Inc. 1989). The CD-ROM contains the AGARD Aerospace Database.

The CD-ROM can be ordered through any member of the Board of Publishers or through NATO-PCO, Overijse, Belgium.

Library of Congress Cataloging-in-Publication Data

```
Acute respiratory distress syndrome : cellular and molecular
mechanisms and clinical management / edited by Sadis Matalon and
Jacob Iasha Sznajder.
      p.   cm. -- (NATO ASI series. Series A, Life sciences ; v.
297)
    "Proceedings of a NATO Advanced Study Institute on Acute
Respiratory Distress Syndrome: Cellular and Molecular Mechanisms and
Clinical Management, held June 15-25, 1997, in Corfu, Greece"--T.p.
verso.
    Includes bibliographical references and index.
    ISBN 978-1-4613-4653-1     ISBN 978-1-4419-8634-4 (eBook)
    DOI 10.1007/978-1-4419-8634-4
    1. Respiratory distress syndrome, Adult--Pathophysiology-
-Congresses.  2. Respiratory distress syndrome, Adult--Treatment-
-Congresses.   I. Matalon, Sadis.   II. Sznajder, Jacob Iasha.
III. NATO Advanced Study Institute on Acute Respiratory Distress
Syndrome: Cellular and Molecular Mechanisms and Clinical Management
(1997 : Kerkyra, Greece)  IV. Series.
    [DNLM: 1. Respiratory Distress Syndrome, Adult--physiopathology
congresses.  2. Respiratory Distress Syndrome, Adult--drug therapy
congresses.   WF 140 A18347 1998]
RC776.R38A2798  1998
616.2'4--dc21
DNLM/DLC
for Library of Congress                             98-17227
                                                        CIP
```

ISBN 978-1-4613-4653-1

© 1998 Springer Science+Business Media New York
Originally published by Plenum Press, New York in 1998
Softcover reprint of the hardcover 1st edition 1998

http://www.plenum.com

10 9 8 7 6 5 4 3 2 1

FOREWORD

Acute lung injury, respiratory failure, and acute respiratory distress syndrome associated with sepsis and multiorgan dysfunction is becoming more common in hospitals throughout the world. Although new classes of antimicrobial medications have been introduced and aggressive intensive life support systems are in use in many hospitals, mortality remains in the range of 35 to 65 percent in even the best equipped medical facilities. However, with new technologies in physiology and cellular and molecular biology, wonderful opportunities now exist for scientists to explore more effective approaches to identification of patients at risk, to develop better insights to pathogenic mechanisms, to use new tools to follow the progression and natural history of disease, and to develop better modes of therapies.

The NATO Advanced Study Institute on *Acute Respiratory Distress Syndrome: Cellular and Molecular Mechanisms and Clinical Management* was an outstanding opportunity for laboratory scientists and clinical investigators to discuss their own research approaches, explore opportunities for bringing new dimensions to their laboratories, and work together to identify possible clinical applications. The conference brought together investigators who have worked with patients of all ages, including those who have so successfully treated infants with respiratory distress syndrome. There is no doubt that this stimulating conference will bring new dimensions, new technologies, and new investigators into the scientific laboratories of each participant. This provides hope that new therapeutic approaches will eventually be available.

It was a special privilege that the NATO organizers included representatives from the National Heart, Lung, and Blood Institute, National Institutes of Health. We were most impressed with the presentations and the scientific interactions that took place between the attendees.

Claude Lenfant, M.D.
Director, National Heart, Lung, and Blood
   Institute
National Heart, Lung, and Blood Institute
National Institutes of Health
Bethesda, Maryland

Suzanne S. Hurd, Ph.D.
Director, Division of Lung Diseases
National Institutes of Health
Bethesda, Maryland

PREFACE

This monograph contains the contributions of invited speakers and participants at the NATO Advanced Study Institute on *Acute Respiratory Distress Syndrome: Cellular and Molecular Mechanisms and Clinical Management,* held in Corfu, Greece from June 15–25, 1997. The Advanced Study Institute provided a unique opportunity for internationally known basic scientists and clinicians with expertise in various aspects of lung injury and repair to exchange ideas on the pathophysiology and treatment of Acute Lung Injury. There were 25 invited speakers and 65 students representing 14 countries. Some of the most notable contributions of this symposium are outlined below:

Role of Nitric Oxide ('NO) in the treatment of Acute Respiratory Distress Syndrome (ARDS). Currently, there is considerable controversy as to whether nitric oxide should be administered to patients with ARDS. There was a lively debate on the use of 'NO in the treatment of ARDS with extensive audience participation. Several studies, mostly from European centers, summarized data on the use of nitric oxide ('NO) in the treatment of ARDS. Although the earlier reports were encouraging, more recent data from several centers show that administration of 'NO did not improve morbidity or mortality of patients with ARDS. There was concern with the the fact that cessation of 'NO treatment led to dramatic increases in the pulmonary circulation pressure, thus making it difficult to wean patients from this type of therapy. Furthermore, a number of presentations from basic scientists pointed out that reactive oxygen-nitrogen intermediates, produced by the reaction of 'NO with oxygen free radicals, released by inflammatory cells, may damage the pulmonary surfactant system and alveolar type II cells. It was agreed that although a short course of 'NO administration may prove beneficial in selective patients by improving oxygenation, its long-term use should be discouraged without additional data providing definitive evidence for decreased mortality and morbidity and lack of toxicity. On the other hand, the use of prostacyclin that also has pulmonary vascular dilation effects appears to show beneficial effects that warrant further investigation in this area.

Importance of active ion transport across the alveolar epithelium in enhancing the clearance of alveolar edema in ARDS. A number of exciting presentations focused on the cellular and molecular mechanisms of ion transport across the alveolar epithelium. Discussions helped to reconcile existing controversies on the contribution of various transporters in fluid clearance across the alveolar epithelium of animals with oxidant lung injury. One of the most exciting aspects of this meeting was the presentation of data indicating that up-regulation of ion transport across the alveolar epithelium of patients with ARDS and in animals with ARDS-type injury, using inotropic drugs such as isoproterenol, dobutamine and dopamine, decreased the amount of pulmonary edema and improved oxygenation. This is an exciting development and may have a significant impact in the treatment of ARDS.

Drug delivery to the injured alveolar epithelium. There is considerable interest in identifying agents that can improve the delivery of various drugs (such as antioxidant enzymes, anti-inflammatory compounds, etc.) to the alveolar epithelium of patients with ARDS. It has been known that surfactant deficiency contributes to the development of ARDS and that surfactant replacement has been efficacious in improving lung mechanics and oxygenation in both animals and patients with ARDS. Exciting new evidence from a number of laboratories indicate that surfactant delivery mixtures, already used for the correction of surfactant deficiency in neonates, greatly improve the delivery of an antioxidant enzyme to the alveolar epithelium of newborn animals. Also, there were a number of important presentations discussing strategies for gene delivery to the alveolar epithelium. Evidence was presented that a new family of polycationic amino polymer compounds is extremely efficient in transferring genes to a number of cell lines. There was considerable interest in testing the efficacy of these compounds in animals with lung injury resembling ARDS. These compounds may circumvent some of the problems associated with adenoviruses, such as lung inflammation, and may render the lung more resistant to oxidant injury.

Controversies in the clinical management of patients with ARDS. There was a lively debate on the advantages and disadvantages of ventilating patients with low versus high tidal volume. Large tidal volumes appear to compound the pre-existing lung injury, causing increased permeability and pulmonary edema. In addition, there were exciting presentations on ventilating both animals and patients with ARDS using perfluorocarbons (liquid breathing). There were several suggestions for additional experiments to better demonstrate the efficacy of this potentially exciting form of treatment in ARDS. Plans were developed for an industry-sponsored multicenter trial to systematically explore the ability to partially ventilate ARDS patients with perfluorocarbons and determine the efficacy of this form of ventilation by measuring physiologic variables and changes in mortality. In addition, studies from several investigators summarized novel approaches to mechanical ventilation of patients with ARDS among which tracheal gas insufflation and prone positioning are being increasingly utilized with encouraging results. Several presentations also dealt with the potentially damaging effects of excessive tidal volume ventilation on the lungs.

Perhaps the best aspects of this symposium were the lively interactions among the participants and the fact that several new collaborations were established among scientists and clinicians who previously knew each other only by reputation.

The organizers would like to take this opportunity to thank the members of the local organizing committee, Drs. Behrakis, Baltopoulos, and Chainis for all their hard work and encouragement. We would also like to thank His Excellency, the Governor of Corfu, Mr. Andreas Pagrates, His Excellency, the Mayor of Corfu, Mr. Chrysanthos Sarles, His Holiness, the Metropolitan of Corfu, Mr. Timotheo, the President of the General Hospital of Corfu, Dr. Kostas Alexandropoulos, The Town Council of Corfu, and the citizens of Corfu for their generosity and hospitality. Special thanks to Mr. Spyros Lychnos for working day and night to ensure that all participants arrived in Corfu safe and sound and for handling the myriad administrative details. Finally, we would like to thank Ms. Nancy Matalon and Ms. Mary Beth Campbell for their superb editorial assistance in putting this volume together.

We want to offer special thanks to our corporate sponsors listed below. Their generous contributions helped make the conference a success and provided support for additional activities.

- Alliance Pharmaceutical Corporation, San Diego, CA, USA
- AmGen, Inc., Thousand Oaks, CA, USA
- AmSouth Bank, Birmingham, AL, USA
- Avanti Polar Lipids, Inc., Alabaster, AL, USA
- Biotechnology General Corporation, Iselin, NJ, USA
- Byk Gulden, Lomberg Chemische Fabrik GmbH, Germany
- Ciba-Geiga, Anthousaa, Greece
- General Hospital of Corfu, Greece

- Glaxo-Wellcome, Halandri, Greece
- Eli Lilly Pharmaceuticals, Athens, Greece
- Hewlett Packard, Andover, MA, USA
- Hoechst Marion Roussel, Bridgewater, NJ, USA
- Rhone Poulence Rorer, Athens, Greece
- Ross Products Division, Abbott Laboratories, Columbus, OH, USA
- SmithKline Beecham Pharmaceuticals, Lisle, IL, USA
- Wyeth Hellas E.P.E., Argyroupolis, Greece

Dr. Sadis Matalon
Professor of Anesthesiology, Physiology
  and Biophysics and Pediatrics
University of Alabama at Birmingham
Birmingham, Alabama

Dr. Jacob Iasha Sznajder
Professor of Medicine Pulmonary and
  Critical Care Medicine
University of Illinois at Chicago and
  Columbia Michael Reese Hospital
Chicago, Illinois

CONTENTS

SECTION 1: SODIUM (Na⁺) TRANSPORT AND FLUID CLEARANCE ACROSS THE NORMAL AND INJURED LUNGS

** indicates short communication from participants.

SECTION 2: ROLE OF THE PULMONARY SURFACTANT IN THE PATHOGENESIS AND TREATMENT OF ARDS

SECTION 3: MECHANISMS AND MODIFICATION OF ACUTE AND CHRONIC LUNG INJURY

SECTION 5: NITRIC OXIDE: FRIEND AND FOE

SECTION 6: HEALTH CARE LEGISLATION

REGULATION OF THE SODIUM PUMP IN HYPEROXIC LUNG INJURY

David H. Ingbar, Joseph M. Lasnier, O. Douglas Wangensteen, and Christine H. Wendt

Departments of Medicine, Physiology and Pediatrics
University of Minnesota School of Medicine
Minneapolis, MN 55455

INTRODUCTION

A major component of the early lesion in the adult respiratory distress syndrome (ARDS) is alveolar edema. Considerable data suggests that decreasing alveolar flooding improves the outcome from acute lung injury (ALI). Studies in humans with ARDS demonstrate that prognosis correlates with the capacity to resorb fluid (Matthay 1990). Retrospective studies indicate that patients with fluid balance out>in, lesser weight gain, lower pulmonary capillary wedge pressures or diuretic treatment have greater survival rates. It is not surprising that decreasing the degree of early alveolar flood could decrease the subsequent need for high pressure ventilation or high levels of oxygen - thus avoiding amplification of the injury.

Active resorption of alveolar sodium and fluid is critical for edema resolution and avoiding further amplification of lung injury (O'Brodovich 1995, Matthay 1996). It also is critical for the normal transition to postnatal breathing (O'Brodovich 1991). In the normal human and rodent lung many substances augment alveolar sodium and fluid resorption. Oxidant stress is a major mechanism of damage during ALI and this stress persists due to the neutrophilic inflammation and use of high levels of inspired oxygen in many patients. Oxidative injury can either inhibit or augment active sodium transport, depending on a number of variables (Ingbar 1997b, 1998). Recently we demonstrated that even the injured alveolar epithelium is capable of active fluid resorption and this can be stimulated by β-adrenergic agonists, such as terbutaline (Lasnier). This chapter examines the effects of oxidants on one component of the sodium resorption system, the sodium pump - Na,K-ATPase; the sodium channel is the other critical component of this system.

Clearance of sodium and water from the alveolar space is primarily mediated by a combination of an apical amiloride-sensitive sodium channel (Matalon 1996) and the basolateral sodium pump, Na,K-ATPase (Matthay 1996). The importance of this pathway is supported by inhibitor and ion substitution studies in living animals, isolated perfused lungs and cultured type II alveolar epithelial cell monolayers (Matalon 1991, Saumon). We have examined the mechanisms by which oxidant stress affects alveolar epithelial solute and fluid resorption, and particularly on the sodium pump, the major

Acute Respiratory Distress Syndrome: Cellular and Molecular Mechanisms and Clinical Management
Edited by Matalon and Sznajder, Springer Science+Business Media New York, 1998

1

basolateral pathway for sodium transport.

Na,K-ATPase STRUCTURE

Na,K-ATPase, is a ubiquitous heterodimeric transmembrane protein composed of $\alpha 1$ and $\beta 1$ subunits in a 1:1 ratio (Horisberger). The $\alpha 1$ subunit is a polypeptide of approximately 96-112 kD with 7 membrane spanning domains that catalyzes the movement of Na+ and K+, is phosphorylated by ATP and binds the specific inhibitor ouabain (Skou). It has 3 mRNA isoforms whose expression varies in different tissues and in development (Jewell). Alpha-1 is the predominant form in epithelial tissues (Sweadner) including ATII cells (Nici, Schneeberger), but cultured type I-like cells may express $\alpha 2$ (Ridge). The β subunit has 3 isoforms and is a glycosylated polypeptide with a peptide core of ~35kD. It targets the assembled molecule to the plasma membrane (Geering) through the C-terminal extracellular domain of the β subunit. The β subunit also is necessary for K-dependent reactions (McDonough 1990a). The $\beta 1$ isoform is ubiquitous, whereas $\beta 2$ is in brain only.

Na,K-ATPase REGULATION

The Na,K-ATPase is regulated at multiple levels and by short and long term mechanisms (Bertorello 1995 a,b), as befits an enzyme that consumes up to 30% of cellular ATP. Regulation of the sodium pump can be achieved rapidly by altering the activity of the enzyme or by moving it to and from the plasma membrane. Pump activity is rapidly increased by lower K^+ or higher Na^+ intracellular ion concentrations, and by hormones such as glucocorticoids (GCs), mineralocorticoids, β-adrenergic agonists and thyroid hormone (McDonough 1990b). Beta agonists increase sodium resorption and Na,K-ATPase activity in alveolar type II (ATII) cells (Suzuki). In some cell types, Na,K-ATPase activity can be inhibited by dopamine, endothelin or PGE2 and stimulated by vasopressin, or growth factors (Berterello). Hypoxia decreased ATII cell sodium pump activity, presumably through impaired calcium homeostasis (Planes).

Post-transcriptional regulation is important at multiple levels (Gick). Unequal ratios of α and β mRNA and protein concentrations are present in many tissues (Lingrel), even though the functional sodium pump enzyme has 1:1 stoichiometry. The mRNA degradation rates can change with cell differentiation and this may contribute to changing ratios of the subunits (Chambers). Translation is regulated and this probably accounts for some of the differences in the ratios of subunit mRNAs and proteins. The mRNA for $\alpha 1$ is translated less efficiently than $\beta 1$ due to the $\alpha 1$ mRNA's 3' untranslated region (UTR) which is extremely G-C rich (Devarajan). Once the subunits are translated, they can not leave the endoplasmic reticulum without assembly into α-β heterodimers (Geering), glycosylation and disulfide bond formation (Beggah). After assembly, they undergo an ill-defined maturation and then are transported into the cytoplasmic membrane over 1-4 hours to generate active pump that can bind ouabain (Caplan, Mircheff). There is an inactive, endogenous pool of Na,K-ATPase that that recycles from plasma membrane sodium pump. Movement of pump molecules between this intracellular pool and the plasma membrane can be stimulated, as by insulin in muscle, but the regulatory factors are not well understood. Mechanical ventilation or stretch may recruit Na,K-ATPase in ATII cells from this pool. The α subunit can be phosphorylated at several amino acids by either protein kinases C or A (Feschenko, Beguin, but the functional consequences vary depending on the cell type, the particular kinase and the stimulant (Berterello). Normally the pump molecules in the plasma membrane do not operate at peak capacity, providing an important mechanism for short term regulation. Hormones such as glucocorticoids (GCs), thyroid hormone, aldosterone and catecholamines stimulate both short term and long term (transcriptional) increases in pump function (Ewart). Degradation rates of the two subunits also can be differentially affected, changing the balance of alpha and beta subunits (Lescale-Matys). The wide repertoire of potential regulatory

2

mechanisms makes studying the response of the Na,K-ATPase to injury much more complex.

TRANSCRIPTIONAL REGULATION OF Na,K-ATPase

Transcriptional regulation is common (reviewed in Lingrel), and likely plays a role during lung development (Ingbar 1996). Pump transcription in cultured cells usually is upregulated by increased intracellular Na or lower K levels (McDonough1990b) and hormones such as GCs, aldosterone and T3 (Horisberger). The Na,K-ATPase isoforms belong to a multigene family and they are expressed in a tissue-specific and developmentally regulated manner (reviewed in Lingrel). Since the subunit genes are on different chromosomes, transcription may be independently regulated. Both subunits have >70% homology between isoforms of each subunit and there is significant interspecies homology of the Na,K-ATPase coding and promoter sequences. For example, rat and sheep have 97% homology of the $\alpha 1$ isoform.

For the rat $\alpha 1$ subunit, the major transcription initiation site is 262 base pairs upstream from the translation initiation site and is preceded by a TATA box at position -32 (Kawakami 1993, 1996). The factors regulating $\alpha 1$ gene transcription are partially defined. Included in the 5' flanking region are 2 Sp1 TF binding sites, two putative GC response element (GRE) half consensus sequences (Yagawa 1990) and a consensus cAMP response element (CRE). The human gene has a TATA box and 5 Sp1-like elements (Shull 1990). Two positive regulatory regions in the rat $\alpha 1$ subunit promoter are (1) at -155 and -49 bp and (2) an "Atp1α1" regulatory element or "ARE" at -94 to -69 bp that binds both common (ARE C1 and C2) and cell type-specific transcription factors (ARE C3) (Yagawa 1992). This sequence is conserved across rat, horse and human genes, but is not present in other Na,K-ATPase subunit promoters. At least 7 ARE-binding factors have been identified by gel mobility shift assays and southwestern cloning (Watanabe). One stimulatory transcription factor (TF), AREC3, is a homologue of the Drosophila sine oculis (Six) gene product that is required for visual system development, but is distinct from other homeodomain proteins. It is present at low levels in adult mouse lung, but otherwise has not been assessed in lung tissue. AREB6 is a zinc finger protein that can regulate Atp1α1 positively or negatively, depending on cell type and conformational state of the protein. It is a 5' extended version of Nil-2, a negative regulator of the interleukin-2 gene. Across multiple cell lines, activating transcription factor-1 (ATF-1)/ CRE binding protein (CREB) heterodimers bound to the -61/-72 region of the $\alpha 1$ promoter and were required for transcription (Kobayashi). Although these TFs are ubiquitous and probably are required for basal transcription of many genes, they can be regulated by cAMP, Ca, and transforming growth factor-α (TGF-α). Recent in vivo footprinting found that rat tissue-specific transcriptional control involved cooperation between a canonical GC box at -57/-49 that binds an Sp1-like TF and the ATF region (-72/-65) (Nomoto). There also were DNAse I hypersensitive sites at -83/-80 in liver and kidney, but not brain. HeLa cells treated with aldosterone after mineralocorticoid receptor transfection had less cAMP-inducible $\alpha 1$ gene transcription of the -103/-57 promoter region (Ahmad), but little is known of mineralocorticoid regulation. The human $\alpha 1$ gene has three potential thyroid hormone response elements, but only one is functional (Feng). In summary a number of potential TFs have been identified, but their presence and function in the lung are virtually undefined.

Much less is known about transcriptional regulation of the $\beta 1$ subunit gene. The rat gene's promoter contains a potential TATA box at position -31, four GC rich boxes, a CRE-like sequence and two sites with putative half consensus thyroid response element sequences TRE-1 and TRE-2 (Liu). The human gene consists of six exons and has two major and three minor transcription initiation sites.

LUNG ION TRANSPORT AND Na,K-ATPase

Although present to some extent in all cells, the sodium pump is found in high density on the basolateral membranes of epithelial cells specialized for Na transport. Net vectoral ion transport across the alveolus resorbs fluid and maintains a dry alveolus. The joint importance of Na,K-ATPase and the sodium channel in salt and water resorption across the alveolar epithelium is supported by studies of intact animals, intact lungs and cultured cells (Saumon, Matalon 1991, Matthay 1996).

The ion transport properties of adult type II cells recently were reviewed (Matalon 1991). Sodium may enter the apical membrane by at least 3 mechanisms: Na channels, Na-glucose cotransport or Na-H+ exchange, but Na channels are quantitatively predominant, at least in adult cells. There likely are several different types of Na channels in alveolar epithelial cells (AECs) (Matalon 1996), since the amiloride affinity and conductance properties differ between freshly isolated and cultured ATII cells. In alveoli, this ubiquitous enzyme is detected primarily on the basolateral surface of type II cells (Nici, Schneeberger), but has not been detected on type I AECs in vivo. One group detected increases in $\alpha2$ isoform and decreases in $\alpha1$ with time that ATII cells are in culture, suggesting that ATI cells may have $\alpha2$ (Ridge), but this has not been established in vivo. The beta subunit has not yet been localized in lung. The density of sodium pump on type II cells supports their importance in maintaining a dry alveolus, although type I cells also may contribute to fluid clearance. The Na,K-ATPase is the primary sodium exit pathway via the basolateral membrane. Both sodium pump and channel are needed for Na transport, but it is not clear which is limiting under normal or pathologic circumstances - especially since their activityusually changes in parallel. In addition to clearing alveoli of fluid, the low intracellular sodium concentration maintained by Na,K-ATPase allows sodium-coupled co-transport of amino acids, glucose and other vital substrates into ATII cells.

REGULATION OF LUNG SODIUM TRANSPORT AND RESPONSE TO INJURY

Many factors can stimulate sodium and fluid resorption in the normal lung (reviewed in Matthay 1996), including catecholamines, dobutamine, dopamine, keratinocyte growth factor (KGF), aldosterone, hepatocyte growth factor (HGF), transforming growth factor (TGFα), cAMP, endotoxin, Pseudomonas exotoxin A, and tumor necrosis factor-α (TNF-α). It is not clear if these factors act directly or through indirect mechanisms and what transporters they influence. Whenever tested, inhibitors of sodium channel or Na,K-ATPase significantly reduced alveolar fluid resorption. In cultured ATII cells, catecholamines and epidermal growth factor (EGF) (Borok) augmented transepithelial Na transport and terbutaline stimulated Na,K-ATPase activity independent of changes in intracellular sodium concentration (Suzuki). Important recent studies demonstrated that β-adrenergic agonists increased alveolar fluid clearance several fold in ex vivo human and rat lungs (Sakuma).

The impact of lung injury on alveolar fluid clearance and sodium transport has been examined in several injury models (Matthay 1996). Sepsis, pneumonia and endotoxin augmented clearance, in part through increased TNFα. Endotoxin-stimulated macrophages impaired lung epithelial transport (Compeau) through a nitric oxide (NO)-dependent mechanism. Intravascular and intratracheal dibutyryl cAMP attenuated edema formation in a phosgene-induced lung injury through activation of prostaglandins. In permeability pulmonary edema induced by napthyl-thiourea, there is increased sodium pump acitivity and alveolar solute resorption during recovery (Zuege).

Na,K-ATPase AND REACTIVE OXIDANT SPECIES

The effects of oxidant stress on Na,K-ATPase and alveolar fluid resorption have varied in different models studied. This also has been demonstrated in many different cell types from other organs (Ingbar 1997). Severe oxidant stress may inhibit general protein synthesis, inhibit sodium pump function or increase protein degradation. Recently we reviewed the multiple reported effects of oxidants on Na,K-ATPase function in different systems (Ingbar 1997). The studies of oxidant effects by other groups will first be reviewed and then the results of our studies on the lung Na,K-ATPase

4

will be summarized.

Acute severe hyperoxia, an injury with little type II cell proliferation, has variable effects on fluid resorption depending on severity of injury. Moderate injury has normal or increased resorption (Zheng, Garat, Carter 1997b), but severe injury can decrease clearance (Olivera 1995). In spite of increased permeability, the lung can still resorb fluid actively (Lasnier, Garat). Short periods of hyperbaric hyperoxia increase rat lung Na,K-ATPase mRNA levels (Harris). KGF is a growth factor for ATII cells in vivo and protects lungs from HCl, hyperoxic, bleomycin and radiation-induced injuries (Panos). The mechanism of KGF protection is not clear - it may be due to type II cell proliferation, increased surfactant decreased apoptosis or, in part, increased fluid clearance.

Less severe or more subacute injury can stimulate type II cell proliferation and may increase resorption through this mechanism. For example, chronic hyperoxia causes ATII proliferation with increased Na channels (Yue 1995) and Na,K-ATPase (Olivera 1994), but fluid clearance increased only in some rats (Yue 1997).

Multiple methods of sodium pump inhibition by oxidants are possible (Boldyrev), including direct protein oxidation (such as sulfhydryl oxidation), peroxidation of surrounding membrane lipid, depletion of intracellular energy stores, uncoupling of energy utilization and ion transport (Elmosehi) or changes in ionic gradients. ROS can aggregate or fragment many proteins and may provide denatured substrates for protease degradation (Davies, Zolotarjan). The sodium pump's sensitivity to ROS may be due to oxidation of critical cysteine residues and disulfide bonds that maintain the enzyme conformation (Gevondyan). While the enzyme functions after all the $\alpha 1$ subunit cysteines have been mutated (Lane), the $\beta 1$ disulfides are critical for α subunit catalytic activity (Ueno). Other amino acids, such as tyrosines, may undergo modification to nitrotyrosine or chlorotyrosine (Heinecke) that could alter function, but has not yet been demonstrated for Na,K-ATPase. The effects of hyperoxia and ROS on sodium pump activity likely vary with different reactive oxygen species, the severity of injury, and cell type. Reactive oxygen species inhibit Na,K-ATPase isolated from canine kidneys (Huang) and myocardiocytes (Haddock) and activated neutrophils inhibit Na,K-ATPase in vitro (Julin), likely through oxidant mechanisms. Nitric oxide inhibits sodium pump activity in mouse proximal tubule epithelial cells. In the brain, lipid peroxidation is the likely mechanism of decreased activity due to modification of sodium pump binding sites for ATP, K, and Na (Mishra). In ATII cells, H_2O_2 or peroxynitrite can inhibit the pump with differential effects from the apical and basal sides (Clerici, Hu, Kim).

In contrast, in at least some circumstances oxidants can stimulate Na,K-ATPase. Reactive oxygen species increase the activity of the sodium pump in intact endothelial cells (Elliott, Meharg). While protein kinases are activated in this model, they do not seem to be causal (Charles). Nitrogen dioxide augments airway epithelial sodium pump (Robison). The apparent contradiction between these results and the inhibition noted above may be partially a function of when the effect is examined. In one study of ATII cells, high levels of hydrogen peroxide initially inhibited the pump, followed shortly by rebound upregulation (Gonzalez-Flecha) - again consistent with homeostatic defense of this vital cell function.

We analyzed the effects of hyperoxia on Na,K-ATPase mRNA, protein and activity in three lung injury models and have ion transport data for two of the three models. The basis for our experimental design was that this comprehensive approach would provide information about the important steps affected in the response to hyperoxia and control of Na,K-ATPase function in the lung. The models used were (1) in vivo hyperoxia of rats with analysis of peripheral lung biochemical parameters and isolated perfused lung or in situ sodium transport; (2) type II cells isolated from rats after 60 hours of in vivo hyperoxia; and (3) rat alveolar epithelial cells exposed in vitro to hyperoxia for 24-48 hours. We discovered that the Na,K-ATPase mRNA-protein-activity parameters did not change in parallel, but there were many consistent responses in the different models. For example, across each model, Na,K-ATPase gene expression consistently increased. The type II cell protein level changes were complex, with maintained β-1, but decreased α subunit protein in acute hyperoxia, followed by return to normal levels. In terms of transport, we were surprised to discover heterogeneity of the response of unidirectional active sodium resorption in the rats (Carter 1997b). In intact type II cells (models 2 and 3) steady state sodium pump transport activity was normal, but the sodium pump Vmax of membrane preparations decreased. Thus these data indicate that sodium pump and transport are affected and regulated in at least several discrete ways during hyperoxia - increased gene expression; differential changes in relative expression of the α-1 and β-1 subunit proteins; and modification of peak enzyme capacity for pump molecules in the plasma membrane with maintained steady state functional activity. It is very likely that there is simultaneous oxidative injury to some pump molecules while there is a homeostatic increase in transcription accompanied by altered intracellular regulation.

In a standard in vivo acute hyperoxic model (60 hour >95% oxygen) (Thet, Crapo), α1 and β1 mRNA increased early and were 3-5 fold elevated at 60 hours when the alveolar septae were widened with interstitial and mild alveolar edema present, but AECs appeared normal (Nici). No new cell population with sodium pump subunit staining was seen. Levels of α1 and β1 subunit antigenic proteins increased significantly by day 1 after O$_2$ exposure ended and remained elevated through 7 days of recovery. We measured the active (amiloride-inhibitable) and total unidirectional sodium resorption from the alveolar space (Carter 1997b). We were surprised to discover heterogeneity in the degree of injury - as reported in chronic hyperoxia (Yue 1997). The more severely injured rats had augmented active Na resorption, whereas the mildly injured rats had decreased active Na resorption - consistent with upregulation with worsening injury.

ATII cells were studied from rats after 60 hours of in vivo 100% oxygen (Carter 1997a). Ouabain-sensitive Rb uptake into intact ATII cells was normal with hyperoxia, however, the sodium pump Vmax of ATII cell membranes was reduced to 75% of control. The ATII cell α1 subunit protein decreased in hyperoxia while the β1 was stable. The mRNA levels of both subunits increased. Thus levels of mRNA, protein and enzyme activity did not respond in parallel to hyperoxia and steady state activity of the intact cells correlated best with the amount of β1 subunit protein.

We developed an in vitro hyperoxic ATII cell model to eliminate interactions with other cell types and hormonal influences, enabling us to study the effects of hyperoxia more directly (Carter 1994). One day after recovery from isolation, normal rat ATII cells were exposed to 95% O$_2$/5% CO$_2$ in vitro. The cells demonstrated ~25% death with increased Erythrosin B uptake and decreased DNA/plate at 48, but not 24, hours. Within 24 hours of hyperoxia, the type II cells increased levels of Na,K-ATPase mRNA for both subunits, with β1 rising more than α1, but the α1 protein level was decreased and the β1 level was unchanged. The sodium pump Rb uptake of the intact cells was normal, but the Vmax of cell membranes again was reduced significantly. Results in this model were almost identical to those in the ATII cells isolated after in vivo injury, supporting the direct effects of hyperoxia in effecting these changes in ATII cells in this much simpler system.

An important general scientific question is the mechanisms through which oxidants and oxygen tension affect gene expression in the lung (Fanburg). To study the detailed mechanism of Na,K-ATPase gene upregulation in hyperoxia, we screened for hyperoxic induction of Na,K-ATPase mRNA in other cell lines. Two alveolar epithelial cell (AEC) lines, A549 and MP48 cells did not increase Na,K-ATPase mRNA in response to hyperoxia; but they do not form tight junctions in cell culture. In contrast, the renal tubular MDCK (Madin-Darby canine kidney) epithelial cells increased Na,K-ATPase α1 and β1 mRNA in response to hyperoxia in vitro (Wendt). Hyperoxia increased the transepithelial resistance of MDCK cells cultured on Millicell HA (Millipore) filters, suggesting that increased functional Na,K-ATPase and consistent with increased dome formation (Dreher). We sought to determine the mechanism of increased Na,K-ATPase steady state mRNA levels. The mRNA half lives of the α1 and β1 subunit mRNAs did not significantly change with hyperoxia, with approximate half-lives for the α1 subunit were 4.4 hrs in normoxia and 4.7 hrs. in hyperoxia and for the β1 subunit were 5.9 hrs. in normoxia and 6.0 hrs. in hyperoxia. In nuclear run-on assays on both MDCK and type II cells after normoxia or 12 hrs. of hyperoxia, hyperoxia-induced transcription of both subunits in both cell types: 1.7 fold for $\alpha 1$ subunit and ~3 fold for $\beta 1$ subunit in type II cells and 1.3 fold for $\alpha 1$ subunit and 3.0 fold for $\beta 1$ subunit in MDCK cells. Thus the increased sodium pump gene expression was primarily due to increased transcription.

To identify hyperoxia-sensitive regulatory regions of the gene promoters, MDCK cells were transfected with wild type and 5' deletions of $\alpha 1$ (1537bp) and $\beta 1$ (817 bp) subunit promoter-luciferase reporter constructs (gifts of K. Kawakami, Dept. Biology, Jichi Medical School and G. Gick, SUNY Brooklyn, respectively) (Wendt). Promoter activity of full length and multiple 5' deletion constructs of the $\alpha 1$ construct did not change with hyperoxia, suggesting that transcriptional control may lie outside of the 5' 1537 bp region. In contrast, hyperoxia increased $\beta 1$ luciferase activity two fold relative to normoxia. Transfection with two $\beta 1$ 5' promoter deletion constructs identified a region between -102 and -41 required for increased promoter activity with hyperoxia. Sequence analysis of this region did not identify consensus sequences for known oxidant-sensitive transcription factors, such as AP-1, ARE (Rushmore), or NFκB. The region is GC rich and has partial sequence homology with an SP-1 site, but this redox-sensitive transcription factor usually is believed to be important for basal transcription..

We confirmed that this 61 bp $\beta 1$ promoter region (-102 to -41) was specific and sufficient for hyperoxia induction, by subcloning oligonucleotides spanning this region into a luciferase vector containing a minimal promoter of the mouse mammary tumor virus (MMTV) that alone was not inducible by hyperoxia. When transfected into MDCK cells, hyperoxia induced promoter activity for both the larger oligonucleotide (-157 to -38) and even more with the smaller dimer of -84 to -44. Thus a 40 bp region between -84 and -44 is necessary and sufficient for hyperoxic induction. The specific transcription factors in addition to SP1 that interact with this region are being determined.

PHYSIOLOGIC RELEVANCE OF ALTERING SODIUM PUMP IN LUNG INJURY

A major question concerning clinical strategies to increase fluid resorption is whether they still work in high permeability injury states. As discussed by Dr. Matthay, this now is being demonstrated in several lung injury models. We developed an in situ method for measuring alveolar fluid resorption in warm dead rats (Lasnier). Fluid absorption reflected changes in alveolar fluid volume. In this model, terbutaline stimulated fluid resorption approximately 2 fold in both normal rat lungs and lungs injured by 60 hours of hyperoxia. In both situations the terbutaline-induced increased absorption was prevented by ouabain. There also was a statistical trend towards greater resorption induced by hyperoxia itself.

SUMMARY

Oxidants have been intensively studied for many years because of their importance as a pathophysiologic mechanism of disease. However, relatively little is known about their impact on certain fundamental cellular processes, such as translation, and on the integrated response of transporting epithelia. A disadvantage of studying oxidants is their complex multiplicity of effects - as demonstrated by the non-parallel changes in Na,K-ATPase mRNA, protein and activity we have observed. Our studies indicate that multiple simultaneous effects likely occur, including inhibition of peak Na,K-ATPase capacity and increased gene transcription. Understanding the mechanisms by which oxidants impact ATII cell monolayer sodium transport and sodium pump oxidation, translation and transcription should significantly augment our ability to design therapeutic strategies likely to succeed in upregulating alveolar fluid resorption in ARDS and other conditions with alveolar flooding. It also will add knowledge about this critical enzyme, important for many cell functions in a variety of organs.

ACKNOWLEDGEMENTS:

The experimental work by the authors reported herein was supported by grants from the NIH (Acute Lung Injury SCOR HL50152 for DHI and ODW, K08HL03114 To CHW), the American Heart Association (Research Grant-In-Aids to DHI and CHW) and the American Lung Association (Career Investigator Award to DHI and a Research Fellowship to JL).

LITERATURE CITED:

Ahmad M, RM Medford. Evidence for the regulation of Na,K-ATPase α-1 gene expression through   the interaction of aldosterone and cAMP-inducible transcriptional factors. Steroids 60:147-152, 1995.

Beggah AT, P Jaunin, K Geering. Role of glycosylation and disulfide bond formation in the β subunit in the folding and functional expression of Na,K-ATPase. *J Biol Chem* 272:10318-10326, 1997.

Beguin P, A Beggah, S Cotecchia, K Geering. Adrenergic, dopaminergic and muscarininc receptor stimulation leads to PKA phophorylation of Na,K-ATPase. *Am J Physiol* 270:C131-C1347, 1996.

Bertorello AM, AI Katz. Regulation of Na-K pump activity: pathways between receptors and effectors. *NIPS* 10:253-259, 1995.

Bertorello AM, Katz AI. Short-term regulation of renal Na,K-ATPase activity: physiological relevance and cellular mechanisms. *Am J Physiol* 265:F743-55, 1995

Berthiaume Y: Effects of exogenous cAMP and aminophylline on alveolar and lung liquid clearance in anesthetized sheep. *J Appl Physiol* 70:2490-2497, 1991.

Boldyrev A, E Kurella: Mechanism of oxidative damage of dog kidney Na/K-ATPase. *Biochem Biophys Res Comm* 222:483-487, 1996.

Borok Z, Hami A, Danto SI, Lubman RL, Kim KJ, Crandall ED: Effects of EGF on alveolar epithelial junctional permeability and active sodium transport. *Am J Physiol* 270:L559-565, 1996.

Caplan MJ, B Forbush, GE Palade, JD Jamieson. Biosynthesis of the Na,K-ATPase in MDCK cells: activation and cell surface delivery. *J Biol Chem* 265:3528-2534, 1990.

Carter EP, Duvick SE, Wendt CH, Dunitz J, Nici L, Wangensteen OD, Ingbar DH. Hyperoxia increases active alveolar Na+ resorption in vivo and type II cell Na,K-ATPase in vitro. *Chest* 105:75S-78S, 1994.

Carter, E.P., Wangensteen, O.D., O'Grady, S.M. and Ingbar, D.H., Effects of hyperoxia on type II cell Na-K-ATPase function and expression. *Am. J. Physiol. (Lung)* 272:L542-L551, 1997a.

Carter E, Wangensteen OD, Carter EP, Dunitz J, Ingbar DH: Hyperoxic effects on alveolar sodium resorption and lung Na,K-ATPase. *Am. J. Physiol.* 273:L1191-L1202, 1997b.

Chambers SK, Gilmore-Hebert M, Kacinski BM, Benz EJ: Changes in Na,K-ATPase gene expression during granulocytic differentiation of HL60 cells. *Blood* 80:1559-1564, 1992

Charles A, Dawicki DD, Oldmixon E, Kuhn C, Cutaia M, Rounds S. Studies on the mechanism of short-term regulation of pulmonary artery endothelial cell Na/K pump activity. J Lab Clin Med 130:157-168, 1997.

Clerch LB, Massoro D. Oxidation-reduction-sensitive binding of lung protein to rat mRNA. *J Biol Chem* 267:2853-2855, 1992

Clerici C, Friedlander G, Amiel C: Impairment of sodium-coupled uptakes by hydrogen peroxide in alveolar type II cells: protective effect of a-Tocopherol. *Am J Physiol* 262:L542-548, 1992

Compeau CG, OD Rotstein, H Tohda, Y Marunaka, B Rafii, AS Slutsky, H O'Brodovich: Endotoxin-stimulated alveolar macrophages impair lung epithelial Na transport by an L-Arg-dependent mechanism. *Am J Physiol* 266:C1330-C1341, 1994.

Crapo JD: Morphologic changes in pulmonary oxygen toxicity. *Ann Rev Physiol* 48:721-731, 1986.

Davies KJA, ME Delsignore, SW Lin: Protein damage and degradation by oxygen radicals, I. General aspects & II. Modification of amino acids *J Biol Chem* 262:9895-9901;9902-9907, 1987.

Devarajan P, Gilmore-Hebert M, Benz EJ: Differential translation of the Na,K-ATPase subunit mRNAs. *J. Biol. Chem.* 267:22435-22439, 1992.

Dreher D, Rochat T. Hyperoxia induces alkalinization and dome formation in MDCK epithelial cells. *Am J Physiol* 262:C358-364, 1992.

Elliot SJ, Schilling WP. Oxidant stress alters Na+ pump and Na+ -K+ -Cl⁻ cotransporter activities in vascular endothelial cells. *Am J Physiol* 263:H96-H102, 1992.

Elmoselhi AB, Butcher A, Samson SE, and Grover AK. Free radicals uncouple the sodium pump in pig coronary artery. *Am J Physiol* 266:C720-C728, 1994.

Ewart HS, A Klip. Hormonal regulation of the Na-KATPase: mechanisms underlying rapid and sustained changes in pump activity. *Am J Physiol* 269:C295-311, 1995.

Fanburg BL, Massaro DJ, Cerutti PA, Gail DB, Berberich MA. Regulation of gene expression by O_2 tension. *Am J Physiol* 263: L235-L241, 1992.

Feng J, Orlowski J, Lingrel JB: Identification of a functional thyroid hormone response element in the upstream flanking region of the human Na,K-ATPase α-1 gene. *Nucl. Acids Res.* 21:2619-2626, 1993.

Feschenko MS, KJ Sweadner: Conformation-dependent phosphorylation of Na,K-ATPase by protein kinase A and protein kinase C. *J Biol Chem* 269:30436-30444, 1994.

Garat C, M Meignan, MA Matthay, DF Luo, C Jayr: Alveolar epithelial fluid clearance mechanisms are intact after moderate hyperoxic lung injury in rats. Chest 111:1381-88, 1997.

Geering K. Subunit assembly and posttranslational processing of Na⁺-pumps. *Acta Physiol Scand* 146:177-181, 1992.

Gevondyan NM, VS Gevondyan, NN Modyanov: Location of disulfide bonds in the Na,K-ATPase α subunit. *Biochem Molecul Biol Intl* 30:347-355, 1993.

Gick GG, Ismail-Beigi F, Edelman IS. Thyroidal regulation of rat renal and hepatic Na,K-ATPase gene expression. *J Biol Chem* 263:16610-16618, 1988.

Gonzalez-Flecha B, Evelson P, Ridge K, Sznajder JI. Hydrogen peroxide increases Na⁻ / K⁻ ATPase function in alveolar type II cells. *Biochim Biophys Acta* 1290:46-52, 1996.

Haddock, PS, MJ Shattock, DJ Hearse. Modulation of cardiac Na-K pump current: role of protein and nonprotein sulfhydryl redox status. *Am J Physiol* 269:H297-H307, 1995.

Harris ZL, Ridge KM, Gonzalez-Flecha B, Gottlieb L, Zucker A, Sznajder JI. Hyperbaric oxygenation upregulates rate lung Na,K-ATPase. *Eur Respir J* 9:472-477, 1996

Heinecke JW, W Li, GA Francis, JA Goldstein: Tyrosyl radical generated by myeloperoxidase catalyzes the oxidative cross-linking of proteins. *J Clin Invest* 91:2866-2872, 1993.

Horisberger, J-D. The Na,K-ATPase: Structure-Function Relationship. R.G. Landes Co., Austin, 1994.

Hu P, Zhu L, Ischiropoulos H, Matalon S. Peroxynitrite inhibition of oxygen consumption and ion transport in alveolar type II cells. *Am J Physiol* 266:L628-34, 1994.

Huang W-H, Wang Y, Askari A. $(Na^+ + K^+)$-ATPase: inactivation and degradation induced by oxygen radicals. *Int J Biochem* 24:621-626, 1992.

Ingbar DH, Duvick S, Burns CA, Jacobsen E, Dowin R, Savik SD, Gilmore-Hebert M, Jamieson JD. Developmental Regulation of Lung Na, K-ATPase. *Am. J. Physiol. (Lung)* 270:L619-629, 1996.

Ingbar DH, Wendt CH: Oxidant effects on Na,K-ATPase: If only it were so simple. (Editorial) *J Lab Clin Med* (In Press) 1997.

Ingbar DH, Wendt CH, Crandell EC: Role of Na,K-ATPase in the resolution of pulmonary edema. in Ingbar DH, Matthay MA Eds.: Pulmonary Edema, volume in Lung Biology in Health & Disease, L'Enfant, C., Ed. M. Dekker (In Press) 1998

Jewell EA, Shamraj OI, Lingrel JB. Isoforms of the α subunit of Na, K-ATPase and their significance.

Acta Physiol Scand 146:161-169, 1992.

Julin CM, Zimmerman JJ, Sundaram V, and Chobanian MC. Activated neutrophils inhibit Na+ -K+ - ATPase in canine renal basolateral membrane. *Am J Physiol* 262:C1364-C1370, 1992.

Kawakami K, Yanagisawa K, Watanabe Y, Tominaga S, Nagano K: Different factors bind to the regulatory region of the Na,K-ATPase α-1 subunit gene during the cell cycle. *FEBS Lett* 335:251-254, 1993.

Kawakami K, K Masuda, K Nagano, Y Ohkuma, RG Roeder. Characterization of the core promoter of the Na/K-ATPase α1 subunit gene: elements required for transcription by RNA polymerase II and III in vitro. *Eur J Biochem* 237:440-446, 1996

Kim KJ, Suh DJ. Asymmetric effects of H2O2 on alveolar epithelial barrier properties. *Am J Physiol.* 264:L308-315, 1993.

Kobayashi M, Kawakami K. ATF-1CREB heterodimer is involved in constitutive expression of the housekeeping Na,K-ATPase α-1 subunit gene. *Nucl Acids Res* 23:2848-2855, 1995.

Lane LK: Functional expression of rat α1 Na,K-ATPase containing substitutions for cysteines 454, 458, 459, 513 and 551. *Biochem Molecul Biol Intl* 31:817-822, 1993.

Lasnier J, OD Wangensteen, LS Schmitz, CR Gross, & DH Ingbar. Terbutaline stimulates alveolar fluid reabsorption in hyperoxic lung injury *J. Appl. Physiol.* 81:1723-1729, 1996

Lescale-Matys L, Putnam DS, McDonough AA. Na$^+$-K$^+$-ATPase α_1- and β_1-subunit degradation: evidence for multiple subunit specific rates. *Am J Physiol* 264:C583-C590, 1993.

Lingrel JB, Orlowski J, Shull MM, Price EM. Molecular genetics of Na,K-ATPase. *Progress in Nucl Acid Res* 38:37-89, 1990

Liu B, Gick G. Characterization of the 5' flanking region of the rat Na+/K+-ATPase β1 subunit gene. *Biochim Biophys Acta* 1130:336-338, 1992.

Matalon S: Mechanisms and regulation of ion transport in adult mammaliam alveolar type II pneumocytes. *Am J Physiol* 261:C727-738, 1991.

Matalon S, DJ Benos, RM Jackson: Biophysical and molecular properties of amiloride-inhibitable Na channels in alveolar epithelial cells. *Am J Physiol* 271:L1-L22,1996.

Matthay MA, Wiener-Kronish JP. Intact epithelial barrier function is critical for the resolution of alveolar edema in humans. *Am Rev Respir Dis* 142:1250-1257, 1990

Matthay MA, Folkesson HG, Verkman AS: Salt and water transport across alveolar and distal airway epithelia in the adult lung. *Am J Physiol* 270:L487-503, 1996.

McDonough AA, Geering K, Farley RA. The sodium pump needs its β subunit. *FASEB J.* 4:1598-1605, 1990a.

McDonough AA, Tang M-J, Lescale-Matys L: Ionic regulation of the biosynthesis of Na,K-ATPase subunits. *Semin Nephrol* 10:400-409, 1990b.

Meharg JV, McGowan-Jordan J, Charles A, Parmelee JT, Cutaia MV, Rounds S. Hydrogen peroxide stimulates sodium-potassium pump activity in cultured pulmonary arterial endothelial cells. *Am J Physiol.* 265:L613-L621, 1993.

Mircheff AK, Bowen JW, Yiu SC, McDonough AA. Synthesis and translocation of Na+-K+-ATPase α-and β-subunits to plasma membrane in MDCK cells. *Am J Physiol* 262:C470-C483, 1992.

Mishra OP, M Delivoria-Papadopoulos, G Cahillane, LC Wagerie. Lipid peroxidation as the mechanism of modification of the affinity of the NaK-ATPase active sites for ATP, K, Na and strophanthidin in vitro. *Neurochem Res* 14:845-851, 1989.

Nici L, Dowin R, Jamieson JD, Ingbar DH: Upregulation of rat type II pneumocyte Na,K-ATPase during hyperoxic lung injury. *Am J Physiol* 261:L307-314, 1991.

Nomoto M, Gonzales FJ, Mita T, Inoue N, Kawamura M. Analysis of cis-acting regions upstream of the rat Na/K-ATPase α-1 subunit gene by in vivo footprinting. *Biochim. Biophys Acta* 1264:35-39, 1995.

O'Brodovich H. Epithelial ion transport in the fetal and perinatal lung. *Am J Physiol* 261:C555-564, 1991.

O'Brodovich HM. The role of active Na transport by lung epithelium in the clearance of airspace fluid. *New Horizons* 3:240-247, 1995.

Olivera WG, Ridge KM, Sznajder JI. Lung liquid clearance and Na, K-ATPase during acute hyperoxia and recovery in rats. *Am J Respir Crit Care Med* 152:1229-34, 1995.

Olivera W, Ridge K, Wood LDH, Sznajder JI. Active sodium transport and alveolar epithelial Na-K-ATPase increase during subacute hyperoxia in rats. *Am J Physiol.* 266:L577-L584, 1994.

Panos R, Bak PM, Simonet WS, Rubin JS, Smith LJ. Intratracheal instillation of keratinocyte growth factor decreases hyperoxia-induced mortality in rats. *J Clin Invest* 96:2026-2033, 1995.

Planes C, Friedlander G, A Loiseau, C Amiel, C Clerici: Inhibition of Na-K-ATPase activity after prolonged hypoxia in an alveolar epithelial cell line. *Am J Physiol* 271:L70-L78,1996.

Ridge KM, Rutschman DH, Factor P, Katz AI, Bertorello AM, Sznajder JI. Differential expression of Na,K-ATPase isoforms in rat alveolar cells. *Am J Physiol* 273:L246-L255, 1997.

Robison T, Kim KJ. Enhancement of airway epithelial Na,K-ATPase activity by NO2 and protective role of nordihydroguaiaretic acid. *Am J Physiol (Lung)* 270:L266-L272, 1996.

Rushmore TH, Morton MR, Pickett CB. The antioxidant responsive element. *J Biol Chem* 266:11632-11639, 1991.

Sakuma T, HG Folkesson, S Suzuki, G Okaniwa, S Fujimura, MA Matthay. Beta-adrenergic agonist stimulated alveolar fluid clearance in ex vivo human and rat lungs. *Am J Respir Crit Car Med* 155:506-512, 1997

Saumon G, Basset G: Electrolyte and fluid transport across the mature alveolar epithelium. *J Appl Physiol* 74:1-15, 1993.

Schneeberger EE, McCarthy KM: Cytochemical localization of Na-K-ATPase in rat type II pneumocytes. *J Appl Physiol* 60:1584-1589, 1986.

Shull MM, Pugh DG, Lingrel JB. The human Na, K-ATPase a1 Gene: characterization of the 5'-flanking region and identification of a restriction fragment length polymorphism. *Genomics* 6:451-460, 1990.

Skou JC: Overview: The Na,K-Pump. *Meth Enzymol* 156:1-25, 1988.

Suzuki S, D Zuege, Berthiaume Y. Sodium-independent modulation of Na,K-ATPase activity by β-adrenergic agonist alveolar type II cells. *Am J Physiol* 268:L983-990, 1995.

Sweadner KJ. Isozymes of the Na,K-ATPase. *Biochim Biophys Acta* 1989; 988:185-220.

Thet LA. Repair of oxygen-induced lung injury in: physiology of oxygen radicals. *Am Physiol Soc*, 87-107, 1986

Ueno S, M Kusaba, K Takeda, M Maeda, M Futai, F Izumi, M Kawamura. Functional consequences of substitution of the disulfide-bonded segment, Cys127-Cys150, located in the extracellular domain of the Na,K-ATPase β subunit: Arg148 is essential for the functional expression of Na,K-ATPase. *J Biochem* 117:591-596, 1995.

Watanable YU, Kawakami K, Hirayama Y, Nagano K: Transcription factors positively and negatively regulating the Na,K-ATPase α-1 subunit gene. *J Biochem* 114:849-855, 1993.

Wendt CH, H Towle, R Sharma, S Duvick, K Kawakami, G Gick & Ingbar DH. Regulation of Na,K-ATPase gene expression in hyperoxia. *Am J Physiol (Cell)* (in press) 1997.

Yagawa Y, Kawakami K, Nagano K. Cloning and analysis of the 5'-flanking region of rat Na^+/K^--ATPase α-1 subunit gene. *Biochim Biophys Acta* 1049:286-292, 1990.

Yagawa YS, Kawakami K, Nagano K. Housekeeping Na,K-ATPase α-1 subunit gene promoter is composed of multiple cis elements to which common and cell type-specific factors bind. *Molec Cell Biol* 12:4046-4055, 1992.

Yue G, Russell WJ, Benos DJ et al. Increased expression and activity of sodium channels in alveolar type II cells of hyperoxic rats. *Proc Natl Acad Sci* 92(18): 8418-22, 1995.

Yue G, S Matalon. Mechanisms and sequelae of increased alveolar fluid clearance in hyperoxic rats. *Am J Physiol* 272:L407-L412, 1997.

Zheng LP, Du RS, Goodman BE. Effects of acute hyperoxic exposure on solute fluxes across the blood-gas barrier in rat lungs. J Appl Physiol 82:240-247, 1997.

Zolotarjova N. Ho C, Mellgren RL, Askari A, Huang Wu-h. Different sensitivities of native and oxidized forms of Na^+ / K^+-ATPase to intracellular proteinases. *Biochim Biophys Acta* 1192:125-131, 1994.

Zuege D, Suzuki S, Berthiaume Y. Increase of lung sodium-potassium-ATPase activity during recovery from high permeability pulmonary edema. *Am J Physiol* 271:L896-L909, 1996.

LUNG EDEMA CLEARANCE DURING HYPEROXIC LUNG INJURY

Jacob Iasha Sznajder and Karen M. Ridge

Michael Reese Hospital and Medical Center
Pulmonary and Critical Care Medicine
University of Illinois at Chicago
Chicago IL 60616

INTRODUCTION:

Prolonged exposure to elevated concentrations of oxygen causes extensive lung injury in all mammalian species studies to date[1]. Lung oxidant damage is characterized by alterations in the alveolo-capillary barrier which increases permeability to solutes and causes interstitial and alveolar edema leading to respiratory failure. These phenomena can be seen in patients with acute hypoxemic respiratory failure (AHRF) when airspace flooding occurs from damaged pulmonary vessels and alveolar membranes, thus interfering with oxygen transfer from the airspaces into the blood [2-4].

Pulmonary edema accumulation results from the imbalance of hydrostatic and oncotic pressure gradients[5] across the alveolar barrier, while resolution of edema is dependent on liquid clearance across the alveolar barrier following osmotic gradients generated by active Na^+ transport[6-10]. The importance of active Na^+ transport in edema clearance has been demonstrated in live sheep, rabbits and dogs, and in normal isolated rat and rabbit lungs[6-10]. Several mechanisms such as Na^+,H exchanger, Na^+Cl co-transport, Na^+/glucose co-transport, and Na^+,bicarbonate co-transport may contribute to active Na^+ transport[11-13]. However, clearance of edema from the alveoli appears to be mostly dependent on the vectorial Na^+ flux via alveolar apical Na^+ channels and the basolaterally located Na,K-ATPases[14-18]. Although it is unclear whether either is the limiting factor we describe here the changes of Na,K-ATPase function in hyperoxic lung injury.

Na,K-ATPase is a membrane-bound heterodimer protein consisting of two polypeptide chains, α and β, in a one-to-one stoichiometry. The Na,K-pump is responsible for maintaining cellular ionic gradients, osmotic balance, and membrane electrical potential[19]. Energy is provided by the hydrolysis of ATP, which is coupled to the transport of Na^+ and K^+ ions

Acute Respiratory Distress Syndrome: Cellular and Molecular Mechanisms and Clinical Management
Edited by Matalon and Sznajder, Springer Science+Business Media New York, 1998

across the plasma membrane[20]. The active enzyme unit is an αβ complex in which the catalytic properties and the binding domains for ATP and cations are associated with the α subunit. The ß-subunit is a glycosylated polypeptide, which apparently controls incorporation of the α-subunit into the plasma membrane[21].

During lung injury, diffuse alveolar damage occurs and is characterized by two sequential phases: an early exudative phase when edema rapidly accumulates, and a later proliferative phase when ATII cells increase in number, repopulate the alveolar epithelium and the reparative processes begins[22,23]. Previous studies have shown that during the early acute-exudative phase of hyperoxic lung injury the alveolar epithelium is disrupted, leading to edema accumulation[24]. Restoration of alveolar epithelial permeability and increased edema clearance in patients with AHRF has been associated with improved clinical outcome [25.]

We will discuss our studies on lung edema clearance and Na,K-ATPase function in the alveolar epithelium during the acute exudative phase and during the reparative phase of hyperoxia. The two models of hyperoxic lung injury described are (A) rats exposed to 100% O_2 for 64 h, representing the acute exudative phase and (B) rats exposed to 85% O_2 for 7 days representing the proliferative-reparative phase.

Acute Hyperoxic Lung Injury

In the first series of experiments, adults rats exposed to 100% O_2 for 64 h (before they develop fatal respiratory distress) were studied at 0, 7, and 14 days after removal from the hyperoxic environment and compared to normoxic controls. We used the isolated-perfused, fluid-filled rat lung model to measure active Na^+ transport and lung liquid clearance from the alveolar space and to assess the passive flux of solutes. In parallel we assessed the changes in Na,K-ATPase function in ATII cells isolated from the same rats[26].

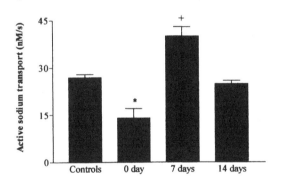

Figure 1. Active Na^+ transport decreased after 64 h of 100% O_2 exposure, increased to above room air control values after 7 d of recovery, and normalized after 14 d of room-air recovery. The bars represent means ± SE. The asterisk and cross represent differences from the other groups (p<0.05). Used with permission[26.]

As shown in Figure 1, during the acute hyperoxic phase active Na+ and lung liquid clearance across the alveolar epithelium decreased by ~45% as compared to controls, whereas after 7days of recovery active Na+ transport increased by ~56% and returned to control values after 14 days of recovery. We isolated ATII cells from rats after the physiologic experiments and found that Na,K-ATPase function decreased after 64 h of 100% O_2 exposure and increased after 7 day of recovery from hyperoxia (Figure 2). These data support the notion that Na,K-ATPase in ATII cells plays a role effecting active Na+ transport and lung liquid clearance during the hyperoxic lung injury. The Na,K-ATPase activity during the acute exudative phase of lung injury was decreased possibly as a result of the oxidative transient inactivation of the pump[27]. The permeability for albumin increased after 64 hours of 100% O_2 exposure and returned to control values after 7 days of recovery while breathing room air. In contrast, passive movement of small solute (e.g. Na+) across the alveolar epithelium increased immediately following oxygen exposure and remained elevated after 7 days of recovery before returning to normal after 14 days. Our data concur with another report where younger rats had similar changes in solute permeability after 60 h of 100% O_2 exposure[28]. These changes in the alveolo-capillary membrane barrier may be caused by the upregulation of metalloproteinases that contribute to hyperoxic lung damage through the degradation of extracellular matrix components[29]. These differential changes in alveolar barrier permeability restoration between small and large solutes may reflect the complexity of the alveolar epithelial reparative process. The reestablishment of the alveolo-capillary barrier and epithelial functional integrity has been helpful in the stratification of patients recovering from AHRF[25].

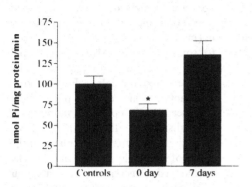

Figure 2. Na,K-ATPase activity decreased in alveolar type 2 cells isolated from rats after 64 h of 100% O_2 exposure as compared with that in control rats and it tended to increase above room air control values in AT2 cells isolated from rats after 7 d of room-air recovery. The bars represent means ± SE. The asterisk denotes differences from the other groups (p< 0.01). Used with permission[26].

Thus, we report that in the early exudative phase of hyperoxic lung injury the alveolar epithelial permeability is increased, possibly due to increased metalloproteinases which increases permeability to solutes, in association with decreased Na,K-ATPase function and thus, lung edema clearance. The acute oxidative lung injury damages the alveolar epithelium and inhibits protective mechanisms against edemagenesis (e.g. Na,K-ATPase) contributing to alveolar flooding and respiratory failure.

Proliferative-reparative hyperoxic phase

In another model of hyperoxic lung injury adult rats were exposed to 85% O_2 for 7 days and studied at 0, 7, 14 and 30 days after removal from the hyperoxic environment and compared to normoxic controls (Figure 3). We found that active Na+ transport and lung liquid clearance increased by 40% in rats exposed to 85% O_2 for 7 days as compared to normoxic controls[18]. This increase was associated with the upregulation of Na,K-ATPase activity and

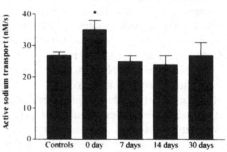

Figure 3. Active Na⁺ transport increased immediately after 85% O_2 exposure for 7 days compared with room air controls and normalized after 7 days of room air recovery. The bars represent means ± SE. The asterisk denotes differences from the other groups (p< 0.05). Used with permission[18].

protein expression in ATII cells immediately following hyperoxic exposure (Figure 4). Seven days after removal from hyperoxic exposure and breathing room air, the active Na+ transport and lung liquid clearance returned to control level and remained unchanged for 30 days[18]. The passive movement of Na^+ across the alveolar epithelium increased in rats immediately after hyperoxic exposure and remained elevated after 7 and 14 days or recovery before returning to control levels by 30 days. Crapo et al.[23] reported no significant changes in the alveolar surface area in rat exposed to 85% O_2 for 7 days, thus we reason that the increased movement of small solutes is due to changes in permeability. These changes are possibly due to the increased proportion of AT2 cells in the alveolar epithelial population increasing the number of intercellular junctions making it a AT2-AT2 cell junction (instead of the normal AT2-AT1 cell junction) which may increase alveolar permeability to small solutes. Alveolar epithelial permeability to small solutes increased after oxygen exposure probably due to injury caused by oxygen free radicals. Furthermore, oxidants increase paracellular permeability in epithelial monolayers by disrupting the normal actin-cytoskeleton pattern particularly in the area of intercellular junctions[30]. The prolonged increase in alveolar epithelial permeability to small solutes from the intravascular and interstitial compartments may be also caused by the early stages of interstitial fibrosis seen post hyperoxic exposure[31].

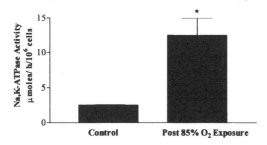

Figure 4. Na,K-ATPase activity was increased in alveolar type II cells from rats exposed to 85% O_2 for 7 days. The bars represent means ± SE. The asterisk denotes differences from the other groups (p< 0.01).

Contrary to the slow normalization of permeability for small solutes, the permeability for albumin increased after 7 days of 85% O_2 exposure and returned to control values in the first week of recovery. These differential changes in alveolar permeability restoration for small and large solutes reflect the complexity of the alveolar epithelial reparative process[31]. It is possible that the small epithelial pores (through which small solute diffuse), increase during hyperoxia and need more time to repair, whereas the epithelial barrier to large solutes such as albumin repairs more rapidly. These observations may have clinical relevance because they suggest that the functional integrity of the alveolar epithelial barrier may be restored rapidly.

We observed that Na,K-ATPase activity and protein expression increased in AT2 cells from rats after 7 days of 85% oxygen exposure and this was associated with an increase in active Na+ transport and lung liquid clearance. The increased Na,K-ATPase expression could be due in part to increased number of AT2 cells -which double in this model[31] and/or to increased number of Na,K-ATPases per cell. An increase of Na,K-ATPase protein expression per cell has been previously observed in other tissues when increased Na+ pumping was needed to cope with threatening extracellular ionic imbalance[32,33].

This association between lung edema clearance and Na,K-ATPase function does not demonstrate necessarily a cause/effect relationship indicating that active Na+ transport and lung liquid clearance are effected by Na,K-ATPase in ATII cells. Therefore, additional studies

16

were conducted demonstrating that in lungs from rats exposed to 85% O_2 for 7 days, amiloride (a Na+ channel blocker) inhibited ~87% of lung liquid clearance, whereas ouabain inhibited ~52%[34]. Both amiloride and ouabain inhibited active Na+ transport by a greater percentage in rats exposed to hyperoxia than in normoxic control rat lungs. The role of amiloride channels is further supported by Haskell et al. who report that rats exposed to 85% O_2 for 7 days had higher levels of Na+ channel protein in ATII cells[35]. These results suggest that upregulation of both amiloride-sensitive Na+ channels and Na,K-ATPase contribute to effective alveolar edema clearance in subacute hyperoxic lung injury.

In summary, during lung injury and specifically in the hyperoxic animal models there are significant changes in permeability and decreased ability of the lungs to clear edema. Studies from other laboratories, and our own, suggest that alveolar epithelial Na+ channels and Na,K-ATPases have an important role contributing to active Na+ transport and thus lung edema clearance. Our challenges are to develop strategies to manipulate these two important mechanisms of vectorial Na+ flux and either prevent the accumulation of, or accelerate lung edema clearance.

References:

1. J. M. Clark, C. J. Lamberstein, *Pharmacol. Rev.* 23, 37 (1971).
2. H. Humphrey, Hall, J., Sznajder, J.I., Silverstein, M., Wood, L.D.H., *Chest* 97, 1176 (1990).
3. J. P. Mitchel, D. Schuller, F. S. Calandrino, D. P. Schuster, *Am. Rev. Respir. Dis.* 145, 990 (1992).
4. J. I. Sznajder, L. D. H. Wood, *Chest* 100, 890 (1991).
5. A. E. Taylor, *Cir. Res.* 49(3), 557 (1981).
6. Y. Berthiaume, Staub, N.C., Matthay, M.A., *J. Clin. Invest* 79, 335 (1987).
7. R. M. Effros, G. R. Mason, J. Hukkanen, P. Silverman, *J. Appl. Physiol.* 66, 906 (1989).
8. B. Goodman, R. Fleisher, E. D. Crandall, *Am. J. Physiol.* 245, C78 (1983).
9. D. H. Rutschman, W. Olivera, J. I. Sznajder, *J. Appl. Physiol.* 75, 1574 (1993).
10. N. Smedira, et al., *J. Appl. Physiol.* 70(4), 1827 (1991).
11. D. G. Oelberg, F. Xu, F. Shabarek, *Biochim. Biophys. Acta* 1194, 92 (1994).
12. B. E. Goodman, Anderson, J.L., Clemens, J.W., Kircher, K.J., Stormo, M.L., Waltz, J.S., Waltz, W.F., White, J.W., *J. Appl. Physiol.* 76, 2578 (1994).
13. E. P. Nord, Brown, S.E.S., Crandall, E.D., *Am. J. Physiol.* 252, C490 (1987).
14. S. Matalon, et al., *Am. J. Physiol.* 262(Cell Physiol. 31), C1228 (1992).
15. H. O'Brodovich, et al., *Am J Physiol* 265 (Cell Physiol 34), C491 (1993).
16. E. E. Schneeberger, K. M. McCarthy, *J. Appl. Physiol.* 60 (5), 1584 (1986).
17. L. Nici, R. Dowin, M. Gilmore-Hebert, J. D. Jamieson, D. H. Ingbar, *Am. J. Physiol.* 261, L307 (1991).
18. W. Olivera, K. Ridge, L. D. H. Wood, J. I. Sznajder, *Am. J. Physiol.* 266, L577 (1994).
19. L. C. Cantley, *Curr. Top. Bionerget.* 11, 201 (1981).
20. J. C. Skou, *FEBS* 268(2), 314 (1990).
21. A. A. McDonough, K. Geering, R. A. Farley, *Faseb J.* 4, 1598 (1990).
22. A. A. Katzenstein, C. M. Bloor, A. A. Leibow, *Am. J. Pathol.* 85(1), 210 (1976).
23. J. D. Crapo, *Annu. Rev. Physiol.* 48, 721 (1986).

24. B. Royston, N. Webster, J. Nunn, *J. Appl. Physiol.* 69, 1532 (1990).
25. M. A. Matthay, J. P. Wiener-Kronish, *Am. Rev. Respir. Dis.* 142, 1250 (1990).
26. W. Olivera, K. Ridge, L.D.H. Wood, and J.I. Sznajder, *Am. J. Physiol.* 266, L577 (1994)
27. B. Gonzalez-Flecha, P. Evelson, K. Ridge, J. I. Sznajder, *Biochim. Biophys. Acta* 1290, 46 (1995).
28. E. P. Carter, S. E. Duvick, C. H. Wendt, J. M. Dunitz, L. Nici, O. D. Wagensteen, and D. H. Ingbar, *Chest* 105, 75S (1994).
29. A. Pardo, M. Selman, K. Ridge, R. Barrios, J. I. Sznajder, *Am. J. Respir. Crit. Care Med.* 154, 1067 (1996).
30. M. J. Welsh, D. M. Shasby, R. M. Husted, *J. Clin. Invest.* 76, 1155 (1985).
31. J. D. Crapo, B. E. Barry, H. A. Foscue, J. Shelburne, *Am Rev. Respir. Dis.* 122, 123 (1980).
32. J. W. Bowen, A. McDonough, *Am. J. Physiol.* 252 (Cell Physiol. 21), C179 (1987).
33. B. A. Woltizky, D. M. Fambrough, *J. Biol. Chem.* 261, 9990 (1986).
34. J. I. Sznajder, W. G. Olivera, K. M. Ridge, D. H. Rutschman, *Am. J. Resp. Crit. Care Med.* 151, 1519 (1995).
35. J. F. Haskell, Yue, G., Benos, D.J., Matalon, S., *Am. J. Physiol.* 266, L30 (1994).

THE ROLE OF LUNG LYMPHATICS IN PULMONARY EDEMA CLEARANCE

Kyriakos D. Chainis[1], Karen M. Ridge[2], Jacob I. Sznajder[2], Dean Schraufnagel [3]

[1]Corfu General Hospital, Pulmonary Department, Corfu, Greece
[2]Michael Reese Hospital and Medical Center, Pulmonary and Critical Care Medicine, Chicago IL 60616
[3]Unversity of Illinois at Chicago, Pulmonary and Critical Care Medicine, Chicago, IL 60680

INTRODUCTION

The acute respiratory distress syndrome (ARDS) is characterized by damage to the alveolar endothelial-epithelial barrier with increased permeability[1] resulting in influx of fluid from the circulation into the lung. The edema is characterized by the presence of protein and cellular elements. The removal of edema from the interstitial and alveolar space of the lung is one of the major goals of ARDS management[2, 3.] Understanding the mechanisms contributing to the clearance of pulmonary edema may improve our ability to treat ARDS. This article will consider morphologic and physiologic factors that have provided some insight into the role of lung lymphatics in pulmonary edema clearance.

Liquid Clearance from Interstitial Spaced

Interstitial pulmonary edema could be cleared by the lymphatics, the pulmonary circulation, or the bronchial circulation[3], although their relative contributions have not been determined[2]. It is probable that the bronchial circulation contributes to the formation rather than the clearance edema because, the brochovascular bundle is made of pulmonary and bronchial arteries, airways and lymphatics, but generally not bronchial veins. Additionally, the higher pressure in the bronchial arteries would cause a flux away from the bronchial circulation[4].

The lung lymphatics play an important role in the interstitial edema clearance. Different studies have established that lymph flow increases in the edematous lung. Drake et al., [5] measured the flow rate from cannulated lung lymph vessels in anaesthetized dogs and found that the lymph flow increased from a base line of 20 µl/min to 388 µl/min during lung edema due to increased left atrial pressure. Mitzner and Sylvester[6], measured lymph flow in isolated sheep lungs during pulmonary hydrostatic edema. They found that during the period of left atrial hypertension lymph flow rose from 3 to 12 ml/h. The role of lung lymphatics is probably more important especially in protein-rich interstitial edema clearance. It is believed that protein-rich interstitial fluid is cleared by lymphatics,

Acute Respiratory Distress Syndrome: Cellular and Molecular Mechanisms and Clinical Management
Edited by Matalon and Sznajder, Springer Science+Business Media New York, 1998

since there is no osmotic or hydrostatic gradient flavoring liquid clearance into either the pulmonary or the bronchial circulation[2].

Interstitial Space Protein Clearance

Clearance of protein from the interstitial space is affected by lymphatic drainage, endocytosis, and metabolic degradation. The route of clearance may depend on the size of the protein and whether it is in solution or not[2].

Different experiments using radiolabeled albumin injected intravenously[7,8] have shown that tracer proteins cross into the lung interstitium and are removed from the lung by lung lymphatics. This route of interstitial protein clearance is effective during the early phase of ARDS, when the protein is in solution. In the late stages of ARDS when the exudate coagulates removal of interstitial protein may depend on a much slower process requiring cellular engulfment and metabolic degradation.

Alveolar Liquid Clearance

The alveolar edema is cleared primarily by the active transport of sodium from the alveoli to the interstitial space with water following passively[2,9,10]. The active transport of sodium depends on the action of the sodium-potassium pump located in the alveolar type II cells, as well as in the type I cells[11]. When the fluid passes to the interstitial space it is removed by lymphatics as well as by the pulmonary and bronchial circulation[2].

Alveolar Protein Clearance

Physiological and morphologic experimental studies using radiolabeled albumin have shown that alveolar protein is removed from the alveolar air spaces through the airways,

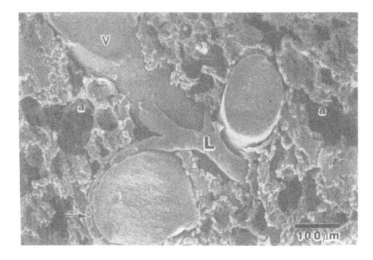

Figure 1 is a scanning micrograph showing a vascular and lymphatic cast of a rat lung that was exposed to hyperoxia. The large vessel on the vein (V) is a conduit lymphatic (L). Initial lymphatics (arrow) are seen around the large vein. Alveolar blood capillaries are in the background (a). All lymphatics are greatly expanded in this model of lung edema. Bar is 100 μm.

20

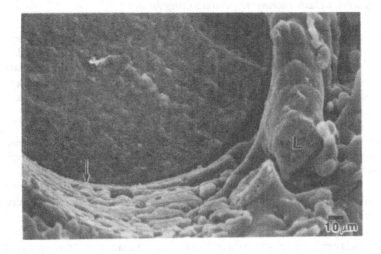

Figure 2 is a scanning micrograph showing the casts of initial lung lymphatics (arrow) around a large blood vessel of a rat exposed to hyperoxia. The initial lymphatics join a conduit lymphatic (L). Bar is 10 μm.

by paracellular diffusion, endocytosis, and the action of alveolar macrophages[2]. The protein needs to pass into the interstitium to be removed by lymphatics.

Alveolar and Interstitial Clearance of Inflammatory Cells

The protein-rich fluid that occurs in the setting of acute lung injury or ARDS is accompanied by an influx of white blood cells and macrophages[1]. Circulating monocytes that enter the interstitium and the airspaces are transformed into alveolar macrophages which clear protein and cells[12].

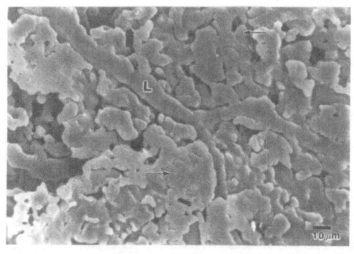

Figure 3 is a scanning micrograph of pleural lymphatic casts of a rat exposed to hyperoxia. The flat structures with quasi-tubular structures are initial lymphatics (arrows) that join with the interstitial space and a conduit lymphatic (L). Bar is 10 μm.

Lung Lymphatics Casts During Hyperoxic Lung Injury

To determine which lymphatics are expanded by edema Schraufnagel and colleagues[13] cast lung lymphatics from rats exposed to 85% O_2 for 7 days. These animals develop a subacute lung injury characterized by increased permeability of alveolar capillary endothelium, pulmonary edema, and proliferation of lung cells. For casting methyl methacrylate resin was injected through the vena cava into the lung vasculature after the animals were removed from hyperoxia.

The animals not exposed to hyperoxia had few lymphatic casts around the blood vessels and airways. In rats that were exposed to hyperoxia, 22% of arteries, 30% of veins, and 51% of indeterminate blood vessels (which could be arteries or veins) were encompassed by saccular lymphatics (Figures 1-2). The amount of lymphatics that could be cast returned to control values fourteen days after recovery from hyperoxia. Lymphatics were also increased on the pleural surface (Figure 3). The median percentage of the pleural surface that was covered with lymphatics was 0 in the animals exposed to ambient air, 65% in animals exposed to hyperoxia and again 0 in animals allowed to recover for fourteen days following hyperoxic exposure (p<0.001). There is also increased active sodium transport, lung edema clearance and alveolar epithelial Na+K-ATPase function associated with this model[9,10]. This study found that all compartments of the lung lymphatics expanded after the injury and edema caused by oxygen and returned to normal with the resolution of lung edema. The increase could have resulted from dilatation of existing lymphatic vessels or the formation of new lymphatic vessels. There is considerable proliferation of many cell types in this model but lung lymphatics can expand quickly in the face of acute edema.

Lung Lymphatics Cast from the Airspaces

In the absence of lung edema the methyl-methacrylate resin injected into the airway does not penetrate the epithelium, but the lymphatics of rats exposed to 85% O_2 for 7 days can

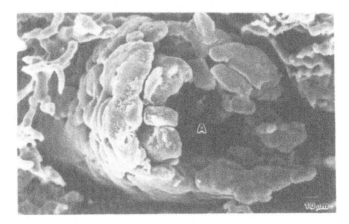

Figure 4 is a scanning micrograph of lymphatic casts around an airway (A) of a rat lung exposed to hyperoxia. The airway has been dissolved away showing 2 layers of lymphatics in its wall. Bar is 10 μm.

be cast by filling the airway with methacrylate[14]. On removal from hyperoxia (day 0) 29% of the bronchioles had saccular lymphatic casts around them and 6% of bronchioles were surrounded by these lymphatics. Twenty-five percent of bronchioles had conduit lymphatic cast. Fourteen percent of arteries had lymphatic casts around them. All were different from control rats breathing room air (p<0.0001). Rats exposed to hyperoxia had lymphatics on the pleural surface near alveoli and alveolar ducts and around veins. The peribronchial (Figure 4) and periarterial saccular lymphatics formed separate groups but communicated with conduit lymphatics. The perivenous lymphatic had their own separate conduit lymphatics. Fourteen days after returning to ambient air the lymphatics were similar to those of control animals. The changes shown in these studies of lymphatics during pulmonary edema and its resolution show the great capacity of the lymphatic system to accommodate excess fluid. Lymph flow increases with pulmonary injury and the increased lymph flow and increased absorptive functions are major protective mechanisms against pulmonary edema.

SUMMARY

- Lung lymphatics can be cast by injecting a resin through the vasculature or airways if the alveolar epithelium is damaged as occurs in animals exposed to hyperoxia.

- Lung edema is cleared via peribronchial and periarterial lymphatics and appears to enter the lymphatics through the alveolar and bronchiolar epithelium.

- The three-dimensional image of lung lymphatics shows that they are enlarged during hyperoxic lung injury and return to normal as the injury resolves.

- Casting lymphatics through the airways shows that the lymphatic structure is similar whether they are cast through the endothelium or epithelium.

- Airway casting provides a better image for the perivascular lymphatics than the pleural lymphatics. The peribronchial lymphatics casts show the communication of saccular peribronchial and perivascular lymphatics to the conduit lymphatics in the bronchovascular bundle.

- The communications between peribronchial and periarterial lymphatics may allow the flux of fluid from one compartment to the other.

- The location of saccular lymphatics around the bronchioles and blood vessels may facilitate draining these structures.

- These lymphatics cover a large surface area and may proliferate or distend to accommodate a large fluid volume. This volume coupled with interstitial fluid can impinge on the airway caliber. This impingement could contribute to the mechanical changes occurring in the airways and blood vessels during pulmonary edema. The cyclical movement of the bronchial and vascular trees may increase the lymph flow from the saccular to the conduit lymphatics.

References

1. Bachofen M. and Weibel E.R.: Structural Alteration of Lung Parenchyma in the ARDS. *Clinics in Chest Medicine* 3:1, (1982)
2. Matthay M.A.: Resolution of Pulmonary Edema. Mechanisms of Liquid, Protein and

Cellular Clearance from the Lung. *Clinics in Chest Medicine* Vol. 6 No. 3 September (1985).

3. Sznajder J.I. and L.D.H. Wood. Beneficial effects of reducing pulmonary edema in patients with hypoxemic respiratory failure. (Editorial) *Chest* 100:890-892, (1991).

4. Staub, N.C., Nagano, H., and Pearce, M.C.: Pulmonary edema in dogs, especially the sequence of fluid accumulation in lungs, *J. Appl. Physiol.* 22:227-240, (1967).

5. Drake RE, Allen SJ, Williams JP. Luine GA, and Gabet JC: Lymph flow from edema dog lungs, *J. Appl. Physiol* 62:2416-2420, (1987)

6. Mitzner W. and JT Sylvester. Lymph flow and lung weight in isolated sheep lungs. *J Appl. Physiol.* 61: 1830-1835, (1986).

7. Gorin, A.B. and Stewart, P.A.: Differential permeability of the endothelial and epithelium barrier to albumin flux. *J. Appl Physiol* 47:1315-1324, (1979).

8. Meyer, E.C., and Ottaviano, R. Right lymphatic duct distribution volume in dogs: Relationship to pulmonary interstitial volume. *Circ. Res* 35:197-203. (1974).

9. Olivera W., Ridge, K. Wood LDH, Sznajder JI: Active sodium transport and alveolar epithelial Na,K-ATPase increase during subacute hyperoxia in rats. *Am J Physiol* 261:L307-L314, (1994).

10. Sznajder J.I., D.H. Rutschman, K.M. Ridge and W. Olivera. Mechanisms of lung liquid clearance during subacute hyperoxia in isolated rat lungs. *Am. J. Respir. Crit. Care Med.* 151:1519-1525, (1995).

11. Ridge K.M., D.H. Rutschman, Factor P., A.I. Katz, A.M. Bertorello and J.I. Sznajder. Differential Na,K-ATPase isoforms in cultured rat alveolar type 2 cells. *Am. J. Physiol.* 273:L246-L255, (1997).

12. Henson, F.M., Larsen, G.L., Hanson, J.E. et al: Resolution of pulmonary inflammation, *Fed. Proc.* 43:2799-2806, (1984).

13. Schraufnagel, D.E., Hainis K., and Sznajder JI: Lung lymphatics increase after hyperoxic lung injury. An ultra structural study of casts. *Am J. Pathol.* 144:1393-1402, (1994)

14. Hainis K.D., Sznajder JI, and Schraufnagel D.E.: Lung lymphatics cast from the airspace. *Am. J. Physiol.* 267: L199-L205, (1994).

Molecular biology of lung Na$^+$ absorption

Pascal Barbry

Institut de Pharmacologie Moléculaire et Cellulaire, CNRS-UPR 411, 660, route des Lucioles, Sophia Antipolis, 06560 Valbonne, France

INTRODUCTION

A transcellular sodium reabsorption through high-resistance *epithelia*: couples passive electrodiffusion of sodium through the apical membrane, and active extrusion of intracellular sodium by basolateral Na$^+$/K$^+$/ATPase, and generates a vectorial transcellular sodium transport. In lung, this mechanism participates to the correct hydration of the luminal compartment. The apical electrodiffusion, which corresponds to the limiting step of transcellular transport, is mediated by an ionic channel, highly selective for sodium and lithium over potassium. This channel is blocked by the diuretics amiloride and triamterene. Molecular identification of the proteins involved in amiloride-sensitive sodium permeation has been achieved. Three homologous subunits, entitled αENaC, βENaC, and γENaC (for epithelial Na$^+$ channel), correspond to the pore-forming subunits. They are distinct from voltage-dependent Na$^+$ channels. Instead, they constitute with more than 20 homologous proteins, a new gene super-family of ionic channels. This family can be divided into three main subfamilies: (1) channels involved in vectorial transport of electrolytes, such as ENaC; (2) degenerins from *Caenorhabditis elegans*, such as DEG-1,

Acute Respiratory Distress Syndrome: Cellular and Molecular Mechanisms and Clinical Management
Edited by Matalon and Sznajder, Springer Science+Business Media New York, 1998

25

MEC-4, MEC-10, UNC-8 and UNC-105, which are likely to correspond to mechanosensitive channels; (3) ligand-gated channels, such as FaNaC (for FMRFamide Na^+ channel), an ionotropic receptor for the cardioexcitatory peptide Phe-Met-Arg-Phe-NH_2 (FMRFamide) found in *Helix aspersa* nervous system, or ASIC (for acid sensing ionic channel), a mammalian H^+-gated channel selective for monovalent and divalent cations. The physiological importance of the epithelial Na^+ channel is highlighted by identification of mutations into αENaC, βENaC, and γENaC genes in families affected by pseudohypoaldosteronism type I (PHA1) or hereditary low-renin hypertension (Liddle's syndrome).

PHARMACOLOGICAL AND BIOPHYSICAL PROPERTIES

Amiloride (3,5-diamino-N-(aminoiminomethyl)-6-chloropyrazinecarboxamide) was discovered in 1964 after screening for nonsteroidal saliuretic agents with antikaliuretic properties in rat. A large amiloride pharmacology has been developed. Amiloride analogs can inhibit distinct Na^+ transport systems: 1) the epithelial Na^+ channel, 2) the Na^+/H^+ exchange system, 3) the Na^+/Ca^{2+} exchange system, which have usually distinct sensitivities for amiloride derivatives. Phenamil, or benzamil, substituted on the guanidino moiety, are potent inhibitors of the epithelial Na^+ channel, but poor inhibitors of the Na^+/H^+ antiporter. Conversely, 5-N-disubstituted derivatives of amiloride, such as ethylisopropylamiloride (EIPA), are the most potent inhibitors of the ubiquitous isoform of the Na^+/H^+ exchanger, while the Na^+ channel is not blocked by these derivatives. The Na^+/Ca^{2+} exchange system is poorly inhibited by amiloride, but some amiloride derivatives that are substituted on the guanidino moiety, such as dichlorobenzamil, inhibit it, although with a low affinity. Importantly, the pharmacological characterization of a Na^+ transport system is not always sufficient to identify the correct Na^+ pathway, since some Na^+/H^+ antiporter isoforms are insensitive to amiloride and to EIPA, while EIPA-sensitive electrogenic Na^+ transport, presumably through Na^+ permeant channels, has been described in cultured cells (Matalon et al., 1996).

Noise analysis induced by submaximal concentrations of amiloride allowed Lindemann and Van Driessche to identify the unitary properties of the epithelial Na^+ channel. Later on, the same channel was characterized by the patch clamp technique in primary cultures of fetal rat lung epithelial cells (Voilley et al., 1994). It is characterized by a low unitary conductance (about 4 pS in 140 mM NaCl), and a high selectivity for sodium and lithium over potassium. Patch clamp or noise analysis also show that the channel is present in A6 cells (derived from amphibian kidney), in apical membrane of cortical collecting tubule cells, in intact epithelium of the toad urinary bladder, in distal segments of colon, in airway epithelium, in granular duct cells of mouse mandibular glands, in *stria vascularis* marginal cells from the cochlea, and in many other high resistance *epithelia* (reviewed by Barbry and Hofman, 1997).

PRIMARY STRUCTURE

A clone encoding a subunit of the epithelial Na$^+$ channel was characterized after functional screening of a rat distal colon cDNA expression library (Lingueglia et al., 1993). It is 3081 nucleotides long and encodes a 699 amino acid protein. This protein was called RCNaCh, for rat colon Na$^+$ channel, or αENaC, for epithelial Na$^+$ channel. Two homologous cDNAs, called βENaC and γENaC, sharing ~35% identity with aENaC were then identified (Canessa et al., 1994; Lingueglia et al., 1994). Co-expression of the three subunits increases the amplitude of the current by two orders of magnitude. The ionic selectivity, gating properties, and pharmacological profile of the channel formed after co-expression of the three subunits in oocyte are similar to those of the native channel (Canessa et al., 1994). The human αENaC, βENaC, γENaC have been subsequently characterized from lung cDNA libraries (Voilley et al., 1994; Voilley et al., 1995). Human αENaC gene maps to chromosome 12p13 (Voilley et al., 1994), while βENaC and γENaC genes are colocalized within a common 400 kilobases fragment on chromosome 16p12-13 (Voilley et al., 1995).

Rat αENaC, βENaC and γENaC mRNAs have been detected by Northern blot analysis as unique bands of 3.6, 2.6, and 3.2 kilobases, respectively (Voilley et al, 1997). They are expressed at a high level in epithelial tissues, such as renal cortex and medulla, distal colon, urinary bladder, lung, placenta, and salivary glands. Low levels of transcripts were also identified in proximal colon, in uterus, in thyroid, and in intestine. Similar patterns of expression were observed for α, β and γ proteins in immunolabelling experiments (Renard et al., 1995)) using specific anti-peptide antibodies raised against each of the three subunits.

Identification of the ENaC subunits has also revealed the existence of similitude with a family of proteins previously identified in the nematode *Caenorhabditis elegans* (Chalfie et al., 1993). While αENaC, βENaC and γENaC share no significant identity with previously cloned ionic channels, a ~12% amino acid identity was observed between them and the proteins MEC-4, MEC-10 and DEG-1, also called degenerins. It is likely that DEG-1, MEC-4 and MEC-10 are subunits of ion channels, but the expression of a channel activity after their heterologous expression has not yet been reported. All these proteins share the same overall organization, characterized by the presence of two large hydrophobic domains, and the presence of one (or two for degenerins) cysteine rich region(s) (Renard et al., 1994).

In *Helix aspersa* neurons, Phe-Met-Arg-Phe-NH$_2$ (FMRFamide) induces a fast excitatory depolarizing response due to direct activation of an amiloride-sensitive Na⁺ channel. A cDNA has been isolated from *Helix aspersa* nervous tissue (Lingueglia et al, 1995). It encodes a FMRFamide-activated Na⁺ channel (FaNaC) that can be blocked by amiloride. FaNaC displays the structural organization of epithelial Na⁺ channel subunits. It corresponds to the first example of an ionotropic receptor for a peptide.

Comparison of ENaCs, FaNaC or degenerin sequences with databases of expressed sequence tags (ESTs) has revealed the existence of other homologues in mammals. The first mammalian protein identified by this approach, called δENaC, corresponds to an "αENaC-like" subunit (Waldmann et al., 1995). The same expressed sequence tags strategy has allowed the identification of others mammalian homologues. Screening of a human brain cDNA library with an expressed sequence tag led to the identification of a 512 residues long protein called MDEG (for mammalian degenerin, (Waldmann et al., 1996). Screening of a rat brain cDNA library revealed the existence of a 67% identical protein (Waldmann et al., 1997). Expression in *Xenopus* oocyte of that latter protein, called ASIC (for acid sensing ionic channel) induced a H⁺-gated cation channel (IH⁺), with permeability ratio: $p_K^+/p_{Na}^+ = 0.077$, $p_{Li}^+/p_{Na}^+ = 0.77$, $p_{Ca}^{2+}/p_{Na}^+ = 0.4$, $p_H^+/p_{Na}^+ = 1.25$ (Waldmann et al., 1997).

Identification of degenerins, ENaC, FaNaC, and ASICs clearly define a large gene super-family that contains not only ionic channels, involved in vectorial transport of electrolytes, characterized by low kinetics of opening and closure such as the amiloride-sensitive Na^+ channel, but also proteins involved in sensory perception, especially mechanosensation, as *C. elegans* MEC-4 and MEC-10, and ligand-gated ionotropic receptors, such as FaNaC and ASIC.

SECONDARY STRUCTURE

Analysis of ENaCs primary structure clearly indicates the presence of two large hydrophobic domains, that divide the proteins into five distinct domains: a 50-100 residues long NH_2-terminal segment (presumably cytoplasmic, due to the absence of a signal peptide), and a 20-100 residues long COOH-terminal segment; two hydrophobic domains, the second one being more conserved over the family than the first one (~30% identity against ~20%); a large domain located between the two hydrophobic domains, that represents more than 50% of the total mass of the protein.

The orientation of the two short NH_2- and COOH-terminal segments toward the cytoplasm. Existence of specific regulation mechanisms for each protein is consistent with the poor conservation of the cytosolic domains among the family. A PPPXY sequence, located into the COOH-terminal segment, is implicated into the hyperactivity of the channel, associated with Liddle's syndrome (reviewed by Barbry and Hofman, 1997).

Many experimental observations confer a high functional importance to the second hydrophobic domain and are consistent with a structural model where it is divided into two distinct parts. The COOH-terminal segment is likely to correspond to a classical transmembrane α-helix. One sector of this helix interacts with ions and with the amiloride molecule, while the others interact with the lipid bilayer. The NH_2-terminal segment, where two putative β-strand structures are linked by a coil-region that contains one (or two) conserved glycine(s), participates in the formation of the ionic pore (Barbry and Hofman, 1997).

ENaCs EXPRESSION IN ALVEOLAR AND AIRWAY *EPITHELIA*

There is no doubt that the highly-amiloride-sensitive Na^+ channel, encoded by αENaC, βENaC and γENaC plays a crucial role for controlling the quantity and composition of the respiratory tract fluid, and during the transition from a fluid-filled to an air-filled lung at the time of birth. ENaCs mRNAs have been identified in the lung (Voilley et al., 1994, 1995, 1997), and the biophysical properties reported by Voilley et al. (1994) appear identical to those of the renal Na^+ channel. Surprisingly, the electrophysiological properties of the channels expressed in pure preparation of type II pneumocytes are strikingly different. A cationic channel, poorly selective for Na^+ over K^+, and with an altered amiloride sensitivity, has been characterized in these cells (Matalon et al., 1996). The link between this channel and the ENaC subunits needs now to be clarified. Around birth, an increase in Na^+ channel transcription and expression results in a switch of the ionic transport in lung from active Cl^- secretion to active Na^+ reabsorption (O'Brodovich, 1993; Voilley et al., 1994; 1997; Tchepichev et al., 1995). This results in clearance of the pulmonary fluid as the lung switches to an air-conducting system. After inactivation of murine αENaC, deficient neonates develop respiratory distress and die within 40 hours of birth from failure to clear their lungs of liquid (Hümmler et al., 1996). A still pending question is the exact site where active Na^+ reabsorption takes place: while *in vitro* and *in vivo* studies have clearly shown the presence of an amiloride-inhibitable alveolar fluid clearance in distal lung, the exact contribution of alveolar and distal airway *epithelia* to reabsorption of fluid by the distal air spaces of the lung is not known (Matthay et al., 1996). Since the alveolar epithelium makes up 99% of the available surface of the lung, it is usually assumed that distal airways only contribute minimally to net fluid absorption. However, the relative contribution of each segment to the active Na^+ transport also depends upon the level of expression of the three ENaC subunits in *alveoli* versus airways. Such an analysis has begun using the different molecular tools developped after ENaCs cloning.

In rat lung, αENaC mRNA has been detected in trachea, bronchi, bronchioles and alveoli (Matsushita et al., 1996). βENaC and γENaC mRNAs were most abundant in the bronchiolar and bronchial epithelium (Farman et al., 1997). ENaC proteins have been detected by Renard et al. (1995) with specific anti-ENaC *antisera*. Airways from trachea to terminal bronchioles express strongly the three subunits. Some αENaC and γENaC co-express with the surfactant protein SP-A in type II pneumocytes. The relative ENaCs expression in alveolar versus airway *epithelia* is not completely established, but the most robust expression was detected at the level of the bronchiolar epithelium (Gaillard et al., manuscript in preparation).

Administration of glucocorticoids has previously been demonstrated to induce an Na^+ absorptive capacity in the immature fetal lung, suggesting that this modification may be due to ENaC stimulation by glucocorticoids. Champigny et al. (1995) and Voilley et al. (1994, 1997) analyzed the mechanisms of this stimulation in primary cultures of fetal rat lung epithelial cells, and showed that the ENaC activity is controlled by corticosteroids. Treatment with dexamethasone, or with RU28362, a synthetic pure glucocorticoid agonist, increases Na^+ channel activity *via* stimulation of the three ENaC subunits transcription, *i.e.* a distinct effect of those observed in kidney and in colon with aldosterone. While the increase in Na^+ channel activity observed in the lung around birth might be related to the raise of corticosteroids, Tchepichev et al. (1995) have observed differences between development and steroid effect, suggesting the existence of others triggers, such as change in P_{O_2}.

In lung, β-adrenergic agonists increase lung fluid clearance and this effect is inhibited by amiloride (Matthay et al., 1996). In rat, cyclic AMP stimulates ENaC activity in primary cultures of fetal lung epithelial cells and in cortical collecting tubules but not in colon, even though the three subunits that make up the channel are present in these three tissue types (Barbry and Hofman, 1997). These observations are consistent with an indirect stimulatory mechanism, which does not require ENaCs phosphorylation by cyclic AMP dependent protein kinase. In accordance, ENaCs do not contain conserved consensus sites for phosphorylation by protein kinase A in their cytoplasmic domains.

CONCLUSION

Molecular biology of colonic Na^+ transport has permitted the identification of the key molecules involved into Na^+ permeation through the apical membrane of alveolar and/or airway *epithelia*. With the different tools which have been developed, several important questions have been, or are near to be solved: distinct mechanisms of regulation by steroids, which depend of the cell type, have been identified; the relationships between the epithelial Na^+ channel and other proteins have been revealed. A lot of questions remain unanswered, about the exact site of expression into the lung, about the mechanisms of regulation by second messengers, such as cyclic AMP, free radicals, or cytokines, or about the molecular structures associated with other Na^+-permeant channels.

Acknowledgements : This work was supported by the Centre National de la Recherche Scientifique (CNRS) and the Association Française de Lutte contre la Mucoviscidose (AFLM).

REFERENCES

(Gastrointest. Liver Physiol. 36): G571-G585.

Canessa, C., L. Schild, G. Buell, B. Thorens, I. Gautschi, J. D. Horisberger and B. C. Rossier, 1994, Amiloride-sensitive epithelial sodium channel is made of three homologous subunits. *Nature* 367: 463-467.

Chalfie, M., M. Driscoll and M. Huang, 1993, Degenerin similarities. *Nature* 361: 504.

Champigny, G., N. Voilley, E. Lingueglia, V. Friend, P. Barbry and M. Lazdunski, 1994, Regulation of expression of the lung amiloride-sensitive Na$^+$ channel by steroid hormones. *EMBO J.* 13: 2177-2181.

Farman, N., C. R. Talbot, R. Boucher, M. Fay, C. Canessa, B. Rossier and J.-P. Bonvalet, 1997, Noncoordinated expression of alpha-, beta-, and gamma-subunit mRNAs of epithelial Na$^+$ channel along rat respiratory tract. *Am. J. Physiol.* 272: C131-C141.

Hümmler, E., P. Barker, J. Gatzy, F. Beermann, C. Verdumo, A. Schmidt, R. Boucher and B. C. Rossier, 1996, Early death due to defective neonatal lung liquid clearance in alpha-ENaC-deficient mice. *Nature Genet.* 12: 325-328.

Lingueglia, E., G. Champigny, M. Lazdunski and P. Barbry, 1995. Cloning of the amiloride-sensitive FMRFamide peptide-gated sodium channel. *Nature* 378: 730-733.

Lingueglia, E., S. Renard, R. Waldmann, N. Voilley, G. Champigny, H. Plass, M. Lazdunski and P. Barbry, 1994, Different homologous subunits of the amiloride-sensitive Na$^+$ channel are differently regulated by aldosterone. *J. Biol. Chem.* 269: 13736-13739.

Lingueglia, E., N. Voilley, R. Waldmann, M. Lazdunski and P. Barbry, 1993, Expression cloning of an epithelial amiloride-sensitive Na$^+$ channel. A new channel type with homologies to *Caenorhabditis elegans* degenerins. *FEBS Lett.* 318: 95-99.

Matalon, S., D. J. Benos and R. M. Jackson, 1996, Biophysical and molecular properties of amiloride-inhibitable Na$^+$ channels in alveolar epithelial cells. *Am. J. Physiol.* 271: L1-L22.

Matsushita, K., P. B. MacCray, Jr, R. D. Sigmund, M. J. Welsh and J. B. Stokes, 1996, Localization of epithelial sodium channel subunit mRNAs in adult rat lung by *in situ* hybridization. *Am. J. Physiol.* 271: L332-L339.

Matthay, M. A., H. G. Folkesson, A. S. Verkman, 1996, Salt and water transport across alveolar and distal airway epithelia in the adult lung. *Am. J. Physiol.* 270 (Lung Cell. Mol. Physiol. 14): L487-L503.

O'Brodovich, H., C. Canessa, J. Ueda, B. Rafii, B. C. Rossier and J. Edelson, 1993, Expression of the epithelial Na$^+$ channel in the fetal rat lung. *Am. J. Physiol.* 265: C491-C496.

Renard, S., E. Lingueglia, N. Voilley, M. Lazdunski and P. Barbry, 1994, Biochemical analysis of the membrane topology of the amiloride-sensitive Na$^+$ channel. *J. Biol. Chem.* 269: 12981-12986.

Renard, S., N. Voilley, F. Bassilana, M. Lazdunski and P. Barbry, 1995, Localization and regulation by steroids of the α, β and γ subunits of the amiloride-sensitive Na$^+$ channel in colon, lung and kidney. *Pflügers Arch. - Eur. J. Physiol.* 430: 299-307.

Tchepichev, S., J. Ueda, C. Canessa, B. C. Rossier and H. O'Brodovich, 1995, Lung epithelial Na channel subunits are differentially regulated during development and by steroids. *Am. J. Physiol.* 269: C805-C812.

Voilley, N., F. Bassilana, C. Mignon, S. Merscher, M.-G. Mattéi, G. F. Carle, M. Lazdunski and P. Barbry, 1995, Cloning, chromosomal localization and physical linkage of the β and γ subunits of the human epithelial amiloride-sensitive sodium channel. *Genomics* 28: 560-565.

Voilley, N., A. Galibert, F. Bassilana, S. Renard, E. Lingueglia, S. Le Néchet, G. Champigny, P. Hofman, M. Lazdunski and P. Barbry, 1997, The amiloride-sensitive Na$^+$ channel: from primary structure to function,. *Comp. Biochem. Physiol.* 118(2): 193-200.

Voilley, N., E. Lingueglia, G. Champigny, M.-G. Mattéi, R. Waldmann, M. Lazdunski and P. Barbry, 1994, The lung amiloride-sensitive Na$^+$ channel : biophysical properties, pharmacology, ontogenesis, and molecular cloning. *Proc. Natl. Acad. Sci. USA* 91: 247-251.

Waldmann, R., G. Champigny, F. Bassilana, C. Heurteaux and M. Lazdunski, 1997, A proton gated cation channel involved in acid sensing. *Nature* 386: 173-177.

Waldmann, R., G. Champigny, F. Bassilana, N. Voilley and M. Lazdunski, 1995, Molecular cloning and functional expression of a novel amiloride-sensitive Na$^+$ channel. *J. Biol. Chem.* 270: 27411-27414.

Waldmann, R., G. Champigny, N. Voilley, I. Lauritzen and M. Lazdunski, 1996, The Mammalian degenerin MDEG, an amiloride-sensitive cation channel activated by mutations causing neurodegeneration in *C. elegans*. *J. Biol. Chem.* 271: 10433-10436.

REGULATION OF ALVEOLAR SODIUM TRANSPORT BY HYPOXIA

Carole Planès,[1] and Christine Clerici[1]

[1]Laboratoire de Physiologie, Faculté de Médecine Paris 13
Bobigny, 93012, France

INTRODUCTION

In normal conditions, alveolar epithelium is exposed to relatively high oxygen (O_2) tension of about 100 mmHg. However, a decrease in alveolar O_2 availability may be observed either in pathologic situations such as respiratory distress syndrome, or in environmental conditions associated with decreased barometric pressure. Generalized alveolar hypoxia caused by ascent to high altitude potentially induces acute lung injury with pulmonary edema in healthy subjects as well as in animals[1]. This observation leads to the hypothesis that hypoxia may be directly implicated in edema formation. In the past years, research in the high altitude pulmonary edema pathogenesis focused on lung hemodynamic changes and pulmonary microvasculature insult induced by alveolar hypoxia. One hypothesis[2] is that the uneven hypoxic pulmonary vasoconstriction may markedly increase capillary pressure in lung regions located downstream of small pulmonary arteries that do not constrict, thereby inducing stress failure of capillary walls and edema formation. An alternative hypothesis[1,3,4] suggests that hypoxia, probably via the local release of mediators, could increase pulmonary microvascular permeability and vascular leakage, independently of hemodynamic effects. Surprisingly, although alveolar epithelial cells are directly exposed to variations in ambient O_2 tension, there has been little concern about the potential role of alveolar epithelium in hypoxia-induced alveolar edema.

Over the last decade, numerous studies[5,6] have shown that alveolar epithelium plays a decisive role in maintaining the airspace free of fluid. In physiologic conditions, fluid absorption from alveoli occurs chiefly as a result of active transepithelial sodium (Na) transport[5]. Sodium ions diffuse passively across the apical membrane of alveolar epithelial cells down a favorable electrochemical gradient and are actively extruded at the basolateral side of the cells by the ouabain-sensitive Na,K-adenosine triphosphatase[7,8] (Na,K-ATPase). Amiloride-sensitive Na channels represent *in vivo* and *in vitro* the major pathway for apical Na entry in alveolar cells, and are considered to be the rate-limiting step in alveolar transepithelial Na transport[5]. Their presence in isolated alveolar type II cells (ATII cells) has been evidenced by functional and pharmacological studies using ^{22}Na flux and

Acute Respiratory Distress Syndrome: Cellular and Molecular Mechanisms and Clinical Management
Edited by Matalon and Sznajder, Springer Science+Business Media New York, 1998

35

short-circuit current measurement or whole cell patch-clamp,[7,9,10] and more recently by molecular studies[11] demonstrating that adult rat ATII cells express mRNA transcripts encoding the three subunits α, β, and γ of the rat epithelial sodium channel[12,13] (rENaC). In pathologic situations associated with alveolar flooding, the capacity of alveolar epithelium to maintain effective transepithelial Na transport could be critical in determining the severity and duration of alveolar edema[6]. When the structure and function of epithelium are intact, alveolar Na reabsorption may participate in the clearance of alveolar edema. In some cases, the upregulation of the expression and activity of Na transport proteins in alveolar cells may in addition enhance alveolar fluid resorption, thereby limiting alveolar edema[14-16]. On the contrary, when alveolar cell Na channel or Na,K-ATPase activity is decreased, the impairment of alveolar Na and fluid clearance could favor the maintenance of pulmonary edema[17].

The effects of hypoxia on alveolar epithelium have been poorly investigated. Both *in vivo* and *in vitro* studies[18,19] suggest that alveolar epithelial cells are quite resistant to the decrease in O_2 tension inasmuch as neither ultrastructural characteristics nor cell viability are altered by prolonged hypoxic exposure. However, the ability of these cells to maintain efficient vectorial Na transport under hypoxia remains questionable since a recent preliminary report[20] showed that subacute hypoxic exposure in rat was associated with a decrease in alveolar Na and fluid clearance. Moreover, we have previously reported[21] that prolonged hypoxia inhibits Na,K-ATPase activity in a rat alveolar epithelial cell line. Whether and to what extent hypoxia may alter expression and activity of proteins involved in Na transport by alveolar cells therefore represents an interesting issue for a better understanding of hypoxia-induced pulmonary edema pathogenesis.

EFFECT OF HYPOXIA ON EPITHELIAL NA CHANNEL EXPRESSION AND ACTIVITY IN ATII CELLS IN CULTURE

Previous studies[22,23] using cultures of type II cells, the most numerous alveolar epithelial cells, have demonstrated that alveolar cells from mature lungs are able *in vitro* to actively transport Na. Cultures of ATII cells isolated from adult rat lungs constitute therefore a useful model to study *in vitro* the effects of hypoxia on Na channel expression and activity. Exposure of ATII cell monolayers to hypoxia induced a decrease in Na channel activity estimated by amiloride-sensitive ^{22}Na (ASNa) influx[24]. This decrease was

Table 1. Effect of decreasing oxygen concentration on amiloride-sensitive ^{22}Na influx in ATII cells.

	Oxygen concentration			
	21%	5%	3%	0%
Amiloride-sensitive^{22}Na influx (nmol / mg prot / 5 min)	38.6 ± 2.4	36.3 ± 7.5	26.3 ± 2.9 *	12.7 ± 2.9 *§

Rat ATII cells grown on plastic dishes for 4 days were exposed for 18 h to either 21%, 5%, 3% or 0% O_2, corresponding to O_2 tension in culture medium of 140, 60, 45 and 30 mmHg respectively. Immediately at the end of exposure, ^{22}Na influx measurements were performed over a 5 min period. AsNa influx was determined as the difference between uptake values in the absence and presence of amiloride (100 μM). Values are means ± SE of 6 separate experiments in which triplicates were obtained. Statistical difference (P < 0.05) of values from 21% group is indicated by *, from 3% group is indicated by §.

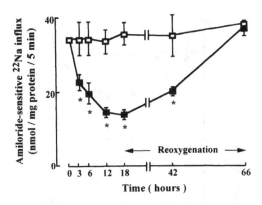

Figure 1. Effect of hypoxia and reoxygenation on AsNa influx in ATII cells. After 4 days in culture. rat ATII cells were exposed after 4 days in culture to either normoxia (21% O_2; open squares) or hypoxia (0% O_2; solid squares) for 3. 6. 12 or 18 h. or to hypoxia followed by reoxygenation (18 h 0% O_2 + 24 or 48 h 21% O_2). Immediately at the end of exposure. ^{22}Na influx measurements were performed over a 5 min period. AsNa influx was determined as the difference between uptake values in the absence and presence of amiloride (100 μM). Results represent means ± SE of 3 to 7 different experiments in which triplicates were obtained. * Significantly different from normoxic control value. P < 0.05.

O_2 concentration-dependent, observed with 3% O_2 and 0% O_2 but not with 5% O_2 (Table 1), and also time-dependent, apparent at 3 h of 0% O_2 exposure and maximal at 12 h and 18 h[25] (Fig. 1). In addition, exposure to 0% O_2 induced a time-dependent decline in α-, β- and γ-rENaC subunit mRNA levels evaluated by RNAse protection assay (Fig. 2), together with a 42% decrease in the rate of α-rENaC protein synthesis as determined by immunoprecipitation with anti-rENaC α subunit antibody. Hypoxia-induced decrease in Na channel expression and activity was not due to a general toxic effect. First, while rENaC mRNA transcripts decreased, the level of other mRNA transcripts like β-actin mRNA remained unchanged. Second, the inhibitory effects of hypoxia were completely reversed when hypoxic cells were transferred to normoxic atmosphere for 48 h (Fig. 1 and 2). Third, neither light nor electron microscopy examination revealed any modification in hypoxic cell aspect as compared with normoxic cells, except the fact that the number of domes (which are thought to be the consequence of active ion transport from the apical to the basolateral side of ATII cells with water following passively)[26] was strikingly reduced in cell monolayers exposed to 0% O_2 hypoxic atmosphere for 18 h. The fact that in these experiments the decrease in ASNa influx roughly paralleled that in rENaC subunit mRNA levels suggests a causative link between reduced mRNA expression and impaired activity, at least for long exposure times. Likewise, it was previously reported[13] that in Xenopus oocytes, maximal channel activity was obtained only when the three rENaC subunit cDNAs were simultaneously coinjected. That the expression of the three subunits is crucial for optimal activity is supported herein by the fact that, during *long-term hypoxia* (18h exposure), maximal inhibition of Na channel activity occurred concomitantly with maximal decrease in α-, β- and γ-rENaC mRNA transcripts and with reduced synthesis of α-rENaC for optimal activity is supported herein by the fact that, during *long-term hypoxia* (18h exposure), maximal inhibition of Na channel activity occurred concomitantly with maximal

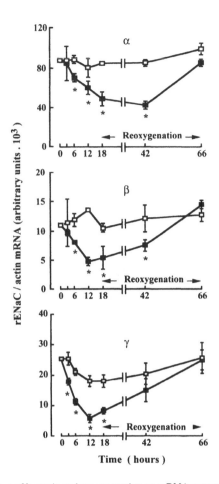

Figure 2. Effect of hypoxia and reoxygenation on mRNA expression of α-, β- and γ-rENaC subunits in ATII cells. Rat ATII cells were exposed after 4 days in culture to either normoxia (21% O_2; open squares) or hypoxia (0% O_2; solid squares) for 3, 6, 12 or 18 h, or to hypoxia followed by reoxygenation (18 h 0% O_2 + 24 or 48 h 21% O_2). At the end of exposure, RNase protection assays were performed on cell lysates (RNA equivalent to 10^6 cells), and α-, β- and γ-rENaC subunit mRNA levels were quantitated using an Instant Imager. Data were normalized for the corresponding actin signal in each lane. Results are expressed as the unitless ratio of α-, β- and γ-rENaC subunit mRNA / actin mRNA, and represent means ± SE of 3 to 6 independent experiments. * Significantly different from normoxic control value. $P < 0.05$.

decrease in α-, β- and γ-rENaC mRNA transcripts and with reduced synthesis of α-rENaC protein. Moreover, the observation that during reoxygenation, the progressive recovery in Na channel activity paralleled the recovery in α-, β- and γ-rENaC mRNA level, and was associated with normalization of α-rENaC protein synthesis suggests that de novo synthesis of rENaC subunits is necessary to restore normal Na channel activity after prolonged hypoxia, and that this protein synthesis requires the restoration of adequate rENaC mRNA levels. As regard to *short-term hypoxia* (3 h exposure), it is noteworthy that reduced channel activity occured at a time when only γ-rENaC mRNA was decreased, whereas α- and β-rENaC mRNA levels were unchanged. One possibility is that reduced amount of γ-rENaC mRNA led to insufficient production of γ-rENaC subunit and subsequently accounted for reduced functional activity, inasmuch as γ-rENaC subunit was previously reported[27] to be necessary for the correct processing of Na channel proteins at the cell surface. The other possibility is that the decrease in Na channel activity is related to translational or post-translational events including decreased efficiency in the translation of rENaC mRNA or in the apical membrane trafficking of rENaC subunits, abnormal degradation or internalization of the channel protein,[16] or hypoxia-induced modification of intracellular signals that modulate Na channel activity[28].

EFFECT OF HYPOXIA ON EXPRESSION AND ACTIVITY OF NA,K-ATPASE IN ATII CELLS IN CULTURE

Along with apical Na channels, basolateral Na,K-ATPase represents the major protein involved in transepithelial Na transport by alveolar cells, and its functional activity is usually tightly coupled with apical Na entry in order to ensure efficient vectoriel Na transport and to maintain cell homeostasis. 0% O_2 hypoxia resulted in a time-dependent decrease in Na,K-ATPase activity estimated by ouabain-sensitive [86]rubidium influx (OsRb influx) in ATII cell monolayers which grossly paralleled that in Na channel activity, being significant but moderate at 3 and 6 h of exposure, and maximal at 18 h hypoxia[29] (Fig. 3). RNase protection assays showed that the levels of mRNA transcripts encoding α_1- and β_1-

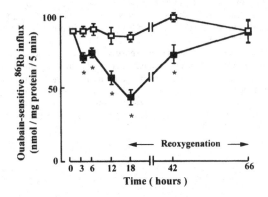

Figure 3. Effect of hypoxia and reoxygenation on OsRb influx in ATII cells. Rat ATII cells were exposed after 4 days in culture to either normoxia (21% O_2; open squares) or hypoxia (0% O_2; solid squares) for 3, 6, 12 or 18 h, or to hypoxia followed by reoxygenation (18 h 0% O_2 + 24 or 48 h 21% O_2). [86]Rb influx studies were performed immediately at the end of exposure over a 5 min period. OsRb influx was determined as the difference between uptake values in the absence and presence of ouabain (1 mM). Results represent means ± SE of 4 to 6 different experiments in which triplicates were obtained. * Significantly different from normoxic control value. P < 0.05.

Na,K-ATPase, the enzyme subunits predominantly expressed in ATII cells,[30,31] were reduced after at least 6 h of 0% O_2 exposure (Fig. 4).

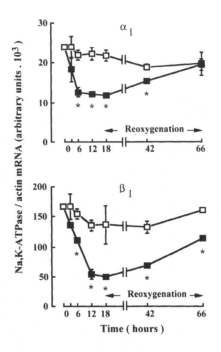

Figure 4. Effect of hypoxia and reoxygenation on mRNA expression of α_1- and β_1-Na,K-ATPase subunits in ATII cells. Rat ATII cells were exposed after 4 days in culture to either normoxia (21% O_2; open squares) or hypoxia (0% O_2; solid squares) for 3, 6, 12 or 18 h, or to hypoxia followed by reoxygenation (18 h 0% O_2 + 24 or 48 h 21% O_2). At the end of exposure, RNase protection assays were performed on cell lysates (RNA equivalent to 10^6 cells), and α_1- and β_1-Na,K-ATPase mRNA levels were quantitated using an Instant Imager. Data were normalized for the corresponding actin signal in each lane. Results are expressed as the ratio of α_1- and β_1-Na,K-ATPase subunit mRNA / actin mRNA, and represent means ± SE of 3 to 6 independent experiments.*Significantly different from normoxic control value. P < 0.05.

The decrease in OsRb influx observed after *short term hypoxia* was likely related to changes in cellular effectors that modulate Na,K-ATPase activity in intact cells rather than to decreased expression of Na,K-ATPase, since the maximal velocity (Vmax) of the enzyme reflecting the number of available functional units of sodium pump was not modified by 3 hours of 0% O_2 exposure (Table 2). The concentration of adenosine

Table 2. Effect of hypoxia and reoxygenation on the maximal velocity of Na,K-ATPase activity in ATII cells.

	Vmax of Na,K-ATPase activity (μmol Pi / mg prot / h)		
	Exposure time		18 h + 48 h reox.
	3 h	18 h	
Normoxia (21% O_2)	4.16 ± 1	4.47 ± 0.49	5.98 ± 0.28
Hypoxia (0% O_2)	4.31 ± 0.65	2.20 ± 0.53 *	6.05 ± 0.13

Rat ATII cells were exposed after 4 days in culture to either normoxia (21% O_2) or hypoxia (0% O_2) for 3 or 18 h, or to hypoxia followed by reoxygenation (18 h 0% O_2 + 48 h 21% O_2). Immediately at the end of exposure, crude cell homogenates were prepared and Na,K-ATPase activity was determined by the rate of ouabain-inhibitable ATP hydrolysis under conditions of maximal velocity (Vmax) determination over a 15 min period. Results represent means $\pm$ SE of 3 or 6 separate experiments in which triplicates were obtained. * Significantly different from normoxic value. $P < 0.05$.

triphosphate (ATP), one of the cellular effectors known to modulate Na,K-ATPase activity[32], may be severely decreased during hypoxia[33]. In our experiments however, ATP content was reduced by 40% only in ATII cells exposed for 3 h to 0% O_2 (Table 3). Such a moderate ATP depletion is likely not responsible for OsRb influx decrease inasmuch as we previously reported[34] that a 90% ATP depletion was not sufficient to impair Na,K-ATPase activity in normoxic ATII cells. Intracellular Na concentration also modulates Na,K-ATPase activity in intact cells. Since AsNa influx was significantly reduced by a 3 h exposure to 0% O_2, the OsRb influx decrease observed concomitantly could reflect *short-term* feedback inhibition of the Na pump in response to decreased intracellular Na concentration due to reduced rate of Na entry through Na channels[35].

Table 3. Effect of hypoxia on alveolar type II cell ATP content.

	Cell ATP content (nmol / mg protein)			
	Exposure time			
	3 h	6 h	12 h	18 h
Normoxia (21% O_2)	35.8 ± 4.6	36.2 ± 2.9	32.6 ± 2.6	27.8 ± 3.3
Hypoxia (0% O_2)	21.3 ± 2.5 *	26 ± 3.7	31.3 ± 3.6	30.9 ± 2.7

Rat ATII cells were exposed after 4 days in culture to 21% O_2 or 0% O_2 for 3, 6, 12 or 18 h. At the end of exposure, cell ATP content was measured using a luciferin-luciferase assay in normoxic and hypoxic cell extracts Results represent means $\pm$ SE of 6 independent experiments. * Significantly different from normoxic control ($p < 0.05$).

During *long term hypoxia*, the decrease in OsRb influx was associated with reduced Vmax of the enzyme, with decreased levels of Na,K-ATPase mRNA transcripts, but with normal cell ATP content. In this case, the impairment of Na pump activity is probably the consequence of a direct effect of hypoxia either on Na,K-ATPase mRNA expression, or at a post-translational level including abnormal cellular routage or increased degradation rate of the protein[36]. It cannot be ruled out, however, that prolonged inhibition of Na channels might also be partly responsible for reduced Vmax in this case, since sustained decrease in intracellular Na was previously shown[37] to reduce the number of pumps inserted in plasmic membrane. Finally during reoxygenation, the normalization of OsRb influx parallels the normalization of α_1-Na,K-ATPase subunit mRNA level, suggesting that the synthesis of new catalytic subunits of the enzyme is necessary to restore normal Na,K-ATPase activity.

DOWNREGULATION OF EPITHELIAL NA CHANNELS AND NA,K-ATPASE BY HYPOXIA IN ATII CELLS : PHYSIOPATHOLOGICAL IMPLICATIONS

The data presented herein clearly show that decrease in O_2 tension downregulates expression and activity of Na transport proteins in ATII cells *in vitro*. During *short term* exposure to hypoxia, only the functional activities - but not the mRNA expression- of both Na channel and Na,K-ATPase are decreased, and it can be hypothetized that the impairment of Na,K-ATPase activity in this case is secondary to the decrease in Na channel activity. By contrast, after *long term* hypoxia, the marked decrease in Na channel and Na,K-ATPase activities is likely related to reduced expression of both proteins. Indeed, the time-dependent decrease in rENaC and Na,K-ATPase mRNA levels under 0% O_2 followed by an increase when the cells were transferred from 0% to 21% O_2 strongly suggests that Na channel and Na,K-ATPase expression is regulated by O_2 tension. Consistent with this hypothesis, others have shown that, in alveolar cells, an increase in O_2 tension upregulated the level of rENaC transcripts: (i) the transfer of rat fetal distal lung epithelial cells in culture from 3% O_2 to higher O_2 concentration induced an increase in rENaC mRNA transcripts as well as in Na channel activity[38], (ii) hyperoxic lung injury in adult rat was associated with increased α-rENaC mRNA expression and Na channel activity in ATII cells[16]. In addition, it was recently reported[39] that α_1- and β_1-Na,K-ATPase subunit mRNA levels both increased in ATII cells from adult rats exposed to hyperoxia. The molecular mechanisms whereby decrease in O_2 tension downregulates rENaC and Na,K-ATPase mRNA transcripts in alveolar cells has not been yet elucidated and it is not known whether this regulation occurs at translational or post-translational level, or both. More generally, the mechanism of O_2-sensing, of hypoxic signal transduction and the nature of transcription factors controlling the expression of O_2-responsive genes in ATII cells has not been yet clearly investigated.

Whatever the mechanism involved, the downregulation of ATII cell Na transport proteins by hypoxia we observed *in vitro* may have important pathophysiological implications *in vivo*, inasmuch as it may alter the capacity of ATII cells to actively transport Na across alveolar epithelium. It is generally assumed that, *in vivo*, hypoxia-induced alveolar edema is mostly related to lung hemodynamic changes or increase in pulmonary microvascular permeability[4]. From our results, we can speculate that decreased expression and activity of Na transport proteins in ATII cells may in addition hamper the resorption of alveolar edema in the early phase of hypoxic lung injury, unless catecholamine-dependent and independent compensatory mechanisms develop to counteract this effect[14-16]. In line with our data, a recent preliminary report[20] showed that subacute hypoxic exposure in rat was associated with a decrease in alveolar Na and fluid clearance. Finally, in our experiments, the recovery in Na channel and Na,K-ATPase expression and activity observed upon reoxygenation suggests that active transepithelial Na transport may be restored during the reparative phase of hypoxic lung injury, and may therefore contribute, along with other adaptive mechanisms, to the clearance of alveolar fluid.

Acknowledgments: the authors thank Dr B. Escoubet and Dr M. Blot-Chabaud for their fruitful collaboration. This work was supported by grants from Recherche et Partage and Fondation pour la Recherche Médicale.

REFERENCES

1. J.-P. Richalet, High-altitude pulmonary edema: still a place for controversy?, *Thorax* 50:923-929 (1995).
2. J. B. West and O. Mathieu-Costello, High altitude pulmonary edema is caused by stress failure of pulmonary capillaries, *Int. J. Sports Med.* 13:S54-S58 (1992).
3. S. Ogawa, H. Gerlach, C. Esposito, A. Pasagian-Macaulay, J. Brett, and D. Stern, Hypoxia modulates the barrier and coagulant function of cultured bovine endothelium, *J. Clin. Invest.* 85:1090-1098 (1990).
4. T. J. Stelzner, R. F. O'brien, K. Sato, and J. V. Weil, Hypoxia-induced increases in pulmonary transvascular protein escape in rats, *J. Clin. Invest.* 82:1840-1847 (1988).
5. G. Basset, C. Crone, and G. Saumon, Significance of active ion transport in transalveolar water absorption : a study on isolated rat lung, *J. Physiol. Lond.* 384:311-324 (1987).
6. M. A. Matthay and J. P. Wiener-Kronish, Intact epithelium barrier function is critical for the resolution of alveolar edema in humans, *Am. Rev. Respir. Dis.* 142:1250-1257 (1990).
7. S. Matalon, R. J. Bridges, and D. J. Benos, Amiloride-inhibitable Na^+ conductive pathways in alveolar type II pneumocytes, *Am. J. Physiol.* 260:L-90-L96 (1991).
8. G. Saumon and G. Basset, Electrolyte and fluid transport across the mature alveolar epithelium, *J. Appl. Physiol.* 74:1-15 (1993).
9. S. Matalon, K. L. Kirk, J. K. Bubien, Y. Oh, P. Hu, G. Yue, R. Shoemaker, E. J. Cragoe Jr, and D. J. Benos, Immunocytochemical and functional characterization of Na^+ conductance in adult alveolar pneumocytes, *Am. J. Physiol.* 262:C1228-C1238 (1992).
10. R. M. Russo, R. L. Lubman, and E. D. Crandall, Evidence for amiloride-sensitive sodium channels in alveolar epithelial cells, *Am. J. Physiol.* 262:L405-L411 (1992).
11. S. Tchepichev, J. Ueda, C. Canessa, B. C. Rossier, and H. O'Brodovich., Lung epithelial Na channel subunits are differentially regulated during development and by steroids, *Am. J. Physiol.* 269:C805-C812 (1995).
12. C. M. Canessa, J.-D. Horisberger, and B. C. Rossier, Epithelial sodium channel related to proteins involved in neurodegeneration, *Nature Lond.* 361:467-470 (1993).
13. C. M. Canessa, L. Schild, G. Buell, B. Thorens, I. Gautschi, J.-D. Horisberger, and B. C. Rossier, Amiloride-sensitive epithelial Na^+ channel is made of three homologous subunits, *Nature Lond.* 367:463-467 (1994).
14. J. F. Pittet, J. P. Wiener-Kronish, M. C. McElroy, H. G. Folkesson, and M. A. Matthay, Stimulation of lung epithelial liquid clearance by endogenous release of catecholamines in septic shock in anesthetized rats, *J. Clin. Invest.* 94:663-671 (1994).
15. W. G. Olivera, K. M. Ridge, L. D. H. Wood, and J. I. Sznajder, Active sodium transport and alveolar epithelial Na-K-ATPase increase during subacute hyperoxia in rats, *Am. J. Physiol.* 266:L577-L584 (1994).
16. G. Yue, W. J. Russell, D. J. Benos, R. M. Jackson, M. A. Olman, and S. Matalon, Increased expression and activity of sodium channels in alveolar type II cells of hyperoxic rats, *Proc. Natl. Acad. Sci. USA.* 92:8418-8422 (1995).
17. W. G. Olivera, K. M. Ridge, and J. I. Sznajder, Lung liquid clearance and Na,K-ATPase during acute hyperoxia and recovery in rats, *Am. J. Respir. Crit. Care Med.* 152:1229-1234 (1995).
18. K. W. M. Scott, G. R. Barer, E. Leach, and I. P. F. Mungall, Pulmonary ultrastructural changes in hypoxic rats, *J. Path.* 126:27-33 (1978).
19. K. K. Graven, L. H. Zimmerman, E. W. Dickson, G. L. Weinhouse, and H. W. Farber, Endothelial cell hypoxia associated proteins are cell and stress specific, *J. Cell. Physiol.* 157:544-554 (1993).
20. S. Suzuki, T. Sakuma, M. Sugita, Y. Hoshikawa, M. Noda, S. Ono, T. Tanita, and S. Fujimura, Subacute hypoxic challenge decreases alveolar fluid clearance in rat, *Am. J. Respir. Crit. Care Med.* 153:A505 Abstr. (1996).
21. C. Planès, G. Friedlander, A. Loiseau, C. Amiel, and C. Clerici, Inhibition of Na,K-ATPase activity after prolonged hypoxia in an alveolar epithelial cell line, *Am. J. Physiol.* 271:L71-L78 (1996).
22. B. E. Goodman, F. Fleisher, and E. D. Crandall, Evidence for active ion transport by cultured monolayers of pulmonary alveolar epithelial cells, *Am. J. Physiol.* 245:C78-C83 (1983).
23. R. J. Mason, M. C. Williams, W. J. H., M. F. Sanders, D. S. Misfeldt, and L. C. Berry JR, Transepithelial transport by pulmonary alveolar type II cells in primary culture, *Proc. Natl. Acad. Sci. USA.* 79:6033-6037 (1982).
24. C. Clerici, S. Couette, A. Loiseau, P. Herman, and C. Amiel, Evidence of Na-K-Cl cotransport in alveolar epithelial cells: effect of phorbol ester and osmotic stress, *J. Membrane Biol.* 147:295-304 (1995).

25. C. Planès, B. Escoubet, C. Amiel, and C. Clerici, Hypoxia decreases expression and activity of amiloride-sensitive sodium channels in adult alveolar type II cells, *Am. J. Respir. Crit. Care Med.* 163:A 505 Abstr. (1996).

26. B. E. Goodman and E. D. Crandall, Dome formation in primary cultured monolayers of alveolar epithelial cells, *Am. J. Physiol.* 243:C96-C100 (1982).

27. E. Lingueglia, S. Renard, R. Waldmann, N. Voilley, G. Champigny, H. Plass, M. Lazdunski, and P. Barbry, Different homologous subunits of the amiloride-sensitive Na^+ channel are differently regulated by aldosterone, *J. Biol. Chem.* 269:13736-13739 (1994).

28. L. G. Palmer and G. Frindt, Effects of cell Ca^{2+} and pH on Na^+ channels from rat cortical collecting tubule, *Am. J. Physiol.* 253:F333-F339 (1987).

29. C. Planès, B. Escoubet, C. Amiel, and C. Clerici, Hypoxia decreases expression and activity of Na,K-ATPase in adult alveolar type II cells, *Am. J. Respir. Crit. Care Med.* 163:A 505 Abstr. (1996).

30. D. H. Ingbar C. Burns Weeks, M. Gilmore-Hebert, E. Jacobsen, S. Duvick, R. Dowin, S. K. Savik, and J. D. Jamieson, Developmental regulation of Na,K-ATPase in rat lung, *Am. J. Physiol.* 270:L619-L629 (1996).

31. S. Suzuki, D. Zuege, and Y. Berthiaume, Sodium-independent modulation of Na,K-ATPase activity by β-adrenergique agonist in alveolar type II cells, *Am. J. Physiol.* 268:L983-L990 (1995).

32. S. P. Soltoff, ATP and the regulation of renal cell function, *Annu. Rev. Physiol.* 48:9-31 (1986).

33. A. V. Tretyakov and H. W. Farber, Endothelial cell tolerance to hypoxia. Potential role of purine nucleotide phosphates, *J. Clin. Invest.* 95:738-744 (1995).

34. C. Clerici, G. Friedlander, and C. Amiel, Impairment of sodium-coupled uptakes by hydrogen peroxyde in alveolar type II cells: protective effect of d-α-tocopherol, *Am. J. Physiol.* 262:L542-L548 (1992).

35. T. A. Pressley, Ion concentration-dependent regulation of Na,K-pump abundance, *J. Membrane Biol.* 105:187-195 (1988).

36. B. A. Molitoris, P. D. Wilson, R. W. Schrier, and R. S. Simon, Ischemia induces partial loss of surface membrane polarity and accumulation of putative calcium ionophores, *J. Clin. Invest.* 76:2097-2105 (1985).

37. D. Kim and T. W. Smith., Effect of growth in low-Na^+ medium on transport sites in cultured heart cells, *Am. J. Physiol.* 250:C32-C39 (1986).

38. O. Pitkänen A. K. Tanswell, G. Downey, and H. O'Brodovich, Increased PO_2 alters the bioelectric properties of fetal lung epithelium, *Am. J. Physiol.* 270:L1060-L1066 (1996).

39. E. P. Carter, O. D. Wangensteen, S. M. O'Grady, and D. H. Ingbar, Effects of hyperoxia on type II cell Na-K-ATPase function and expression, *Am. J. Physiol.* 272:L542-L551 (1997).

VECTORIAL MOVEMENT OF SODIUM IN LUNG ALVEOLAR EPITHELIUM: ROLE AND REGULATION OF Na^+,K^+-ATPase

Alejandro M. Bertorello[1], Karen Ridge[2], Goichi Ogimoto[1], Guillermo Yudowski[1], Adrian I. Katz[3] and J. Iasha Sznajder[2]

[1]Department of Molecular Medicine, Karolinska Institute, 171 76 Stockholm, Sweden

[2]Michael Reese Hospital and Medical Center, Pulmonary and Critical Care Medicine and University of Illinois at Chicago, Chicago IL 60616

[3]Department of Medicine, University of Chicago Pritzker School of Medicine, Chicago, Illinois 60637

INTRODUCTION

The Na^+,K^+-ATPase (Na:Kpump) is an integral membrane protein responsible for maintaining the gradients of sodium and potassium across the plasma membrane[1]. These gradients constitute the basis for resting membrane potential permitting membrane excitability in neurons[2], and control indirectly diverse secretory processes, such as that of insulin from pancreatic β-cells[3].

In most transporting epithelia, including those from the kidney and lung alveoli, the Na^+,K^+-ATPase is confined to the basolateral domain of the cells[4], and this polarized distribution is critical for the vectorial transport of sodium followed isosmotically by that of water. A decrease in renal tubule Na^+,K^+-ATPase activity induced by dopamine is an important adaptive mechanism responsible for the control of sodium reabsorption during high salt diet[5], and impaired Na^+,K^+-ATPase regulation under these circumstances has been associated with the

Acute Respiratory Distress Syndrome: Cellular and Molecular Mechanisms and Clinical Management
Edited by Matalon and Sznajder, Springer Science+Business Media New York, 1998

45

development of salt-sensitive hypertension[6,7]. Equally important, changes in Na[+],K[+]-ATPase activity in alveolar epithelial cells has been associated with increased clearance of alveolar fluid and improvement of pulmonary edema during development[8].

VECTORIAL TRANSPORT OF SODIUM IN THE ALVEOLAR EPITHELIUM

Recent investigations have demonstrated the importance of transcellular sodium transport and its role in edema clearance in the survival of mechanically ventilated patients with respiratory failure[9,10]. Resolution of edema is dependent predominantly on liquid clearance across the alveolar barrier following osmotic gradients generated by active sodium transport[11-17], which has been shown in live animal models and in isolated lungs[12-17]. Transport of sodium across the alveolar epithelium, and hence edema clearance, is predominantly dependent on its entry into cells through apical sodium channels and its active extrusion mediated by the basolateral Na[+],K[+]-ATPase[18-22]. Supporting the importance of active sodium transport in effecting lung edema clearance are studies in rat lungs showing that lung liquid clearance was completely stopped by hypothermia, possibly by inhibiting metabolic processes on which solute transport is highly dependent[23,24], and that it was significantly decreased by both amiloride, a sodium channel inhibitor and ouabain, a Na[+],K[+]-ATPase inhibitor[13,14,25-27].

Catecholamines are important modulators of fluid transport in several transporting epithelia. There is increasing evidence that β-adrenergic stimulation doubles the basal levels of lung liquid clearance in isolated and in vivo rat lungs[14,17]. Administration of β-agonists has variable but mostly stimulatory effects on lung liquid clearance in different species[15,16,26]. Recent reports suggest that these changes in alveolar fluid movement elicited by β-adrenergic agonists are mediated by stimulation of both amiloride-sensitive sodium channels and Na[+],K[+]-ATPase - mediated active sodium transport[27,28]. This synergy is also seen in several models of hyperoxia[29-32].

REGULATION OF Na[+],K[+]-ATPase ACTIVITY

Changes in Na[+],K[+]-ATPase activity in response to agonists could be due to a conformational change in the enzyme's structure with the consequent alteration in its catalytic activity or, alternatively, be the result of changes in the number of active Na[+],K[+]-ATPase units within the plasma membrane. While the former is representative of short-term regulation (within minutes), the latter represents a more sustained response that may require several hours, or longer if pumps are to be synthesized *de novo*.

Cellular mechanisms. Acute changes in Na^+,K^+-ATPase activity have been extensively documented[33], and in recent years it has been clearly established that they require a cell- and agonist-specific signaling system that ultimately leads to activation of protein kinases[33]. To provide a concise description of these mechanisms, this brief survey will review, only the effects of various agonists in the alveolar epithelium, which will be compared to those in renal tubule cells.

Second messengers. Distinct and cell-specific signaling pathways, set in motion after membrane receptor activation, participate in both the short-term and sustained regulation of the Na:K-pump. A great variety of responses have been observed involving mostly the activation of adenylyl cyclase, phospholipase C, phospholipase A_2 as well as the lipid kinase, phosphatydilinositol 3-kinase (for review see Bertorello & Katz[33]; Eward & Klip[34]). Many of these stimuli seem to converge to steps that involve the increased production of arachidonic acid and of its various metabolites, principally those generated by the action of cytochrome P450-dependent monooxygenase. Of the latter, 20-hydroxyeicosatetraenoic acid (20-HETE) plays an important role in cell signalling in proximal tubule cells and elsewhere[33].

Protein phosphorylation. *In vitro* experiments have demonstrated that the Na^+,K^+-ATPase catalytic (α) subunit is a good substrate for protein kinases, in particular protein kinase C (PKC) and the cAMP-dependent protein kinase (cAMP-K), which phosphorylate distinct amino acid residues, namely Ser^{11} and Ser^{18} by PKC, and Ser^{943} by cAMP-K[35]. Phosphorylation of the α subunit is translated into different effects on its catalytic activity, e.g., while PKC-dependent phosphorylation is generally associated with a decrease in its hydrolytic activity[36,37], phosphorylation by cAMP-K often results in an increased activity[38]. These effects are presumably the consequence of a change in the conformational structure of the enzyme[39].

While in cell-free preparations inhibition or stimulation of Na^+,K^+-ATPase activity is associated with activation of PKC or cAMP-K, experiments performed in intact cells demonstrate conflicting effects that complicate their interpretation. For example, the effects of PKC activators on enzyme activity and their correlation to α subunit phosphorylation require a more thorough examination. Many of these discrepancies stem from the fact that activation of

PKC is commonly induced with phorbol esters. One must therefore take into account the possibility that other PKC-dependent processes are activated simultaneously, and that PKCs participating in the regulation of Na^+,K^+-ATPase activity may be phorbol ester-insensitive. Another potential variable when evaluating the role of PKC in the regulation of Na^+,K^+-ATPase is when intact- or modified (mutated) subunit cDNAs are transfected into cells (amphibian or mammalian) that may not fully exhibit the specific physiological functions linked to the regulation of Na^+,K^+-ATPase activity.

Subunits endocytosis. Certain cardiac glycoside, such as ouabain, bind to the Na^+,K^+-ATPase α subunit and act as specific inhibitors of the enzyme. Because exposure of cells to protein kinase C activators leads to a decrease in ouabain binding, it has been postulated that regulation of Na^+,K^+-ATPase activity may be also achieved by removal of the pump's component units from the plasma membrane. Supporting this observation, when renal epithelial cells are treated with PKC activators and plasma membranes are then prepared, the total Na^+,K^+-ATPase activity is reduced, suggesting that this decrease in hydrolytic activity (which persists after cell disruption) may have been the consequence of reduced number of copies in the plasma membrane. Using a physiological agonist such as dopamine in intact renal proximal tubule cells, where it is produced during high salt diet we recently demonstrated that a fraction of Na^+,K^+-ATPase α/β subunits migrate from their location in the plasma membrane to intracellular compartments, notably endosomes[40].

Endocytosis of integral plasma membrane proteins occurs through a step-wise translocation into early- and late endosomes, from where they may recycle to the plasma membrane or eventually proceed to lysosomes, where they are degraded. The endocytic process is initiated at the plasma membrane after formation of a membrane pit by accumulation of clathrin[41], which is pinched off by the action of a small GTP-binding protein termed dynamin[42]. Clathrin vesicles fuse with, and deliver their contents to early endosomes.

Using immunoblotting techniques we have observed that in proximal tubule cells under basal conditions, clathrin vesicles and early- and late endosomes all contain considerable amounts of Na^+,K^+-ATPase α- and β subunits[40], suggesting the existence of a constitutive endocytic process. Incubation of renal tubule cells with dopamine results in decreased Na^+,K^+-ATPase activity[40,43] associated with an increased abundance of α- and β subunits in clathrin vesicles, early- and late endosomes[40]. Inhibition of enzyme activity, as well as its endocytosis, are blocked by PKC inhibitors. It appears that under the influence of dopamine the Na^+,K^+-ATPase α subunit is phosphorylated (via PKC), and that this phosphorylation constitutes the triggering signal for its endocytosis.

The above conclusions are also supported by studies performed in opossum kidney (OK) cells transfected with the full rat α subunit DNA or with the DNA lacking the putative PKC phosphorylation sites. In the latter, dopamine failed to phosphorylate the α subunit and did not increase its endocytosis (A.V. Chibalin, C.H. Pedemonte, A.I. Katz, E. Féraille, P.-O. Berggren, and A.M. Bertorello, unpublished observations). Although it is clear that the mechanisms involved are far more complex, identification of the triggering signal and of the traffic sequence in α subunit endocytosis represent a significant advance in our understanding of how the regulation of some membrane transport proteins activity is accomplished.

Shuttle of Na⁺,K⁺-ATPase subunits to the plasma membrane. In the renal and alveolar epithelia, both α and β adrenergic agonists stimulate Na⁺,K⁺-ATPase activity. In renal epithelial cells the stimulation (by α-adrenergic agonists) appears to be due to a direct effect on the enzyme[44] as well as to changes in sodium permeability[45], while in the alveolar epithelium this effect (β-adrenergic stimulation) seems to be chiefly by modification of the number of pump molecules in the plasma membrane[46].

In kidney proximal tubules oxymetazoline, and α-adrenergic agonist, stimulated Na⁺,K⁺-ATPase activity[44]. Because this effect was prevented by calcineurin inhibitors, the authors conclude that the phenomenon is mediated by protein phosphatases, and in particular by protein phosphatase 2B. Those studies, however, did not assess the possibility that this effect could have been secondary to changes in sodium permeability, a documented result of these agonists' action in renal proximal tubule cells.

The β-adrenergic agonist-induced increase of Na⁺,K⁺-ATPase activity in alveolar epithelial cells[48] is associated with an increased number of Na⁺,K⁺-ATPase copies in the plasma membrane (A.M. Bertorello, K.M. Ridge, A.V. Chibalin, A.I. Katz, and J.I. Sznajder, unpublished observations). Moreover, this effect parallels a significant decrease of α subunits within endosomal compartments (late endosomes), which strongly suggests that the subunits are recruited from these intracellular pools to the plasma membrane. Neither the changes in subunit abundance nor in pump activity were affected by amiloride, lending further support to a postulated direct regulation of Na⁺,K⁺-ATPase activity, independently of changes in sodium permeability and intracellular sodium concentration.

Subunits traffic and cytoskeletal organization. The Na^+,K^+-ATPase is anchored in the plasma membrane by a complex interplay of several cytoskeletal proteins[46]. Subunits' removal from, or insertion in the membrane is therefore likely be accompanied by reorganization of the actin/microtubule cytoskeleton.

Endocytosis: Dopamine-induced endocytosis of Na^+,K^+-ATPase α/β subunits is affected by compounds that selectively modify the actin/microtubule network[40]. Overnight treatment of intact cells with the fungal toxin phallacidin stabilizes the cortical actin cytoskeleton without affecting the state of cytoplasmic actin, which results in impaired depolymerization and reorganization of actin filaments. The aforementioned increase in subunits' sequential accumulation into clathrin vesicles, early- and late endosomes was blunted by pretreatment with phallacidin. This observation suggests that stabilization of the cortical actin has prevented first clathrin-coated pit formation, and thereby clathrin vesicle-dependent transport. Although microtubules are responsible for endosomal transport, they do not affect clathrin vesicle formation[47]. In accord with this formulation, in our studies two microtubule depolymerizing agents (nocodazole and colchicine) prevented α/β subunit transport into early- and late endosomes, but did not affect the subunits' accumulation in clathrin vesicles[40].

Reinsertion: The actin/microtubule system also plays a critical role in the traffic of Na^+,K^+-ATPase subunits to the plasma membrane. We have already mentioned that isoproterenol-stimulated Na^+,K^+-ATPase activity is associated with an increased number of pump units in the plasma membrane. Changes in pump activity as well as in α subunit abundance were blocked by overnight treatment with the actin stabilizing agent, phallacidin[49]. Additionally, the increase in lung fluid clearance in response to isoproterenol was blocked by pretreatment with colchicine, a microtubule depolymerizing drug[29].

REFERENCES

1. I.M. Glynn. The Na,K-transporting adenosine triphosphatase. In: The Enzymes of Biological Membranes, edited by A.N. Martonosi. New York: Plenum, 3: 28 (1985).
2. J.W. Phillis, and P.H. Wu. Catecholamines and the sodium pump in excitable cells. *Prog. Neurobiol.* 17: 141 (1981).
3. B. Ribalet, C.J. Mirell, D.G. Johnson, and S.R. Levin. Sulfonylurea binding to a low-affinity site inhibits the Na/K-ATPase and the K_{ATP} channel in insulin-secreting cells. *J. Gen. Physiol.* 107: 231 (1996).
4. R.K. Kinne. Polarity, diversity, and plasticity in proximal tubule transport systems. *Pediatr. Nephrol.* 2: 477 (1988).
5. A.M. Bertorello, T. Hökfelt, M. Goldstein, and A. Aperia. Proximal tubule Na-K-ATPase activity is inhibited during high salt diet: evidence for a DA-mediated effect. *Am. J. Physiol.* 254: F795 (1988).
6. C. Chen, R.E. Beach, and M.F. Lokhandwala. Dopamine fails to inhibit renal tubular sodium pump in hypertensive rats. *Hypertension* 21: 364 (1993).
7. A. Nishi, A.C. Eklöf, A.M. Bertorello, and A. Aperia. Dopamine regulation of renal Na^+,K^+-ATPase activity is lacking in Dahl salt-sensitive rats. *Hypertension* 21: 767 (1993)

8. R.D. Bland, and D.W. Nielson. Developmental changes in lung epithelial ion transport and liquid movement. *Annu Rev Physiol* 54: 373 (1992).

9. M.A. Matthay, and J.P. Wiener-Kronish. Intact epithelial barrier function is critical for the resolution of alveolar edema in humans. *Am. Rev. Respir. Dis.* 142:1250 (1990).

10. H. O'Brodovich. When the alveolus is flooding, it's time to man the pumps. *Am. Rev. Respir. Dis.* 142: 1247 (1990).

11. M.A. Matthay, C.C. Landolt, and N.C. Staub. Differential liquid and protein clearance from alveoli of anesthetized sheep. *J. Appl. Physiol.* 53: 96 (1982).

12. B.E. Goodman, R.S. Fleishner and E.D. Crandall. Evidence for active Na^+ transport by cultured monolayers of pulmonary alveolar epithelial cells. *Am. J. Physiol* 254: C78 (1983).

13. R.M. Effros, G.R. Mason, K. Sietsema, P. Silverman, and J. Hukkanen. Fluid reabsoption and glucose consumption in edematous rat lungs. *Circ Res.* 60: 708 (1987).

14. E.D. Crandall, T.A. Heming, R.L. Palombo, and B.E. Goodman. Effects of terbutaline on sodium transport in isolated perfused rat lung. *J. Appl. Physiol.* 60: 289 (1986).

15. Y. Berthiaume, N.C. Staub, and M. Matthay. Beta-adrenergic agonists increase lung liquid clearance in anesthetized sheep. *J. Clin. Invest.* 79: 335 (1987).

16. Y. Berthiaume, V.C. Broadus, M.A. Grooper, T. Tanita, and M. Matthay. Alveolar liquid and protein clearance from normal dog lungs. *J. Appl. Physiol.* 65: 585 (1988)

17. R.M. Effros, G.R. Mason, J. Hukkanen, and P. Silverman. New evidence for active sodium transport from fluid filled rat lungs. *J. Appl. Physiol.* 66: 906 (1989).

18. S. Matalon, K.L. Kirk, J.K. Bubien, Y. Oh, P. Hu, G. Yue, R. Shoemaker, E.J. Cragoe, and D.J. Benos. Immunocytochemical and functional characterization of Na^+ conductance in adult alveolar pneumocytes. *Am. J. Physiol.* 262: C1228 (1992).

19. N. Voilley, E. Lingueglia, G. Champigny, M. Mattei, R. Waldman, M. Lazdunski, and P. Barby. The lung amiloride-sensitive Na^+ channel: Biophysical properties, pharmacology, ontogenesis, and molecular cloning. *Proc. Natl. Acad. Sci, USA* 91: 247 (1994).

20. E.E. Schneeberger, and K.M. McCarthy. Cytochemical localization of Na,KATPase in rat type II pneumocytes. *J. Appl. Physiol.* 20: 1584 (1986).

21. L. Nici, R. Dowin, M. Gilmore-Hevert, J.D. Jamieson, D.H. Ingbar. Response of rat Type II pneumocyte Na,K-ATPase to hyperoxic injury. *Am. J. Physiol.* 261: L307 (1991).

22. K.M. Ridge, D.H. Rutschman, P. Factor, A.I. Katz, A.M. Bertorello, and J.I. Sznajder. Differential Na,K-ATPase isoforms in cultured rat alveolar type 2 cells. *Am. J. Physiol.* 273: L246 (1997).

23. D.H. Rutschman, W. Olivera, and J.I. Sznajder. Active transport and pasive fluid movement in isolated perfused rat lungs. *J. Appl. Physiol.* 75: 1574 (1993).

24. V.B. Serikov, M. Grady, and M. Matthay. Effect of temperature on alveolar liquid and protein clearance in an in situ perfused goat lung. *J. Appl. Physiol.* 75: 940 (1993).

25. W. Olivera, K. Ridge, L.D.H. Wood and J.I. Sznajder. Atrial natriuretic factor decreases active sodium transport and increases alveolar epithelial permeability. *J. Appl. Physiol.* 75: 1581 (1993).

26. G. Saumon and G Basset. Electrolyte and fluid transport across the mature alveolar epithelium. *J. Appl. Physiol.* 74: 1 (1993).

27. F.A. Tibayan, A. Chesnutt, H. Folkesson, J. Eandi and M.A. Matthay. Dobutamine increases alveolar liquid clearance in ventilated rats by beta-2 receptor stimulation. *Am. J. Respir. Crit. Care Med.* 156: 438 (1997).

28. F. Saldias, E. Friedman, S. Rajesh, M.L. Barnard and J.I. Sznajder. Isoproterenol increases alveolar lung liquid clearance in isolated perfused rat lungs. *Am. J. Respir. Crit. Care Med.* A646 (1997).

29. J.I. Sznajder, D.H. Rustchman, K.M. Ridge, and W. Olivera. Mechanisms of lung liquid clearance during subacute hyperoxia in isolated rat lungs. *Am. J. Respir. Crit. Care Med.* 151:1519 (1995).

30. W. Olivera, K.M. Ridge, L.D.H. Wood, and J.I. Sznajder. Active sodium transport and alveolar epithelial Na,K-ATPase increase during subacute hyperoxia in rats. *Am. J. Physiol.* 266: L577 (1994).

31. W. Olivera, K.M. Ridge, and J.I. Sznajder. Lung liquid clearance and alveolar epithelial Na,K-ATPase during hyperoxia and recovery in rats. *Am. J. Respir. Crit. Care Med.* 152: 1229 (1995).

32. J.F. Haskell, G. Yue, D.J. Benos, and S. Matalon. Upregulation of sodium conductive pathways in alveolar type 2 cells in sublethal hyperoxia. *Am. J. Physiol.* 266: L30 (1994).

33. A.M. Bertorello, and A.I. Katz. Short-term regulation of Na-K-ATPase activity: physiological relevance and cellular mechanisms. *Am. J. Physiol.* 265: F743 (1993).

34. H.S. Ewart, and A. Klip. Hormonal regulation of the Na^+-K^+-ATPase: mechanisms underlying rapid and sustained changes in pump activity. *Am. J. Physiol.* 269, C295 (1995).

35. M.S. Feschenko, and K.J. Sweadner. Structural basis for species-specific differences in the phosphorylation of Na,K-ATPase by protein kinase C. *J. Biol. Chem.* 270: 14072 (1995).

36. A.M. Bertorello, A. Aperia, S.I. Walaas, A.C. Nairn, and P. Greengard. Phosphorylation of the catalytic subunit of Na$^+$,K$^+$-ATPase inhibits the activity of the enzyme. *Proc. Natl. Acad. Sci. USA* 88: 11359 (1991).

37. Nishi, A.M. Bertorello, and A. Aperia. Renal Na$^+$,K$^+$-ATPase in Dahl salt-sensitive rats: K+ dependence, effect of cell environment and protein kinases. *Acta. Physiol. Scand.* 149, 377 (1993).

38. F. Cornelious, and N. Logvinenko. Functional regulation of reconstituted Na,K-ATPase by protein kinase A phosphorylation. *FEBS Lett.* 380: 277 (1996).

39. N.S. Logvinenko, I. Dulubova, N. Fedosova, S.H. Larsson, A.C. Nairn, M. Esman, P. Greengard, and A. Aperia. Phosphorylation by protein kinase C of serine-23 of the α-1 subunit of rat Na$^+$,K$^+$-ATPase affects its conformational equilibrium. *Proc. Natl. Acad. Sci. USA* 93:9132 (1996).

40. A.V. Chibalin, A.I. Katz, P.-O. Berggren, and A.M. Bertorello. Receptor-mediated inhibition of renal Na$^+$-K$^+$-ATPase is associated with endocytosis of its α- and β-subunits. *Am. J. Physiol.* In press.

41. B.M.F. Pearse, and M.S. Robinson. Clathrin, adaptors, and sorting. *Annu. Rev. Cell Biol.* 6: 151 (1990).

42. M.S. Robinson. The role of clathrin, adaptors and dynamin in endocytosis. *Curr. Opin. Cell Biol.* 6: 538 (1994).

43. A.M. Bertorello. Diacylglycerol activation of protein kinase C results in a dual effect on Na$^+$,K$^+$-ATPase activity from intact renal proximal tubule cells *J. Cell Sci.* 101: 343 (1992).

44. F. Ibarra, A. Aperia, L-B Svensson, A.-C. Eklöf, and P. Greengard. Bidirectional regulation of Na$^+$,K$^+$-ATPase activity by dopamine and β-adrenergic agonist. *Proc. Natl. Acad. Sci. USA* 90: 21 (1993).

45. H. Singh, and S. Linas. β2-adrenergic function in cultured rat proximal tubule epithelial cells. *Am. J. Physiol.* 271: F71 (1996).

46. E. Rodriguez-Boulan, W.J. Nelson. Morphogenesis of the polarized epithelial cell phenotype. Science 107: 718 (1989).

47. J. Gruenberg, and F.R. Maxfield. Membrane transport in the endocytic pathway. *Curr. Opin. Cell Biol.* 7: 552 (1995).

48. S. Suzuki, D. Zuege, and Y. Berthiaume. Sodium-independent modulation of Na,K-ATPase activity by β-adrenergic agonist in alveolar type II cells. *Am. J. Physiol.* 268:L983 (1995).

49. K. Ridge, A. Bertorello, A. Chibalin, A.I. Katz, and J.I. Sznajder. Short-term upregulation of the Na,K-ATPase in rat alveolar epithelial cells by isoproterenol. *Am. J. Respir. Crit. Care Med.* A650, (1997).

52

THE PREMATURE INFANT'S INABILITY TO CLEAR FETAL LUNG LIQUID RESULTS IN RESPIRATORY DISTRESS SYNDROME (RDS)

Hugh M. O'Brodovich MD, FRCP(C)

MRC Group in Lung Development (Respiratory Research Division of the Hospital for Sick Children), Department of Paediatrics of the University of Toronto

INTRODUCTION:

Prior to the era of modern cellular and molecular biology, it was believed that the epithelium lining the conducting airways and distal lung units merely protected the lung from external noxious particulate matter and provided a thin barrier for gas exchange. However, it is now known that the epithelium is extremely active metabolically and has many functions, including the active transport of solutes and fluid. In concert with the overriding theme of this Conference, the objective of this paper will be to present a paradigm to help us understand the pathogenesis of neonatal acute lung injury. Specifically, that if the liquid that fills the fetal lung is cleared normally and efficiently at the time of birth, there is a normal transition from intrauterine to post-natal life. In contrast, when the liquid clearance is abnormal and delayed, respiratory distress ensues. This delayed clearance of lung liquid is one of two main factors, the other being relative surfactant deficiency, contributing to the pathogenesis of acute respiratory distress syndrome in the full-term and prematurely born human infant. To explain this paradigm, one requires an understanding of the sites and physiologic importance of lung epithelial ion transport during fetal and perinatal life and how developmental immaturity and endogenous hormones alter ion transport. A more detailed overview of this area and appropriate original references are available in published reviews (28,29).

PHYSIOLOGICAL IMPORTANCE OF ION TRANSPORT IN THE DEVELOPING FETAL LUNG.

From early gestation onward, the lungs are filled with a protein poor and Cl^- rich fetal lung liquid. In the early portion of this century it was questioned whether this liquid represented aspirated amniotic fluid or was derived from the body itself. Jost and Policard (22) proved that the fluid actually arose from the lung. In the 1960's investigators in the United States noted that the fluid in the developing fetal lamb lung was Cl^- rich (1) but it was the group at University College Hospital (UCH) in London England who demonstrated that the fluid resulted from the active transport of Cl^-, with Na^+ and water following, by the developing lung epithelia (2). The

Acute Respiratory Distress Syndrome: Cellular and Molecular Mechanisms and Clinical Management
Edited by Matalon and Sznajder, Springer Science+Business Media New York, 1998

53

secretion of fluid by the developing lung is indeed remarkable. In the developing fetal lamb lung the average secretion rate is ~5 mls/kg/hr which, if extrapolated to an adult male, would represent almost 8 liters of fluid secretion per day. The UCH investigators documented that the liquid was essentially protein-free with a protein concentration comparable to that of normal cerebral spinal fluid. Although the Na^+ concentrations in the fetal lung liquid were comparable to that of plasma, the Cl^- concentration was markedly elevated, and the K^+ concentration was approximately 50% greater than the plasma levels. The pH of the fluid varies between species: in the guinea pig and lamb, the Cl is extremely low resulting in an acidic pH, whereas in the canine and primate lung it is comparable to plasma levels (34).

The cellular mechanism resulting in active Cl^- secretion is in many ways comparable to that of other Cl^- and fluid secreting epithelium, including the presence of apical tight junctions, and a basolateral membrane located Na^+/K^+ ATPase, which in concert with K^+ channels results in the electrochemical driving force for Cl^- secretion. It is likely that the basolateral entry step for Cl^- secretion is the bumetanide sensitive $Na^+/K^+/2Cl^-$ co-transporter, although it is possible that there are other paired cotransporters such as the Na^+/HCO_3^- symport which is SITS inhibitable (36) with an accompanying HCO_3^-/Cl^- exchanger. In the apical membrane there is likely more than one type of Cl^- permeant ion channel. It is clear that the CFTR Cl^- channel is present in the apical membrane of the developing lung epithelium; however, transgenic mouse experiments have indicated that there must be alternative chloride channels (6). These may include the Ca^{++} activated Cl^- channel or the ClC_2 channel (27). It is presumed that Na^+ follows the Cl^- secretion through a paracellular route. The pathway for water movement is uncertain and it may occur via a paracellular route or across cell membranes via the aquaporin water channels (see Chapter by Michael Matthay). There appear to be low or undetectable levels of expression of the different aquaporin water channels during early development; however, aquaporins are highly expressed as full-term gestation approaches.

The secretion of Cl^- with Na^+ and fluid following has important biologic implications as normal lung development only occurs in the presence of adequate amounts of intralumenal fluid. A classic article was published by Alcorn et al (3) who either ligated, or drained fluid from, the trachea of fetal lambs *in utero*. They demonstrated that when the intra-pulmonary lumenal content of fluid was increased there was hypertrophy and an accelerated development of the lung; whereas, when the fluid was decreased there was hypoplasia. It appears that it is the volume of liquid within the developing lung, rather than the rate of secretion which is critical for lung development.

Although the presence of fluid within the fetal lung is critical for its normal development, the airspaces must become fluid free and air-filled at the time of birth to enable a normal transition to extrauterine life. It is unknown whether the Cl^- secretion continues, is down regulated, or is merely overwhelmed by a fluid absorptive response in the distal lung at the time of birth. *In-vivo* and *in-vitro* experiments have indicated the airspace fluid clearance takes place in the distal regions of the lungs utilizing apical membrane Na^+ channels and a basolateral Na^+/K^+ ATPase (Figure 1). The Na^+/K^+ ATPase and the relative abundance of Na^+, K^+, and other ion channels result in the intracellular membrane potential being negative relative to the external fluid and there being a very low (~10 mM) intracellular Na^+ concentration. This sets up a large favourable electrochemical gradient for Na^+ entry into the cell. The presence of apical membrane Na^+ permeant ion channels allow Na^+ movement into the cell and subsequent vectorial movement of Na^+ from the air to interstitial spaces. The Cl^- likely moves in a paracellular route, whereas water may follow between or through cells.

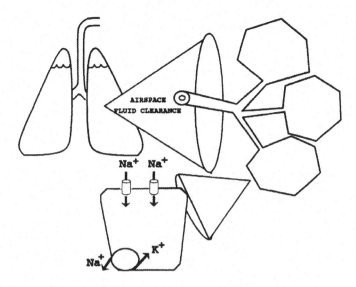

Figure 1: See text for details. Two Na$^+$ permeant ion channels are illustrated to indicate that data suggests the presence of Na$^+$ channels with different biophysical and pharmacologic properties.

Studies in lambs (*in utero*) had demonstrated that the mature fetal lung could be converted from Cl$^-$ with fluid secretion to Na$^+$ with fluid absorption within minutes of instituting a β-agonist stimulus (11). However, it was unknown whether this process had physiologic implications at the time of birth. Experiments conducted in guinea pigs (33) demonstrated that the clearance of fluid from the fetal lung at the time of birth was sensitive to the Na$^+$ transport inhibitor - amiloride. In these experiments, term, mature fetal guinea pigs were delivered by Caesarean section and prior to the first breath had amiloride or its vehicle instilled in a randomized blinded fashion into their lungs filled with fetal lung liquid. The umbilical cord was subsequently clamped, and following an extremely brief period of assisted ventilation, the animals were allowed to breathe room air. Newborn guinea pigs who had amiloride instilled into their lung liquid developed respiratory distress and hypoxemia, and had a dose-dependent failure in clearance of their lung liquid (33). Subsequent experiments evaluated the time course of the fluid clearance from the lungs of newborn guinea pigs. Surprisingly there were two components to this post-natal lung water clearance, the initial component being amiloride-insensitive while the second component was amiloride-sensitive (Figure 2).

Although experiments have demonstrated that the amiloride sensitive component is consistent with an inhibition of active transport of Na$^+$ by the pulmonary epithelium, there are at least two possible explanations for the amiloride-insensitive component of newborn lung water clearance. One explanation is that this amiloride-insensitive component is related to oncotic forces where the protein-rich interstitial fluid promotes the movement of water and salt from the protein-poor lumenal fluid through a transient increase in the permeability of the tight junctions (Note that in the normal lung the inter-epithelial junctions are so tight that they restrict the movement of electrolytes). A transient increase in epithelial tight junction permeability has been demonstrated *in-vivo* at birth (16) and *in-vitro* when primary cultures of fetal distal lung epithelium are switched from fetal (3%) to post-natal (21%) oxygen concentrations (38). A second explanation for the amiloride-insensitive fluid clearance is that Na$^+$ transport is occurring via amiloride-insensitive Na$^+$ conductive pathways. This explanation also has some support from the published literature regarding lung Na$^+$ and fluid transport where Na$^+$ dependent but amiloride

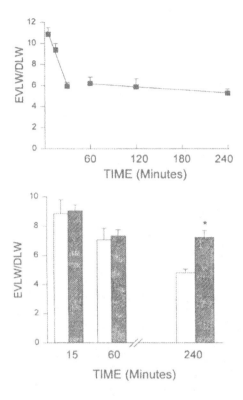

Figure 2:

Upper Panel: Extravascular lung water (EVLW) corrected to blood free dry lung weight (DLW) in newborn guinea pigs decreases after birth in two distinct phases. (Note that the normal EVLW/DLW in pups killed at 9 days of age is 4.4 ± 0.24 SEM EVLW and adult EVLW/DLW is 3.6 ± 0.12 (33).

B) EVLW/DLW in newborn guinea pigs receiving 0.3 ml saline (open bars) or 0.3 ml of a 10^{-3}M amiloride (closed bar) solution into their fluid filled lungs before their first breath (final estimated amiloride concentration = 10^{-4}M). The Na^+ transport blocker amiloride did not alter the early rapid phase of lung water clearance (n = 6 at 15 min and n = 7 at 60 minutes) after birth. Amiloride did delay the slower second phase of lung water clearance between 60 and 240 minutes (n = 11, $*p < 0.05$). Reproduced with permission from Ref. #28.

insensitive short-circuit current has been observed in primary cultures of fetal distal lung epithelium (32). *In vivo* there is also amiloride insensitive $^{22}Na^+$ and water clearance from the lung (35). In addition, patch-clamp electrophysiologic and biochemical experiments have demonstrated the presence of at least 3 amiloride-sensitive sodium conductive channels in the apical membrane of the developing distal lung epithelium (26,43,45).

An important question is whether the lung transports Na^+ at a comparable rate during juvenile and adult life. Recent work has demonstrated that although the total amount of $^{22}Na^+$ active transport increases with growth of the postnatal lung, the amount of amiloride-sensitive $^{22}Na^+$ transport per unit of distal lung epithelial surface area *in vivo* is comparable at both 7 and 30 days after birth (35). Similarly, primary cultures of the distal lung epithelium from the fetal (36) and adult (25) lung have comparable amiloride sensitive Na^+ transport.

MECHANISMS INVOLVED IN THE CONVERSION OF FETAL Cl⁻ SECRETION TO POST-NATAL Na⁺ ABSORPTION

Although it is well documented that the intact fetal lung can rapidly convert from Cl⁻ and fluid secretion to Na⁺ and fluid absorption, it is presently uncertain whether this represents the same cell(s) switching from Cl⁻ secretion to Na⁺ absorption or whether there are Cl⁻ secreting cells which are overwhelmed by Na⁺ absorbing epithelia after they receive the appropriate stimuli. It is also possible that a combination of these events occur. There are convincing data that there is a decrease in the amount of Cl⁻ chloride channel expression and functional activity in the developing lung. In the fetal human trachea, *in-situ* hybridization studies have shown a diminution of CFTR expression(42). At 12 weeks of gestation the human fetal tracheal epithelium has high levels of expression within all tracheal epithelial cells, however, by 28 weeks of gestation, there are only very few cells which express CFTR, albeit at a high level in these infrequently seen cells. This would be consistent with electrophysiologic studies of the human trachea showing that it is, after birth, a Na⁺ absorptive tissue (10). Other studies (24) have also demonstrated that CFTR Cl⁻ channel functional activity in primary cultures of rat lung epithelium decreases markedly during the latter parts of gestation. In addition, there is a developmental regulation of the ClC₂ Cl⁻ channel in the fetal and perinatal lung (27), the pattern of which is compatible with our understanding of the physiologic changes.

Concurrent with the reduction in the gene expression and functional activity of the Cl⁻ channels, there is an increase in Na⁺/K⁺ ATPase expression and activity in the fetal distal lung epithelium (37), the likely site of Na⁺ and fluid absorption in the perinatal lung. It is assumed, that apical membrane Na⁺ conductive channels are the rate limiting step in fluid absorption. Thus it is of course important to know the ontogeny of the epithelial Na⁺ channel (ENaC). Although there may be additional Na⁺ conductive pathways in the developing lung, the amiloride-sensitive ENaC (12,13,23) is critical to the Na⁺ transport which occurs at birth. The α subunit of ENaC is

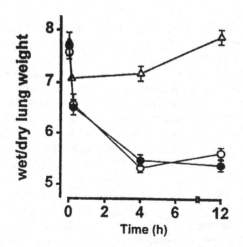

Figure 3:

Lung water content of normal and ENaC knock out neonatal mice. Lungs were excised from fetuses (0 h) or from neonates at different time points after birth (15 min, 4 hr and 12 h) and the wet/dry lung weight ratio was determined for all three genotypes at different times after birth: ● = ENaC (+/+ normal mice), 0 = ENaC (+/- heterozygote mice) and Δ = ENaC (-/- knock-out mice). Reproduced with permission (19)

developmentally regulated in rat (30) and human (45) lungs, however, it was genetic experiments which conclusively demonstrated the importance of the α subunit of ENaC in the transition from fetal to postnatal life. Hummler et al (19) made a knock-out mouse deficient for the α subunit of ENaC. They demonstrated that the bioelectric properties of these mice's tracheal epithelial cysts were insensitive to amiloride and that the newborn mice were unable to clear their lung liquid, in a manner comparable to the experiments in newborn guinea pigs who had received airspace instillation of amiloride (33)

The homozygote knock-out mice died shortly after birth, whereas heterozygotes survived with a normal pattern of lung liquid clearance. Regrettably detailed examinations of the lung, including surfactant analysis, were not performed in these studies (19), however on a gross and light microscopic basis the homozygote lungs appeared to have developed normally.

Other factors, such as the matrix of the lung, can modulate the ability of the premature and mature lung to transport Na^+ from its airspaces. Pitkänen et al (39) performed experiments in which mature late-gestation monolayers of distal lung epithelium were seeded and grown upon a matrix secreted by an immature mixed lung cell population that had been grown under the normal fetal (3%) oxygen concentration. This underlying matrix altered the ion transporting phenotype of the epithelium and induced a more immature mode with less amiloride-sensitive Na^+ transport and more bumetanide sensitive Cl^- secretion. Compatible with these findings, the amounts of mRNA for the α, β, and γ subunits of ENaC were decreased to between 10 and 30% of control values (40). This immaturity in ion transport is consistent with previous experiments that documented that mature lambs *in utero* could, whereas immature lambs could not convert from Cl^- secretion to Na^+ absorption in response to a β agonist or cAMP stimulus, (11). Thus, assuming that there are similarities between the fetal lamb and human infant, an infant born after approximately 26-32 weeks of gestation would not only have their lungs filled with fluid, they would not convert to fluid and Na^+ absorption and thus would continue to secrete liquid into their airspaces.

Thyroid and corticosteroid hormones can also markedly affect lung epithelial ion transport. Barker et al (7) have demonstrated that fetal lambs who had undergone thyroidectomy at an early fetal age and received exogenous thyroid and corticosteroid hormones were able to induce a Na^+ absorptive capability. Wallace et al (46) subsequently showed that a continuous low-dose cortisol infusion starting at 30 days gestation could similarly, by itself, induce a Na^+ absorptive response. A potential cellular mechanism for these changes have recently been provided by *in-vivo* experiments which demonstrated that dexamethasone by itself can induce a 3-fold increase in the αENaC subunit within 8 hours of infusion (41). Similarly, glucocorticoids can increase the Na^+ transport of rat distal lung epithelia (14) and the mRNA levels coding for the α, β, and γ subunits ENaC in cultured fetal human (44) distal lung epithelia.

The mechanisms responsible for the stable induction of Na^+ transport are incompletely understood. One promising candidate is the physiologic change in oxygen tension that occurs at birth. The fetal distal lung epithelium is exposed to the equivalent of ~3% oxygen and after birth there is an approximate 7-fold increase in oxygen concentration to 21%. Barker and Gatzy first observed (8) the oxygen tension could affect the ability of the late gestation fetal lung explants to secrete fluid. Pitkänen et al (38) maintained primary cultures of distal lung epithelia isolated from late gestation fetal lungs under a 3% oxygen environment. These monolayers had little amiloride sensitive Na^+ transport and very low levels of mRNA coding for α, β and γ rENaC. Exposure of comparable monolayers to 21% oxygen resulted in marked increases in both mRNA levels and amiloride sensitive Na^+ transport. These data (38) and those of Barker and Gatzy (8) in the intact lung strongly suggests that the lung uses oxygen as one indicator of postnatal life and as a signal for increasing Na$^+$ transport.

IMPLICATIONS FOR NEWBORN RESPIRATORY DISTRESS SYNDROMES.

For several decades it was thought that the sole factor that initiated acute respiratory distress in the prematurely born human infant was a surfactant deficiency (4,17,21) relative to the full term infant.   There is much scientific data to support this hypothesis, however, there is increasing evidence that an additional second factor plays a role in the initiation of RDS in the prematurely born infant, specifically the inability to initiate effective Na$^+$ transport and liquid clearance from the lungs at the time of birth. deSa (15) demonstrated that the lung-liquid content of newborn human infants was greater when the infant was more premature.   In addition, he demonstrated that the clearance of liquid from the lungs took many hours and if the infant developed RDS there was a marked increase rather than reduction in the airspace liquid.   These findings were subsequently demonstrated in animal and primate experiments by several investigators (see review (28)).   Indeed, flash frozen sections of a primate model of respiratory distress syndrome demonstrated widespread filling of the airspaces with fluid and that this fluid was the major cause of reduction in lung gas volumes in these animals (20).

Exogenous surfactant dramatically improves the oxygen exchanging abilities of the lungs of premature infants with RDS.   This likely results from an improvement in the air-liquid interface surface tension related to the infant's surfactant deficiency relative to the full term infant.   To test whether exogenous surfactant may also have a similar beneficial effect in fluid filled lungs, experiments were done in two-week old juvenile rabbits (31).   These animals had lungs which must be surfactant replete. When 20 ml/kg of saline was instilled into their lungs there was, not surprisingly, a marked reduction in oxygenation.   As illustrated in Figure 4, delivering surfactant to the animal when the lungs were filled with saline resulted in a dramatic improvement in oxygenation, comparable to that seen in premature infants with RDS.

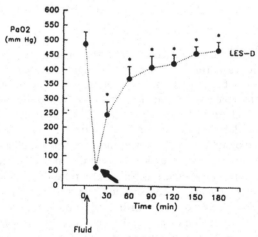

Figure 4:

Juvenile healthy rabbits receiving assisted ventilation with an F_IO_2 = 1.0 had normal gas exchange at time 0 as indicated.   The PaO_2 fell dramatically when their airspaces were filled with 20 ml/kg saline (fluid) to simulate the fluid filled airspaces of a newborn animal.   Subsequent delayed administration of a natural lipid extracted surfactant (LES-D) resulted in a rapid and marked increase in the PaO_2 similar to that seen when premature infants with RDS are given the same surfactant.   * indicates $p < 0.05$ relative to 15 minute (min) time point. Figure is reproduced with permission (31).

Human lungs obtained from the immature fetuses show undetectable levels of the α subunit of hENaC when analyzed by northern analysis (45), and primary cultures of early gestation human fetal distal lung epithelium have very low levels of the α, β, and γ subunits of hENaC (44). The functional correlate of these low levels of hENaC mRNA were recently demonstrated in studies of the bioelectric properties of human fetal distal lung epithelium (5) and *in vivo* (9). In this latter study (9), Barker et al measured the electrical potential difference (PD) between the surface of nasal respiratory epithelium of human infants relative to these subcutaneous space. They demonstrated that infants with RDS, relative to infants of comparable gestational age who did not have RDS, had lesser amounts of amiloride-sensitive PD.

As discussed above (Figure 3), genetic experiments in mice demonstrated that the lack of the α subunit of mENaC resulted in respiratory distress at birth with death ensuing several hours later (19). This is very strong evidence to support the hypothesis that Na^+ transport is critical at the time of birth. It was therefore surprising that humans with pseudo-hypoaldosteronism type I with N-terminal stop codon mutations of the α subunit of hENaC apparently did not suffer RDS at birth (although they do experience respiratory symptomatology during later life). Although several possibilities exist, one explanation arises from cystic fibrosis genetic knockout mouse experiments. Although the human with a mutation in CFTR almost always develops a lethal lower airway respiratory disease, the typical CFTR knockout mouse has no lower airway disease, likely because of the presence of an alternative calcium activated Cl^- channel which can replace the function of the CFTR Cl^- channel. The opposite may occur with αhENaC. In the human there is not severe neonatal disease, perhaps because of expression of an alternative Na^+ conductive pathway. Whereas, when the α subunit of mENaC is knocked out, the mouse gets the lethal expression of disease.

In summary, there is much *in-vivo* and *in-vitro* animal and human experimental data to support the critical role that epithelial active Na^+ transport plays in fluid clearance at the time of birth during a normal transition to post-natal life. This, along with the well established surfactant deficiency supports the notion that one has to have the correct amount of "soap and water" before one can prevent the initiation of RDS in the prematurely born infant (Figure 5, reproduced with permission Ref. #28).

Speculation regarding the initial mechanisms responsible for the pathogenesis of non-infective respiratory distress syndromes in the newborn infant. In the mature term infants the "normal" amount of surfactant is approximately 10 fold greater than the amount in preterm fetuses or adults (21). This enables the infant to survive the "salt water drowning" that results from the fetal lung liquid that normally fills the airspaces at birth. A normal fluid (Na^+) reabsorptive capacity then clears the lung liquid. In the transient tachypnea syndrome there is an abnormally slow clearance of lung liquid, however, the large amount of surfactant still results in normal surface tension at the air-liquid interface. This results in mild to moderate respiratory distress which resolves within 24 - 48 hours as Na^+ transport rapidly matures (18). In contrast, when there is both inadequate fluid (Na^+) reabsorption an a relative surfactant deficiency the premature infant develops a severe respiratory distress syndrome (hyaline membrane disease) that is typically characterised by severe lung injury. A recent report demonstrating abnormally low amounts of amiloride sensitive PD in respiratory epithelial of infants with hyaline membrane disease supports this hypothesis (9).

The on-going injury in infants who do not recover from RDS and develop BPD likely involves many other factors including inadequate amounts of antioxidant protective mechanisms, structural immaturity of the collagen and others supporting elements within the lungs, and other immature cellular physiologic functions.

ACKNOWLEDGMENTS

This research was supported by grants-in-aid from the Medical Research Council of Canada (Group in Lung Development) and the Heart and Stroke Foundation of Ontario. Dr. O'Brodovich was a career investigator of the Heart & Stroke Foundation of Ontario.

REFERENCES

1. Adams, F. H., A. J. Moss, and L. Fagan. The tracheal fluid in the fetal lamb. *Biol. Neonate* 5: 151-158, 1963.

2. Adamson, T. M., R. D. H. Boyd, H. S. Platt, and L. B. Strang. Composition of alveolar liquid in the foetal lamb. *J. Physiol. (Lond.)* 204: 159-168, 1969.

3. Alcorn, D., T. M. Adamson, T. F. Lambert, J. E. Maloney, B. C. Ritchie, and P. M. Robinson. Morphological effects of chronic tracheal ligation and drainage in the fetal lamb lung. *J. Anat.* 123: 649-660, 1977.

4. Avery, M. E. and J. Mead. Surface properties in relation to atelectasis and hyaline membrane disease. *Am. J. Dis. Child.* 97: 517-523, 1959.

5. Barker, P. M., R. C. Boucher, and J. R. Yankaskas. Bioelectric properties of cultured monolayers from epithelium of distal human fetal lung. *Am. J. Physiol. (Lung Cell Mol. Physiol.)* 268: L270-L277, 1995.

6. Barker, P. M., K. K. Brigman, A. M. Paradiso, R. C. Boucher, and J. T. Gatzy. Cl⁻ secretion by trachea of CFTR (+/-) and (-/-) fetal mouse. *Am. J. Respir. Cell Mol. Biol.* 13: 307-313, 1995.

7. Barker, P. M., M. J. Brown, C. A. Ramsden, L. B. Strang, and D. V. Walters. The effect of thyroidectomy in the fetal sheep on lung liquid reabsorption induced by adrenaline or cyclic AMP. *J. Physiol. (Lond.)* 407: 373-383, 1988.

8. Barker, P. M. and J. T. Gatzy. Effect of gas composition on liquid secretion by explants of distal lung of fetal rat in submersion culture. *Am. J. Physiol. (Lung Cell Mol. Physiol.)* 265: L512-L517, 1993.

9. Barker, P. M., C. W. Gowen, E. E. Lawson, and M. Knowles. Decreased sodium ion absorption across nasal epithelium of very premature infants with respiratory distress syndrome. *J. Pediatr.* 130: 373-377, 1997.

10. Boucher, R. C., J. Narvarte, C. Cotton, M. J. Stutts, M. Knowles, A. L. Finn, and J. T. Gatzy. Sodium absorption in mammalian airways. In: *Fluid Electrolyte Abnormalities in Exocrine Glands in Cystic Fibrosis*, edited by P. Quinton, J. R. Martinex, and U. Hopfer. San Francisco: San Francisco Press, 1982, p. 271-287.

11. Brown, M. J., R. A. Olver, C. A. Ramsden, L. B. Strang, and D. V. Walters. Effects of adrenaline and of spontaneous labour on the secretion and absorption of lung liquid in the fetal lamb. *J. Physiol. (Lond.)* 344: 137-152, 1983.

12. Canessa, C. M., J. D. Horisberger, and B. C. Rossier. Functional cloning of the epithelial sodium channel: relation with genes involved in neurodegeneration. *Nature* 361: 467-470, 1993.

13. Canessa, C. M., L. Schild, G. Buell, B. Thorens, I. Gautschi, J.-D. Horisberger, and B. C. Rossier. Amiloride-sensitive epithelial Na⁺ channel is made of three homologous subunits. *Nature* 367: 463-467, 1994.

14. Cott, G. R. and A. K. Rao. Hydrocortisone promotes the maturation of Na+-dependent ion transport across the fetal pulmonary epithelium. *Am. J. Respir. Cell Mol. Biol.* 9: 166-171, 1993.

15. deSa, D. J. Pulmonary fluid content in infants with respiratory distress. *J. Pathol.* 97: 469-479, 1969.

16. Egan, E. A., R. E. Olver, and L. B. Strang. Changes in non-electrolyte permeability of alveoli and the absorption of lung liquid at the start of breathing in the lamb. *J. Physiol. (Lond.)* 244: 161-179, 1975.

17. Farrell, P. M. and M. E. Avery. State of the Art: Hyaline membrane disease. *Am. Rev. Respir. Dis.* 111: 657-658, 1975.

18. Gowen, C. W., E. E. Lawson, J. Gingras, R. Boucher, J. T. Gatzy, and M. Knowles. Electrical potential difference and ion transport across nasal epithelium of term neonates: correlation with mode of delivery, transient tachypnea of the newborn, and respiratory rate. *J. Pediatr.* 113: 121-127, 1988.

19. Hummler, E., P. Barker, J. Gatzy, F. Beermann, C. Verdumo, A. Schmidt, R. Boucher, and B. C. Rossier. Early death due to defective neonatal lung liquid clearance in $\alpha ENaC$-deficient mice. *Nature Genet.* 12: 325-328, 1996.

20. Jackson, C., A. Mackenzie, E. Chi, T. Standaert, W. Truog, and W. Hodson. Mechanisms for reduced total lung capacity at birth and during hyaline membrane disease in premature newborn monkeys. *Am. Rev. Respir. Dis.* 142: 413-419, 1990.

21. Jobe, A. and M. Ikegami. State of the Art: Surfactant for the treatment of respiratory distress syndrome. *Am. Rev. Respir. Dis.* 136: 1256-1275, 1987.

22. Jost, P. A. and A. Policard. Contribution experimentale a L'etude du developpement prenatal du poumon chez le lapin. *Arch. D'Anatomie Microscopique* 37: 323-332, 1948.

23. Lingueglia, E., N. Voilley, R. Waldmann, M. Lazdunski, and P. Barbry. Expression cloning of an epithelial amiloride-sensitive Na^+ channel: A new channel type with homologies to *Caenorhabditis elegans* degenerins. *FEBS* 318: 95-99, 1993.

24. MacLeod, R. J., J. R. Hamilton, H. Kopelman, and N. Sweezey. Developmental differences of cystic fibrosis transmembrane conductance regulator functional expression in isolated rat fetal distal airway cells. *Pediatr. Res.* 35: 45-49, 1994.

25. Mason, R., M. C. Williams, J. H. Widdicombe, M. J. Sanders, D. S. Misfeldt, and L. C. J. Berry. Transepithelial transport by pulmonary alveolar type II cells in primary culture. *Proceedings of the National Academy of Sciences U. S. A.* 79: 6033-6037, 1982.

26. Matalon, S., M. Bauer, D. Benos, T. Kleyman, C. Lin, E. J. J. Cragoe, and H. M. O'Brodovich. Fetal lung epithelial cells contain two populations of amiloride-sensitive Na+ channels. *Am. J. Physiol. (Lung Cell Mol. Physiol.)* 264: L357-L364, 1993.

27. Murray, C. B., M. M. Morales, T. R. Flotte, S. A. McGrath-Morrow, W. B. Guggino, and P. L. Zeitlin. ClC-2: A developmentally dependent chloride channel expressed in the fetal lung and downregulated after birth. *Am. J. Respir. Cell Mol. Biol.* 12: 597-604, 1995.

28. O'Brodovich, H. Immature epithelial Na^+ channel expression is one of the pathogenetic mechanisms leading to human respiratory distress syndrome. *Proc. Assoc. Am. Phys* 108: 1-12, 1996.

29. O'Brodovich, H. M. Epithelial ion transport in the fetal and perinatal lung. *Am. J. Physiol. (Cell Physiol.)* 261: C555-C564, 1991.

30. O'Brodovich, H. M., C. M. Canessa, J. Ueda, B. Rafii, B. C. Rossier, and J. Edelson. Expression of the epithelial Na+ channel in the developing rat lung. *Am. J. Physiol. (Cell Physiol.)* 265: C491-C496, 1993.

31. O'Brodovich, H. M. and V. Hannam. Exogenous surfactant rapidly increases PaO2 in mature rabbits with lungs that contain large amounts of saline. *Am. Rev. Respir. Dis.* 147: 1087-1090, 1993.

32. O'Brodovich, H. M., V. Hannam, and B. Rafii. Sodium channel but neither Na+-H+ nor Na-glucose symport inhibitors slow neonatal lung water clearance. *Am. J. Respir. Cell Mol. Biol.* 5: 377-384, 1991.

33. O'Brodovich, H. M., V. Hannam, M. Seear, and J. B. M. Mullen. Amiloride impairs lung water clearance in newborn guinea pigs. *J. Appl. Physiol.* 68(4): 1758-1762, 1990.

34. O'Brodovich, H. M. and T. A. Merritt. Bicarbonate concentration in rhesus monkey and guinea pig fetal lung liquid. *Am. Rev. Respir. Dis.* 146: 1613-1614, 1992.

35. O'Brodovich, H. M., J. B. M. Mullen, V. Hannam, and B. E. Goodman. Active $^{22}Na^+$ transport by the intact lung during early postnatal life. *Can. J. Physiol. Pharmacol.* 75: 431-435, 1997.

36. O'Brodovich, H. M., B. Rafii, and M. Post. Bioelectric properties of fetal alveolar epithelial monolayers. *Am. J. Physiol.* 258: L201-L206, 1990.

37. O'Brodovich, H. M., O. Staub, B. Rossier, K. Geering, and J.-P. Kraehenbühl. Ontogeny of alpha1 and beta1 isoforms of Na/K ATPase in fetal distal lung epithelium. *Am. J. Physiol. (Cell Physiol.)* 264: C1137-C1143, 1993.

38. Pitkänen, O., A. K. Tanswell, G. Downey, and H. O'Brodovich. Increased PO_2 alters the bioelectric properties of fetal distal lung epithelium. *Am. J. Physiol. Lung Cell. Mol. Physiol.* 270: L1060-L1066, 1996.

39. Pitkänen, O., A. K. Tanswell, and H. M. O'Brodovich. Fetal lung cell-derived matrix alters distal lung epithelial ion transport. *Am. J. Physiol. (Lung Cell Mol. Physiol.)* 268: L762-L771, 1995.

40. Ruddy, M. K., J. M. Drazen, O. Pitkänen, H. M. O'Brodovich, and H. W. Harris. Aquaporin-4 is expressed in rat fetal distal lung epithelial cells (FDLE) where it may function in Na^+ mediated water reabsorption during the perinatal period. *Am. J. Respir. Crit. Care Med.* 153: A2321996.(Abstract)

41. Tchepichev, S., J. Ueda, C. M. Canessa, B. C. Rossier, and H. O'Brodovich. Lung epithelial Na channel subunits are differentially regulated during development and by steroids. *Am. J. Physiol. (Cell Physiol.)* 269: C805-C812, 1995.

42. Tizzano, E. F., H. M. O'Brodovich, D. Chitayat, and M. Buchwald. Regional expression of CFTR in developing respiratory tissues: the Lung Paradox. *Am. J. Respir. Cell Mol. Biol.* 10: 355-362, 1994.

43. Tohda, H. and Y. Marunaka. Insulin-activated amiloride-blockable nonselective cation and Na^+ channels in the fetal distal lung epithelium. *Gen. Pharmacol.* 26: 755-763, 1995.

44. Venkatesh, V. C. and H. D. Katzberg. Glucocorticoid regulation of epithelial sodium channel genes in human fetal lung. *Am. J. Physiol. (Lung Cell Mol. Physiol.)* 273: L227-L233, 1997.

45. Voilley, N., E. Lingueglia, G. Champigny, M.-G. Mattéi, R. Waldmann, M. Lazdunski, and P. Barbry. The lung amiloride-sensitive Na^+ channel: Biophysical properties, pharmacology, ontogenesis, and molecular cloning. *Proc. Natl. Acad. Sci. USA* 91: 247-251, 1994.

46. Wallace, M. J., S. B. Hooper, and R. Harding. Effects of elevated fetal cortisol concentrations on the volume, secretion, and reabsorption of lung liquid. *Am. J. Physiol. (Regulatory Integrative Comp. Physiol.)* 269: R881-R887, 1995.

ROLE OF MATRIX MACROMOLECULES IN CONTROLLING LUNG FLUID BALANCE: TRANSITION FROM A DRY TISSUE TO EDEMA

Daniela Negrini,[1] Alberto Passi,[2] Giancarlo de Luca,[2] and Giuseppe Miserocchi,[1]

[1]Istituto di Fisiologia Umana
Università degli Studi, Milano, ITALY
[2]Dipartimento di Biochimica "A.Castellani"
Università degli Studi, Pavia , ITALY

Alveolar gas exchange is guaranteed by the structure of the alveolo-capillary barrier, whose extreme thinness essentially depends upon the hydration state of the extracellular matrix (ECM) laying between the capillary endothelium and the alveolar epithelium. In normal conditions, lung interstitial fluid volume is minimized, as indicated by the subatmospheric pulmonary interstitial pressure [P_{ip}, ~ -10 cmH$_2$O at right atrium level[1,2]]; since pulmonary microvasculature provides fluid filtration into surrounding interstitium[3], the negative interstitial pressure is set by lymphatic drainage. The main fibril constituents of pulmonary ECM are collagen and elastin that provide a mechanical scaffold for proteoglycans (PGs), large macromolecules consisting of a core protein covalently linked to glycosaminogycan (GAG) chains. Three main PG families are present in the lung parenchyma: (1) chondroitinsulphate PGs (versican) that form aggregates with hyaluronic acid in the interstitial matrix; (2) heparansulphate PGs (perlecan) of epithelial and endothelial basal membrane, and (3) small dermatansulphate PGs (decorin) associated with collagen fibrils. PGs contribute to determine and maintain tissue mechanical features and, due to the high anionic charge of their lateral GAG chains, they display high hydrophilic properties, thus controlling the hydration of interstitial tissue. The latter depends upon: (1) a selective pulmonary endothelium, limiting transcapillary fluid fluxes and restricting plasma protein escape; (2) an efficient tissue drainage through pulmonary lymphatics; (3) a very low interstitial compliance.

We related the time course of P_{ip} to the interstitial fluid dynamic, to the biochemical structure of the matrix and to its mechanical properties when a perturbation of the steady state was induced during the development of hydraulic and lesional pulmonary edema.

Acute Respiratory Distress Syndrome: Cellular and Molecular Mechanisms and Clinical Management
Edited by Matalon and Sznajder, Springer Science+Business Media New York, 1998

P_{ip} was measured through the micropuncture technique performed in anesthetized supine rabbits with lungs physiologically expanded in the intact pleural space. A glass micropipette, driven by a micromanipulator under stereomicroscopic view, was advanced through the intact parietal pleura exposed after removal of the intercostal muscles. P_{ip} was recorded in control and during development of pulmonary edema induced by either intrajugular infusion of 0.5 ml/(kg·min) of saline solution (hydraulic edema) or by an i.v. bolus of 200 µg of pancreatic elastase (lesional edema) that would mimic the effect of acute pancreatitis, often complicated by development of Adult Respiratory Distress Syndrome (ARDS). Elastase is an omnivorous proteolytic enzyme with broad affinity for a variety of soluble and insoluble protein substrates, including ECM components.

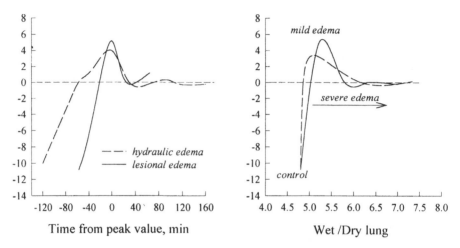

Figure 1. Left panel: time course of pulmonary interstitial pressure (P_{ip}) during development of hydraulic[4] or lesional[5] lung edema. Time is expressed relative to the attainment of a positive peak pressure; data are mean ± 1 SD. **Right panel:** P_{ip} data shown in the left panel are plotted as a function of the corresponding wet weight to dry weight lung ratio.

As shown in the left panel of Figure 1, in both hydraulic and lesional edema P_{ip} initially increases from a control value of ~ -10 cmH$_2$O to a positive peak pressure of ~ +5 cmH$_2$O, decreasing thereafter to zero during development of severe edema[4,5]. In the right panel of Figure 1 one can appreciate that the marked P_{ip} change during the initial phase of lesional and hydraulic edema is coupled with a negligible increase of the wet weight to dry weight lung ratio (W/D). Hence, in normal lung and up to the attainment of mild edema (characterized by the positive P_{ip} peak), the mechanical compliance of the interstitial matrix is very low [estimated at ~ 0.5 ml/(mmHg·100g wet weight)[4]]; this mechanical property represents an important "tissue safety factor" to counteract further progress of pulmonary edema. Severe edema develops when P_{ip} drops back to zero with no further changes in face of a marked increase in W/D, suggesting a substantial increase of interstitial tissue compliance in this late phase. From the standpoint of microvascular fluid exchange, since the net capillary to interstitium transendothelial pressure gradient is smaller in severe edema compared to control[6], the increased fluid filtration leading to accumulation of extravascular lung water, necessarily implies an increase of the

66

endothelial membrane permeability. Hence, two factors seem to act simultaneously in determining severe edema: (1) the loss of the "tissue safety factor" provided by the structural integrity of pulmonary tissue matrix and (2) an increase in microvascular permeability.

To relate P_{ip} to tissue matrix organization, we studied the structure of pulmonary interstitial PGs in control and at various stages of development of hydraulic or lesional pulmonary edema. After extraction with 0.4 M and 4 M guanidil chloride (GuHCl), PGs were cleared of hyaluronic acid and proteins and they were analyzed through gel-filtration chromatography and electrophoresis; in addition, radiolabelled PGs were tested to assess their binding properties with other matrix macromolecules. In both hydraulic and lesional edema the extractability of PGs, evaluated from the concentration of hexuronate (a specific marker of GAG chains) in the tissue samples, increases from control to mild and severe edema: this suggests a progressive weakening of the non covalent bonds of PGs with other macromolecules of the matrix. Furthermore, the electrophoretic profile of PGs from control lungs identifies a family of slowly migrating large PGs and another family of fast migrating small PGs; during development of hydraulic mild and severe edema a progressive decrease of the large PG component, with a marked increase of the smaller PG fraction is observed, indicating a fragmentation of native PGs.

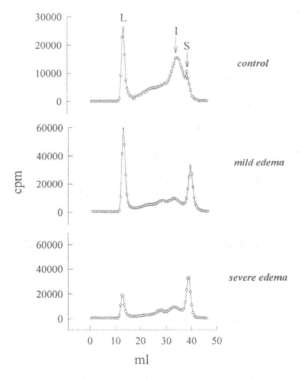

Figure 2. Gel filtration chromatography of radiolabelled proteoglycans isolated from 0.4 M lung extracts in control, mild and severe lesional edema. The relative content of large molecular size PGs (peak L, eluted in the void volume of the column) decreased markedly only in severe edema. This material might be identified as versican, the large chondroitinsulphate containing PG of the lung ECM. The relative content of heterogeneous smaller size PGs (peak I) decreased both in mild and severe edema; these PGs were likely to contain heparansulphate chains including those of the capillary and alveolar basement membrane. Peak S, including the smallest PGs (decorin) and PG fragments, progressively increases during edema development.

Similarly, gel filtration chromatography of radiolabelled PGs (Figure 2, top panel) shows that in control condition 23% of total PGs are large chondroitinsulphate PGs (peak L), 66% consist of smaller heparansulphate PGs (peak I) and the remaining 11% are represented by even smaller PG fragments (peak S). During lesional mild edema, the heparansulphate PG fraction is greatly reduced, while PG fragments are increased, the larger PG fraction being essentially unaltered. The indication is that in lesional elastase-induced mild edema there is an initial fragmentation of endothelial basement membrane PGs (perlecan). Progression to lesional severe edema shows a fragmentation of the larger matrix chondroitinsulphate PGs (versican). An important functional consequence of PGs sequential fragmentation is that they progressively loose their capability to bind with other tissue macromolecules like collagen type I, II or IV, fibronectin and hyaluronic acid as assessed by solid binding assay on microplates.

It is a common clinical observation that progression from mild interstitial edema, usually undetectable with common diagnostic procedure, to accelerated alveolar edema occurs very rapidly. In light of our results, it seems that a cascade of sequential events, likely common to both hydraulic and lesional edema, might be involved at the onset and during the development of pulmonary edema: (a) an increased endothelial membrane permeability to water and solutes, depending upon the loss of the structural integrity of heparansulphate PGs; (b) a decreased interstitial tissue compliance due to fragmentation of chondroitinsulphate PGs and disorganization of the matrix macromolecules. Hence, the integrity of pulmonary interstitial matrix seems to play a crucial role in controlling lung fluid balance.

REFERENCES

1. G. Miserocchi, D. Negrini, and C. Gonano. Direct measurement of interstitial pulmonary pressure in in situ lung with intact pleural space. *J.Appl.Physiol.* 69:2168 (1990).
2. G. Miserocchi, D. Negrini, and C. Gonano. Parechymal stress affects interstitial and pleural pressure in in situ lung. *J.Appl.Physiol.* 71: 1967 (1991).
3. D. Negrini, C. Gonano, and G. Miserocchi. Microvascular pressure profile in intact in situ lung. *J.Appl.Physiol.* 72: 332 (1992).
4. G. Miserocchi, D. Negrini, M. Del Fabbro and D. Venturoli. Pulmonary interstitial pressure in in situ lung: transition to interstitial edema. *J.Appl.Physiol.* 74: 1171 (1993).
5. D. Negrini, A. Passi, G. De Luca, and G. Miserocchi. Pulmonary interstitial pressure and proteoglycans during development of pulmonary edema. *Am.J.Physiol.* 270 (*Heart Circ. Physiol.* 39): H2000 (1996).
6. D. Negrini. Pulmonary microvasculare pressure profile during development of pulmonary edema. *Microcirculation* 2:173 (1995).

Role of the Na,K-ATPase α_2 subunit in lung liquid clearance

Karen M. Ridge, Walter Olivera, David H. Rutschman, Stuart Horowitz,
Philip Factor and Jacob Iasha Sznajder.

Pulmonary and Critical Care Medicine Division
Michael Reese Hospital
University of Illinois at Chicago
Chicago, Illinois 60616

Department of Mathematics
Northeastern Illinois University
Chicago, Illinois 60625

CardioPulmonary Research Institute
Winthrop-University Hospital
Mineola, New York 11501

INTRODUCTION

An important function of alveolar epithelial cells is to keep the airspace free of liquid and preserve gas exchange. Previous studies have suggested that Na,K-ATPase may play a role contributing to alveolar epithelial vectorial sodium transport and thus lung liquid clearance[1-3]. In the lungs the α1 and β1 isoforms have been localized to alveolar type 2 (AT2) cells [4,5]. The α2 isoform has been described in whole lung homogenates from adult rats [6,7]. The normal lung has two types of alveolar epithelial cells. Alveolar type 2 cells (AT2) are cuboidal cells that occupy about 5% of the alveolar surface area, and synthesize and secrete pulmonary surfactant [8]. Alveolar type 1 cells (AT1) are large flattened cells that cover >90% of the alveolar surface and are thought to be terminally differentiated[8]. Expression of Na,K-ATPase α1, α2 and β1 subunits was investigated in rat AT2 cells cultured for seven days, a period during which they lose their phenotypic markers and differentiate toward an alveolar type 1-like cell phenotype.

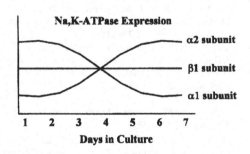

Na,K-ATPase Expression

α2 subunit

β1 subunit

α1 subunit

1 2 3 4 5 6 7
Days in Culture

RESULTS

Differentiation of AT2 cells to an AT1-like phenotype resulted in a decrease of α1 and an increase of α2 mRNA and protein abundance without changes in the β1 subunit as represented schematically in Figure 1[9]. By *in situ* hybridization we established that a strong hybridization signal, corresponding to α2 isoform mRNA, was evident throughout the alveolar epithelium of normal rat lung tissue. Although the data suggests that alveolar epithelial cells express the $\alpha2$ subunit, the resolution of light microscopy does not permit the unambiguous assignment of cell type in the alveolar epithelium [10]. Definitive localization of the Na,K-ATPase $\alpha2$ isoform within the alveolar epithelium was established by immuno-electron microscopy. Isoform specific monoclonal antibody to the Na,K-ATPase $\alpha2$ subunit was used in conjunction with a 5 nm gold-labeled secondary antibody, localizing the $\alpha2$ protein to AT1 cells.

In rodent tissue, it is possible to quantitate the relative amounts of $\alpha1$ and $\alpha2$ isoforms of Na,K-ATPase due to the large difference in their affinity for ouabain. Previous studies have established that in the range of 10^{-6}-10^{-5} M ouabain, $\alpha2$ is about 90% saturated with ouabain, whereas less that 10% of the $\alpha1$ subunit is inhibited. Thus in the alveolar epithelium, at a concentrations of 10^{-7} and 10^{-5} M ouabain, the Na,K-ATPase $\alpha2$ subunit is almost exclusively inhibited [5,9]. To determine whether there is a role for the $\alpha2$ subunit in active Na+ transport and lung liquid clearance we perfused the pulmonary circulation of the isolated lung model with increasing concentrations of ouabain. A solution containing Evans blue dye-bound albumin (EBD-albumin) was instilled into the alveoli; changes in the concentration of EBD-albumin within the alveolus were used to calculate lung liquid clearance [3,11,12]. Control rat lungs and lungs perfused with 10^{-9} M ouabain cleared ~10% of the total airspace instillate in one hour. Perfusion of the pulmonary circulation with 10^{-7} M ouabain resulted in a 29% reduction of basal lung edema clearance ($p<0.05$) from rat lungs, and perfusion with 10^{-5} M ouabain resulted in a 58% reduction in lung edema clearance as compared to controls ($p<0.05$).

These results demonstrate the presence of Na,K-ATPase $\alpha2$ subunit in rat alveolar epithelial cells *in vitro* and *in situ* and suggest that the Na,K-ATPase $\alpha2$ subunit contributes to the vectorial Na$^+$ flux across the lung epithelium.

REFERENCES:

1. W. Olivera, K. Ridge, LDH Wood, and JI Sznajder. Active sodium transport and alveolar epithelial Na,K-ATPase increase during subacute hyperoxia in rats. *Am. J. Physiol.*; 266:L577-L584 (1994).
2. MA Matthay, JP Wiener-Kronish. Intact epithelial barrier function is critical for the resolution of alveolar edema in humans. *Am. Rev. Respir. Dis.*; 142:1250-1257 (1990)
3. JI Sznajder, W Olivera, KM Ridge, DH Rutschman. Mechanisms of lung liquid clearance during hyperoxia in isolated rat lungs. *Am. J. Respir. Crit. Med.*; 151:1519-1525 (1995)
4. EE Schneeberger, KM McCarthy. Cytochemical localization of Na+-K+-ATPase in rat type II pneumocytes. *J. Appl. Physiol*; 60 (5):1584-1589 (1986)
5. S Suzuki, D Zuege, Y Berthiaume. Sodium-independent modulation of Na,K-ATPase activity by β-adrenergic agonist in alveolar type II cells. *Am. J. Physiol.* 268:L983-L990. (1995)
6. AW Shyjan, R Levenson. Antisera specific for the $\alpha1,\alpha2,\alpha3$, and β subunits in rat tissue membranes. *Biochem.* 28:4531-4535 (1989)
7. J Orlowski, JB Lingrel. Tissue specific and developmental regulation of rat Na,K-ATPase catalytic α and β subunit mRNAs. *J. Biol. Chem.*; 263:10436-10442 (1988).
8. IR. Adamson, DH Bowden. Derivation of type 1 epithelium from type 2 cells in the developing rat lung. *Lab Investigation.* 32:736 (1975).
9. K Ridge, DH Rutschman, P Factor, AI Katz, AM Bertorello, JI Sznajder. Differential expression of Na,K-ATPase isoforms in rat alveolar epithelial cells. *Am. J. Physiol.* 273:L246-L255 (1997)
10. B. Piedboeuf, J. Frenette, P. Petrov, SE Welty, JA Kazzaz, and S Horowitz. In vivo expression of intercelular adhesion molecule 1 in type II pneumocytes during hyperoxia. *Am. J. Respir. Cell Mol. Biol.* 15:71-77 (1996)
11. G. Saumon and G Basset. Electrolyte and fluid transport across the mature alveolar epithelium. *J. Appl. Physiol.* 74:1-15 (1993)
12. DH Rutschman, W Olivera, and JI Sznajder. Active transport and passive liquid movement in isolated perfused rat lungs. *J. Appl. Physiol*; 75(4):1574-1580 (1993)

ALVEOLAR EPITHELIAL FLUID TRANSPORT UNDER NORMAL AND PATHOLOGICAL CONDITIONS

Michael A. Matthay, Colleen Horan, Chun-Xue Bai, and Yibing Wang

Cardiovascular Research Institute
University of California, San Francisco

This chapter focuses on the mechanisms of salt and water transport across alveolar and distal airway epithelium of the adult lung. The first section presents evidence for active sodium transport as a mechanism for regulating *in vivo* alveolar fluid clearance, including a discussion of catecholamine and non-catecholamine dependent mechanisms for stimulating fluid transport. The second section reviews new evidence for involvement of transcellular water channels in alveolar and distal airway fluid transport, and the third section describes how the normal capacity of the alveolar epithelial barrier to transport salt and water is altered by exposure to clinically relevant pathological conditions).

EVIDENCE FOR ACTIVE SODIUM TRANSPORT AS A MECHANISM FOR *IN VIVO* ALVEOLAR FLUID CLEARANCE

For many years, it was generally believed that differences in hydrostatic and protein osmotic pressures (Starling forces) accounted for removal of excess fluid from the air spaces of the lung.[1] This misconception persisted in part because experiments that were designed to measure solute flux across the epithelial and endothelial barriers of the lung were done at room temperature.[2] Also, these studies were done in dogs, a species that has a very low rate of alveolar epithelial sodium and fluid transport.[3] However, in the early 1980's, experimental work from both *in vivo* and *in vitro* studies provided direct evidence that active sodium transport drives alveolar fluid transport across the alveolar barrier. The principal findings of the *in vivo* studies are summarized below).

In Vivo Studies of Alveolar Fluid Clearance

Experiments in anesthetized or unanesthetized sheep over 4, 12, and 24 hours indicated that spontaneous alveolar fluid clearance occurs in the face of a rising alveolar protein concentration.[4,5] The final alveolar protein concentration exceeded plasma protein concentration by 3-6 g/100 ml. The same pattern was documented subsequently in humans in the resolution phase of pulmonary edema.[6] The final alveolar protein concentration in some patients exceeded 10 g/100 ml with a simultaneous plasma protein concentration of 5-6 g/100 ml. These observations indicated an active ion transport mechanism was responsible for the removal of alveolar fluid).

If active ion transport were responsible for alveolar fluid clearance, then alveolar fluid clearance should be temperature dependent. In an *in situ* goat lung preparation, the rate of alveolar fluid clearance progressively declined as temperature was lowered from 37 to 18° C.[7] Similar results were obtained in an isolated rat lung preparation in which hypothermia inhibited alveolar sodium transport.[8] In the isolated human lung, alveolar fluid clearance

Acute Respiratory Distress Syndrome: Cellular and Molecular Mechanisms and Clinical Management
Edited by Matalon and Sznajder, Springer Science+Business Media New York, 1998

ceased when temperature was lowered to 20° C.[9] In addition, if active ion transport were primarily responsible for alveolar fluid clearance, then the elimination of transpulmonary hydrostatic pressure generated by ventilation should not alter the rate of alveolar fluid clearance. Studies in rabbits and sheep indicated that rate of alveolar fluid clearance was unchanged in the absence of ventilation.[10]

Additional evidence for active fluid transport was obtained in intact animals with the use of amiloride, an inhibitor of sodium uptake by the apical membrane of alveolar epithelium and distal airway epithelium. Amiloride inhibited 40-70% of basal alveolar fluid clearance in sheep, rabbits, rats, and in the human lung,[4,9,11,12] similar to the data obtained in isolated rat lung preparations.[13,14,15] Amiloride also inhibits sodium uptake in distal airway epithelium from sheep and pigs.[16,17] To further explore the role of active sodium transport, experiments were designed to inhibit the Na,K-ATPase in alveolar type II cells. It is difficult to study the effect of ouabain in intact animals because of cardiac toxicity. However, in the isolated rat lung, ouabain was shown to inhibit >90% of clearance.[18] Subsequently, with the development of an *in situ* sheep preparation for measuring alveolar fluid clearance in the absence of blood flow, it was demonstrated that ouabain inhibited 90% of alveolar fluid clearance over a 4-hour period.[19]

Important species differences in the basal rates of alveolar fluid clearance have been identified. In order to normalize for differences in lung size or the available surface area, different instilled volumes were used ranging from 1.5 to 6.0 ml/kg. The slowest alveolar fluid clearance was measured in dogs,[3] intermediate rates of alveolar fluid clearance in sheep and goats,[4,5,7,20] and the highest basal alveolar fluid clearance rates have been measured in rabbits and rats.[11,12] The basal rate of alveolar fluid clearance in the human lung has been difficult to estimate, but based on the isolated, non-perfused human lung model, basal clearance rates appear to be intermediate-to-fast.[9] In fact, recent data indicates that the clearance in the *ex vivo* human lung is approximately half of the rate in the *ex vivo* rat lung.[21] The explanation for the species differences is not apparent, although may be related to the number or activity of sodium channels or the density of Na,K-ATPase pumps in alveolar epithelium in different species. However, morphometric studies[22,23] show no significant difference in the number of alveolar type II cells in different species).

Catecholamine-Dependent Upregulation of Alveolar Fluid Clearance

Studies in newborn lambs suggested that endogenous release of epinephrine might stimulate reabsorption of fetal lung fluid from the lung.[24] Recent work in newborn guinea pigs provides conclusive evidence that elevated endogenous epinephrine drives alveolar fluid clearance at birth.[25] In addition, studies of isolated alveolar type II cells indicated that sodium transport could be augmented with ß-adrenergic agonists, probably by cAMP dependent mechanisms.[26,27,28] The enhancement of alveolar fluid clearance by catecholamines was confirmed in isolated perfused lung studies.[14,29,30] Subsequent experiments in isolated lungs[31,32] and *in vivo* studies provided further evidence that cAMP is the second messenger for the ß-adrenergic effects, whereas activation of protein kinase C is not involved.[33,34]

In vivo studies over 4 hours were carried out in sheep to examine the physiologic factors that might influence alveolar fluid clearance, including systemic and pulmonary hemodynamics, pulmonary blood flow, and lung lymph flow. Terbutaline (10^{-5} M) was instilled with autologous serum into the distal air spaces of the lung.[20] Terbutaline nearly doubled alveolar fluid clearance over 4 hours in sheep, and the increase was 90% prevented by co-administration of amiloride in the instilled solution. Although terbutaline increased pulmonary blood flow, this factor was not important since studies with nitroprusside, an agent that increased pulmonary blood flow to an equivalent degree, did not increase alveolar fluid clearance. There was an increase in lung lymph flow, a finding that reflected removal of some of the excess alveolar fluid volume that had been transported to the lung interstitium. The ß-adrenergic agonist effect was prevented by co-administration of propranolol into the air spaces. Terbutaline also doubled alveolar fluid clearance in the dog lung.[3] Subsequent studies demonstrated that alveolar fluid clearance is markedly increased in the intact rat lung by ß-adrenergic agonists.[11] Interestingly, ß-adrenergic agonist therapy does not increase alveolar fluid clearance in rabbits and hamsters.[12,35] The explanation for this lack of effect is unclear, particularly since there are ß-receptors in rabbit type II cells that stimulate surfactant secretion.[36]

Do ß-adrenergic agonists increase alveolar fluid clearance in the human lung? Based on studies of the resolution of alveolar edema in humans, it has been difficult to quantify the contribution of endogenous catecholamines to the basal alveolar fluid clearance rate.[6] However, recent studies of alveolar fluid clearance in the isolated human lung have demonstrated that ß-adrenergic agonist therapy increases alveolar fluid clearance, and the increased clearance can be inhibited with propranolol or amiloride.[9,37] The magnitude of the effect is similar to that observed in other species, with a ß-agonist-dependent doubling of alveolar fluid clearance over baseline levels. This data is particularly important based on recent evidence indicating that aerosolized ß-agonist therapy with salmeterol, a long acting lipid soluble ß-agonist, markedly increased the rate of alveolar fluid clearance in sheep.[38] Also, there is recent data that the long acting lipid soluble ß-agonists are more potent than hydrophilic ß-agonists in the *ex vivo* human lung (**Figure 1**).[21] Thus, aerosolized ß-agonist treatment in some patients with pulmonary edema might accelerate the resolution of alveolar edema).

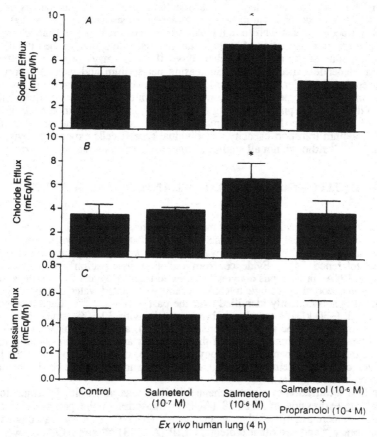

Figure 1. Effects of salmeterol on sodium, chloride, and potassium transport from the alveolar spaces in the *ex vivo* human lung over 4 hours. A. The values of net sodium efflux. B. The values of net chloride efflux. C. The values of net potassium influx. *$P<0.05$ vs corresponding values in the control experiments. The effect of salmeterol was significantly greater than terbutaline in augmenting the rate of alveolar fluid clearance (with permission, reference 21).

Catecholamine Independent Upregulation of Alveolar Fluid Clearance

In addition to the well studied effects of ß-adrenergic agonists, there is recent evidence that several non-catecholamine dependent pathways can increase the rate of alveolar fluid clearance. Incubation of isolated alveolar type II cells with epidermal growth factor for 24-48 hours increases their capacity to transport sodium.[39,40] Transforming growth factor-α (TGF-α) recently has been reported to increase alveolar fluid clearance acutely in anesthetized, ventilated rats.[41,42] Compared to controls, 50 ng/ml of TGF-α in the instilled fluid increased alveolar liquid clearance by 45% over 1 hour and by 53% over 4 hours. This increase was similar to the 50% increase in alveolar fluid clearance in rats treated with a ß-agonist.[42] Interestingly, since cAMP was only minimally increased in isolated alveolar type II cells exposed to TGF-α, it is likely that the TGF-α effect is mediated by an alternative signal transduction pathway).

New evidence suggests that cytokines may stimulate sodium uptake and alveolar fluid clearance. It is well known that alveolar instillation of endotoxin or exotoxin releases several proinflammatory cytokines from alveolar macrophages. For example, exotoxin A from *P. aeruginosa* can stimulate alveolar fluid clearance in rats by a non-catecholamine dependent pathway.[43] In a recent study, instillation of endotoxin from *E. Coli* into the distal air spaces of rats upregulated epithelial sodium transport and alveolar fluid clearance for up to 48 hours by catecholamine independent mechanisms.[44] The mechanisms for the endotoxin effect may depend on release of tumor necrosis factor-α (TNF-α), since a monoclonal antibody against TNF-α inhibited the increase in alveolar fluid clearance that occurred 24 hours after instillation of bacteria into the distal air spaces of the rat lung.[45]

Proliferation of alveolar epithelial type II cells may provide another non-catecholamine dependent mechanism for increasing net sodium and water transport across the alveolar epithelial barrier. Recent work with bleomycin injured rat lungs indicate that hyperplasia of alveolar type II cells contributes to increased alveolar fluid clearance, especially in the sub-acute phase following acute lung injury.[46] In addition to an increase in the number of alveolar type II cells, there may also be an oxidant-dependent mechanism that increases the sodium transport capacity of individual type II cells exposed to hyperoxia for several days[47,48,49,50] although not all studies of hyperoxia demonstrate this effect.[51,52,53]

ROLE OF AQUAPORINS IN ALVEOLAR AND DISTAL AIRWAY TRANSPORT

The existence of specialized water transporting proteins had been proposed for many years based on biophysical measurements showing that osmotic water permeability in erythrocytes and certain kidney tubules was high and weakly temperature dependent (for review, see reference 54). Evidence from radiation inactivation[55] and expression of heterologous mRNAs in Xenopus oocytes[56] suggested that the putative water channel was an ~30 kDa protein encoded by a single mRNA. A family of related water transporting proteins (aquaporins) was subsequently identified over the past 4 years.[57,58,59] Each member of the family is a small (~30 kDa) integral membrane protein with 30-50% amino acid sequence identity to the major intrinsic protein of lens fiber (MIP), and related proteins from plants, bacteria and yeast.[60] Hydropathy plots of these proteins are similar, suggesting up to six transmembrane helical segments. Homology in amino acid sequence between the first and second halves of each protein suggests genesis from tandem, intragenic duplication of a three-transmembrane segment.

The first localization of a water channel in lung was an in situ hybridization study showing diffuse expression of CHIP28 (AQP-1) transcript in the peri-alveolar region.[61] Subsequently, a mercurial-insensitive water channel (MIWC, AQP-4) was cloned from a lung cDNA library,[62] and two other proteins, GLIP (AQP-3)[63,64] and AQP-5,[65] were cloned from other sources and then found to be expressed in trachea and/or lung (**Table 1**).

Structure and Function of CHIP28 (AQP-1)

The majority of molecular-level information comes from studies on CHIP28, an integral membrane protein identified initially in the plasma membrane of erythrocytes.[66] CHIP28 is easily purified in milligram quantities from blood cell membranes, and forms mercurial-sensitive water selective channels when reconstituted into liposomes[67,68] or when expressed in Xenopus oocytes[69,70] or mammalian cells.[71] Residue cysteine 189 has been shown to be the site at which mercurials inhibit CHIP28 water permeability.[72,73] Fifty percent of CHIP28 monomers are modified by N-linked glycosylation with a polylactosamine sugar at residue Asn42.[74] The single channel water permeability of CHIP28 is low (5×10^{-14} cm^3/s), so that the membrane density of CHIP28 must be quite high ($>10^3$ µm^{-2}) to increase water permeability above the background level related to water diffusion across membrane lipids. CHIP28 is expressed in many fluid-transporting tissues, including kidney proximal tubule and thin descending limb of Henle, choroid plexus, iris, ciliary body, placental synctiotrophoblast, gallbladder, colonic crypt and several reproductive tissues.[75,76,77,78,79]

Table 1. Properties of Mammalian Water Channels that are Found in the Lung

Original Names Alternate Names	MIP26	CHIP28 AQP-1	MIWC AQP-4	AQP-5
Protein size (amino acids)	263	268	301	282
mRNA size (kb)	1.4	2.8	5.5	1.6
N-linked glycosylation	-	+	-	+
Mercurial inhibition	-	+	-	+
Transport function	?	water	water	water
Human gene locus	12q13	7p14	18q22	?
Tissue expression	lens	wide	wide	wide

Major intrinsic protein (MIP) family members expressed in mammalian tissues are listed (left to right) in the order in which they were identified and cloned. Data on mRNA and protein are given for rat and chromosomal locus for human).

Mercurial-insensitive Water Channel (MIWC or AQP-4)

MIWC was the first mercurial-insensitive water channel identified[80] and is expressed in multiple tissues, including the basolateral plasma membrane of tracheal and bronchial epithelia, kidney collecting duct, ependymal and astroglial cells in brain, iris, ciliary body and nuclear layer of retina in eye, gastric parietal cells, colon surface epithelium, skeletal muscle and various secretory glands.[81,82] MIWC has greatest homology to the big brain protein of Drosophila. MIWC forms mercurial-insensitive water-selective channels when expressed in Xenopus oocytes.[80] The lack of mercurial sensitivity is due to absence of cysteine residues in the vicinity of the two conserved NPA motifs.[83] Analysis of MIWC biogenesis and topology indicated that MIWC contains 6-membrane spanning domains with N- and C- termini in the cytoplasm.[84] Recently a human homolog (hMIWC) of rat MIWC was cloned.[85] Two distinct hMIWC cDNAs were found: clone hMIWC1, encoding 301 amino acids, and hMIWC2, which contains a distinct 5'-sequence upstream from bp -34 in clone hMIWC1 and two additional in-frame translation start codons. Analysis of hMIWC gene structure indicated 2 distinct but overlapping transcription units from which multiple hMIWC mRNAs are transcribed. There were 3 introns (lengths 0.9, 0.2 and 6 kb) in the hMIWC1 coding sequence. Genomic Southern blot analysis and in situ hybridization indicated the presence of a single copy hMIWC gene at location 18q2).

Aquaporin-5

Another related mercurial-sensitive water channel, AQP-5, was cloned from a rat submandibular cDNA library.[86] Northern blot analysis showed transcript expression in salivary gland, lacrimal gland, eye, trachea and lung. Recent data indicates that aquaporin-5 is localized to the apical portion of alveolar epithelial type I cells.[87]

We recently obtained data on water channel transcript expression in the developing and post-natal lung using the RNase protection assay.[88] Transcripts encoding CHIP28, MIWC and AQP-5 are expressed at relatively low levels in lung prior to birth. CHIP28 mRNA increases strongly just after birth and remains elevated. MIWC mRNA increases strongly between days 1 and 2 after birth, and decreases slightly over the first week. In contrast, AQP-5 mRNA slowly and progressively increases over first week. Functional studies of water permeability and analysis of protein expression will be needed to define the physiological implications of these observations).

Measurement of Water Permeability in the Lung

There have been few measurements of water permeability in lung. More than 20 years ago, Effros demonstrated rapid translocation of pure solute-free water into the vascular space following injection of a hypertonic solution into the perfusate of isolated perfused lungs.[89] However, direct evidence for the existence of specific transcellular water pathways in the lung was not available until recently, when a combination of molecular, cellular, isolated airway, and whole lung studies were utilized to test the hypothesis that osmotically driven water movement in the lung occurs across plamsa membrane water channels.[90,91]

An *in situ* perfused sheep lung model was utilized to measure transalveolar osmotic water permeability. Intact lungs were perfused continuously with an isosmolar dilute blood solution. Hypertonic fluid (900 mOsm) was instilled bronchoscopically into the airspaces, and the time course of water movement from capillary to airspace was deduced from the dilution of instilled radiolabeled-albumin and from air space fluid osmolality.[91] In control lungs, osmotically-induced water movement was rapid (equilibration half-time ~45 s) and had an apparent P_f of ~0.02 cm/s, similar to that in erythrocytes. Water permeability in the contralateral lung was inhibited reversibly by ~70% by $HgCl_2$. These results indicated that mercurial-sensitive water channels facilitated the transcellular movement of water between the airspace and capillary compartments in lung. High osmotic water permeability was found recently in the mouse lung utilizing a novel fluorescence method described earlier.[92]

We recently developed a strategy to measure both osmotic and diffusional water permeability in intact airways.[93] Small airways (100-200 μm diameter, 1-2 mm length) from guinea pig lung were microdissected and perfused in vitro with fluorescent markers in the lumen. Osmotic water permeability (P_f) was 4-5 x 10^{-3} cm/s at 23° C and was independent of lumen flow rate and osmotic gradient size and direction. Temperature dependence measurements gave an activation energy of 4.3 kcal/mol, suggesting the passage of water through molecular water channels, consistent with the water channel immunolocalization studies that were done. Osmotic water permeability has not yet been measured in epithelia of trachea and larger airways).

Recently, it has been possible to measure water permeability across the alveolar barrier in perinatal rabbit lungs.[88] Interestingly, osmotic water permeability increased immediately after birth, potentially consistent with the need to reabsorb alveolar fluid in the newborn lung).

Despite the strong expression of several water channels in pulmonary tissues, the exact physiological relevance of these proteins remains to be proven. Mutations in water channel CHIP28 have been detected in humans.[94] Genomic DNA analysis on three rare individuals who do not express CHIP28-associated Colton blood group antigens demonstrated that two of them carried a CHIP28 pseudogene with different nonsense mutations, and another had a missense mutation encoding a nonfunctional CHIP28 molecule. Although red blood cells from these individuals had low osmotic water permeability, the subjects were phenotypically normal, raising questions about the physiological importance of CHIP28. Naturally occurring mutations in MIWC (AQP-4), GLIP (AQP-3) and AQP-5 have not as yet been identified. Definition of the physiologic role of these proteins will require evaluation of knock-out transgenic animals or other suitable models and/or the identification of specific, non-toxic inhbitors of water channel function).

ALVEOLAR FLUID TRANSPORT UNDER PATHOLOGICAL CONDITIONS

The fluid transport capacity of the alveolar epithelial barrier under pathological conditions, particularly in patients with pulmonary edema and acute lung injury, is of major interest to both basic and clinical investigators. More than ten years ago, clinical studies indicated that protein-rich pulmonary edema can be collected from patients with acute lung injury, whereas patients with cardiogenic or hydrostatic pulmonary edema have a significantly lower protein concentration in the edema fluid.[95,96] However, there was no direct information in these clinical studies regarding the contribution of the epithelial barrier to the development or resolution of the alveolar edema. Until recently, pathological studies provided the only direct information regarding the status of the alveolar epithelial barrier in patients with acute lung injury. For example, post-mortem studies of patients who die with acute lung injury report diffuse alveolar damage to both the endothelial and epithelial barriers of the lung with protein-rich edema, inflammatory cells, and intra-alveolar exudate, pathological hallmarks of the pulmonary response to acute lung injury.[97] Ultrastructural studies indicate widespread necrosis and denuding of alveolar epithelial type I cells, usually with some evidence of alveolar epithelial type II cell hyperplasia.[98] However, these post-mortem studies represent a biased sampling of only the most severe cases of acute lung injury. Recent clinical studies indicate that there is considerable heterogeneity in the fluid transport and barrier properties of the alveolar epithelial barrier of patients with acute lung injury.[99,100,101,102]

Two properties of the epithelial barrier can be assessed clinically. First, since the epithelial barrier is normally impermeable to protein, the quantity of protein that accumulates in the distal air spaces is a good index of epithelial permeability. Secondly, since concentration of protein in alveolar fluid reflects net clearance of alveolar fluid, measurement of protein concentration in sequential alveolar edema fluid samples provides a physiologic index of the ability of the alveolar epithelial barrier to remove edema fluid. In a recent study, patients who died with acute lung injury had 3-fold more protein lavaged from their distal air spaces than patients who survived.[103] In another study, approximately 40% of patients were able to reabsorb some of the alveolar edema fluid within 12 hours of intubation and acute lung injury.[102] These patients had a more rapid recovery from respiratory failure and a lower mortality. In contrast, the patients who did not reabsorb any of the alveolar edema fluid in the first 12 hours following acute lung injury had protracted respiratory failure and a higher mortality. Based on clinical studies, the ability of the alveolar epithelial barrier to reabsorb alveolar edema fluid from acute lung injury within the first 12 hours after acute lung injury is preserved in 30-40 % of patients.[102]

As illustrated in **Table 2**, many experimental studies have provided new insights into the function of the alveolar epithelial barrier under clinically relevant pathological conditions. In each of the studies, the primary focus was to assess the net fluid transport capacity of the alveolar and distal airway epithelium under specific physiologic stresses as well as well defined pathological insults. Interestingly, the results indicate that the alveolar and distal airway epithelium is remarkably resistant to injury, particularly compared to the adjacent lung endothelium. Even when mild to moderate alveolar epithelial injury occurs, the capacity of the alveolar epithelium to transport salt and water is often preserved. In addition, several mechanisms may result in an upregulation of the fluid transport capacity of the distal pulmonary epithelium, even after moderate to severe epithelial injury).

The first evidence demonstrating the resistance of the alveolar epithelial barrier to injury evolved from studies in which large numbers of neutrophils and monocytes crossed the tight alveolar epithelial barrier without inducing a significant change in either permeability to protein or the transport capacity of the alveolar epithelium. Instillation of autologous serum or plasma into the distal air spaces of sheep was associated with an influx of neutrophils and monocytes. Despite the influx of inflammatory cells, there was no increase in epithelial permeability to plasma protein; in addition, alveolar fluid clearance was normal.[4] In a subsequent study in normal human volunteers, large numbers of neutrophils were recruited to the distal air spaces by instillation of the potent neutrophil chemotactic factor, leukotriene B_4, without influx of plasma protein into the air spaces.[104] Instillation of a hyperosmolar solution (sea water) into rabbit lungs caused a rapid translocation of a large volume of water into the distal air spaces as well as the influx of large numbers of neutrophils.[105] However, there was only a transient change in epithelial permeability to protein. Moreover, after osmotic equilibaration occurred, the rate of alveolar sodium and fluid transport was normal in rabbits[105] and in one well-described clinical case.[106] Finally, recent data indicates that alveolar epithelial transport mechanisms in the human lung are not altered by 6-8 hours of severe hypothermia (7° C) followed by rewarming to 37° C.[107]

Table 2. Effect of Pathological Conditions on Alveolar Epithelial Fluid Clearance in Several Species

Pathological Condition	Species	Severity of Lung Injury		Alveolar Fluid Clearance
		Endothelium	Epithelium	
Endotoxin, IV	Sheep	Mild	None	Normal
Endotoxin, alveolar	Sheep	None	None	Normal
	Rat	None	None	Increased
Exotoxin, alveolar	Rat	None	None	Increased
Bacteria, IV[1]	Sheep	Moderate	Mild	Increased
	Rat	Mild	None	Increased
	Rabbit	Mild	None	Normal
Bacteria, alveolar[1]	Sheep	Mild	Mild	Normal
	Rat	Mild	Mild	Increased
Oleic acid[2]	Sheep	Moderate	Moderate	Normal
Acid aspiration	Rabbit	Severe	Severe	Decreased
Salt water aspiration	Rabbit	Mild	Mild	Normal
Hyperoxia[3]	Rat	Moderate	Mild	Normal
Drug induced (Bleomycin)	Rat	Moderate	Mild	Increased
Drug induced (ANTU)	Rat	Moderate	None	Unknown

[1] The extent of injury to the epithelial and endothelial barriers is dependent on the dose and virulence of bacteria that are administered. In some sheep, clearance was negligible).

[2] Initially the injury after intravenous oleic acid is severe, but the alveolar epithelium recovers after 4 hours sufficiently to remove some of the excess edema fluid).

[3] The extent of injury to the epithelial and endothelial barrier depends on the oxygen concentration and duration of exposure of the rats).

Even when lung endothelial injury occurs, the alveolar epithelial barrier may remain normally impermeable to protein and retain its normal fluid transport capacity (**Table 2**). For example, intravenous endotoxin or bacteria have been used to produce lung endothelial injury in sheep[108] or rats,[109] but permeability to protein across the lung epithelial barrier was not increased. When septic shock was produced in rats, there was a marked increase in plasma epinephrine levels. Even though there was endothelial injury and mild interstitial pulmonary edema, alveolar epithelial fluid transport was increased from 45% over 4 h in control rats to 75% over 4 h in the septic rats. The effect was inhibited with instillation of amiloride (10^{-4} M) or propranolol (10^{-4} M) into the distal air spaces, proving that the stimulated clearance depended on ß-agonist stimulation of alveolar epithelial sodium transport. When more severe septic shock was produced in sheep, the alveolar epithelial barrier was resistant to injury in the majority of sheep with confinement of the edema to the pulmonary interstitium (**Table 2**).[110] In some sheep, however, more severe systemic and pulmonary endothelial injury was associated with alveolar flooding, a marked increase in epithelial permeability to protein, and the inability to transport fluid from the air spaces of the lung. The inability to remove excess fluid from the air spaces in these sheep was probably related more to a marked increase in paracellular permeability from injury to the epithelial tight junctions rather than to a loss of salt and water transport capacity of alveolar epithelial cells).

In some experimental studies, such as acid aspiration induced lung injury,[111] the injury to the epithelial barrier is so severe that recovery does not occur (**Table 2**). In other types of severe lung injury, as occurs from intravenous oleic acid, the initial injury to the tight junctions results in severe alveolar flooding, although recovery may occur within a few hours, presumably from re-establishment of the normal tight barrier characteristics of the alveolar barrier.[112] A similar pattern of severe injury may develop in animal models of pneumonia in which large numbers of virulent bacteria are used,[113] although studies with less virulent bacteria are associated with less epithelial injury and a preserved capacity to transport fluid from the distal air spaces of the lung.[114]

Although much has been learned about the resistance of the alveolar epithelial barrier to injury and its capacity for preserved transport function after injury, more work is needed to understand the local and systemic factors that regulate sodium and water transport across distal airway and alveolar epithelium under pathological conditions. Upregulation of alveolar fluid clearance from endogenous or exogenous beta-adrenergic stimulation has been clearly demonstrated in several clinically relevant animal models. New evidence suggests that alveolar epithelial type II cell hyperplasia can be associated with a sustained upregulation of alveolar epithelial fluid clearance, an important potential mechanism in the recovery phase following acute lung injury).

FUTURE DIRECTIONS

Recent studies have established that transport of sodium from the air spaces to the lung intersititum is the primary mechanism driving alveolar fluid clearance. While there are significant differences among species in the basal rates sodium and fluid transport, the basic mechanism depends on sodium uptake by channels on the apical membrane of alveolar type II cells followed by extrusion of sodium on the basolateral surface by the Na,K-ATPase.[76] This process can be upregulated by several catecholamine dependent and independent mechanisms. The identification of water channels expressed in lung, together with the high water permeabilities, suggest a role for channel-mediated water movement between the airspace and capillary compartments. Direct experimental evidence utilizing transgenic knock-out and specific water channel inhibitors is required to define the roles of specific water channels in lung physiology.[76]

The application of this new knowledge regarding salt and water transport in distal airway and alveolar epithelium to pathological conditions has been successful in clinically relevant experimental studies, as well as in a few clinical studies. The studies of exogenous and endogenous catecholamine regulation of alveolar fluid clearance are a good example of how new insights into the basic mechanisms of alveolar sodium and fluid transport can be translated to clinically relevant experimental studies. For example, it is clear that exogenous catecholamines can increase the rate of alveolar fluid clearance in several species including the human lung, and it is also apparent that release of endogenous catecholamines can upregulate alveolar fluid clearance in animals with septic or hypovolemic shock. It is possible that therapy with beta adrenergic agonists might be useful in hastening the resolution of alveolar edema in some patients, but the use of exogenous beta adrenergic agonists in some clinical conditions may be unnecessary if endogenous catecholamines are already elevated by the clinical condition, such as shock. In some patients, the extent of injury to the alveolar epithelial barrier may be too severe for beta adrenergic agonists to enhance the resolution of alveolar edema, although several experimental studies indicate that alveolar fluid clearance can be augmented in the presence of moderately severe lung injury. Clinical research is needed to evaluate the use of exogenous beta adrenergic agonists in patients with pulmonary edema and the conditions that predispose to the development of acute lung injury).

Acknowledgements

This work was supported in part by NIH grants HL51854, DK35124, and HL42368).

REFERENCES

1. N.C. Staub, Pulmonary edema, *Physiol. Rev*. 54:678-811 (1974).
2. A.E. Taylor, A.C. Guyton, and V.S. Bishop, Permeability of the alveolar epithelium to solutes, *Circ. Res*. 16:353-362 (1965).
3. Y. Berthiaume, V.C. Broaddus, M.A. Gropper, T. Tanita, and M.A. Matthay, Alveolar liquid and protein clearance from normal dog lungs, *J. Appl. Physiol*. 65:585-593 (1988).
4. M.A. Matthay, Y. Berthiaume, and N.C. Staub, Long-term clearance of liquid and protein from the lungs of unanesthetized sheep, *J. Appl. Physiol*. 59:928-934 (1985).
5. M.A. Matthay, C.C. Landolt, and N.C. Staub, Differential liquid and protein clearance from the alveoli of anesthetized sheep, *J. Appl. Physiol*. 53:96-104 (1982).
6. M.A. Matthay and J.P. Wiener-Kronish, Intact epithelial barrier function is critical for the resolution of alveolar edema in humans, *Am. Rev. Respir. Dis*. 142:1250-1257 (1990).
7. V.B. Serikov, M. Grady, and M.A. Matthay, Effect of temperature on alveolar liquid and protein clearance in an *in situ* perfused goat lung, *J. Appl. Physiol*. 75:940-947 (1993).
8. D.H. Rutschman, W. Olivera, and J.I. Sznajder, Active transport and passive liquid movement in isolated perfused rat lungs, *J. Appl. Physiol*. 75:1574-1580 (1993).
9. T. Sakuma, G. Okaniwa, T. Nakada, T. Nishimura, S. Fujimura, and M.A. Matthay, Alveolar fluid clearance in the resected human lung, *Am. J. Respir. Crit. Care Med*. 150:305-310 (1994).
10. C. Jayr and M.A. Matthay, Alveolar and lung liquid clearance in the absence of pulmonary blood flow in sheep, *J. Appl. Physiol*. 71:1679-1687 (1991).
11. C. Jayr, C. Garat, M. Meignan, J.-F. Pittet, M. Zelter, and M.A. Matthay, Alveolar liquid and protein clearance in anesthetized ventilated rats, *J. Appl. Physiol*. 76:2636-2642 (1994).
12. N. Smedira, L. Gates, R. Hastings, C. Jayr, T. Sakuma, J.-F. Pittet, and M.A. Matthay, Alveolar and lung liquid clearance in anesthetized rabbits, *J. Appl. Physiol*. 70:1827-1835 (1991).
13. G. Basset, C. Crone, and G. Saumon, Fluid absorption by rat lung *in situ*: pathways for sodium entry in the luminal membrane of alveolar epithelium, *J. Physiol. (London)* 384:325-345 (1987).
14. E.D. Crandall, T.H. Heming, R.L. Palombo, and B.E. Goodman, Effect of terbutaline on sodium transport in isolated perfused rat lung, *J. Appl. Physiol*. 60:289-294 (1986).
15. R.M. Effros, G.R. Mason, J. Hukkanen, and P. Silverman, New evidence for active sodium transport from fluid-filled rat lungs, *J. Appl. Physiol*. 66:906-919 (1988).
16. F.J. Al-Bazzaz, Regulation of Na and Cl transport in sheep distal airways, *Am. J. Physiol*. 267:L193-L198 (1994).
17. S.T. Ballard, S.M. Schepens, J.C. Falcone, G.A. Meininger, and A.E. Taylor, Regional bioelectric properties of porcine airway epithelium, *J. Appl. Physiol*. 73:2021-2027 (1992).
18. G. Basset, C. Crone, and G. Saumon, Significance of active ion transport in transalveolar water absorption: a study on isolated rat lung, *J. Physiol. (London)* 384:311-324 (1987).
19. T. Sakuma, J.F. Pittet, C. Jayr, and M.A. Matthay, Alveolar liquid and protein clearance in the absence of blood flow or ventilation in sheep, *J. Appl. Physiol*. 74:176-185 (1993).
20. Y. Berthiaume, N.C. Staub, and M.A. Matthay, Beta-adrenergic agonists increase lung liquid clearance in anesthetized sheep, *J. Clin. Invest*. 79:335-343 (1987).
21. T. Sakuma, H. Folkesson, S. Suzuki, K. Usuda, M. Handa, G. Okaniwa, S. Fujimura, and M.A. Matthay, Salmeterol increases alveolar epithelial fluid clearance in both in vivo and ex vivo rat lungs, as well as in ex vivo human lungs, *Am. J. Resp. Crit. Care Med*. (1996).

22. J.D. Crapo, S.L. Young, E.K. Fram, K.E. Pinkerton, B.E. Barry, and R.O. Crapo, Morphometric characteristics of cells in the alveolar region of mammalian lungs, *Am. Rev. Respir. Dis.* 128:S42-S46 (1983).

23. K.E. Pinkerton, B.E. Barry, J.J. O'Neil, J.A. Raub, P.C. Pratt, and J.D. Crapo, Morphologic changes in the lung during the lifespan of Fisher 344 rats, *Am. J. Anat.* 164:155-174 (1982).

24. M.J. Brown, R.E. Olver, C.A. Ramsden, L.B. Strang, and D.V. Walters, Effects of adrenaline and of spontaneous labour on the secretion and absorption of lung liquid in the fetal lamb, *J. Physiol. (London)* 344:137-152 (1983).

25. N. Finley, A. Norlin, D.C. Baines, and H.G. Folkesson, Alveolar epithelial fluid clearance is mediated by endogenous catecholamines at birth in guinea pigs, *J. Clin. Invest.*, in press (1998).

26. B.E. Goodman, S.E. Brown, and E.D. Crandall, Regulation of transport across pulmonary alveolar epithelial cell monolayers, *J. Appl. Physiol.* 57:703-710 (1984).

27. R.J. Mason, M.C. Williams, J.H. Widdicombe, M.J. Sanders, D.S. Misfeldt, and L.C.J. Berry, Transepithelial transport by pulmonary alveolar type II cells in primary culture, *Proc. Natl. Acad. Sci. USA* 79:6033-6037 (1982).

28. S. Matalon, Mechanisms and regulation of ion transport in adult mammalian alveolar type II pneumocytes, *Am. J. Physiol.* 261:C727-C738 (1991).

29. R.M. Effros, G.R. Mason, K. Sietsema, P. Silverman, and J. Hukkanen, Fluid reabsorption and glucose consumption from edematous rat lungs, *Circ. Res.* 60:708-719 (1987).

30. B.E. Goodman, K.J. Kim, and E.D. Crandall, Evidence for active sodium transport across alveolar epithelium of isolated rat lung, *J. Appl. Physiol.* 62:2460-2466 (1987).

31. B.E. Goodman, J.L. Anderson, and J.W. Clemens, Evidence for regulation of sodium transport from airspace to vascular space by cAMP, *Am. J. Physiol.* 257:L86-L93 (1989).

32. G. Saumon and G. Basset, Electrolyte and fluid transport across the mature alveolar epithelium, *J. Appl. Physiol.* 74:1-15 (1993).

33. Y. Berthiaume, Effect of exogenous cAMP and aminophylline on alveolar and lung liquid clearance in anesthetized sheep, *J. Appl. Physiol.* 70:2490-2497 (1991).

34. Y. Berthiaume, M. Sapijaszko, J. MacKenzie, and M.P. Walsh, Protein kinase C activation does not stimulate lung liquid clearance in anesthetized sheep, *Am. Rev. Respir. Dis.* 144:1085-1090 (1991).

35. B.E. Goodman, Lung fluid clearance, in: *Fluid and Solute Transport in the Air Spaces of the Lungs*, R.M. Effros and H.K. Chang, eds., Marcel Dekker, New York (1993).

36. J.V. McDonald, Jr., L.W. Gonzales, P.L. Ballard, J. Pitha, and J.M. Roberts, Lung beta-adrenoreceptor blockade affects perinatal surfactant release but not lung water, *J. Appl. Physiol.* 60:1727-1733 (1986).

37. T. Sakuma, G. Okaniwa, T. Nakada, T. Nishimura, S. Fugimura, and M.A. Matthay, Terbutaline, a betaadrenergic agonist, increases alveolar liquid clearance in the resected human lung (abstract), *FASEB J.* 7:A436 (1993).

38. A.R. Campbell, H.G. Folkesson, O. Osorio, J.M. Cohen-Solal, and M.A. Matthay, Alveolar fluid clearance can be accelerated in ventilated sheep with an aerosolized beta-adrenergic agonist (salmeterol) (abstract), *Am. J. Respir. Crit. Care Med.* 151:A620 (1995).

39. Z. Borok, S.J. Danto, K.J. Kim, R.L. Lubman, and E.D. Crandall, Effects of EGF and dexamethasone on bioelectric properties of alveolar epithelial cell monolayers, *Am. Rev. Respir. Dis.* 147:1005A (1993).

40. H.A. Jaffe, W. Olivera, K. Ridge, D. Yeates, and J.I. Sznajder, *In vivo* administration of EGF increases Na^+,K^+-ATPase activity in alveolar type II cells, *FASEB J.* 8:A141 (1994).

41. H.G. Folkesson, J.-F. Pittet, G. Nitenberg, and M.A. Matthay, Transforming growth factor-α increases alveolar liquid clearance in anesthetized, ventilated rats, *Am. J. Physiol.* (Submitted).

42. M.A. Matthay, J.-F. Pittet, G. Nitenberg, and H.G. Folkesson, Transforming growth factor-α (TGF-α) and beta adrenergic agonist therapy:Comparative studies of

their capacity to increase alveolar liquid clearance (abstract), *FASEB J.* 9:A568 (1995).

43. J.F. Pittet, S. Hashimoto, M. Pian, M. McElroy, G. Nitenberg, and J.P. Wiener-Kronish, Exotoxin A stimulates fluid reabsorption from distal airspaces in anesthetized rats, *Am. J. Physiol. (Lung Cell. Mol. Physiol.)* 270:L232-L241 (1996).

44. C. Garat, S. Rezaiguia, M. Meignan, M.P. d'Ortho, A. Harf, M.A. Matthay, and C. Jayr, Alveolar endotoxin increases alveolar liquid clearance in rats, *J. Appl. Physiol.* 79:2021-2028 (1995).

45. S. Rezaiguia, C. Garat, M. Meignan, and C. Jayr, Acute bacterial pneumonia increases alveolar epithelial fluid clearance by a tumor necrosis factor-a dependent mechanism in rats (abstract), *Am. J. Respir. Crit. Care Med.* 151:A763 (1995).

46. G. Nitenberg, H.G. Folkesson, O. Osorio, J.M. Cohen-Solal, and M.A. Matthay, Alveolar epithelial liquid clearance is markedly increased 10 days following acute lung injury from bleomycin (abstract), *Am. J. Respir. Crit. Care Med.* 151:A620 (1995).

47. R.H. Hastings, M. Grady, T. Sakuma, and M.A. Matthay, Clearance of different-sized proteins from the alveolar space in humans and rabbits, *J. Appl. Physiol.* 73:1310-1316 (1992).

48. L. Nici, R. Dowin, M. Gilmore-Hebert, J.D. Jamieson, and D.H. Ingbar, Upregulation of rat lung Na-K-ATPase during hyperoxic injury, *Am. J. Physiol.* 261:L307-L314 (1991).

49. W. Olivera, K. Ridge, L.D. Wood, and J.I. Sznajder, Active sodium transport and alveolar epithelial Na-K-ATPase increase during subacute hyperoxia in rats, *Am. J. Physiol.* 266:L577-L584 (1994).

50. G. Yue, W.J. Russell, D.J. Benos, R.M. Jackson, M.A. Olman, and S. Matalon, Increased expression and activity of sodium channels in alveolar type II cells of hyperoxic rats, *Proc. Natl. Acad. Sci. USA* 92:8418-8422 (1995).

51. E.P. Carter, S.E. Duvick, O.D. Wangensteen, and D.H. Ingbar, Rat lung Na,K-ATPase activity is decreased following 60 hours of hyperoxia, *Am. J. Respir. Crit. Care Med.* 149:A588 (1994).

52. J.D. Crapo, B.E. Barry, H.A. Foscue, and J. Shelburne, Structural and biochemical changes in rat lungs occurring during exposures to lethal and adaptive doses of oxygen, *Am. Rev. Resp. Dis.* 122:123-143 (1980).

53. J.I. Sznajder, W.G. Olivera, K.M. Ridge, and D.H. Rutschman, Mechanisms of lung liquid clearance during hyperoxia in isolated rat lungs, *Am. J. Respir. Crit. Care Med.* 151:1519-1525 (1995).

54. A.S. Verkman. *Water Channels*, Landes, Austin (1993).

55. A.N. van Hoek, M.L. Hom, L.H. Luthjens, M.D. de Jong, J.A. Dempster, and C.H. van Os, Functional unit of 30 kDa for proximal tubule water channels as revealed by radiation inactivation, *J. Biol. Chem.* 226:16633-16635 (1991).

56. R.B. Zhang, K.A. Logee, and A.S. Verkman, Expression of mRNA coding for kidney and red cell water channels in Xenopus oocytes, *J. Biol. Chem.* 265:15375-15378 (1990).

57. P. Agre, G.M. Preston, B.L. Smith, J.S. Jung, S. Raina, C. Moon, W.B. Guggino, and S. Nielsen, Aquaporin CHIP: the archetypal molecular water channel, *Am. J. Physiol.* 265:F463-F476 (1993).

58. C.H. van Os, P.M.T. Deen, and J.A. Dempster, Aquaporins: water selective channels in biological membranes; molecular structure and tissue distribution, *Biochim. Biophys. Acta* 1197:291-309 (1994).

59. A.S. Verkman, A.N. van Hoek, T. Ma, A. Frigeri, W.R. Skach, A. Mitra, B.K. Tamarrappoo, and J. Farinas, Water transport across mammalian cell membranes, *Am. J. Physiol. (Cell Physiol.)* 270:C12-C30 (1996).

60. J. Reizer, A. Reizer, and M.H. Saier, The MIP family of integral membrane channel proteins: sequence comparisons, evolutional relationships, reconstructed pathway of evolution, and proposed functional differentiation of two repeated halves of the protein, *Crit. Rev. Biochem. Mol. Biol.* 28:235-257 (1993).

61. J.F. Haskell, G. Yue, D.J. Benos, and S. Matalon, Upregulation of sodium conductive pathways in alveolar type II cells in sublethal hyperoxia, *Am. J. Physiol.* 266:L30-L37 (1994).

62. H. Hasegawa, R. Zhang, A. Dohrman, and A.S. Verkman, Tissue-specific expression of mRNA encoding rat kidney water channel CHIP28k by *in situ* hybridization, *Am. J. Physiol.* 264:C237-C245 (1993).

63. K. Ishibashi, S. Sasaki, K. Fushimi, S. Uchida, M. Kuwahara, H. Saito, T. Furukawa, K. Nakajima, Y. Yamaguchi, T. Gojobori, and F. Marumo, Molecular cloning and expression of a member of the aquaporin family with permeability to glycerol and urea in addition to water expressed at the basolateral membrane of kidney collecting duct cells, *Proc. Natl. Acad. Sci. USA* 91:6269-6273 (1994).

64. T. Ma, A. Frigeri, H. Hasegawa, and A.S. Verkman, Cloning of a water channel homolog expressed in brain meningeal cells and kidney collecting duct that functions as a stilbene-sensitive glycerol transporter, *J. Biol. Chem.* 269:21845-21849 (1994).

65. S. Raina, G.M. Preston, W.B. Guggino, and P. Agre, Molecular cloning and characterization of an aquaporin cDNA from salivary, lacrimal, and respiratory tissues, *J. Biol. Chem.* 270:1908-1912 (1995).

66. G.M. Preston and P. Agre, Isolation of the cDNA for erythrocyte integral membrane protein of 28 kilodaltons:member of an ancient channel family, *Proc. Natl. Acad. Sci. USA* 88:11110-11114 (1991).

67. A.N. van Hoek and A.S. Verkman, Functional reconstitution of the isolated erythrocyte water channel CHIP28, *J. Biol. Chem.* 267:18267-18269 (1992).

68. M.L. Zeidel, S.V. Ambudkar, B.L. Smith, and P. Agre, Reconstitution of functional water channels in liposomes containing purified red cell CHIP28 protein, *Biochemistry* 31:7436-7440 (1992).

69. G.M. Preston, T.P. Carroll, W.B. Guggino, and P. Agre, Appearance of water channels in Xenopus oocytes expressing red cell CHIP28 protein, *Science* 256:385-387 (1992).

70. R. Zhang, W. Skach, H. Hasegawa, A.N. van Hoek, and A.S. Verkman, Cloning, functional analysis and cell localization of a kidney proximal tubule water transporter homologous to CHIP28, *J. Cell Biol.* 120:359-369 (1993).

71. T. Ma, A. Frigeri, S.T. Tsai, J.M. Verbavatz, and A.S. Verkman, Localization and functional analysis of CHIP28k water channels in stably transfected Chinese hamster ovary cells, *J. Biol. Chem.* 268:22756-22764 (1993).

72. G.M. Preston, J.S. Jung, W.B. Guggino, and P. Agre, The mercury-sensitive residue at cysteine 189 in the CHIP28 water channel, *J. Biol. Chem.* 268:17-20 (1993).

73. R. Zhang, A.N. van Hoek, J. Biwersi, and A.S.Verkman, A point mutation at cysteine 189 blocks the water permeability of rat kidney water channel CHIP28k, *Biochemistry* 32:2938-2941 (1993).

74. A.N. van Hoek, M.C. Wiener, J.M. Verbavatz, D. Brown, P.H. Lipniunas, R.R. Townsend, and A.S. Verkman, Purification and structure-function analysis of native, PNGase F-treated, and endo-b-galactosidase-treated CHIP28 water channels, *Biochemistry* 34:2212-2219 (1995).

75. H. Hasegawa, S.C. Lian, W.E. Finkbeiner, and A.S. Verkman, Extrarenal tissue distribution of CHIP28 water channels by *in situ* hybridization and antibody staining, *Am. J. Physiol.* 266:C893-C903 (1994).

76. M.A. Matthay, H.G. Folkesson, A.S. Verkman, Salt and water transport across alveolar and distal airway epithelia in the adult lung, *Am. J. Physiol. (Lung Cell. Mol. Physiol.)* 270:L487-L503 (1996).

77. S. Nielsen, B.L. Smith, E.I. Christensen, and P. Agre, Distribution of the aquaporin CHIP in secretory and resorptive epithelia and capillary endothelia, *Proc. Natl. Acad. Sci. USA* 90:7275-7279 (1993).

78. S. Nielsen, B.L. Smith, E.I. Christensen, M.A. Knepper, and P. Agre, CHIP28 water channels are localized in constitutively water-permeable segments of the nephron, *J. Cell Biol.* 120:371-383 (1993).

79. I. Sabolic, G. Valenti, J.M. Verbavatz, A.N. van Hoek, A.S. Verkman, D.A. Ausiello, and D. Brown, Localization of the CHIP28 water channel in rat kidney, *Am. J. Physiol.* 263:C1225-C1233 (1992).

80. H. Hasegawa, T. Ma, W. Skach, M.A. Matthay, and A.S. Verkman, Molecular cloning of a mercurial-insensitive water channel expressed in selected water-transporting tissues, *J. Biol. Chem.* 269:5497-5500 (1994).

81.   A. Frigeri, M.A. Gropper, C.W. Turck, and A.S. Verkman, Immunolocalization of the mercurial-insensitive water channel and glycerol intrinsic protein in epithelial cell plasma membranes, *Proc. Natl. Acad. Sci. USA* 92:4328-4331 (1995).

82.   A. Frigeri, M.A. Gropper, F. Umenishi, M. Katsura, D. Brown, and A.S. Verkman, Localization of MIWC and GLIP water channel homologs in neuromuscular, epithelial, and glandular tissues, *J. Cell Sci.* 108:2993-3002 (1995).

83.   L.B. Shi and A.S. Verkman, Selected cysteine point mutations confer mercurial sensitivity to the mercurial-insensitive water channel MIWC, *Biochemistry* 35:538-544 (1996).

84.   L.B. Shi, W.R. Skach, T. Ma, and A.S. Verkman, Distinct biogenesis mechanisms for the water channels MIWC and CHIP28 at the endoplasmic reticulum, *Biochemistry* 34:8250-8256 (1995).

85.   B. Yang, T. Ma, and A.S. Verkman, cDNA cloning, gene organization, and chromosomal localization of a human mercurial-insensitive water channel:evidence for distinct transcriptional units, *J. Biol. Chem.* 270:22907-22913 (1995).

86.   S. Raina, G.M. Preston, W.B. Guggino, and P. Agre, Molecular cloning and characterization of an aquaporin cDNA from salivary, lacrimal, and respiratory tissues, *J. Biol. Chem.* 270:1908-1912 (1995).

87.   Z. Borok, R.L. Lubman, S.I. Danto, X.L. Zhang, S.M. Zabski, L.S. King, D.M. Lee, P. Agre, and E.D. Crandall, Keratinocyte growth factor modulates alveolar epithelial cell phenotype in vitro: expression of aquaporin 5 (AQP5), *Am. J. Physiol.*, in press (1998).

88.   E.P. Carter, F. Umenishi, M.A. Matthay, and A.S. Verkman, Developmental changes in water permeability across the alveolar barrier in perinatal rabbit lung, *J. Clin. Invest.* 100:1071-1078 (1997).

89.   R.M. Effros, Osmotic extraction of hypotonic fluid from the lungs, *J. Clin. Invest.* 5:935-947 (1974).

90.   H.G. Folkesson, F. Kheradmand, and M.A. Matthay, The effect of salt water on alveolar epithelial barrier function, *Am. J. Respir. Crit. Care Med.* 150:1555-1563 (1994).

91.   H.G. Folkesson, M.A. Matthay, H. Hasegawa, F. Kheradmand, and A.S. Verkman, Transcellular water transport in lung alveolar epithelium through mercury-sensitive water channels, *Proc. Natl. Acad. Sci. USA* 91:4970-4974 (1994).

92.   E.P. Carter, M.A. Matthay, J. Farinas, and A.S. Verkman, Transalveolar osmotic and diffusional water permeability in intact mouse lung measured by a novel surface fluorescence method, *J. Gen. Physiol.* 108:133-142 (1996).

93.   H.G. Folkesson, M.A. Matthay, A. Frigeri, and A.S. Verkman, Transepithelial water permeability in microperfused distal airways: evidence for channel-mediated water transport, *J. Clin. Invest.* 97:664-671 (1996).

94.   G.M. Preston, B.L. Smith, M.L. Zeidel, J.J. Moulds, and P. Agre, Mutations in *aquaporin-1* in phenotypically normal humans without functional CHIP water channels, *Science* 265:1585-1587 (1994).

95.   A. Fein, R.F. Grossman, J.G. Jones, E. Overland, J.F. Murray, and N.C. Staub, The value of edema fluid protein measurements in patients with pulmonary edema, *Am. J. Med.* 67:32-39 (1979).

96.   M.A. Matthay, W.L. Eschenbacher, and E.J. Goetzl, Elevated concentrations of leukotriene D_4 in pulmonary edema fluid of patients with the adult respiratory distress syndrome, *J. Clin. Immunol.* 4:479-483 (1984).

97.   Y.M. Fukuda, M. Ishizaki, Y. Masuda, G. Kimura, O. Kawanami, and Y. Masugi, The role of intra-alveolar fibrosis in the process of pulmonary structural remodeling in patients with diffuse alveolar damage, *Am. J. Pathol.* 126:171-182 (1987).

98.   M. Bachofen and E.R. Weibel, Alterations of the gas exchange apparatus in adult respiratory insufficiency associated with septicemia, *Am. Rev. Respir. Dis.* 116:589-615 (1977).

99.   J.G. Clark, J.A. Milberg, K.P. Steinberg, and L.D. Hudson, Type III procollagen peptide in adult respiratory distress syndrome: association of increased peptide levels in bronchoalveolar lavage with increased risk for death, *Ann. Intern. Med.* 122:17-23 (1995).

100.   R.H. Hastings, J.R. Wright, K.H. Albertine, R. Ciriales, and M.A. Matthay, Effect of endocytosis inhbitors on alveolar clearance of albumin, immunoglobulin G, and

SP-A in rabbits, *Am. J. Physiol. (Lung Cell. Mol. Physiol.)* 266:L544-L552 (1994).

101. M.A. Matthay, G. Nitenberg, and C. Jayr, The critical role of the alveolar epithelial barrier in acute lung injury, in: *Yearbook of Intensive Care and Emergency Medicine*, J.-L. Vincent, ed., Berlin: Springer-Verlag, Berlin (1995).

102. M.A. Matthay and J.P. Wiener-Kronish, Intact epithelial barrier function is critical for the resolution of alveolar edema in humans, *Am. Rev. Respir. Dis.* 142:1250-1257 (1990).

103. J.G. Clark, J.A. Milberg, K.P. Steinberg, and L.D. Hudson, Type III procollagen peptide in adult respiratory distress syndrome: association of increased peptide levels in bronchoalveolar lavage with increased risk for death, *Ann. Intern. Med.* 122:17-23 (1995).

104. T.R. Martin, B.P. Pistorese, E.Y. Chi, R.B. Goodman, and M.A. Matthay, Effects of leukotriene B$_4$ in the human lung: recruitment of neutrophils into the alveolar spaces without a change in protein permeability, *J. Clin. Invest.* 84:1609-1619 (1989).

105. H.G. Folkesson, F. Kheradmand, and M.A. Matthay, The effect of salt water on alveolar epithelial barrier function, *Am. J. Respir. Crit. Care Med.* 150:1555-1563 (1994).

106. D.S. Cohen, M.A. Matthay, M.G. Cogan, and J.F. Murray, Pulmonary edema associated with salt water near-drowning: new insights, *Am. Rev. Respir. Dis.* 146:794-796 (1992).

107. T. Sakuma, S. Suzuki, K. Usuda, M. Handa, G. Okaniwa, T. Nakada, S. Fujimura, and M.A. Matthay, Alveolar epithelial fluid transport mechanisms are preserved in the rewarmed human lung following severe hypothermia, *J. Appl. Physiol.* 80:1681-1686 (1996).

108. J.P. Wiener-Kronish, K.H. Albertine, and M.A. Matthay, Differential responses of the endothelial and epithelial barriers of the lung in sheep to *Escherichia coli* endotoxin, *J. Clin. Invest.* 88:864-875 (1991).

109. J.-F. Pittet, J.P. Wiener-Kronish, M.C. McElroy, H.G. Folkesson, and M.A. Matthay, Stimulation of lung epithelial liquid clearance by endogenous release of catecholamines in septic shock in anesthetized rats, *J. Clin. Invest.* 94:663-671 (1994).

110. J.-F. Pittet, J.P. Wiener-Kronish, V. Serikov, and M.A. Matthay, Resistance of the alveolar epithelium to injury from septic shock in sheep, *Am. J. Respir. Crit. Care Med.* 151:1093-1100 (1995).

111. H.G. Folkesson, M.A. Matthay, C.A. Hébert, and V.C. Broaddus, Acid aspiration induced lung injury in rabbits is mediated by interleukin-8 dependent mechanisms, *J. Clin. Invest.* 96:107-116 (1995).

112. J.P. Wiener-Kronish, V.C. Broaddus, K.H. Albertine, M.A. Gropper, M.A. Matthay, and N.C. Staub, Relationship of pleural effusions to increased permeability pulmonary edema in anesthetized sheep, *J. Clin. Invest.* 82:1422-1429 (1988).

113. I. Kudoh, J.P. Wiener-Kronish, S. Hashimoto, J.-F. Pittet, and D. Frank, Exoproduct secretions of *Pseudomonas aeruginosa* strains influence of alveolar epithelial injury, *Am. J. Physiol.* 267:L551-L556 (1994).

114. J.P. Wiener-Kronish, T. Sakuma, I. Kudoh, J.F. Pittet, D. Frank, L. Dobbs, M.L. Vasil, and M.A. Matthay, Alveolar epithelial injury and pleural empyema in acute *P. aeruginosa* pneumonia in anesthetized rabbits, *J. Appl. Physiol.* 75:1661-1669 (1993).

HOST DEFENSE CAPACITIES OF PULMONARY SURFACTANT

Ulrich Pison and Sylvia Pietschmann

Department of Anesthesiology and Intensive Care Medicine, Virchow-Klinikum, Humboldt-University, Berlin, Germany

INTRODUCTION

The most well characterized function of pulmonary surfactant is its ability to modify surface tension at the alveolar air-liquid interface in a manner that depends on alveolar surface area. Thereby surfactant helps to keep the gas exchange surface available during breathing. In addition, several lines of evidence suggest that surfactant may also interact with pulmonary host defense systems. It seems reasonable to consider pulmonary surfactant defending the host's gas exchange surface against infectious agents and irritants, because many of the lung diseases that are associated with surfactant alterations are either primarily pneumonia or frequently complicated by lung infection (Pison, Max et al. 1994). Although homozygous SP-A-deficient mice were without detectably altered postnatal survival or pulmonary function (Korfhagen, Bruno et al. 1996), these animals were more susceptible to group B streptococcal infection (LeVine, Bruno et al. 1997).

The following interactions between surfactant and host defenses are possible, though more questions than answers exist, regarding a complete understanding of the various role(s) of surfactant components in modulating host defenses in lungs. (I) The specific components of surfactant (proteins and phospholipids) react with different alveolar cells affecting their host defense function. (II) Surfactant and their components react with pathogens and particles in a sense that helps phagocytes to clear them from the alveolar space. (III) Single surfactant components or whole surfactant macromolecular super-structures scavenge bacterial toxins and/or neutralize endogenous mediators of inflammation. (IV) Surfactant by lowering surface tension at end expiration, enhances the removal of particulate substances and damaged cells from the alveoli via the mucociliary transport system.

SURFACTANT COMPONENTS REACT WITH ALVEOLAR CELLS

Surfactants main cellular target in alveoli is the type II cell (Dobbs 1990). Surfactant and its components, however, could also interact with alveolar macrophages and immune competent cells affecting their host defense function. In the 1960s, several investigators demonstrated a major bactericidal role for alveolar macrophages in vivo (Green and Kass 1964), but suprisingly these macrophages lacked such activity in vitro (reviewed in (Juers, Rogers et al. 1976)). In 1973 LaForce and coworkers demonstrated that lavage fluid from rat lungs increased intracellular killing of microbes by alveolar macrophage, regardless of whether the organism had been ingested in vitro or in vivo (LaForce, Kelly et al. 1973). This finding somewhat resolved the controversy between in vivo and in vitro properties of alveolar macrophages. Juers extended this observation to human alveolar lining material (Juers, Rogers et al. 1976), and in 1984 O'Neill et al.

Acute Respiratory Distress Syndrome: Cellular and Molecular Mechanisms and Clinical Management
Edited by Matalon and Sznajder, Springer Science+Business Media New York, 1998

87

demonstrated that pulmonary surfactant is the active principle that increases killing of bacteria by alveolar macrophages (O'Neill, Lesperance et al. 1984). Ansfield and associates showed that surfactant was also capable of modulating lymphocyte functions in vitro (Ansfield, Kaltreider et al. 1979). Numerous studies followed, addressing the question of whether and how the alveolar lining fluid and surfactant prepared from this fluid modulates alveolar cell defense functions in vitro (Juers, Rogers et al. 1976; Coonrod and Yoneda 1983; O'Neill, Lesperance et al. 1984; Sitrin, Ansfield et al. 1985; Coonrod, Jarrells et al. 1986; Webb and Jeska 1986; Hoffman, Claypool et al. 1987; van Iwaarden, Welmers et al. 1990). These studies suggest that surfactant enhances macrophage defense capacities probably via its proteins, and suppresses that of lymphocytes through its phospholipid components. Controversies, however, exist regarding the specific surfactant components involved and the potential effects of surfactant on macrophages in general. Surfactant from human donors, e.g., was unable to stimulate human alveolar macrophages antimicrobial defenses (Jonsson, Musher et al. 1986), and high amounts of surfactant decreased candidacidal (Zeligs, Nerurkar et al. 1984) and bactericidal activity (Sherman, D'Ambola et al. 1988) of rabbit macrophages. Species differences in general, differences in preparing surfactant, and differences of the cell populations used or the conditions under which the macrophages were obtained might explain some of these controversies. In addition, phagocytes that become overfed with surfactant could loose their functional capacities, as was shown for blood monocytes (Wiernik, Curstedt et al. 1987).

SP-A, the most abundant surfactant protein, seems to be substantial for surfactant macrophage interactions. Evidence for this hypothesis is that SP-A binds with high affinity to alveolar macrophages through a specific cell-surface receptor, probably via its collagen like tail (Malhotra, Haurum et al. 1992; Pison, Wright et al. 1992), and that some (van Iwaarden, Welmers et al. 1990) but not all studies (Juers, Rogers et al. 1976) found that the active principle of surfactant regarding macrophages was associated with SP-A. Despite these controversies, the uptake of phospholipids by alveolar macrophages is specifically accelerated in the presence of SP-A (Wright, Wager et al. 1987).

SP-A binding to alveolar macrophages may effect macrophage host defense capacities. One of these capacities is the release of reactive oxygen species (Klebanoff 1992). One study using soluble human SP-A prepared from alveolar proteinosis patients demonstrated an increase in oxygen radical release from macrophages (van Iwaarden, Welmers et al. 1990). Experimental data with soluble dog SP-A, however, showed the opposite effect (Weber, Heilmann et al. 1990). Furthermore, the inflammatory responsiveness of alveolar macrophages, e.g., measured as oxygen radical release, is down regulated in the presence of surfactant (Hayakawa, Myrvik et al. 1989), depends on postnatal development (Zeligs, Nerurkar et al. 1984), starvation (Sakai, Kweon et al. 1992), and is impaired after exogenous surfactant application (Sherman, D'Ambola et al. 1988). From a teleological point of view it seems reasonable to speculate that SP-A (or whole surfactant) which excists in close proximity to alveolar macrophages does not activate these cells under normal conditions. The experimental data from our group indicate that SP-A indeed does not simply enhance the release of reactive oxygen species from resident alveolar macrophages in vitro, which might result in unrestricted alveolar inflammation in vivo. Only surface bound, but not soluble, SP-A enhances oxygen radical release from resident alveolar macrophages (Neuendank, Weißbach et al. 1992; Weißbach, Neuendank et al. 1994). Our data indicate that either a conformational change induced by surface binding, or a multivalent presentation of the SP-A molecule is required for functional activity regarding macrophage activation. A similar finding was demonstrated recently for the C1q mediated neutrophil (PMN) activation (Goodman and Tenner 1992).

Other macrophage host defense capacities are phagocytosis of pathogens and the directed cell migration towards chemotactive substances. SP-A enhances both phagocytosis of pathogens by alveolar macrophages and chemotaxis of this cell type. It is not clear, however, whether SP-A modulated phagocytosis requires serum as an opsonin and what the mechanism really is (see also following section). One study demonstrated increased phagocytosis due to SP-A but only for serum-opsonized *Staphylococcus aureus* (van Iwaarden, Welmers et al. 1990). Manz-Keinke and coworkers showed that different natural and recombinant SP-A molecules enhance serum-independent phagocytosis of various bacteria including *Staphylococcus aureus* as well. The functional activity of natural SP-A was most effective, probably due to its higher molecular structure, and phagocytosis depends on the growth phase of bacteria (Manz-Keinke, Plattner et al. 1992).

We found that SP-A could substitute for serum as an opsonin regarding phagocytosis of *Candida tropicalis* (Weißbach, Neuendank et al. 1992). The phospholipids of surfactant had no effect on phagocytosis of *Candida albicans* by blood monocytes (Speer, Götze et al. 1991). SP-A is structurally homologous to C1q and may be functionally interchangeable with this molecule. This hypothesis was tested for FcR and CR1-mediated phagocytosis by Tenner et al. (Tenner, Robinson et al. 1989). SP-A can indeed enhance the phagocytosis of immonoglobulin and complement coated erythrocytes by blood monocytes and culture-derived macrophages, activities shown to be modulated through the collagen-like region of C1q. Functions which require interactions of both the collagenous and the non-collagenous regions (i.e. initiation of the classical complement cascade for C1q, and phopsholipid uptake by type II cells and macrophages for SP-A) are not interchangeable (Tenner, Robinson et al. 1989). Regarding the experiments by Tenner et al. both proteins, allthough not opsonins in the classical sense, may enhance the opsonic capacity of other proteins and stimulates phagocytosis mediated by other receptors (Graham, Gresham et al. 1989). SP-A also stimulates chemotaxis of alveolar macrophages (Wright and Youmans 1993). Through this function SP-A might regulate the amount of macrophages in the alveolar compartment of lungs, recruiting more cells into the alveoli during infection challenges (Hoffman, Claypool et al. 1987).

All experimental results reported so far for the SP-A macrophage interactions need to be interpreted with caution. SP-A is a highly organized quarterly structured protein with distinct binding domains for phospholipids (Ross, Notter et al. 1986), calcium (Haagsman, Sargeant et al. 1990), carbohydrates (Haagsman, Hawgood et al. 1987), glycolipids (Childs, Wright et al. 1992), and probably some lipopolysaccharides like endotoxin. SP-A has a tendency to self-aggregate depending on calcium and other ion concentrations. The three-dimensional structure of various recombinant SP-A's, SP-A as obtained from alveolar proteinosis patients, and SP-A as prepared from healthy mammalian lungs all differ from each other (Voss, Eistetter et al. 1988; Voss, Melchers et al. 1991). It might well be that some of the SP-A preparations used for functional studies were contaminated with substances like platelet activating factor, lipopolysaccharides, or lysophospholipids, which could be toxic to cells. Further studies will need to exclude such contamination and give values for endogenous SP-A ligands like glycospingolipids and divalent ions. The available experimental data together with what we know about the biological requirements in alveoli makes it reasonable to speculate that a multivalent presentation of the SP-A molecule to macrophages is essential for a functional response. Such multivalent presentation is imaginable for the simultaneous adhesion of SP-A to a pathogen or particle together with binding to a macrophage cell surface receptor (see also next section) or by a more self aggregated SP-A molecule itself as in alveolar proteinosis.

SP-D, another hydrophobic surfactant associated protein, interact with alveolar macrophages as well (Kuan, Persson et al. 1994), stimulating the release of oxygen radicals (van Iwaarden, Shimizu et al. 1992), while for the hydrophilic surfactant proteins SP-B and SP-C no macrophage interactions have been described so far.

While some surfactant proteins appear to stimulate certain macrophage defense functions, surfactant phospholipids seem to inhibit those of lymphocytes. Suppressed proliferation of blood lymphocytes in response to various mitogens and alloantigens were found after incubating cells with surfactant (Ansfield, Kaltreider et al. 1979). These studies were extended using lung lymphocytes, revealing comparable results (Ansfield, Kaltreider et al. 1980). Lymphocytes responded to surfactant phospholipids, whereas surfactant associated proteins had no effect (Ansfield and Benson 1980). Concerning surfactant's phospholipid composition phosphatidylglycerol is more suppressive than phosphatidylcholine on a molar basis. In addition, surfactant phospholipids suppress the B cell immunoglobulin production and natural killer cell cytotoxicity in normal, undifferentiated spleen lymphocytes (Sitrin, Ansfield et al. 1985). These in vitro capacities of surfactant on lymphocytes were tested in vivo using an animal model of hypersensitivity pneumonitis (Richman, Batcher et al. 1990). Guinea pigs were immunized using complete Freund's adjuvant containing a protein derivative of *Mycobacterium tuberculosis*. An aerosol challenge with the purified protein derivative was significantly more pronounced in animals which were surfactant deficient, suggesting that surfactant may have an in vivo role in modulating the inflammatory response that accompanies immune lung injury.

The surface active material of alveolar lining fluid has been shown to have immunologic activities. The effect of some phospholipids of pulmonary surfactant on blood monocyte or alveolar macrophage cytotoxicity against a tumor cell line revealed enhanced tumor cell killing by macrophages or monocytes, when phagocytes were preincubated with lipopolysaccaride alone, or coincubating tumor cells and phagocytes with surfactant phospholipids (Baughman, Mangels et al. 1987). Other effects of surfactant phospholipids on phagocytes include suppression of tumor necrosis factor a (TNF) secretion by resting and lipopolysaccaride-stimulated blood monocytes (Speer, Götze et al. 1991). The endotoxin induced secretion of TNF, interleukin-1ß (IL-1), interleukin-6 (IL-6), and interleukin-8 (IL-8) by alveolar macrophages could also be inhibited by a synthetic surfactant used for therapy (Exosurf). This surfactant includes mainly phosphatidylcholine but no proteins (Thomassen, Meeker et al. 1992). TNF inhibits the expression of SP-A and SP-B in a human pulmonary adenocarcinoma cell line (Wispe, Clark et al. 1990), whereas interferon gamma stimulates SP-A production in human fetal lung explants (Ballard, Liley et al. 1990). The effects of surfactant on the cytokine network modifying alveolar cell capacities are not well understood and the biological significance of such interactions need to be elucidated.

SURFACTANT REACTS WITH PATHOGENS AND PARTICLES

Surfactant and its components may react with pathogens and particles in a sense that helps phagocytes clearing them from the alveolar space. In 1903 Wright and Douglas showed that the serum of immune animals contains two factors which bind to bacteria and enhance their phagocytic uptake by leukocytes (Wright and Douglas 1903). One of the factors was stable to heating and is now known to be IgG. The second factor was heat-labile and is now known to be complement protein C3. These factors were shown to bind to a bacterium and thereby bridge the bacterium to a phagocyte. To describe the two factors, Wright and Douglas coined the word "opsonin" from the greek opsono, meaning "to prepare food for". Since that time, IgG, C3 and the receptors for these proteins (FcR, CR3) have been extensively characterized and represent the best-studied models for understanding phagocytosis (see (Wright 1992) for review). In addition to the "classical" opsonins (IgG and C3) several additional proteins serve an opsonic function, or play a role in opsono-phagocytosis without being opsonins per se. Two of these proteins are the surfactant associated proteins SP-A and SP-D. Together with conglutinin, C1q and mannose-binding proteins, they belong to a group of structurally homologous proteins called "collectins". This group of proteins comprise a collagen tail or spoke and a globular head, which contains a carbohydrate recognition domain. The presence of a collagen and lectin domain prompted the term collectin (Malhotra, Haurum et al. 1992). The collectins are pattern recognition molecules sharing the ability to discriminate between the patterns of oligosaccharides that decorate the cell wall of pathogens from normal self-glycoproteins (Sastry and Ezekowitz 1993). Collectins may provide an early line of defense in circumstances in which antibody formation may be slower. It is easy to envision such a scenario in the lungs where inhaled immunogens and irritants meet with alveolar cells and actually have the opportunity to penetrate the epithelial/endothelial barrier. Thus it seems reasonable to hypothesize that there may exist in the alveolar lining fluid a first line of defense against infection that would act quickly before enough time had elapsed to acquire specific immunity. SP-A and SP-D are candidates for this function in the alveolar compartment of mammalian lungs.

SP-A binds to immobilized D-mannose, L-fucose, D-glucose, and D-galactose in the presence of calcium (Haagsman, Hawgood et al. 1987), a characteristic feature of C-type lectins (Drickamer 1988). In agreement with its lectin characteristic rat SP-A binds to gp120, a mannose-rich surface membrane glycoprotein of *Pneumocystis carinii* (Zimmerman, Voelker et al. 1992). SP-A from alveolar proteinosis patients binds to *Staphylococcus aureus* and type 25 *Streptococcus pneumoniae* as well, but facilitates only the attachment/ingestion of opsonized *Staphylococcus aureus* to rabbit alveolar macrophages (McNeely and Coonrod 1993). In addition to binding to bacteria (Tino and Wright 1996), rat and dog SP-A could substitute for serum as an opsonin regarding phagocytosis of *Candida tropicalis* and zymosan by rat alveolar macrophages, indicating binding to some fungi (Weißbach, Neuendank et al. 1992). Furthermore, SP-A is an opsonin for herpes simplex virus type 1 infected cells, promoting its phagocytosis by rat alveolar macrophages (van Iwaarden, van Strijp et al. 1991). SP-A binding to herpes simplex virus infected cells is mediated by its carbohydrate moiety (van Iwaarden, van Strijp et al. 1992).

SP-D is a new member of the ever-growing C-type lectin family as it has the ability to bind various sugars in the presence of calcium, with high affinity for maltose and a slightly lower affinity for glucose (Persson, Chang et al. 1990). SP-D mediates the agglutination of gram-negative bacteria, such as *Escherichia coli, Salmonella paratyphi* and *Klebsiella pneumonia.* SP-D did not bind *Staphylococcus aureus,* the only gram-positive bacteria tested (Kuan, Rust et al. 1992). Binding of SP-D to *E. coli* could be inhibited by competing cold SP-D or saccharides, and agglutination was inhibited by EDTA, a-glycosyl-containing saccharides, antisera to the carbohydrate-binding domain of SP-D, and *E. coli.* lipopolysaccharide. Opsonization of influenza A virus with collectins like SP-D protects neutrophils against the deactivation effects of the virus on cellular respiratory burst responses in vitro (Hartshorn, Reid et al. 1996).

SP-A and SP-D serve as pattern recognition molecules in alveoli, binding to certain structural elements that decorate the cell wall of pathogens and probably to surface motifs of inhaled particles. Through different binding domains they could interact concurrently with phagocytes helping to clear pathogens from the alveolar space. The precise mechanism of such mutivalent interaction is not completly understood, neither are the receptors involved on phagocytes identified. It is likely that a family of related molecules mediates the overlapping and distinct cellular effects induced by collectin ligand complexes (Acton, Resnick et al. 1993; Sastry and Ezekowitz 1993).

SURFACTANT SCAVENGES TOXINS

Single surfactant components or whole surfactant and its macromolecular super-structures may scavenge bacterial toxins and/or neutralize endogenous mediators of inflammation. Such reactions may restrict inflammation in the alveolar space of the lung. The experimental data supporting this concept are scare. Whole surfactant from sheep lung complexes lipopolysaccharides prepared from the gram-negative bacteria *Escherichia coli, Salmonella typhi, Klebsiella pneumonia, Seratia marcescens,* and *Pseudomonas aeruginosa* (Brogden, Cutlip et al. 1986). The toxicity of *E. coli* lipopolysaccharide was enhanced after complexing with surfactant. The specific surfactant components complexing bacterial lipopolysaccharides are unknown from this study. Iodinated SP-D could bind lipopolysaccharides from several gram-negative bacteria, suggesting that it might be involved in complexing these saccharides (Kuan, Persson et al. 1994). Platelet-activating factor (PAF), a potent phospholipid mediator was recently shown to contaminate natural surfactant preparations used for replacement therapy (Moya, Hoffman et al. 1993). The role of PAF in natural surfactant preparations is unknown. We need to await further studies to learn more about the beneficial and harmful aspects of surfactant scavenging of exogenous toxins and inflammatory mediators.

SURFACTANT FACILITATES MUCOCILIARY MOVEMENTS

It might well be that surfactant, by lowering surface tension at end expiration, also enhances the removal of particulate substances and damaged cells from the alveoli and bronchi via the mucociliary transport system. In vitro studies demonstrated unidirectional transport of particles floating on surfactant monolayers during compression and expansion of the film (Rensch and von Seefeld 1984). A substantial microbial pathogenicity factor is bacterial adherence to epithelial surfaces. Adherence of a microbe to the mucosal surface preceeds its colonization and subsequent proliferation. Surfactant could help impede microbial adherence by coating their surfaces, accelerating cilia beat frequency, and conditioning the viscosity of mucus. Most if not all mechanisms await experimental proof. The surfactant covering conducting airways displaces inhaled particles towards the epithelium where they are retained. Surface-balance studies showed that pulmonary surfactant promotes the displacement of spherical particles from air to the aqueous phase, and electron microscopy demonstrated that particles in peripheral airways and alveoli are found below the surfactant film. From here they can be cleared within 24 hours by mucociliary activity or more slowly by macrophage phagocytosis (Gehr and Schürch 1992; Schürch, Lee et al. 1992). Ciliary beat frequency and the regeneration of guinea pig tracheal explants after hydrogen peroxide injury was accelerated by artificial surfactants (Kakuta, Sasaki et al. 1991).

SUMMARY

The most well characterized function of pulmonary surfactant is its ability to reduce surface tension at the alveolar air-liquid interface, thereby preventing lung collapse. However, several lines of evidence suggest that surfactant may also have "non-surfactant" functions: specific components of surfactant (proteins and phospholipids) may interact with different alveolar cells, inhaled particles and microorganisms modulating pulmonary host defense systems.

SP-A the most abundant surfactant protein binds to alveolar macrophages via a specific surface receptor with high affinity (Pison, Wright et al. 1992). Such binding effects the release of reactive oxygen species from resident alveolar macrophages if SP-A is properly presented to the target cell. SP-A also stimulates chemotaxis of alveolar macrophages (Wright and Youmans 1993), and serves as an opsonin in the phagocytosis of herpes simplex virus (van Iwaarden, van Strijp et al. 1991) *Candida tropicalis* (Weißbach, Neuendank et al. 1992) and various bacteria (Manz-Keinke, Plattner et al. 1992). In addition, SP-A enhances the uptake of particles by monocytes and culture-derived macrophages (Tenner, Robinson et al. 1989) and improves bacterial killing. SP-D, another hydrophobic surfactant associated protein, might interact with alveolar macrophages as well, stimulating the release of oxygen radicals (van Iwaarden, Shimizu et al. 1992) while for the hydrophilic surfactant proteins SP-B and SP-C no macrophage interactions have been described so far. SP-A and SP-D are members of the so called "collectins", pattern recognition molecules involved in first line defense.

While some surfactant proteins appear to stimulate certain macrophage defense functions, surfactant phospholipids seem to inhibit those of lymphocytes. Suppressed lymphocyte functions include lymphoproliferation in response to mitogens and alloantigens, B cell immunoglobulin production and natural killer cell cytotoxicity. Concerning surfactant's phospholipid composition phosphatidylglycerol is more suppressive than phosphatidylcholine on a molar basis (Ansfield and Benson 1980). Bovine surfactant has an immunosuppresive effect on the development of hypersensitivity pneumonitis in a guinea pig model (Richman, Batcher et al. 1990).

Despite these interesting observations, several important questions concerning the interactions of surfactant components with pulmonary host defense systems remain unanswered: Sufficient host defense in the lungs works through various humoral-cellular systems in conjunction with the specific anatomy of the airways and the gas exchange surface -- how does the surfactant system fit into this network? Surfactant and alveolar cells are both altered during lung injury -- is there a relationship between alveolar cells from RDS patients and the endogenous surfactant isolated from such patients? How does exogenous surfactant as used for substitution therapy modulate the defense system of the host? Some of those artificial surfactants have been shown to inhibit the endotoxin induced release of inflammatory cytokines from alveolar macrophages, PMN's and monocytes including IL-1, IL-6, and TNF (Speer, Götze et al. 1991; Thomassen, Meeker et al. 1992).

The available experimental data support the following hypotheses: Pulmonary surfactant has ambivalent physiological roles regarding host defenses. It may inhibit cell mediated specific immune responses, e.g., to inhaled antigens, while augmenting both the numbers of macrophages and their ability to ingest and kill foreign invaders. It might well be that surfactant by lowering surface tension at end expiration, also enhance the removal of particulate substances and damaged cells from the alveoli through the mucociliary transport system.

ACKNOWLEDGEMENTS

Research reviewed in this article was financially supported by Deutsche Forschungsgemeinschaft.

REFERENCES

Acton, S., D. Resnick, et al. (1993). "The collagenous domains of macrophage scavenger receptors and complement component C1q mediate their similar, but not identical, binding specificities for polyanionic ligands." J Biol Chem 268(5): 3530-7.

Ansfield, M. J. and B. J. Benson (1980). "Identification of the immunosuppressive components of canine pulmonary surface active material." J. Immunol. 125: 1093-1098.

Ansfield, M. J., H. B. Kaltreider, et al. (1979). "Immunosuppressive activity of canine pulmonary surface active material." J. Immunol. 122(3): 1062-1066.

Ansfield, M. J., H. B. Kaltreider, et al. (1980). "Canine surface active material and pulmonary lymphocyte function studies with mixed-lymphocyte culture." Exp. Lung Res. 1: 3-10.

Ballard, P. L., H. G. Liley, et al. (1990). "Interferon-gamma and synthesis of surfactant components by cultured human fetal lung." Am. J. Respir. Cell Mol. Biol. 2(2): 137-143.

Baughman, R. P., D. J. Mangels, et al. (1987). "Enhancement of macrophage and monocyte cytotoxicity by the surface active material of lung lining fluid." J Lab Clin Med 109(6): 692-7.

Brogden, K. A., R. C. Cutlip, et al. (1986). "Complexing of bacterial lipopolysaccharide with lung surfactant." Infect Immun 52(3): 644-9.

Childs, R. A., J. R. Wright, et al. (1992). "Specificity of lung surfactant protein SP-A for both the carbohydrate and the lipid moieties of certain neutral glycolipids." J Biol Chem 267(14): 9972-9.

Coonrod, J. D., M. C. Jarrells, et al. (1986). "Effect of rat surfactant lipids on complement and Fc receptors of macrophages." Infect. Immun. 54(2): 371-378.

Coonrod, J. D. and K. Yoneda (1983). "Effect of rat alveolar lining material on macrophage receptors." J. Immunol. 130(6): 2589-2596.

Dobbs, L. G. (1990). "Isolation and culture of alveolar type II cells." Am. J. Physiol. 258(Lung Cell. Mol. Physiol. 2): L134-L147.

Drickamer, K. (1988). "Two distinct classes of carbohydrate-recognition domains in animal lectins." J. Biol. Chem. 263(20): 9557-9560.

Gehr, P. and S. Schürch (1992). "Surface forces displace particles deposited in airways toward the epithelium." News Physiol. Sci. 7: 1-5.

Goodman, E. B. and A. J. Tenner (1992). "Signal transduction mechanisms of C1q-mediated superoxide production. Evidence for the involvement of temporally distinct staurosporine-insensitive and sensitive pathways." J Immunol 148(12): 3920-8.

Graham, I. L., H. D. Gresham, et al. (1989). "An immobile subset of plasma membrane CD11b/CD18 (Mac-1) is involved in phagocytosis of targets recognized by multiple receptors." J Immunol 142(7): 2352-8.

Green, G. M. and E. H. Kass (1964). "The role of the alveolar macrophage in the clearance of bacteria from the lung." J. Exp. Med. 119: 167-175.

Haagsman, H. P., S. Hawgood, et al. (1987). "The major lung surfactant protein, SP 28-36, is a calcium-dependent, carbohydrate-binding protein." J. Biol. Chem. 262(29): 13877-13880.

Haagsman, H. P., T. Sargeant, et al. (1990). "Binding of calcium to SP-A, a surfactant-associated protein." Biochemistry 29(38): 8894-900.

Hartshorn, K. L., K. B. Reid, et al. (1996). "Neutrophil deactivation by influenza A viruses: mechanisms of protection after viral opsonization with collectins and hemagglutination-inhibiting antibodies." Blood 87(8): 3450-61.

Hayakawa, H., Q. N. Myrvik, et al. (1989). "Pulmonary surfactant inhibits priming of rabbit alveolar macrophage. Evidence that surfactant suppresses the oxidative burst of alveolar macrophages in infant rabbits." Am. Rev. Respir. Dis. 140(5): 1390-1397.

Hoffman, R. M., W. D. Claypool, et al. (1987). "Augmentation of rat alveolar macrophage migration by surfactant protein." Am. Rev. Respir. Dis 135(6): 1358-1362.

Jonsson, S., D. M. Musher, et al. (1986). "Human alveolar lining material and antibacterial defenses." Am. Rev. Respir. Dis. **133**(1): 136-140.

Juers, J. A., R. M. Rogers, et al. (1976). "Enhancement of bactericidal capacity of alveolar macrophages by human alveolar lining material." J. Clin. Invest. **58**: 271-275.

Kakuta, Y., H. Sasaki, et al. (1991). "Effect of artificial surfactant on ciliary beat frequency in guinea pig trachea." Respir. Physiol. **83**: 313-322.

Klebanoff, S. J. (1992). Oxygen metabolites from phagocytes. Inflammation: Basic Principles and Clinical Correlates. J. I. Gallin, I. M. Goldstein and R. Snyderman. New York, Raven Press: 541-588.

Korfhagen, T. R., M. D. Bruno, et al. (1996). "Altered surfactant function and structure in SP-A gene targeted mice." Proc Natl Acad Sci U S A **93**(18): 9594-9.

Kuan, S.-F., K. Rust, et al. (1992). "Interactions of surfactant protein D with bacterial lipopolysaccharides." J. Clin. Invest. **90**: 97-106.

Kuan, S. F., A. Persson, et al. (1994). "Lectin-mediated interactions of surfactant protein D with alveolar macrophages." Am J Respir Cell Mol Biol **10**(4): 430-6.

LaForce, F. M., W. J. Kelly, et al. (1973). "Inactivation of staphylococci by alveolar macrophages with preliminary observations on the importance of alveolar lining material." Am. Rev. Respir. Dis. **108**: 784-790.

LeVine, A. M., M. D. Bruno, et al. (1997). "Surfactant protein A-deficient mice are susceptible to group B streptococcal infection." J Immunol **158**(9): 4336-40.

Malhotra, R., J. Haurum, et al. (1992). "Interaction of C1q receptor with lung surfactant protein A." Eur J Immunol **22**(6): 1437-45.

Manz-Keinke, H., H. Plattner, et al. (1992). "Lung surfactant protein A (SP-A) enhances serum-independent phagocytosis of bacteria by alveolar macrophages." Eur J Cell Biol **57**(1): 95-100.

McNeely, T. B. and J. D. Coonrod (1993). "Comparison of the opsonic activity of human surfactant protein A for Staphylococcus aureus and Streptococcus pneumoniae with rabbit and human macrophages." J Infect Dis **167**(1): 91-7.

Moya, F. R., D. R. Hoffman, et al. (1993). "Platelet-activating factor in surfactant preparations [see comments]." Lancet **341**(8849): 858-60.

Neuendank, A., S. Weißbach, et al. (1992). "Surfactant protein A (SP-A) enhances the production of reactive oxygen species in alveolar macrophages." Am. Rev. Respir. Dis. **145**: A876.

O'Neill, S., E. Lesperance, et al. (1984). "Rat lung lavage surfactant enhances bacterial phagocytosis and intracellular killing by alveolar macrophages." Am. Rev. Respir. Dis. **130**: 225-230.

O'Neill, S. J., E. Lesperance, et al. (1984). "Human lung lavage surfactant enhances staphylococcal phagocytosis by alveolar macrophages." Am. Rev. Respir. Dis. **130**: 1177-1179.

Persson, A., D. Chang, et al. (1990). "Surfactant protein D (SP-D) is a divalent cation-dependent carbohydrate binding protein." J. Biol. Chem. **265**: 5755-5760.

Pison, U., M. Max, et al. (1994). "Host defence capacities of pulmonary surfactant: evidence for 'non-surfactant' functions of the surfactant system." Eur J Clin Invest **24**(9): 586-99.

Pison, U., J. R. Wright, et al. (1992). "Specific binding of surfactant apoprotein SP-A to rat alveolar macrophages." Am. J. Physiol. **262**(4): L412-L417.

Rensch, H. and H. von Seefeld (1984). Surfactant-mucus interacrion. Pulmonary Surfactant. B. Robertson, L. M. G. van Golde and J. J. Batenburg. Amsterdam, Elsevier: 204-214.

Richman, P. S., S. Batcher, et al. (1990). "Pulmonary surfactant suppresses the immune lung injury response to inhaled antigen in guinea pigs." J. Lab. Clin. Med. **116**: 18-26.

Ross, G. F., R. H. Notter, et al. (1986). "Phospholipid binding and biophysical activity of pulmonary surfactant-associated protein (SAP)-35 and its non-collagenous COOH-terminal domains." J. Biol. Chem. **261**(30): 14283-14291.

Sakai, K., M. N. Kweon, et al. (1992). "Effects of pulmonary surfactant and surfactant protein A on phagocytosis of fractionated alveolar macrophages: relationship to starvation." Cell Mol Biol **38**(2): 123-30.

Sastry, K. and R. A. Ezekowitz (1993). "Collectins: pattern recognition molecules involved in first line host defense." Curr Opin Immunol **5**(1): 59-66.

Schürch, S., M. Lee, et al. (1992). "Pulmonary surfactant: surface properties and function of alveolar and airway surfactant." Pure & Appl. Chem. **64**(11): 1745-1750.

Sherman, M. P., J. B. D'Ambola, et al. (1988). "Surfactant therapy of newborn rabbits impairs lung macrophage bactericidal activity." J. Appl. Phys. **65**: 137-145.

Sitrin, R. G., M. J. Ansfield, et al. (1985). "The effect of pulmonary surface-active material on the generation and expression of murine B- and T-lymphocyte effector functions in vitro." Exp. Lung. Res. **9**: 85-97.

Speer, C. P., B. Götze, et al. (1991). "Phagocytic functions and tumor necrosis factor secretion of human monocytes exposed to natural porcine surfactant (Curosurf)." Pediatr. Res. **30**: 69-74.

Tenner, A. J., S. L. Robinson, et al. (1989). "Human pulmonary surfactant protein (SP-A), a protein structurally homologous to C1q, can enhance FcR- and CR1-mediated phagocytosis." J. Biol. Chem. **264**(23): 13923-13928.

Thomassen, M. J., D. P. Meeker, et al. (1992). "Synthetic surfactant (Exosurf) inhibits endotoxin-stimulated cytokine secretion by human alveolar macrophages." Am. J. Respir. Cell Mol. Biol. **7**: 257-260.

Tino, M. J. and J. R. Wright (1996). "Surfactant protein A stimulates phagocytosis of specific pulmonary pathogens by alveolar macrophages." Am J Physiol **270**(4 Pt 1): L677-88.

van Iwaarden, F., B. Welmers, et al. (1990). "Pulmonary surfactant protein A enhances the host-defense mechanism of rat alveolar macrophages." Am. J. Respir. Cell Mol. Biol. **2**: 91-98.

van Iwaarden, J. F., H. Shimizu, et al. (1992). "Rat surfactant protein D enhances the production of oxygen radicals by rat alveolar macrophages." Biochem. J. **286**: 5-8.

van Iwaarden, J. F., J. A. van Strijp, et al. (1992). "Binding of surfactant protein A (SP-A) to herpes simplex virus type 1-infected cells is mediated by the carbohydrate moiety of SP-A." J Biol Chem **267**(35): 25039-43.

van Iwaarden, J. F., J. A. G. van Strijp, et al. (1991). "Surfactant protein A is opsonin in phagocytosis of herpes simplex virus type 1 by rat alveolar macrophages." Am. J. Physiol. **261**(Lung Cell. Mol. Physiol. 5): L204-L209.

Voss, T., H. Eistetter, et al. (1988). "Macromolecular organization of natural and recombinant lung surfactant protein SP 28-36. Structural homology with the complement factor C1q." J. Mol. Biol. **201**(1): 219-227.

Voss, T., K. Melchers, et al. (1991). "Structural comparison of recombinant pulmonary surfactant protein SP-A derived from two human coding sequences: implications for the chain composition of natural human SP-A." Am J Respir Cell Mol Biol **4**(1): 88-94.

Webb, D. S. A. and E. L. Jeska (1986). "Enhanced luminol-dependent chemiluminescence of stimulated rat alveolar macrophages by pretreatment with alveolar lining material." J. Leukocyte. Biol. **40**(1): 55-64.

Weber, H., P. Heilmann, et al. (1990). "Effect of canine surfactant protein (SP-A) on the respiratory burst of phagocytic cells." FEBS Lett. **270**(1-2): 90-94.

Weißbach, S., A. Neuendank, et al. (1992). "Surfactant protein A (SP-A) stimulates phagocytosis of candida trop. by alveolar macrophages." FASEB J. **6**(4): A1270.

Weißbach, S., A. Neuendank, et al. (1994). "Surfactant protein A (SP-A) modulates the release of reactive oxygen species from alveolar macrophages." Am. J. Physiol. **267**(6): L660-L666.

Wiernik, A., T. Curstedt, et al. (1987). "Morphology and function of blood monocytes after incubation with lung surfactant." Eur J Respir Dis **71**(5): 410-8.

Wispe, J. R., J. C. Clark, et al. (1990). "Tumor necrosis factor-alpha inhibits expression of pulmonary surfactant protein." J Clin Invest **86**(6): 1954-60.

Wright, A. E. and S. R. Douglas (1903). "An experimental investigation of the role of the body fluids in connection with phagocytosis." Proc R Soc Lond **72**: 357-370.

Wright, J. R., R. E. Wager, et al. (1987). "Surfactant apoprotein Mr = 26,000-36,000 enhances uptake of liposomes by type II cells." J Biol Chem **262**(6): 2888-94.

Wright, J. R. and D. C. Youmans (1993). "Pulmonary surfactant protein A stimulates chemotaxis of alveolar macrophages." Am. J. Physiol. **264**: L338-L344.

Wright, S. D. (1992). Receptors for complement and the biology of phagocytosis. Inflammation: Basic Principles and Clinical Correlates. J. I. Gallin, I. M. Goldstein and R. Snyderman. New York, Raven Press: 477-495.

Zeligs, B. J., L. S. Nerurkar, et al. (1984). "Chemotactic and candidacidal responses of rabbit alveolar macrophages during postnatal development and the modulating roles of surfactant in these responses." Infect Immun 44(2): 379-85.

Zimmerman, P. E., D. R. Voelker, et al. (1992). "120-kD surface glycoprotein of Pneumocystis carinii is a ligand for surfactant protein A." J Clin Invest 89(1): 143-9.

ALTERATION OF PULMONARY SURFACTANT IN ARDS - PATHOGENETIC ROLE AND THERAPEUTIC PERSPECTIVES

A. Günther, D. Walmrath, F. Grimminger and W. Seeger

Department of Internal Medicine
Justus-Liebig-University
Klinikstrasse 36
35392 Giessen, Germany

INTRODUCTION

The adult respiratory distress syndrome (ARDS) characterizes different states of acute impairment of pulmonary gas exchange. Underlying noxious events may directly affect lung parenchyma from the alveolar side (e.g. gastric acid aspiration), or - more classically - the lung vasculature may be the primary target site of circulating humoral or cellular mediators activated under conditions of systemic inflammatory events such as sepsis or severe polytrauma. Key pathophysiological features of the initial "exsudative" phase of ARDS include an increase in capillary endothelial and alveolar epithelial permeability, leakage of protein rich edema fluid into interstitial and alveolar spaces, increase in pulmonary vascular resistance with maldistribution of pulmonary perfusion, alveolar instability with formation of atelectases and ventilatory inhomogeneities as well as severe disturbances of gas exchange characterized by ventilation-perfusion mismatch and extensive shunt flow.

This exsudative phase may persist for days to weeks, and full recovery without persistent loss of lung function is well possible during this period of acute respiratory distress. New inflammatory events, such as recurrent sepsis or acquisition of secondary (nosocomial) pneumonia, may repetitively worsen the state of lung function and then progressively trigger proliferative processes with mesenchymal cell activation and rapidly on-going lung fibrosis. Thus, within few weeks, the lung architecture may become dominated by thickened fibrotic alveolar septae and large interposed airspaces ("honeycombing"). Prognosis is very poor during this phase of ARDS, and only partial recovery of lung function may be achieved in the few survivors from this late phase of disease.

Deficiency of pulmonary surfactant has been established as the primary cause of the respiratory distress syndrome in preterm infants (IRDS), and transbronchial application of natural surfactant preparations has been proven to be beneficial in this disease. Surfactant abnormalities may also be involved in the sequelae of pathogenetic events in ARDS, however, due to the diversity of underlying triggering mechanisms and the complexity of pathophysiological events in adult respiratory distress, any evaluation

Acute Respiratory Distress Syndrome: Cellular and Molecular Mechanisms and Clinical Management
Edited by Matalon and Sznajder, Springer Science+Business Media New York, 1998

97

of the role of surfactant in this disease is much less certain. This review focuses on three questions:

- *What is the present evidence for surfactant abnormalities in patients with ARDS?*
- *Which pathophysiological events encountered in the course of ARDS may be ascribed to surfactant abnormalities?*
- *What are the acute effects of a transbronchial surfactant administration in ARDS in view of gas exchange and compliance?*

These aspects aim to provide a rational basis for the question, whether transbronchial surfactant application might become a safe and general therapeutic approach in patients with ARDS as it is in IRDS.

ALTERATION OF SURFACTANT IN ARDS

In early post-mortem investigations in lungs from patients who had died in ARDS, first evidence for severe impairment of surfactant function was obtained[1]. More direct proof was provided by studies of this group and of other investigators with biophysical analysis of bronchoalveolar lavage fluids (BAL) obtained by flexible bronchoscopy during the active state of the disease[2-5]. Compared to normal volunteers, BAL-samples of these patients showed increased minimal surface tension and decreased hysteresis of the surface tension-surface area relationship, two presumably most critical variables of surfactant function in vivo (Fig. 1). Recently, elevated minimal surface tension values were also determined for surfactant samples obtained from patients being at risk for ARDS[2]. Several factors may underlay such loss of surface activity in ARDS.

LACK OF SURFACE-ACTIVE COMPOUNDS AND CHANGE IN PHOSPHOLIPID, FATTY ACID AND APOPROTEIN PROFILES

Clinical studies addressing the phospholipid composition of surfactant samples obtained from patients with ARDS revealed three important features. *Firstly,* the overall content of phospholipids was found to be decreased in two of the four studies performed yet. In addition, this decrease of total phospholipids appeared to be dependent on the severity of ARDS[2]. *Secondly,* the relative amounts of the two functionally most important phospholipids, PC and PG, were markedly depressed in all three studies. Most strikingly, the PG levels decreased by > 80°% in three of the studies; the decrease in the percentage of PC was more moderate in all investigations. However, the relative content of palmitic acid within the phospholipid fatty acids, in particular the relative amount of dipalmitoylated PC (DPPC), was found to be severely reduced in ARDS patients (not given in detail). *Thirdly,* all studies demonstrated an increase in the relative amounts of PI, PE and Sph.

Due to the late detection and - in case of SP-B and SP-C - the extreme hydrophobic nature of the surfactant apoproteins, appropriate analytical techniques for measurement of these essential surfactant compounds have only recently become available; SP-C quantification in BAL samples is still an unresolved problem. Two recent studies with measurement of SP-A and SP-B in BAL samples from patients with ARDS demonstrated an impressive decline of SP-A and SP-B[2,5]. Again, some decrease of these functionally important compounds was also observed in patients being at risk for ARDS[2,5].

ALTERATION OF SURFACTANT SUBTYPE DISTRIBUTION

Under physiological conditions nearly 80-90% of the extracellular surfactant material is recovered among the fraction of large surfactant aggregates and this subfraction imposes with high relative SP-B content and superior surface activity. Under conditions of ARDS, however, increase of the small surfactant aggregates at the expense

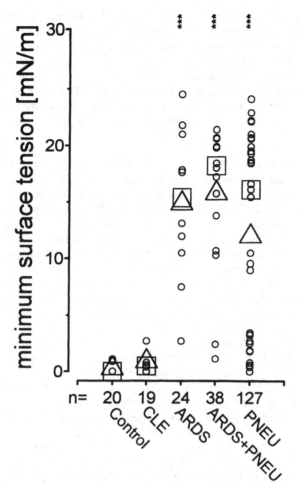

Figure 1.
Biophysical surfactant properties of isolated large surfactant aggregates. For controls and the different groups of patients, all single events (o), means (Δ) and median (□) values are indicated. Surface tension [mN/m] at minimum bubble size after 5min of film oscillation (γmin) is given. Phospholipid concentration was 2 mg/ml; n-numbers are given in the x-axis. All groups were compared to controls; p is indicated by * (p < 0.05), ** (p < 0.01) or *** (p < 0.001). CLE - cardiogenic lung edema; ARDS - Acute Respiratory Distress Syndrome; PNEU - Severe pneumonia necessitating mechanical ventilation; ARDS+PNEU - ARDS and lung infection. Reprinted from [5] with permission.

of the large surfactant aggregates is encountered and is paralleled by a loss of SP-B and surface activity within the large surfactant aggregates.

INHIBITION OF SURFACTANT FUNCTION BY PLASMA PROTEIN LEAKAGE

Leakage of plasma proteins into the alveolar space may substantially contribute to surfactant alterations in ARDS. Measurements of the protein content in BAL samples from these patients persistently showed markedly increased levels as compared to normal controls. Protein leakage is an early event in the sequence of pathogenetic events in ARDS, and is related to the severity of the disease [see e.g. [5]]. Experimental studies in vitro and in vivo have demonstrated that admixture of blood, serum, plasma or alveolar washings obtained during states of plasma leakage may severely compromise biophysical surfactant function. Among different proteins involved, albumin[6], hemoglobin[7] and in particular fibrinogen or fibrinmonomer[6,8,9] possess strong surfactant inhibitory properties. Concerning fibrinogen, it has been demonstrated that its potency to inhibit surfactant function depends on the surfactant apoprotein profile. Surfactant preparations lacking hydrophobic apoproteins are extremely sensitive to fibrinogen inhibition, and least sensitivity is noted in the presence of both SP-C and SP-B in near physiological quantities[8]. In addition, a further reduction of surfactant sensitivity to fibrinogen is achieved by supplementation of phospholipid- and hydrophobic apoprotein-based surfactants with SP-A[10].

"INCORPORATION" OF SURFACTANT IN FIBRIN/HYALINE MEMBRANES

Intraalveolar accumulation of clot material, characterized as `hyaline membranes', is commonly found in ARDS and other acute or chronic inflammatory diseases of the lung. In the alveolar milieu, the extrinsic coagulation pathway represents the predominant clotting sequence. Alveolar macrophages express and shed a procoagulant activity, which is mainly attributable to tissue factor in compound with factor VII[11]. This alveolar procoagulant activity was found to be markedly increased in ARDS patients. In contrast, concentrations of urokinase-type plasminogen activator, representing the predominant fibrinolysis pathway within the alveolar spaces[11,12], were noted to be decreased in lavage fluids from patients with ARDS; concomitantly, increased levels of plasminogen-activator-inhibitor-1 and α2-antiplasmin were detected[11,12]. Moreover, surfactant phospholipid-mixtures were found to inhibit plasmin-, trypsin- or elastase-induced fibrino(geno)lysis, in particular when combined with the surfactant apoproteins SP-B and SP-C[13]. Thus, the hemostatic balance within the alveolar milieu appears to be shifted towards predominance of procoagulant and antifibrinolytic activities in acutely or chronically inflamed lung regions, in particular in ARDS. Recent investigations performed by this group[14] demonstrated loss of surfactant phospholipids from the soluble phase due to binding to/within fibrin strands, when the process of fibrin polymerization occurred in the presence of surfactant material. In parallel, virtually complete loss of surface activity was noted, with fibrin dose-effect curves ranging two orders of magnitude below the corresponding efficacy range of soluble fibrinogen. Overall, the findings obviously suggest "incorporation" of phospholipids (and possibly hydrophobic apoproteins) into nascent fibrin strands, with severe loss of functionally important surfactant compounds in areas with alveolar fibrin and hyaline membrane formation. Interestingly, surface activity may be largely restored by application of fibrinolytic agents in vitro[13] and in vivo [unpublished], with release of formerly incorporated surfactant material into the soluble phase.

DAMAGE OF SURFACTANT COMPOUNDS BY INFLAMMATORY MEDIATORS

A variety of inflammatory processes are assumed to underlie microcirculatory disturbances in ARDS, and mediator generation has also been demonstrated for the alveolar compartment. Free elastase and collagenase activities were repeatedly detected in BAL fluids of patients in ARDS, oxidative inhibition of the alveolar α1-proteinase inhibitor indicated oxygen radical generation in this compartment, and increased levels of lysophospholipids (predominantly lyso-PC)[3] suggested increased phospholipolytic activity in the alveolar space under conditions of ARDS. A variety of in-vitro studies have addressed putative direct inhibitory effects of inflammatory mediators on biophysical surfactant functions. Inhibitory potencies were demonstrated for phospholipases, proteases, oxygen radicals, free fatty acids and activated granulocytes (via release of oxygen radicals). Presently, however, no data are available to quantify the contribution of such surfactant-inhibitory effects of inflammatory mediators to the impairment of surfactant function in patients with ARDS.

PATHOPHYSIOLOGICAL CONSEQUENCES OF SURFACTANT ALTERATIONS IN ARDS

As outlined above, there is strong evidence for severe impairment of the alveolar surfactant system under conditions of ARDS, and several mechanisms may underlay this finding. Thus the question arises, whether and to what extent such surfactant alterations contribute to the sequence of pathogenetic events and the loss of lung functional integrity encountered in this disease. The main issues to be addressed in this context are discussed in the following section.

ALTERATION OF LUNG MECHANICS

Loss of alveolar surface activity will increase surface tension and thereby cause alveolar instability with formation of atelectasis. These features must be expected to result in a marked decrease of lung compliance. This basic finding was, indeed, already described in the very early reports on altered mechanics of post-mortem analysed lungs from patients dying in ARDS[1]. In addition, in a variety of experimental approaches using animal models of ARDS, induction of acute lung injury resulted in significant decrease of compliance [e.g. 15]. Accordingly, transbronchial application of surfactant was shown to completely or partially restore physiological lung compliance in some of these models [e.g. 15]. In patients with severe ARDS, however, reliable measurements of lung compliance are still difficult to perform, mostly because of uncertainties concerning lung volumes (at which part of the pressure-volume curve does the lung actually range?) and transpulmonary pressures.

IMPAIREMENT OF GAS EXCHANGE - V/Q MISMATCH AND SHUNT FLOW

Lack of surface active material has been established as primary cause of severe gas exchange disturbances in IRDS, and dramatic improvement of arterial oxygenation is achieved by transbronchial surfactant application under these conditions. Similarly, experimental approaches in adult animals with removal of alveolar surfactant (lung lavage models [e.g. 16]) or surfactant inactivation by detergent application[17] and subsequent transbronchial re-application of surface active material have underscored the fundamental significance of the alveolar surfactant system for ventilation-perfusion matching also in adult lungs. In more realistic models of ARDS, starting with induction of microvascular or alveolar injury, matters are more complex. Shunt-flow (perfusion of atelectatic regions) and blood flow through lung areas with low ventilation-perfusion ratios (partial closure of alveolar units or small airways) may well be related to an acute impairment of the alveolar surfactant system in such experiments, and transbronchial surfactant application was found to improve gas exchange conditions in models with protein-rich edema formation due to cervical vagotomy, acid aspiration, pneumonia, hyperoxic lung injury and application of N-nitroso-N-methylurethane (NNNMU) or oleic acid [see e.g. 15]. The efficacy of surfactant replacement in these models with induction of lung inflammation is, however, lower than in models with primary surfactant depletion (lavage, preterm newborns), which is most probably attributable to inhibitory capacities of leaked plasma proteins and inflammatory mediators, as discussed above. Larger amounts of surfactant material are apparently needed under these conditions, in order to surpass, at least partially, such inhibitory capacities.

LUNG EDEMA FORMATION

Interstitial and alveolar edema is a key finding in ARDS, primarily ascribed to increased endothelial and epithelial permeability in the diseased lungs. Surfactant alterations may, however, well contribute to the disturbances in fluid balance in ARDS. Any increase in alveolar surface tension must be expected to result in a decrease of interstitial and thus perivascular pressures and, according to Starling's law, increase transendothelial fluid fluxes into septal and interstitial spaces. Similarly, increased alveolar surface tension will favour transepithelial fluid movement into the alveolar spaces. Several experimental studies have indeed demonstrated extensive lung edema formation due to inhibition of surfactant function in vivo by transbronchial detergent application, intratracheal injection of bile acid as well as cooling and ventilating at low FRC or plasma lavage. Concerning patients with ARDS, however, there is presently no conclusive study to evaluate the impact of surfactant abnormalities on lung fluid balance and alveolar epithelial permeability characteristics under clinical conditions.

REDUCTION OF HOST-DEFENSE COMPETENCE ?

Next to the reduction of surface tension within the alveolar compartment, the pulmonary surfactant system is involved in many host defense properties of the lung. Although not fully understood at the present time there is evidence that the hydrophilic apoproteins SP-A and SP-D might act as highly effective opsonins, thereby enhancing the phagocytosis of several strains of bacteria and viruses. In contrast, the lipid fraction of pulmonary surfactant is capable of suppressing the activation and proliferative response of lymphocytes, granulocytes and alveolar macrophages. However, these aspects are at best mosaics of a complex alveolar host defense system, which largely remains to be defined. The marked decrease in SP-A levels in lungs of ARDS patients (see above) may suggest loss of opsonizing capacity and increased susceptibility to nosocomial infections.

"COLLAPSE INDURATION", MESENCHYMAL CELL PROLIFERATION AND FIBROSIS

The proliferative phase of ARDS is characterized by progressive mesenchymal cell activation and proliferation, predominantly in atelectatic regions, and may result in widespread lung fibrosis and honeycombing within few weeks. Underlying mechanisms may well include a major role of the alveolar surfactant system and of alveolar fibrin deposition. A corresponding sequence of events was suggested for the pathogenesis of lung fibrosis in general by Burkhardt[18] and termed "collapse induration". Basically, this concept starts with persistent atelectasis at sites of extensive loss of alveolar surfactant function, in particular regions with fibrin deposition. Alveolar wall apposition and the fibrin matrix represent a nidus for fibroblast activation, and the concerned alveolar space is definitely lost by deposition of fibrous tissue (collapse induration). Thus, thick indurated septae (or conglomerates of several septae) may exist adjacent to widened (remaining) alveoli to provide the typical morphological image of fibrosis and honeycombing.

SURFACTANT REPLACEMENT IN ARDS - CURRENT STATUS

Against this background of surfactant abnormalities, improvement of alveolar surfactant function appears to be a reasonable approach to improve gas exchange in ARDS patients. Such attempts may include pharmacological approaches to stimulate the secretion of intact surfactant material from pneumocytes type II, but clear evidence that this approach may be effectively used under conditions of acute respiratory failure is still missing. In addition, transbronchial administration of exogenous (natural) surfactant preparations, commonly used in IRDS, may also be employed in ARDS, but will clearly demand larger quantities of material to overcome the surfactant inhibitory capacities in the alveolar space under these conditions. Two pilot studies in this field have yet been completed. Performing repetitive intratracheal application of Survanta, with cumulative doses between 300 and 800 mg/kg b.w., T. Gregory and colleagues noted a significant improvement of gas exchange and even obtained some preliminary evidence for an increase in survival in adults with acute respiratory failure[19]. Our group investigated the safety and efficacy of a bronchoscopic application of a natural surfactant extract (Alveofact) in patients with severe ARDS. All patients fulfilled ECMO criteria (mean Murray lung injury score ≈ 3.3) and were treated within the first 5 days of disease, i.e. before onset of major fibrotic processes. Underlying diseases were mostly sepsis and severe pneumonia; at the present time the study includes 26 patients. 300 mg/kg Alveofact was delivered bronchoscopically in divided doses to each segment of both lungs (total dose 22.5 ± 1.4 g in 375 ml saline), followed by a second application of 200 mg/kg (total dose 15.5 ± 1.05 g in 260 ml saline) 18 - 24 h later in selected patients. Assessment of gas exchange including ventilation-perfusion characteristics, hemodynamic measurements and BAL was performed before and after surfactant application. As obvious from an example given in Figure 2, an acute and impressive improvement of the gas exchange was encountered already within the application procedure. When analyzing the course of gas exchange among all patients, the first surfactant application resulted in an immediate increase of mean paO_2/FiO_2 from < 90 mmHg to ≈ 200 mmHg (Figure 3), mainly due to a decrease in shunt-flow (from ≈ 40 % to ≈ 20 %). Approximately 2/3 of the patients „responded" with a paO_2/FiO_2 increase of at least 25 %. The effect was partially lost within the following hours in some of the responders, but restored with prolonged improvement of arterial oxygenation by the second application. Initial BAL showed severe alteration of surfactant composition and impairment of biophysical surfactant function. Surfactant application resulted in a marked, but still incomplete restoration of surfactant properties, with a profound improvement of the phospholipid (PL)-protein-ratio, relative content of large surfactant aggregates, relative content of phosphatidylcholine, and minimum surface tension in absence as well as in presence of the inhibitory BALF proteins. Analysis of the ventilation/perfusion characteristics revealed that in response to the bronchoscopic surfactant application, formerly collapsed alveoli were re-aerated, yielding a reduction of the intrapulmonary shunt flow and an increase in regions with low and normal ventilation/perfusion ratios.

CONCLUSION - PERSPECTIVE

In conclusion, profound alterations of the alveolar surfactant system are encountered in ARDS. There is now good evidence that these abnormalities contribute to the severe impairment in gas exchange under these conditions. Transbronchial surfactant application may offer a feasible and safe approach to improve biochemical and biophysical properties of the endogenous surfactant pool and, by this, the gas exchange conditions in most severe early-stage ARDS. However, a high and/or repetitive dosage regimen appears to be necessary to overcome inhibitory capacities in the alveolar space of these patients and to achieve sustained alveolar recruitment. Forthcoming studies will have to rule out the optimum timing and dosage regimen of such intervention and will have to address the question, whether this therapy is capable of reducing the still high mortality of patients with most severe ARDS. The impact of surfactant apllication on inflammation, host-defense and mesenchymal proliferation in the alveolar compartment will have to be addressed critically.

REFERENCES

1. Petty, T.L., Silvers, G.W., Paul, G.W., Stanford, R.E., (1979), Abnormalities in lung elastic properties and surfactant function in adult respiratory distress syndrome, Chest 75(5):571-574.

2. Gregory, T.J., Longmore, W.J., Moxley, M.A., Whitsett, J.A., Reed, C.R, Fowler, A.A., Hudson, L.D., Maunder, R.J., Crim, C., Hyers, T.M., (1991), Surfactant chemical composition and biophysical activity in acute respiratory distress syndrome, J. Clin. Invest. 88:1976-1981.

3. Hallman, M., Spragg, R., Harrell, J.H., Moser, K.M., Gluck, L., (1982), Evidence of lung surfactant abnormality in respiratory failure, J. Clin. Invest. 70:673-683.

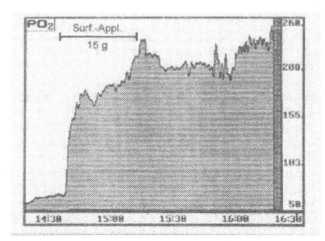

Fig. 2: Example of the course of the arterial pO2 in response to a transbronchial surfactant administration in a 18y old female with severe ARDS. Given is the original on-line registration of the paO2 (Paratrend) at a constant FiO2 of 1.0. 300mg/kg b.w. of a bovine surfactant extract (Alveofact) were given. The paO2 increased from ~ 60mmHG (base-line) to ~ 220mmHG after surfactant application (from [20] with permission).

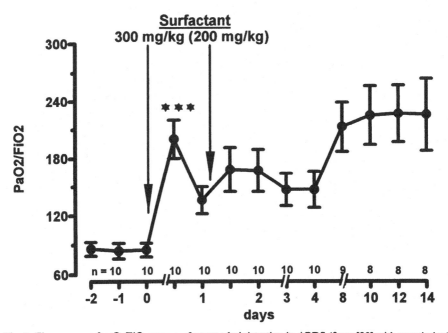

Fig. 3: Time course of paO$_2$/FiO$_2$ upon surfactant administration in ARDS (from [20] with permission)

4. Pison, U., Seeger, W., Buchhorn, R., Joka, T., Brand, M., Obertacke, U., Neuhof, H., Schmit-Neuerburg, K.P., (1989), Surfactant abnormalities in patients with respiratory failure after multiple trauma, Am. Rev. Respir. Dis. 140:1033-1039.

5. Günther, A.G., Siebert, C., Schmidt, R., Ziegler, S., Grimminger, F., Yabut, M., Temmesfeld, B., Walmrath, D., Morr, H., Seeger, W., (1996), Surfactant alterations in severe pneumonia, acute respiratory distress syndrome, and cardiogenic lung edema. Am. J. Respir. Crit. Care Med. 153:176-184.

6. Fuchimukai, T., Fuchiwara, T., Takahashi, A., Enhorning, G., (1987), Artificial pulmonary surfactant inhibited by proteins, J. Appl. Physiol. 62(2):429-437.

7. Holm, B.A., Notter, R.H., (1987), Effects of hemoglobin and cell membrane lipids on pulmonary surfactant activity, J. Appl. Physiol. 63(4):1434-1442.

8. Seeger, W., Günther, A., Thede, C., (1992), Differential sensitivity to fibrinogen-inhibition of SP-C versus SP-B based surfactants, Am. J. Physiol. (Lung Cell. Mol. Physiol. 5):L286-L291.

9. O'Brodovich, H.M., Weitz, J.I., Possmayer, F., (1990), Effect of fibrinogen degradation products and lung ground substance on surface function, Biol. Neonate. 57:325-333.

10. Cockshutt, A.M., Weitz, J., Possmayer, F., (1990), Pulmonary surfactant-associated protein A enhances the surface activity of lipid extract surfactant and reverses inhibition by blood proteins in vitro, Biochemistry 29(36):8424-8429.

11. Idell, S., James, K.K., Levin, E.G., Schwartz, B.S., Manchanda, N., Maunder, R.J., Martin, T.R., McLarty, J., Fair, D.S., (1989), Local abnormalities in coagulation and fibrinolytic pathways predispose to alveolar fibrin deposition in the adult respiratory distress syndrome, J. Clin. Invest. 84:695-705.

12. Bertozzi, P., Astedt, B., Zenzius, L., Lynch, K., LeMaire, F., Zapol, W., Chapman, H., (1990), Depressed bronchoalveolar urokinase activity in patients with adult respiratory distress syndrome, N. Engl. J. Med. 322:890-89713.

13. Günther, A., Kalinowski, M., Elssner, A., Seeger, W., (1994), Clot-embedded natural surfactant: kinetics of fibrinolysis and surface activity, Am. J. Physiol. 267:L618-L624.

14. Seeger, W., Elssner, A., Günther, A., Kraemer, H.-J., Kalinowski, H.O., (1993), Lung surfactant phospholipids associate with polymerizing fibrin: loss of surface activity, Am. J. Respir. Cell Mol. Biol. 9:213-220.

15. Lewis, J.F., Ikegami, M., Higuchi, R., Jobe, A., Absolom, D., (1991), Nebulized vs. instilled exogenous surfactant in an adult injury model, J. Appl. Physiol. 71(4):1270-1276.

16. Hall, S.B., Venkitaraman, A.R., Whitsett, J.A., Holm, B.A., Notter, R.H., (1992), Importance of hydrophobic apoproteins as constituents of clinical exogenous surfactants, Am. Rev. Respir. Dis. 145:24-30.

17. Schermuly, R., Schmehl, T., Günther, A., Seeger, W., Walmrath, D., (1997), Aerosolized surfactant improves gas exchange in detergent treated, isolated rabbit lungs. In press: Am. J. Respir. Crit. Care Med.

18. Burkhardt, A, (1989), Alveolitis and collapse in the pathogenesis of pulmonary fibrosis, Am. Rev. Respir. Dis. 140:513-524.

19. Gregory, T.J., Steinberg, K.P., Spragg, R., Gadek, J.E., Hyers, T.M., Longmore, W.J., Moxley, M.A., Cai, G.Z., Hite, R.D., Smith, R.M., Hudson, L.D., Grim, C., Newton, P., Mitchell, B.R., Gold, A.J., (1997), Bovine surfactant therapy for patients with acute respiratory distress syndrome, Am. J. Respir. Crit. Care Med. 155:1309-1315.

20. Walmrath, D., Günther, A., Ghofrani, A.G., Schermuly, R., Schneider, T., Grimminger, F., Seeger, W., (1996), Bronchoscopic surfactant administration in patients with severe adult respiratory distress syndrome and sepsis, Am. J. Respir. Crit. Care Med, 154:57-62.

SURFACTANT REPLACEMENT IN PATIENTS WITH ARDS: RESULTS OF CLINICAL TRIALS

Roger G. Spragg and Robert M. Smith

Department of Medicine
University of California, San Diego
VA Medical Center
San Diego, CA 92161

INTRODUCTION

Since the first medical description of the acute respiratory distress syndrome (ARDS) three decades ago, there has been growing evidence that disturbance of the lung surfactant system may contribute to the pathophysiology of that syndrome. Initial descriptions of patients with ARDS documented abnormalities of lung surfactant biophysical function. Understanding of the complexity of the lung surfactant system has matured, and it is now possible to describe both biophysical and non-biophysical functions of that system, how the various components of lung surfactant may be altered in disease states, and how those alterations may contribute to lung surfactant dysfunction. Such understanding forms a rational basis for design and evaluation of studies of lung surfactant administration to patients with ARDS.

BIOPHYSICAL AND NON-BIOPHYSICAL PROPERTIES OF LUNG SURFACTANT

These activities may viewed in the context of the composition of lung surfactant - a complex of lipids and apoproteins that is distributed over the air-liquid interface of the pulmonary alveoli and distal airways. Surfactant minimizes alveolar wall surface tension at low lung volumes, thereby stabilizing alveolar volume during tidal breathing and preventing the movement of fluid into the alveolus[1]. Surfactant is synthesized in the alveolar type II cell, stored in lamellar bodies of that cell, and secreted into the alveolar hypophase in which it undergoes complex conformational changes. Current understanding suggests that the structure identified in the alveolar hypophase by electron microscopy as tubular myelin is a transitional state between the lamellar body and dispersed surfactant. Extracellular surfactant

Acute Respiratory Distress Syndrome: Cellular and Molecular Mechanisms and Clinical Management
Edited by Matalon and Sznajder, Springer Science+Business Media New York, 1998

107

exists predominately as large and small aggregates possessing different biophysical activities, and as a monolayer (or near monolayer) covering the air-liquid interface.

The composition of lung surfactant of adult humans and animals, summarized in Table 1, is very constant, although it may change in disease states[2]. Surfactant isolated by endobronchial lavage followed by differential and density gradient centrifugation consists of approximately 90% lipid (of which the majority is phospholipid), with the balance composed of surfactant-associated apoproteins. The lipid composition of lung surfactant is well described[1-3]. Of note, in adult humans, phosphatidylcholine accounts for approximately 80% of the phospholipids. It has been reported that, of this, approximately 74% is dipalmitoylphosphatidylcholine (DPPC)[3]. Phosphatidylglycerol (PG) is the next most abundant phospholipid, but its importance to surfactant biophysical function is unclear; exogenous surfactants devoid of PG and endogenous surfactant with minimal PG present demonstrate excellent biophysical function. Approximately 70% of the phospholipid acyl groups in surfactant recovered from the airway of the mature lung are fully saturated[4].

Table 1. Composition of lung surfactant by weight[1-4]

Lipid		90%		
	Neutral lipids		9%	
	Phospholipids		91%	
	Saturated acyl groups		68%	
	Unsaturated acyl groups		32%	
	Monounsaturated			87%
	Diunsaturated			8%
	Polyunsaturated			5%
Protein		10%		
	SP-A		50%	
	SP-B		10-20%	
	SP-C		10-20%	
	SP-D		10-20%	

Currently recognized surfactant apoproteins include a group of relatively hydrophilic acidic glycoproteins of Mr = 30,000 - 35,000 (SP-A), two markedly hydrophobic nonglycosylated proteins of Mr = 7,500 - 10,000 and Mr = 3,000 -6,000 (SP-B and SP-C, respectively), and a multimeric complex of disulfide-bonded trimers comprised of identical 43 kD collagenous glycoprotein subunits (SP-D). The amino acid sequences and cDNA sequences of the lung surfactant apoproteins are known and the genes for human SP-A, SP-B, and SP-C have been cloned. SP-A, synthesized in alveolar Type II cells, Clara cells, and alveolar macrophages, contains an N-terminal collagen-like sequence, a lipid-binding domain, and a carboxy-terminal lectin-like region sequence. SP-A has been considered to be of importance in phospholipid uptake and reutilization by type II cells[5]. However, the normal pulmonary function and histology of SP-A knockout mice have suggested the lack of a critical biophysical or metabolic role for SP-A[6]. Both SP-B and SP-C are rich in the hydrophobic amino acids valine and leucine and promote adsorbtion of phospholipid to the air-liquid interface. Combining SP-B or SP-C with DPPC and PG results in a surfactant with good biophysical properties that is effective in vivo.

Coordinate regulation of the synthesis of surfactant protein and lipid components is poorly understood. Experiments with transgenic and knockout mice suggest little or no influence on lung surfactant phospholipid synthesis of absent SP-A[6] or of either increased or decreased SP-B levels[7,8]. In contrast, infants with hereditary SP-B deficiency have low levels of surfactant phospholipids, suggesting coordinate regulation[9]. That various hormones may independently modulate the relative abundance of surfactant components suggests an absence of coordinate regulation, although the panoply of hormonal effects makes interpretation difficult. Also, metabolic pathways for surfactant lipids and apoprotein appear to be distinct[10]. The degree to which enhanced synthesis and intracellular level of phospholipid affects surfactant apoprotein synthesis and level in the absence of exogenous hormones is unknown.

Although lung surfactant was initially recognized for its biophysical properties (e.g. the ability to lower alveolar surface tension), recent investigations have highlighted a host of additional non-biophysical properties, including host defense functions, which may be of equal importance in the pathogenesis of acute lung injury. SP-A and SP-D both contain lectin-binding domains that allow them to be recognized by phagocytic cells including neutrophils and alveolar macrophages. SP-D is markedly chemotactic for neutrophils and may act as an opsonin, a role that has been firmly established for SP-A[11,12]. While SP-A or SP-D may function as opsonins or chemotaxins in isolation, phagocytic cells in the alveolus are in contact with all components of lung surfactant including the lipid fraction and the hydrophobic apoproteins SP-B and SP-C. Exposure of phagocytes to lipid-containing surfactant preparations results in: marked diminution of superoxide production by neutrophils and macrophages[13,14], stimulated cytokine production by alveolar macrophages[15,16], phagocytosis by monocytes or neutrophils[17,18], and migration by neutrophils[19,20]. Studies of the mechanisms of surfactant-mediated suppression of cytokine release have demonstrated that exposure of mononuclear phagocytes to either synthetic or modified natural surfactant (Survanta®)resulted in decreased inflammatory cytokine mRNA levels as well as their protein products including TNF-α, IL-1β, and IL-6[21]. Exploration of the mechanism of this effect found that both synthetic and natural surfactants significantly reduce NFκB activation.

In addition to these effects on transcriptional regulation of cytokine production, there is a marked reduction in respiratory burst oxidase activity in human neutrophils exposed to natural or to synthetic surfactant preparations[14]. This inhibitory effect is not related to oxidant scavenging by surfactant, and is not associated with alterations in intracellular calcium mobilization or movement of protein kinase C to the membrane after cell stimulation. Thus, the surfactant effect appears to be mediated through alterations in the activity of the respiratory burst oxidase itself. Detailed analysis of polymorphonuclear leukocytes stimulated with agonist in the presence of lung surfactant indicates that translocation of cytosolic components of the respiratory burst oxidase to the plasma or phagosomal membrane is inhibited[14,22].

In summary, both the biophysical and non-biophysical properties of the lung surfactant may be of importance in maintaining normal pulmonary function. To the extent that properties of lung surfactant are altered in patients with lung injury, questions related to the relationship between those alterations and lung pathophysiology become relevant.

THE LUNG SURFACTANT SYSTEM IN PATIENTS WITH ARDS

The earliest medical reports of patients with the acute respiratory distress syndrome (ARDS) suggested that dysfunction of the lung surfactant system might be of pathophysiologic importance[23]. Early analysis of lung surfactant from patients with ARDS disclosed significant biochemical and functional abnormalities[2]. These findings were subsequently confirmed and extended by studies in which bronchoalveolar lavage fluids (BAL) recovered from patients with ARDS were analyzed for changes in surfactant composition and function[24-27].

Abnormalities of lung surfactant (which is recovered in sedimented pellets of BAL from patients with ARDS) include a 80% fall in the content of total PL (with decline in the fractional contents of PC and PG of 17% and 60%, respectively), and loss of 90-95% of SP-A and SP-B content[24]. Surfactant biophysical function is also profoundly altered, and minimum and maximum surface tensions (as measured with a pulsating bubble surfactometer at 1.5 mM phosphorus) rise from values of approximately 4 and 35 dyne/cm, respectively, to values of approximately 24 and 54 dyne/cm. Qualitatively similar changes of lesser magnitude occur in surfactant recovered from patients with risk factors for development of ARDS and with lung injury not meeting the criteria for ARDS[24]. The variety of mechanisms that may cause a loss of lung surfactant biophysical function in patients with acute lung injury is detailed in Table 2.

Table 2: Possible mechanisms that may cause loss of lung surfactant biophysical function in patients with acute lung injury.

Inhibition by plasma proteins[28]
Phospholipid cleavage by phospholipases[29]
Formation of inactive nitration products[30,31]
Formation of inactive lipid peroxidation products[32]
Apoprotein cleavage by neutrophil enzymes[33]
Oxidative modification by reactive oxygen species[34]
Enhanced conversion to poorly functional small aggregate forms[35]

The finding of abnormal surfactant function in patients with acute lung injury is not surprising. As reviewed by Lachmann[36], a wide variety of agents induce an acute lung injury in experimental animals that is accompanied by alterations in the lung surfactant system. Of particular interest are those injuries in which subsequent surfactant administration appears efficacious. These include: massive lavage[37], hyperoxia[38-41], N-nitroso-N-methylurethane exposure[42,43], acid instillation[44], bilateral cervical vagotomy[45], viral pneumonia[46], and exposure to antibodies directed against surfactant components[47]. In many injuries, surfactant components are variably decreased.

The finding of abnormal surfactant function in patients with acute lung injury and in animal models of acute lung injury, together with apparent benefit associated with surfactant replacement, has provided rational for studies of administration of exogenous surfactant. Further support has come from the success that neonatologists have had in treating infants with established respiratory distress syndrome, for in that syndrome elements of acute lung injury are also present[48].

CLINICAL TRIALS OF SURFACTANT REPLACEMENT

Several uncontrolled and controlled trials of surfactant administration to patients with acute lung injury have been conducted over the past decade. Each of these had had to deal with a variety of issues that are common to such trials.

Issues Relevant to Trials of Surfactant Replacement

When designing trials of surfactant replacement, there are a variety of questions that must be answered: When in the course of the clinical disease should surfactant be administered? What preparation of exogenous surfactant should be administered? What volume and concentration of surfactant should be administered? Should that preparation be administered by instillation or by aerosol? Should repeated doses be administered? What ventilation strategies should be employed?

Clear answers to these questions are not yet available. In general, investigators have opted to define entry criteria for clinical trials so that patients are treated within the first few days of meeting criteria for acute lung injury. This strategy is based on observations that suggest onset after 3-5 days of processes that lead to pulmonary fibrosis and on hope that avoidance of barotrauma early in the course of disease may be of therapeutic efficacy.

A variety of surfactant preparations may be considered for therapeutic use. Whereas production of surfactant for use in treatment of neonatal respiratory distress syndrome can be accomplished in the laboratory of a single investigator, vastly larger amounts are needed to treat adults. This requirement has been a barrier to performance of clinical trials. Of the agents listed in Table 3, only two, rSP-C based surfactant and KL$_4$-surfactant are currently undergoing clinical trial in adults.

Table 3. Surfactants of possible therapeutic use for patients with acute lung injury

Class	Origin	Example	Composition
Modified natural	Bovine	Survanta® Alveofact® CLSE® BLES®	Phospholipids and hydrophobic apoproteins
	Porcine	Curosurf®	
Synthetic		Exosurf®	DPPC, hexadecanol, tyloxapol
		KL$_4$-surfactant	(Lys-Leu-Leu-Leu)$_4$Lys, DPPC, PG, palmitate
		rSP-C	rSP-C, DPPC, PG, palmitate

The volume and concentration in which to administer exogenous surfactants is liable to be a function of the properties of the specific preparations. There is some evidence that larger volumes are associated with more homogeneous distribution[49]. The mode of delivery may also affect surfactant distribution, with the heterogeneity of the underlying disease process an important independent variable[50,51]. Instillation has the advantage of allowing delivery of a large quantity of surfactant in a brief period of time and has been associated with good distribution (in animal studies) and good clinical effect. Disadvantages include lack of standardization of the procedure and the possibility of short-term impairment of ventilation. Aerosolization has the advantages of continuous administration and good effect in laboratory studies in which relatively low doses were administered. Disadvantages include slow delivery, lack of optimized devices, and minimal evidence, in clinical studies, of therapeutic efficacy.

Summary of Clinical Trials

The clinical trials reported to date include those listed in Table 4. Of these, the trials of Exosurf®, Survanta®, and Alveofact® deserve special mention.

Table 4. Trials of suractant replacement for patients with acute lung injury.

Trial	Surfactant	Method	Treated	Controls	Response	Ref
Uncontrolled	Mod. Natural	Instilled	3		Positive	52
	Surfactant TA®	Instilled	2		Positive	53
	BLES®	Instilled	7		Positive	54
	Alveofact®	Instilled[1]	10		Positive	55
	CLSE®	Instilled	29		Positive	56
Controlled	Curosurf®	Instilled	6	6[2]	Positive	57
	Exosurf®	Aerosolized	35	17	Trend positive	58
	Exosurf®	Aerosolized	33	16	Positive	59
	Exosurf®	Aerosolized	364	361	No effect	60
	Survanta®	Instilled	43	16	Positive	61

[1] Surfactant was delivered by bronchoscopic instillation sequentially to lung segments
[2] Subjects served as their own controls

All but one of the trials listed suggests therapeutic benefit. However, the largest of these trials, a phase III trial of Exosurf® delivered by aerosol, showed no benefit[60]. In this trial, 725 patients were randomized to receive either Exosurf® (13.5 mg DPPC/ml) or placebo (0.45% saline) by continuous aerosol administration for up to five days. Although this trial followed two promising phase II trials, it was unable to showed evidence of benefit on overall survival, survival in groups stratified by injury severity score, days of mechanical ventilation, days in the ICU, days on oxygen, or days to death. This negative result raises several interesting questions.

First, what are the explanations for the failure to see benefit, given that the experience with a variety of surfactants in a variety of clinical settings suggests strongly that there may be therapeutic benefit? Certainly it is possible that the underlying hypothesis may be incorrect, and that surfactant supplementation may not benefit patients with acute lung injury. However, no evidence was presented to support the hypothesis that Exosurf® reached or was retained in the gas-exchanging regions of the lung, and it is highly likely that insufficient surfactant was delivered to those regions. Pilot studies of natural surfactants suggest benefit from instilled doses of approximately 300 mg/kg/day. Patients in the phase III Exosurf trial are likely to have received less than 5 mg/kg/day – less than one fifth the delivery rate associated with detectable effect of aerosolized surfactant in sheep with homogeneous lung injury[50]. However, the lung injury in patients with ARDS is often heterogeneous, a setting in which aerosolized surfactant is delivered preferentially to well-ventilated lung parenchyma and therefore may be relatively ineffective. In addition, Exosurf® has an unfavorable concentration-activity profile[32] and is particularly sensitive to protein inhibition; the latter characteristic suggests it may be of little value in the setting of a high-permeability lung injury such as ARDS.

Secondly, what factors might explain the contrast between the findings in the initial phase II studies, which suggested therapeutic benefit, and the negative phase III study? The simplest explanation is that statistically insignificant trends were not substantiated in the larger phase III trial. However, the search for beneficial therapeutic interventions for patients with ARDS contains many similar stories. Initial promises of an IL-1 receptor antagonist, of anti-TNF, and of PGE_1 have all been lost when those agents have been subjected to phase III trials. It is possible in these trials that selection criteria – either explicit or implicit – are different in the phase II and phase III trials. Whereas benefit may be truly seen by patients in the phase II trials, that benefit is not seen when selection criteria are broadened in phase III trials.

Results of a phase II trial of Survanta® delivered by instillation are more promising[61]. In this trial, patients fulfilling criteria for ARDS were randomized to receive surfactant by endotracheal instillation in addition to standard therapy (43 patients) or standard therapy only (16 patients). Patients received one of three treatment regimens: up to eight doses of 50 mg phospholipids/kg, up to eight doses of 100 mg phospholipids/kg, or up to four doses of 100 mg phospholipids/kg. Surfactant was administered by a catheter at the level of the carina with rotation of the patient to the left or right lateral decubitus positions as different dose aliquots were administered. Mortality of control patients and those receiving 50 mg phospholipid/kg was 44% and 50%, respectively, while patients receiving 100 mg phospholipid/kg for 4 or 8 doses had mortality of 19% and 21%, respectively. Patients receiving 100 mg phospholipid/kg for 4 doses also had a significantly decreased F_IO_2 and p_aO_2/F_IO_2 at 120 hours after treatment. Analysis of BAL fluids obtained prior to instillation of surfactant and at 120 hrs after instillation disclosed an increase in phospholipid content in 120 hr samples from patients receiving high dose surfactant. In addition, a trend toward lower minimum surface tensions was seen in those samples[61]. This study provides support for the hypothesis that surfactant supplementation to patients with acute lung injury may be beneficial and provides rationale for further trials. Such trials have not taken place with Survanta®, however, perhaps because of concern by the manufacturers of natural surfactants that surfactants composed entirely of synthetic components will be of equivalent efficacy and less expensive to produce.

Finally, in a recent uncontrolled trial, a modified natural surfactant (Alveofact) was delivered bronchoscopically to each individual lung segment of patients with ARDS. This novel delivery technique was accompanied by an immediate increase in p_aO_2/F_IO_2 from 85 ± 7 to 200 ± 20 mmHg, mainly due to a decrease in shunt flow. The initial administration of 300 mg phospholipids/kg was followed, in a subset of patients who did not maintain full improvement in oxygenation, by a repeat administration of 200 mg phospholipids/kg. Eight of 10 patients survived the 14 day observation period with progressive improvement in gas exchange[62].

FUTURE DIRECTIONS

Clinical trials have not yet demonstrated convincingly whether or not administration of exogenous surfactant will be of benefit to patients with acute lung injury. Given the industrial commitment necessary to produce sufficient surfactant to mount additional phase II and III clinical trials, it is likely that only a few additional such trials will be forthcoming. Optimal trial design, use of the most favorable surfactant preparation available, and selection of clinically significant endpoints will be necessary. The search for efficacious treatments for acute lung injury has been frustrating; hopefully, administration of exogenous surfactant will prove to benefit substantively the large number of patients who now rely only on supportive care.

REFERENCES

1. R. J. King, Pulmonary surfactant, *J.Appl.Physiol.* 53:1 (1982).
2. M. Hallman, R.G. Spragg, J.H. Harrell, K.M. Moser, and L. Gluck, Evidence of lung surfactant abnormality in respiratory failure: study of bronchoalveolar lavage phospholipids, surface activity, phospholipase activity, and plasma myoinositol, *J.Clin.Invest.* 70:673 (1982).
3. S. A. Shelley, J.U. Balis, J.E. Paciga, C.G. Espinoza, and A.V. Richman, Biochemical composition of adult human lung surfactant, *Lung* 160:195 (1982).
4. S. Yu, G.R. Harding, M. Smith, and F. Possmayer, Bovine pulmonary surfactant: chemical composition and physical properties, *Lipids* 18:522 (1983).
5. J. R. Wright, R.E. Wager, R.L. Hamilton, M. Huang, and J.A. Clements, Uptake of lung surfactant subfractions into lamellar bodies of adult rabbit lungs, *J.Appl.Physiol.* 60:817 (1986).
6. T. R. Korfhagen, M.D. Bruno, G.F. Ross, K.M. Huelsman, M. Ikegami, A.H. Jobe, S.E. Wert, B.R. Stripp, R.E. Morris, S.W. Glasser, C.J. Bachurski, H.S. Iwamoto, and J.A. Whitsett, Altered surfactant function and structure in SP-A gene targeted mice, *Proc Natl.Acad.Sci.U.S.A.* 93:9594 (1996).
7. H. T. Akinbi, J.S. Breslin, M. Ikegami, H.S. Iwamoto, J.C. Clark, J.A. Whitsett, A.H. Jobe, and T.E. Weaver, Rescue of SP-B knockout mice with a truncated SP-B proprotein. Function of the C-terminal propeptide, *J Biol.Chem.* 272:9640 (1997).
8. J. C. Clark, T.E. Weaver, H.S. Iwamoto, M. Ikegami, A.H. Jobe, W.M. Hull, and J.A. Whitsett, Decreased lung compliance and air trapping in heterozygous SP-B- deficient mice, *Am.J Respir.Cell Mol.Biol.* 16:46 (1997).
9. L. M. Nogee, D.E. de Mello, L.P. Dehner, and H.R. Colten, Brief report: deficiency of pulmonary surfactant protein B in congenital alveolar proteinosis, *N.Engl.J Med.* 328:406 (1993).
10. M. Ikegami and A. Jobe, Surfactant metabolism, *Semin.Perinatol.* 17:233 (1993).
11. K. Miyamura, L.E. Leigh, J. Lu, J. Hopkin, A. Lopez Bernal, and K.B. Reid, Surfactant protein D binding to alveolar macrophages, *Biochem.J* 300:237 (1994).
12. H. Manz-Keinke, H. Plattner, and J. Schlepper-Schäfer, Lung surfactant protein A (SP-A) enhances serum-independent phagocytosis of bacteria by alveolar macrophages, *Eur.J Cell Biol.* 57:95 (1992).
13. H. Weber, P. Heilmann, B. Meyer, and K.L. Maier, Effect of canine surfactant protein (SP-A) on the respiratory burst of phagocytic cells, *FEBS Lett.* 270:90 (1990).
14. A. Ahuja, N. Oh, W. Chao, R.G. Spragg, and R.M. Smith, Inhibition of the human neutrophil respiratory burst by native and synthetic surfactant, *Am.J Respir.Cell Mol.Biol.* 14:496 (1996).
15. M. J. Thomassen, J.M. Antal, M.J. Connors, D.P. Meeker, and H.P. Wiedemann, Characterization of exosurf (surfactant)-mediated suppression of stimulated human alveolar macrophage cytokine responses, *Am.J Respir.Cell Mol.Biol.* 10:399 (1994).
16. M. J. Thomassen, J.M. Antal, L.T. Divis, and H.P. Wiedemann, Regulation of human alveolar macrophage inflammatory cytokines by tyloxapol: a component of the synthetic surfactant Exosurf, *Clin.Immunol Immunopathol* 77:201 (1995).
17. C. P. Speer, B. Gotze, B. Robertson, and T. Curstedt, [Effect of natural porcine surfactant (Curosurf) on the function of neutrophilic granulocytes] Einfluss von naturlichem porcinen Surfactant (Curosurf) auf die Funktion neutrophiler Granulozyten, *Monatsschr.Kinderheilkd.* 138:737 (1990).
18. C. P. Speer, B. Gotze, T. Curstedt, and B. Robertson, Phagocytic functions and tumor necrosis factor secretion of human monocytes exposed to natural porcine surfactant (Curosurf), *Pediatr.Res* 30:69 (1991).
19. D. C. Anderson, B.J. Hughes, and C.W. Smith, Abnormal mobility of neonatal polymorphonuclear leukocytes. Relationship to impaired redistribution of surface adhesion sites by chemotactic factor or colchicine, *J Clin.Invest.* 68:863 (1981).
20. A. Suwabe, K. Otake, N. Yakuwa, H. Suzuki, and K. Takahashi, [Effects of surfactant TA on adherence and structure of human peripheral blood neutrophils], *Nippon.Kyobu.Shikkan.Gakkai.Zasshi.* 34:290 (1996).
21. J. M. Antal, L.T. Divis, S.C. Erzurum, H.P. Wiedemann, and M.J. Thomassen, Surfactant suppresses NF-kappa B activation in human monocytic cells, *Am.J Respir.Cell Mol.Biol.* 14:374 (1996).

114

22. W. Chao, R.G. Spragg, and R.M. Smith, Inhibitory effect of porcine surfactant on the respiratory burst oxidase in human neutrophils. Attenuation of p47^{phox} and p67^{phox} membrane translocation as the mechanism, *J.Clin.Invest.* 96:2654 (1995).

23. T. L. Petty, O.K. Reiss, G.W. Paul, G.W. Silvers, and N.D. Elkins, Characteristics of pulmonary surfactant in adult respiratory distress syndrome associated with trauma and shock, *Am.Rev.Respir.Dis.* 115:531 (1977).

24. T. J. Gregory, W.J. Longmore, M.A. Moxley, J.A. Whitsett, C.R. Reed, A.A.I. Fowler, L.D. Hudson, R.J. Maunder, C. Crim, and T.M. Hyers, Surfactant chemical composition and biophysical activity in acute respiratory distress syndrome, *J.Clin.Invest.* 88:1976 (1991).

25. U. Pison, W. Seeger, R. Buchhorn, T. Joka, M. Brand, U. Obertacke, H. Neuhof, and K.P. Schmit-Neuerburg, Surfactant abnormalities in patients with respiratory failure after multiple trauma, *Am.Rev.Respir.Dis.* 140:1033 (1989).

26. U. Pison, U. Overtacke, M. Brand, W. Seeger, T. Joka, J. Bruch, and K.P. Schmit-Neuerburg, Altered pulmonary surfactant in uncomplicated and septicemia-complicated courses of acute respiratory failure, *J.Trauma* 30:19 (1990).

27. W. Seeger, U. Pison, R. Buchhorn, U. Obertacke, and T. Joka, Surfactant abnormalities and adult respiratory failure., *Lung* 168 Suppl:891 (1990).

28. B. A. Holm, G. Enhorning, and R.H. Notter, A biophysical mechanism by which plasma proteins inhibit lung surfactant activity, *Chem.Phys.Lipids* 49:49 (1988).

29. B. A. Holm, L. Keicher, M.Y. Liu, J. Sokolowski, and G. Enhorning, Inhibition of pulmonary surfactant function by phospholipases, *J.Appl.Physiol.* 71:317 (1991).

30. J. Cifuentes, J. Ruiz-Oronoz, C. Myles, B. Nieves, W.A. Carlo, and S. Matalon, Interaction of surfactant mixtures with reactive oxygen and nitrogen species. *J Appl.Physiol* 78:1800 (1995).

31. I. Y. Haddad, S. Zhu, H. Ischiropoulos, and S. Matalon, Nitration of surfactant protein A results in decreased ability to aggregate lipids, *Am.J Physiol* 270:L281 (1996).

32. N. Gilliard, G.P. Heldt, J. Loredo, H. Gasser, H. Redl, T.A. Merritt, and R.G. Spragg, Exposure of the hydrophobic components of porcine lung surfactant to oxidant stress alters surface tension properties, *J Clin.Invest.* 93:2608 (1994).

33. L. G. Fong, S. Parthasarathy, J.L. Witztum, and D. Steinberg, Nonenzymatic oxidative cleavage of peptide bonds in apoprotein B-100, *J.Lipid Res.* 28:1466 (1987).

34. W. Palinski, S. Yla-Herttuala, M.E. Rosenfeld, S.W. Butler, S.A. Socher, S. Parthasarathy, L.K. Curtiss, and J.L. Witztum, Antisera and monoclonal antibodies specific for epitopes generated during oxidative modification of low density lipoprotein, *Arteriosclerosis* 10:325 (1990).

35. R. A. Veldhuizen, L.A. McCaig, T. Akino, and J.F. Lewis, Pulmonary surfactant subfractions in patients with the acute respiratory distress syndrome, *Am.J Respir.Crit.Care Med.* 152:1867 (1995).

36. Lachmann, B. and van Daal, G. J. Adult respiratory distress syndrome: Animal models. Robertson, B., Van Golde, L. M. G., and Batenburg, J. J. Pulmonary surfactant: From molecular biology to clinical practice. (26), 635-663. 1992. New York, Elsevier.
Ref Type: Book Chapter

37. B. Lachmann, B. Robertson, and J. Vogel, In vivo lung lavage as an experimental model of the respiratory distress syndrome, *Acta Anaesthesiol.Scand.* 24:231 (1980).

38. S. Matalon, B.A. Holm, and R.H. Notter, Mitigation of pulmonary hyperoxic injury by administration of exogenous surfactant, *J.Appl.Physiol.* 62:756 (1987).

39. P. C. Engstrom, B.A. Holm, and S. Matalon, Surfactant replacement attenuates the increase in alveolar permeability in hyperoxia, *J.Appl.Physiol.* 67:688 (1989).

40. G. M. Loewen, B.A. Holm, L. Milanowski, L.M. Wild, R.H. Notter, and S. Matalon, Alveolar hyperoxic injury in rabbits receiving exogenous surfactant, *J.Appl.Physiol.* 66:1087 (1989).

41. Y. C. Huang, S.P. Caminiti, T.A. Fawcett, R.E. Moon, P.J. Fracica, F.J. Miller, S.L. Young, and C.A. Piantadosi, Natural surfactant and hyperoxic lung injury in primates. I. Physiology and biochemistry, *J Appl.Physiol* 76:991 (1994).

42. J. F. Lewis, M. Ikegami, and A.H. Jobe, Metabolism of exogenously administered surfactant in the acutely injured lungs of adult rabbits, *Am.Rev.Respir.Dis.* 145:19 (1992).

43. J. D. Harris, F.J. Jackson, M.A. Moxley, and W.J. Longmore, Effect of exogenous surfactant instillation on experimental acute lung injury, *J.Appl.Physiol.* 66:1846 (1989).

44. W. J. Lamm and R.K. Albert, Surfactant replacement improves lung recoil in rabbit lungs after acid aspiration, *Am.Rev Respir.Dis.* 142:1279 (1990).

45. D. Berry, M. Ikegami, and A. Jobe, Respiratory distress and surfactant inhibition following vagotomy in rabbits, *Am.J.Physiol.* 61:1741 (1986).

46. G. J. van Daal, K.L. So, D. Gommers, E.P. Eijking, R.B. Fievez, M.J. Sprenger, D.W. van Dam, and B. Lachmann, Intratracheal surfactant administration restores gas exchange in experimental adult respiratory distress syndrome associated with viral pneumonia, *Anesth.Analg.* 72:589 (1991).

47. E. P. Eijking, D.S. Strayer, G.J. van Daal, R. Tenbrinck, T.A. Merritt, E. Hannappel, and B. Lachmann, In vivo and in vitro inactivation of bovine surfactant by an anti-surfactant monoclonal antibody, *Eur.Respir.J* 4:1245 (1991).

48. J. F. Lewis and A.H. Jobe, Surfactant and the adult respiratory distress syndrome, *Am.Rev.Respir.Dis.* 147:218 (1993).

49. N. Gilliard, P.M. Richman, T.A. Merritt, R.G. Spragg, Effect of volume and dose on the pulmonary distribution of exogenous surfactant administered to normal rabbits or to rabbits with oleic acid lung injury, *Am. Rev. Respir. Dis.* 141;743 (1990).

50. J. Lewis, M. Ikegami, R. Higuchi, A. Jobe, and D. Absolom, Nebulized vs. instilled exogenous surfactant in an adult lung injury model, *J.Appl.Physiol.* 71:1270 (1991).

51. J. Lewis, M. Ikegami, B. Tabor, A. Jobe, and D. Absolom, Aerosolized surfactant is preferentially deposited in normal versus injured regions of lung in a heterogenous lung injury model, *Am.Rev.Respir.Dis.* 145:A184 (1992).

52. Lachmann, B. Surfactant replacement in acute respiratory failure: Animal studies and first clinical trials. Lachmann, B. Surfactant replacement therapy. 212-223. 1987. New York, Springer-Verlag. Ref Type: Book Chapter

53. S. Nosaka, T. Sakai, M. Yonekura, and K. Yoshikawa, Surfactant for adults with respiratory failure, *Lancet* 336:947 (1990).

54. Lewis, J., Dhillon, J., and Frewen, T. Exogenous surfactant therapy in pediatric patients with ARDS. Am.J Respir.Crit.Care Med. 149, A125-A125. 1994. Ref Type: Generic

55. D. Walmrath, A. Gunther, H.A. Ghofrani, R. Schermuly, T. Schneider, F. Grimminger, and W. Seeger, Bronchoscopic surfactant administration in patients with severe adult respiratory distress syndrome and sepsis, *Am.J Respir.Crit.Care Med.* 154:57 (1996).

56. D. F. Willson, J.H. Jiao, L.A. Bauman, A. Zaritsky, H. Craft, K. Dockery, D. Conrad, and H. Dalton, Calf's lung surfactant extract in acute hypoxemic respiratory failure in children, *Crit.Care Med.* 24:1316 (1996).

57. R. G. Spragg, N. Gilliard, P. Richman, R.M. Smith, R.D. Hite, D. Pappert, B. Robertson, T. Curstedt, and D. Strayer, Acute effects of a single dose of porcine surfactant on patients with the adult respiratory distress syndrome, *Chest* 105:195 (1994).

58. J. Weg, H. Reines, R. Balk, R. Tharratt, P. Kearney, T. Killian, D. Scholten, D. Zaccardelli, J. Horton, E. Pattishall, and and the Exosurf-ARDS Sepsis Study Group, Safety and efficacy of an aerosolized surfactant (Exosurf) in human sepsis-induced ARDS, *Chest* 100:137S (1992).

59. H. Wiedemann, R. Baughman, B. deBoisblanc, D. Schuster, E. Caldwell, J. Weg, R. Balk, S. Jenkinson, J. Wiegelt, R. Tharratt, J. Horton, E. Pattishall, W. Long, and and the Exosurf ARDS Sepsis Study Group, A multicenter trial in human sepsis-induced ARDS of an aerosolized synthetic surfactant (Exosurf), *Am.Rev.Respir.Dis.* 145:A184 (1992).

60. A. Anzueto, R. Baughman, K.K. Guntupalli, J.G. Weg, H.P. Wiedemann, A.A. Raventos, F. Lemaire, W. Long, D.S. Zaccardelli, and E.N. Pattishall, Aerosolized surfactants in adults with sepsis-induced acute respiratory distress syndrome, *N.Engl.J.Med.* 334:1417 (1996).

61. T. J. Gregory, K.P. Steinberg, R. Spragg, J.E. Gadek, T.M. Hyers, W.J. Longmore, M.A. Moxley, G.Z. Cai, R.D. Hite, R.M. Smith, L.D. Hudson, C. Crim, P. Newton, B.R. Mitchell, and A.J. Gold, Bovine surfactant therapy for patients with acute respiratory distress syndrome, *Am.J Respir.Crit.Care Med.* 155:1309 (1997).

62. D. Walmrath, A. Gunther, H.A. Ghofrani, R. Schermuly, T. Shneider, F. Grimminger, W. Seeger, Bronchoscopic surfactant administration in patients with severe adult respiratory distress syndrome and sepsis, *Am. J. Respir. Crit. Care Med.* 154;57, (1996).

POTENTIAL ROLE FOR PULMONARY SURFACTANT
IN LUNG TRANSPLANTATION

Fred Possmayer,[1] Richard J. Novick,[2] Ruud A.W. Veldhuizen,[3] John Lee,[2]
David Bjarneson[4] and Jim F. Lewis[3]

[1]Departments of Obstetrics & Gynaecology and Biochemistry and
 MRC Group in Fetal and Neonatal Health and Development
[2]Division of Cardiovascular/Thoracic Surgery, Department of Medicine
[3]Division of Respirology, Department of Medicine
 The University of Western Ontario
 London, ON, Canada, N6A 5A5
[4]BLES Biochemicals, Inc., London, ON, Canada, N5V 3K4

INTRODUCTION

Since its introduction in 1954 by Murray and colleagues, allotransplantation has greatly improved the quality, as well as the duration of life for those involved.[1] The discovery of immunosuppressant drugs advanced long term efficacy of transplantation and led to a further improvement in the quality of life. Unfortunately, maximal potential benefits have never been achieved. As with a number of other organs, particularly kidneys, livers and hearts, there are lengthy waiting lists for suitable lungs. This lack of sufficient organs results, not only from the limited number of donors, but in part because many of the lungs available for transplant are judged unsuitable, due to edema, aspiration or contusion. The short period over which lungs remain clinically viable, presently 6-8 hours, also limits the availability of these organs. Furthermore, the practice of maintaining brain-damaged potential organ donors on respirators using high oxygen levels and vigorous mechanical ventilation contributes to the paucity of available lungs.

The above considerations led to the studies described in the present paper. Initial studies documented alterations in pulmonary surfactant composition and function after short (2 h) and long (12 h) storage prior to transplantation. The observed alterations in the surfactant system prompted further studies which showed that surfactant treatment could alleviate many of the detrimental effects arising from prolonged lung storage, even after close to 40 h cold storage. The observation that exogenous surfactant treatment enhanced lung preservation led to continuing studies on the possibility that surfactant administration could mitigate the effects of lung injury, thereby expanding the potential pool of donor lungs.

PULMONARY SURFACTANT

Pulmonary surfactant is a complex mixture of lipids and proteins that stabilizes the terminal airways, particularly during expiration.[2] Surfactant contains approximately 90%

Acute Respiratory Distress Syndrome: Cellular and Molecular Mechanisms and Clinical Management
Edited by Matalon and Sznajder, Springer Science+Business Media New York, 1998

lipids and 10% surfactant-associated proteins. The major lipid components of surfactant are the disaturated lecithin dipalmitoylphosphatidylcholine (DPPC) (~35%), unsaturated phosphatidylcholine (PC) (~33%), the acidic phospholipids, phosphatidylglycerol and phosphatidylinositol (~15%) and small amounts of phosphatidyletholamine, phosphatidylserine and lyso-bis phosphatidic acid.[3,4] Small amounts of neutral lipids, up to approximately 10% of the total lipids, consisting mainly of cholesterol are also found. The surfactant-associated proteins (SP-) consist of SP-A and SP-D, which are water soluble oligomeric glycoproteins and two low-molecular weight, hydrophobic proteins, SP-B and SP-C, which dissolve readily in organic solvents. The structural and functional properties of surfactant apoproteins have been extensively described in recent reviews.[2,5,6]

Pulmonary surfactant is synthesized in the type II cells of the alveoli and secreted into the alveolar hypophase as dense, tightly packed lamellar bodies.[7] In the hypophase, lamellar bodies convert to tubular myelin, closely packed rectangular tubes which have a lattice-like appearance in cross section. Tubular myelin is thought to be the major source of the surface film. With time, possibly due to repeated compression and decompression, small vesicles are generated, which can be taken up by the type II cells for degradation and recycling into lamellar bodies.[2,8]

Pulmonary surfactant isolated through bronchoalveolar lavage consists of a number of different structural forms. These structural forms can be readily separated into two major subtypes, the large aggregates and the small aggregates, through differential or gradient centrifugation. The large aggregate subtype, which consists of lamellar bodies, tubular myelin and large, multilaminated vesicular forms, can be fractionated by centrifuging 40,000 g for 15 min. The small aggregates, composed of small, mainly unilamellar vesicles, remain in the 40,000 g supernatant. Studies conducted with neonates and radioactive pulse-chase experiments have shown that the large aggregate subtypes act as a precursor of the small aggregate subtype.

LUNG TRANSPLANTATION MODEL

Although certain alterations were necessitated by the nature of the questions being addressed, the following experimental left lung transplantation model was employed throughout the investigations described in this article. Two dogs of approximately the same weight, designated as donor and recipient, were used. The donor's left and right lungs were removed, the pulmonary arteries flushed, normally with Euro-Collins (EC) solution. The lungs were inflated with air to approximately residual functional capacity and stored for various periods depending on the study. After the storage period, the donor right lung was lavaged for surfactant analysis. The donor left lung was transplanted into weight matched recipients.

The recipient's left lung, removed to accomodate the donated lung, served as a control. The recipient dog was then ventilated at 100% oxygen for 6 hours in a supine position while blood gas and other measurements were conducted. At the end of reperfusion, the right pulmonary artery was snared for 10 min by a preinserted heavy silk tie, in order to obtain blood-gas measurements fully dependent on the isolated donor left lung. The animals were then euthanized with an overdose of sodium pentobarbital, the double lung block removed and the transplanted left lung and the recipient's native right lung lavaged for analysis.

SURFACTANT ALTERATIONS DURING LUNG TRANSPLANTATION

Initial experiments investigated the effect of short (2 h) and long (12 h) ischemic intervals on gaseous exchange and surfactant biochemistry.[4] In this, but not in subsequent studies, left and right pulmonary artery snares were applied for 10 min every 2 h so that oxygenation dependent on either lung could be assessed independently. Figure 1 shows that there was little alteration in PO_2 during reperfusion after 2 h ischemia. However, with 12 h ischemia, there was a rapid decline in gaseous exchange such that PO_2 levels were

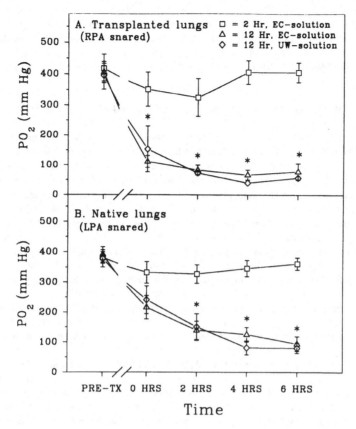

Figure 1. Effect of Storage on PO₂ values on 100 % O₂ with left or right pulmonary artery snared (*PC 0.05 vs 2 h stored lung). (From Veldhuizen et al 1993[4]).

significantly lower than the 2 h stored values immediately after reanastomosis and declined further thereafter. Interestingly, PO₂ levels also declined with the native right lungs, although this was slower than in the transplanted lungs. Similar declines were observed with the standard EC solution flushed lungs as with University of Wisconsin (UW) solution which has a higher potassium concentration.

Examination of pulmonary surfactant from the stored donor right, the transplanted left, the recipient's native right compared to the recipient's excised left lung, which served as control, revealed a number of alterations. No difference was observed in the total phospholipid pools. On the other hand, determination of the small/large aggregate ratio showed a small nonsignificant increase in lavage from transplanted and native right lungs with the 2 h storage and a larger significant increase with the 12 h stored lung groups, except for EC-flushed native right lungs. Compositional analysis revealed a small significant decrease in total phosphatidylcholine, and phosphatidylglycerol in transplanted and native right lung with the 12 h group, compared to surfactant from the control lungs. In addition, there was a tendency for the concentration of phosphatidylethanolamine to rise, while sphingomyelin levels increased significantly in transplanted and native right lung surfactants in the 12 h group.

Further analysis revealed a 50 % or greater fall in SP-A levels in large aggregates from the 12 h, but not the 2 h stored donor, transplanted, and native right lungs relative to control lungs. SP-B levels were not significantly altered. Total protein levels were significantly elevated. Examination of surfactant large aggregates on the pulsating bubble surfactometer revealed small, nonsignificant elevations in surface tension at minimum bubble radius with

119

large aggregates from the 12 h transplanted and native lungs groups. Not surprisingly, lipid extracts from large aggregates retained the ability to reduce surface tension of the pulsating bubble, while lipid extracts of the small aggregates were not very surface active. No differences in lipid extract surface activity were observed between the experimental groups.

EFFECT OF EXOGENOUS SURFACTANT ON LUNG PRESERVATION

The noted similarity between the surfactant alterations included by pulmonary ischemia-reperfusion and the effects reported for acute lung injury [9,10] prompted examination of the possibility that lung preservation times could be extended by administering exogenous surfactant. BLES, the exogenous surfactant used, is a chloroform:methanol extract of neutral surfactant collected from mature cows that has been treated to remove natural lipids.[11] BLES has proven effective in treating neonatal respiratory distress[2] and acute lung injury.[12] Initial studies in which the donor left lungs were treated with BLES (50 mg/kg body weight) immediately after transplantation of 38 h stored donor left lungs revealed a variable response with three of the eight treated animals demonstrating a marked improvement in gaseous exchange compared to controls. In addition, the responders had superior left pulmonary artery perfusion, lower total protein contents and lower small aggregate/large aggregate ratios than nonresponders or the controls.[13]

These promising, albeit variable, results promoted studies attempting to optimize surfactant treatment strategies. Three different conditions; 1) donor treatment with BLES by aerosolization for 3 h prior to graft removal; 2) recipient treatment with instilled BLES at 50 mg/kg; 3) Combined donor and recipient BLES therapy were compared to no treatment control.[14] Figure 2 shows that while recipient BLES treatment had no observable effect, donor aerosol-treated lungs maintained good gaseous exchange. It should be mentioned that the aerosol groups received approximately 35 mg BLES/kg to both lungs prior to graft collection, a dose considerably lower than the 50 mg/kg treatment of the recipients' single transplanted lung. The beneficial effects of donor BLES treatment were further indicated by comparison of transplanted left lung PO_2 values obtained with right pulmonary artery snaring which demonstrated higher PO_2 levels relative to control in the donor aerosol group, although this did not achieve significance. Significantly higher left lung PO_2 values were observed with the combined therapy.

Total lavage protein showed significant increases relative to donor control lungs with the nontreated controls and recipient instilled groups. Protein levels were also elevated in

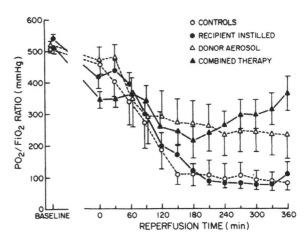

Figure 2. Effect of BLES treatment strategy on PO_2/FiO_2 during reperfusion (From Novick et al., 1996[14]).

recipient native right lungs with donor aerosol but not in the transplanted lungs. Protein levels were not increased in lungs from the combined therapy. The small/large aggregate ratio was elevated in the untreated controls relative to stored donor left lungs. No increases in aggregate ratio were observed with the surfactant-treated groups. It should be noted that this latter effect could be due in part to the fact that BLES is composed of large aggregates.

MITIGATION OF LUNG INJURY BY EXOGENOUS SURFACTANT

The ability of BLES to improve pulmonary function after 38 h storage led to studies examining the possibility that exogenous surfactant could mitigate lung damage.[15] Patients possessing cerebral electrographic patterns incompatible with normal brain function are often maintained on high ventilatory volumes for long periods to maintain function of transplantable organs other than the lungs. Therefore, studies were conducted with dogs subjected to 8 h mechanical ventilation at 45 ml/kg, a volume 3-5 times normal. BLES, 100 mg/kg, was instilled into half of the dogs prior to ventilation. Bronchoalveolar lavage (BAL) samples revealed increased leucocytes, total protein and small/large aggregate ratios in the untreated controls. Furthermore, control animals experienced a decline in PO_2/FiO_2 ratio, which reached significance by 2.5 h with no difference in the BLES group (Figure 3A).

After storage at 4 °C for 17 h, donor left lungs were transplanted into recipients followed by reperfusion for 6 h. Figure 3B shows that the BLES-treated group stabilized at a significantly higher PO_2 than the control group. In addition, this group maintained a significantly lower small/large aggregate ratio. Approximately 65 % of the administered radioactive surfactant was recovered. Of this surfactant 47 % was in the stored donor right lungs, 43 % in the transplanted left lungs while 10 % apparently spilled over into the recipient animals' native right lungs.

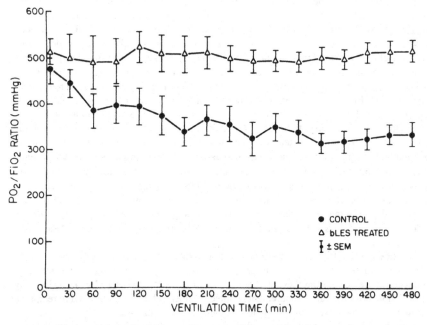

Figure 3/A. Effect of hyperventilation on PO_2/FiO_2 ratio in control and BLES-treated dogs.(Novick et al., 1997[15]).

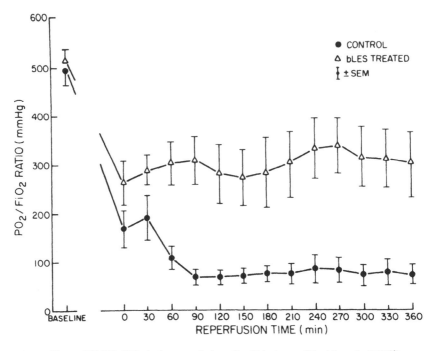

Figure 3/B. PO$_2$/FiO$_2$ during reperfusion after 12 h storage.(Novick et al., 1997[15]). ·

DISCUSSION

The initial studies described in this review documented distinct alterations in the pulmonary surfactant system with 12 h, but not 2 h, storage and 6 h reperfusion. Prolonged lung storage resulted in a rapid deterioration in the ability of the transplanted left lungs to support oxygenation at levels compatible with survival. Interestingly, a marked decline in gaseous exchange occurred with left pulmonary artery snaring demonstrating transplantation of 12 h stored lungs had a pronounced effect on the native right lung. Alterations in surfactant chemistry and surfactant subtype ratio were observed in both lungs. In retrospect, it appears possible that the repeated snaring of left and right pulmonary arteries, thereby subjecting the individual lungs to the entire cardiac output, may have contributed in part to the decline in pulmonary function. Nevertheless, despite this caveat it is evident deterioration in pulmonary function occurred with prolonged lung storage prior to transplant. Prolonged lung storage was also associated with increases in serum protein, which can inhibit surfactant function *in vitro* and *in vivo*. [2,9,16,17] Lung storage resulted in an increase in surfactant small/large aggregate ratio. Since small aggregates possess very poor surface activity, the elevated ratio is indicative of surfactant dysfunction.[2,10] Similar overall alterations in surfactant have been observed after storage with a left lung transplantation model in the rat.[18,19] As indicated earlier, similar alterations occur with the Adult Respiratory Distress Syndrome.[9,10]

Treatment with BLES, a modified natural surfactant, produced improvements in pulmonary function and surfactant properties depending on the particular strategy employed. It is evident from these studies that donor treatment, whether by aerosolization or instillation, prior to organ procurement resulted in superior benefits compared to recipient treatment. Combined donor and recipient treatment produced the best results. Although the mechanisms involved remain unknown, they are likely extremely complex.[20] Nevertheless, it is apparent surfactant treatment somehow prevents induction of an inflammatory cascade in the stored left lungs during reperfusion, thereby protecting the native right lung. Instillation of BLES prior to ventilation proved effective not only in limiting injury due to 8 h mechanical ventilation at 45 ml/kg, but also resulted in satisfactory lung function after 17 h storage. Most of the BLES administered before the insult was still present in the transplanted graft after the 6 h reperfusion period.

An unanticipated finding of these investigations was that recipients' native right lung can be subjected to a marked inflammatory-like response which resembles that in the stored transplanted left lung. The basis of this response needs to be investigated further. The observed effects could be related to cytokines elicited by the stored, transplanted lung with prolonged ischemia prior to reperfusion. It also seems possible that decreased compliance of the stored transplanted lung could result in overextension of the recipients native right lung during the 6 h reperfusion period. Both factors could contribute to the functional decline. Recent studies conducted by Seeger's group in Germany suggest separate ventilation of the lungs post transplantation minimizes the deleterious effects on the recipients' native right lung.[21]

Preliminary studies conducted by our group have shown that delaying BLES installation for 4 h after hyperventilation was initiated did not diminish the salutary effects of surfactant. In these latter studies, there was no decrease in oxygenation during hyperventilation in either delayed-instilled or control animals. The reason for this difference is not known, but it is evident the control lungs were damaged and could no longer support adequate gaseous exchange after transplantation.

CONCLUSION

These studies show pulmonary surfactant was altered during reperfusion following long periods of lung ischemia. Exogenous surfactant greatly improved lung preservation. While donor treatment prior to storage proved superior to recipient treatment, combined donor-recipient treatment was best. Recipient treatment with BLES diminished the deleterious effects of 8 h hyperventilation followed by 17 h storage. While further studies are required, it appears likely surfactant treatment could expand the limited pool of transplantable lungs.

REFERENCES

1. R.P. Lanza, D.K.C. Cooper, and W.L. Chick, Xenotransplantation, *Sci Am.* 277:54 (1997).
2. F. Possmayer, Physicochemical aspects of pulmonary surfactant, in: *Fetal and Neonatal Physiology*, R.A. Poulin and W.W. Fox, eds., WB Saunders Company, Boca Raton (1997). (in press)
3. R.J. King and J.A. Clements, Surface active materials from dog lung. II. Composition and physiological correlations, *Am J Phys.* 223:715 (1972).
4. R.A.W. Veldhuizen, J. Lee, D. Sandler, W. Hull, J.A. Whitsett, F. Possmayer and R.J. Novick, Alterations in pulmonary surfactant composition and activity after experimental lung transplantation, *Am Rev Respir Dis.* 148:208 (1993).
5. L.A.J.M. Creuwels, L.M.G. van Golde and H.P. Haagsman, The Pulmonary surfactant system: biochemical and clinical aspects, *Lung.* 175:1 (1997).
6. J. Johansson, and T. Curstedt, Molecular structures and interactions of pulmonary surfactant components, *Eur J Biochem.* 244:675 (1997).
7. L.M.G. van Golde, J.J. Batenburg and B. Robertson, The pulmonary surfactant system, *Int Union Physiol Sci/Am Physiol Soc.* 9:13 (1994).
8. N.J. Gross, Extracellular metabolism of pulmonary surfactant: the role of a new serine protease, *Annu Rev Physiol.* 57:135 (1995).
9. J.F. Lewis and A.H. Jobe, Surfactant and the adult respiratory distress syndrome, *Am Rev Respir Dis.* 147:218 (1993).

10. J.F. Lewis and R.A.W. Veldhuizen, Factors influencing efficacy of exogenous surfactant in acute lung injury, *Biol Neonate*. 67: 48 (1995).

11. S. Yu, N. Smith, P.G.R. Harding and F. Possmayer, Bovine pulmonary surfactant: Chemical composition and physical properties, *Lipids*. 18:522 (1983).

12. J.F. Lewis, J.S. Dhillon, R. Singh, C.C. Johnson and T.C. Frewen, Exogenous surfactant therapy for pediatric patients with the acute respiratory distress syndrome, *Can Resp J*. 4: 21(1997).

13. R.J. Novick, R.A.W. Veldhuizen, F. Possmayer, J. Lee, D. Sandler, and J.F. Lewis, Exogenous surfactant therapy in thirty-eight hour lung graft preservation for transplantation, *J Thorac Cardiovasc Surg*. 108:259 (1994).

14. R.J. Novick, J. MacDonald, R.A.W. Veldhuizen, J. Duplan, L. Denning, F. Possmayer, A. Gilpin, L.J. Yao, D. Bjarneson and J.F. Lewis.,Evaluation of surfactant treatment strategies after prolonged graft storage in lung transplantation, *Am J Respir Crit Care Med*. 154:98 (1996).

15. R.J. Novick, A.A. Gilpin, K.E. Gehman, I.S. Ali, R.A.W. Veldhuizen, J. Duplan, L. Denning, F. Possmayer, D. Bjarneson and J.F. Lewis, Mitigation of injury in canine lung grafts by exogenous surfactant therapy, *J Thorac Cardio Vasc Surg*. 113:342 (1997).

16. D.S. Strayer, E. Herting, B. Sun and B. Robertson, Antibody to surfactant protein A increases sensitivity of pulmonary surfactant inactivation by fibrinogen in vivo, *Am J Resp Crit Care Med*. 153:1116 (1996).

17. K. Yukitake, C.L. Brown, M.A. Schlueter, J.A. Clements and S. Hawgood, Surfactant apoprotein A modifies the inhibitory effect of plasma proteins on surfactant activity in vivo, *Pediatr Res*. 37:21 (1995).

18. M.E. Erasmus, A.H. Petersen, G. Hofstede, H.P. Haagsman, S.B. Oetomo and J. Prop Surfactant treatment before reperfusion improves the immediate function of lung transplants in rats, *Am J Respir Crit Care Med*. 153:665 (1996).

19. M.E. Erasmus, A.H. Petersen, S.B. Oetomo and J. Prop, The function of surfactant is impaired during the reimplantation response in rat lung transplants, *J Heart Lung Transplan*. 13:791 (1994).

20. Novick, R.J., Gehman, K.E., Ali, I.S., and Lee, J., 1996 Lung preservation - the importance of endothelial and alveolar type II cell integrity, *Ann Thorac Surg*. 62:302.

21. A. Günther, I. Friedrich, F.H. Splittgerber, J. Börgermann, M. Brinkmann, R. Schmidt, M. Yabut, C. Reidemeister and W. Seeger, Attenuation of reperfusion-induced alteration of surfactant properties by application of bovine surfactant extract in a model of single lung transplantation, *Am J Resp Crit Care Med*. 155: A214.(1997).

PREPARATION OF SURFACTANT AND LIPID VECTORS FOR DELIVERY OF PROTEINS AND GENES TO TISSUE

Stephen W. Burgess and Walter A. Shaw

Avanti Polar Lipids, Inc.
700 Industrial Park Drive
Alabaster, AL 35007

INTRODUCTION

Gene therapy, the newest and mostly unexplored frontier of medical research, offers the promise of treatment and perhaps cures for many dreadful afflictions of humanity, many of which are inflicted upon the most innocent of victims, infants and children who are powerless to defend themselves from the devastating effects of disease and illness. While the technology exists to determine the affected gene and manufacture a replacement, a suitable delivery vehicle for the genetic material which allows incorporation into the cell and utilization by the cellular machinery to manufacture the gene product has yet to be developed. Currently there are two approaches to this problem, commonly grouped into the classes referred to as viral vectors, modified viral particles capable of infecting cells and delivering genetic material, and non-viral vectors, primarily cationic amphiphilic lipid-like molecules which complex naked genetic material and facilitate its transport across the cell membrane. Viral vectors, while successful at delivering genetic material to cells, presents a number of severe limitations. Many of these limitations can be circumvented by the use of non-viral vectors, although the efficiency of gene delivery is dramatically decreased. This article will concentrate on discussing the factors affecting the handling and use of surfactant and lipid vectors, as well as the preparation of lipid suspensions for the delivery of proteins and genetic material to cells. We will conclude with a brief discussion of an alternative to the classical approaches to protein and gene delivery.

PROPERTIES OF LIPIDS AFFECTING PARTICLE FORMATION

The major class of non-viral vectors are surfactant or amphiphilic molecules possessing a single or multiple positive charges. Amphiphilic compounds are distinguished by the general structural feature of having regions within the molecule with different solubilities. Generally, they contain a hydrophobic region composed of a non-polar hydrocarbon chain, and a hydrophilic region containing the charged polar group(s). This duality of "water loving" and "water hating" regions establishes the driving force for formation of three dimensional

Acute Respiratory Distress Syndrome: Cellular and Molecular Mechanisms and Clinical Management
Edited by Matalon and Sznajder, Springer Science+Business Media New York, 1998

structures when the compounds are dispersed in water. These structures can be grouped into two major classes, micelles and bilayers. Micelles, usually formed by single chain surfactants whose polar region is larger than the non-polar region, are characterized by there relatively small size and hydrocarbon core with a hydrophilic surface. Bilayers, typically composed of double chain surfactants whose polar region is similar in size to the non-polar region, are distinguishable from micelles since they contain a hydrophobic layer with a hydrophilic surface on each side. This structure is unstable due to the exposed hydrocarbon at the ends of the layer, and will close upon itself to form a lipid vesicle having a trapped internal aqueous compartment.

Polar (Headgroup) Region Properties

Geometry and composition of polar and non-polar regions dramatically affects the size and type of particles formed. The polar or headgroup region of the molecule plays a primary role in defining molecular shape, which predisposes the compound to a particular particle type. Lipids with large headgroups relative to the hydrocarbon chains (*e.g.*, soaps, detergents, and lysophospholipids) are generally described as inverted cones and typically form micelles, while lipids with smaller headgroups relative to the hydrocarbon chains (*e.g.*, phosphatidyl-ethanolamine, phosphatidic acid, cholesterol) are described as cones and form inverted micelles. Lipids with headgroups having cross sectional areas equivalent to the hydrocarbon region (*e.g.*, phosphatidyl-choline, -serine, -inositol, sphingomyelin) are described as cylinders and form lamellar phases or bilayers.

In addition to this function, the lipid headgroup influences the surface characteristics of the lipid particles. Charged headgroups affect the interaction of lipid particles with each other, as well as with other biological components (*e.g.*, complexation of DNA by cationic lipids). The presence of charged functional groups in the headgroup region whose ionized state is pH dependent (pK_a) can induce physical state changes in response to pH changes. Some headgroups are capable of extensive hydrogen bonding, establishing a hydration layer (layer of ordered water molecules) surrounding the lipid particle which can hinder the approach of other particles.

Non-Polar (Hydrocarbon) Region Properties

The non-polar or hydrocarbon region of the compound is crucial to the stability of the particle and integrity of the membrane. Shorter hydrocarbon chains allow the polar headgroup region to dominate. The molecules are usually more water soluble, with the critical micelle concentration (CMC) increasing as the length of the hydrocarbon chain decreases. Consequently, for more stable particles, longer hydrocarbon chain lengths are desirable (carbon lengths ≥14). Size and saturation of the hydrocarbon chain influences the transition temperature of the lipid, which affects the formation of the lipid particle. The phase transition temperature is defined as the temperature required to induce a change in the lipid physical state from the ordered gel phase, where the hydrocarbon chains are fully extended and closely packed, to the disordered liquid crystalline phase, where the hydrocarbon chains are randomly oriented and fluid.[1] There are several factors which directly affect the phase transition temperature including hydrocarbon length, unsaturation, charge, and headgroup species. As the hydrocarbon length is increased, van der Waals interactions become stronger requiring more energy to disrupt the ordered packing, thus the phase transition temperature increases. Likewise, introducing a double bond into the acyl group puts a kink in the chain which requires much lower temperatures to induce an ordered packing arrangement. Using high transition lipids produces lipid particles with highly ordered, non-leaky, impermeable, and generally non-fusogenic membranes that may not be readily taken up by cellular systems. Alternatively, low transition lipids are fluid, release their contents more readily, and may more easily be taken up by cellular systems. By varying the composition of the hydrocarbon region, one has the ability to dramatically alter the kinetics of uptake and delivery of components into the cell.

PREPARATION OF LIPID SUSPENSIONS

The general method for preparation of lipid suspensions includes: 1) dissolution and mixing of lipids in organic solvent, 2) drying of lipid solution, 3) hydration of lipid film, and 4) sizing of lipid suspension to provide a homogenous population of particles. We will describe the procedure and factors affecting each step in the process in the sections below.

Mixing of Lipids

Lipids suspensions are prepared by hydration of a lipid powder or film using distilled or deionized water, or a suitable buffer. When using a single lipid component system containing a long, saturated hydrocarbon chain, the lipid powder can be hydrated directly. For multi-component systems or unsaturated lipids, a solution of the lipid should be prepared in a suitable organic solvent. Unsaturated lipids are hygroscopic and cannot be aliquoted efficiently without moisture uptake and significant loss of material on transfer. An organic solution of the material allows accurate transfer without the inherent problems associated with hygroscopic powders. Lipid preparations composed of two or more lipids should be mixed in an organic solvent to provide a homogenous lipid mixture for hydration. The practice of combining aqueous dispersions of separate lipid components to yield a population of lipid particles containing a homogenous distribution of lipids should be avoided. This technique produces a mixture of multiple populations containing separate lipid components.

Lipid components should be dissolved in an organic solvent at a concentration of 10-100mg/mL and mixed in a glass, stainless steel, or teflon container, until a homogenous solution is obtained. Typical solvents used for dissolving lipids are chloroform, methylene chloride (dichloromethane), methanol, ethanol, and diethyl ether. The most versatile of these solvents is chloroform or mixtures of chloroform and methanol (C:M, 2:1 or 1:1, v/v). Most lipids are completely soluble in these solvents at concentrations ≥ 100mg/mL. When working with organic solutions of lipids, transfer and store organic solutions using glass, stainless steel or teflon. Organic solutions should never be stored in polymer or plastic containers (polystyrene, polyethylene, polypropylene, etc.) as this will leach impurities out of the container.[2]

Drying of Lipids

The lipid solution should be transferred to a glass vessel for drying. Removal of organic solvent is necessary before the aqueous phase can be added to the lipid for hydration. The vessel should be designed to maximize the surface area of the lipid film produced. Round bottom evaporation flasks are suitable for removal of organic solvent using a rotary evaporator with mild heating (35-40°C). After the solvent has been removed, the evaporation system should be backfilled with nitrogen or argon to prevent oxidation of unsaturated lipids and moisture uptake. Alternatively, if the sample is too small to efficiently dry by rotary evaporation, the sample may be dried by evaporating the solvent using a stream of nitrogen or argon.

Once the organic solvent has been removed, two procedures have been utilized to prepare the sample for hydration. The first of these procedures involves redissolving the lipid film using an organic solvent that will freeze, followed by lyophilizing the lipid solution using a high vacuum system. Suitable solvents for lyophilization include t-butyl alcohol (or mixtures of t-butyl alcohol and water), cyclohexane, and cyclohexane:ethanol (97:3, v/v). The lipid solution should be <300mg/mL for freezing and lyophilization. The lipid solution can be frozen in a thin shell using a dry ice/ethanol or dry ice/acetone bath, or frozen in a solid block by placing the lyophilization vessel on dry ice until completely frozen (2 hours or longer depending on volume). If the solution is frozen in a solid block, the block should not be thicker than its diameter to prevent thawing during lyophilization. The vacuum should be <1000mtorr during the lyophilization to prevent thawing and insure thorough drying. The sample should be allowed to dry for a minimum of 12 hours to a maximum defined by the size and dryness of the sample (typically 72-96 hours). This method is generally useful for lipid samples that form a white powder upon lyophilization. Lipid samples which form an oily or gummy film should be handled according the second procedure described below. White powder samples are easily suspended in aqueous medium for hydration and typically form smaller particles upon hydration.

The alternative method to the organic solvent lyophilization procedure involves placing the evaporation vessel used to remove the organic solvent on a high vacuum lyophilization system for a minimum of 12 hours to remove residual organic solvents, producing a dry lipid film or residue. Hydration of the lipid film formed by this procedure is sometimes difficult and typically yields particles with a larger mean diameter.

Regardless of the procedure employed for final drying of the lipid sample, the vessel containing the lipid should be backfilled with nitrogen or argon to prevent oxidation and either hydrated immediately or sealed and stored frozen at -20°C.

Hydration of Lipids

Solutions used for hydration of lipids include distilled or deionized water, buffer solutions, and sodium chloride (saline) or sugar solutions. For *in vivo* applications, the osmolality of the solution should closely match the physiological osmolality of 290 mOsm/kg. Typical solutions which match this criteria are 0.9% saline, 5% dextrose, and 10% sucrose. The amount of hydration solution added to the lipid film should be adjusted to achieve a final lipid concentration of 5-50mM. If the concentration of the lipid suspension needs to be diluted, use an isotonic solution (typically the hydration solution). If the hydration solution and the dilution solution differ greatly in osmolarity, the membranes could swell, causing the vesicles to rupture, or collapse, depending on the osmolarity gradient created.

The most important detail necessary for successful hydration and formation of lipid particles is that the temperature of the hydration buffer must be above the transition temperature of the lipid (see above for discussion of transition temperature). The lipid must be in the liquid crystalline or fluid phase to properly form lipid particles. Lipids hydrated below their transition temperature form partially hydrated lipid aggregates that do not disperse efficiently in the suspension. When using high transition lipids (lipids having a transition temperature above room temperature), warm the buffer solution above the transition temperature of the lipid before adding to the lipid film or powder. For multi-component systems, warm the buffer solution above the transition temperature of the lipid in the mixture having the highest transition temperature.

The warmed hydration solution is added to the lipid film in the drying vessel and vigorously agitated during hydration. The lipid/buffer suspension should be kept above the transition temperature of the lipid during the entire period of hydration. For high transition lipids, the most convenient method for accomplishing hydration at elevated temperatures is to hydrate the lipid film in a round bottom rotary evaporation flask, attaching the flask to the rotary evaporator and spinning the flask in a water bath at a temperature above the transition temperature of the lipid. Small samples can be hydrated by vortexing the solution periodically during the hydration period.

Hydration periods vary among lipids but can be generalized to approximately 60 minutes. Aging of hydrated lipid suspensions for several hours or overnight can ease the task of downsizing the particles. At this point in the procedure, LMV or large, multilamellar vesicles are formed. This structure is characterized by its onion-like features, having several lipid layers separated by aqueous compartments in a single particle. If this the desired particle size and structure required (see Table 1), store the lipid suspension in a refrigerator (4°C) until used. If a smaller particle size is required, follow the procedures outlined in the following section.

Sizing of Lipid Suspensions

Lipid suspensions can be transferred to a flat bottomed tube or bottle and sonicated to form SUV or small, unilamellar vesicles with particle sizes <50nm. Sonication is generally carried out in a bath sonicator, although many prefer the higher power generated by probe sonicators. However, probe sonicators have a number of disadvantages: 1) they generate higher heat which can lead to rapid hydrolysis of the lipid, 2) they release metal ions (titanium) into the solution, and 3) they require larger volumes. SUV's can also be formed spontaneously by injecting an organic solution of the lipid directly into a vigorously stirring aqueous phase held at a temperature above the lipid transition temperature. The organic solvent used is typically ethanol since it is miscible with water. Other water miscible organic solvents could be substituted provided the lipid is completely soluble. Once the lipid solution has been injected, the solvent is removed by evaporation, dialysis, or diafiltration. This technique suffers from the inherent problems associated with complete removal of the organic solvent.

Large, unilamellar vesicles or LUV can be formed by a number of methods. The currently popular method involves forcing the lipid suspension through filters with defined pore sizes to yield particles similar in size to the pore size used. Briefly, the LMV suspension is prefiltered through a 1μm pore size, followed by 5 times through a 0.2-0.4μm membrane, and finally 5-10 times through a 0.1μm membrane. This yields particles having sizes of 110-140nm. Smaller sizes can be obtained although higher curvature particles (<80nm) are inherently unstable and will spontaneously fuse at temperatures below their transition temperature to form larger particles. Suspensions of particles formed by this extrusion technique have a narrow size distribution.

Another class of LUV can be formed by solubilizing the lipid using a detergent solution. The particles formed are mixed micelles of lipid and detergent. Larger particles are formed spontaneously as the detergent is removed by dialysis, gel filtration, or detergent absorbing beads. The population of vesicles formed has a broad size distribution and the membranes typically contain residual detergent which affects the membranes physical properties.

Finally, lipid suspension can be forced at high pressures through a small orifice, colliding with a wall, small ball, or tip of a pyramid. This technique, known simply as homogenization or microemulsification, produces particles of a defined size, however, the size cannot be accurately predetermined by adjusting instrument parameters.

Table 1. Typical types and sizes of lipid vesicles

Vesicle type	Size (μm)	Size (nm)	Size (Å)
SUV	0.05	50	500
LUV	0.12	120	1,200
LMV	5.00	5,000	50,000

STORAGE AND STABILITY OF LIPID SUSPENSIONS

Once lipid particles have been formed, maintaining the physical properties of the particles can be difficult. Size distribution can change on storage due to degradation of the components. Permeabilization of the membrane can lead to leakage of encapsulated material. Stability issues due to hydrolytic degradation is a general problem with lipid products. Aqueous formulations of drug products tend to be less stable since the presence of excess or bulk water leads to rapid hydrolytic degradation in lipid preparations.[3,4,5,6]

After the sizing process is complete, lipid suspensions should be stored at close to pH 7 as possible. Lipids containing ester-linked hydrocarbon chains are susceptible to acid and base hydrolysis.[3] Hydrolysis rate is dramatically affected by temperature,[3,5] therefore lipid suspensions should be kept refrigerated during storage. Lipid suspensions should not be frozen if possible since the freezing process could fracture or rupture the vesicles leading to a change in size distribution and loss of internal contents. The use of cryoprotectants such as dextrose, sucrose, and trehalose may increase stability from hydrolysis. Also, samples may experience oxidation upon storage. The addition of small amounts of antioxidants during processing may stabilize the suspension and limit oxidation of the product.

NON-VIRAL STRATEGIES FOR PROTEIN AND GENE DELIVERY

The emerging technology of gene therapy, while offering the promise of genetic defect treatment or cure, requires a breakthrough to produce a scientifically and commercially viable alternative to conventional drug therapy. Viral-based vectors, currently the most efficient DNA delivery vehicle, suffer several weaknesses, including low packaging capacity and immune reactions that prevent repeat dosing. Alternative approaches involving non-viral vectors such as liposomes may circumvent some of these issues, but their transfecting efficiency does not achieve that of the viral systems. A new class of cationic vectors seeks to overcome the deficiencies of liposomes, including toxicity and serum inhibition, while matching the DNA delivery efficiencies of viral-based systems. The polycationic amino polymer, known as PolyCat 57, was introduced by Avanti Polar Lipids, Inc., a leader in lipid-based products for pharmaceutical applications. The *in vivo* feasibility study, in which the polymer was compared to a liposomal and viral vector for delivery of a model gene to a human brain tumor in mice, was conducted by researchers in the gene therapy program at the University of Alabama at Birmingham.[7] The study, reported in *Nature Biotechnology*, demonstrates the equivalency of the polymer and recombinant adenoviral vector to deliver a plasmid to the tumor. The study also demonstrates the overwhelming superiority of the polymer to a commercially available liposome carrier, both in *in vitro* and *in vivo* transfection efficiency and resistance to serum inhibition. In addition, the polymer offers advantages of water solubility, chemical stability, and low toxicity.

REFERENCES

1. D.M. Small. *Handbook of Lipid Research: The Physical Chemistry of Lipids, From Alkanes to Phospholipids*, Vol. 4, Plenum Press, New York (1986).
2. C. Pidgeon, G. Apostol, and R. Markovich, Fourier transform infrared assay of liposomal lipids, *Anal. Biochem.*, 181:28 (1989).
3. S. Frøkjaer, E.L. Hjorth, and O. Wørts, Stability and storage of liposomes, in: *Optimization of Drug Delivery*, H. Bundgaard, A. Bagger Hansen, and H. Kofod, eds., Munksgaard, Copenhagen (1982).

4. C.R. Kensil and E.A. Dennis, Alkaline hydrolysis of phospholipids in model membranes and the dependence on their state of aggregation, *Biochemistry*, 20:6079 (1981).

5. M. Grit, J.H. de Smidt, A. Struijke, and D.J.A. Crommelin, Hydrolysis of phosphatidylcholine in aqueous liposome dispersions, *Int. J. Pharm.*, 50:1 (1989).

6. M. Grit, N.J. Zuidam, and D.J.A. Crommelin, Analysis and hydrolysis kinetics of phospholipids in aqueous liposome dispersions, in: *Liposome Technology: Liposome Preparation and Related Techniques*, Vol. 1, 2nd edn, G. Gregoriadis, ed., CRC Press, Ann Arbor (1993).

7. C.K. Goldman, L. Soroceanu,N. Smith, G.Y. Gillespie, W. Shaw, S. Burgess, G. Bilbao, and D.T. Curiel, In vitro and in vivo gene delivery mediated by a synthetic polycationic amino polymer, *Nature Biotechnology*, 15: 462 (1997).

RECOMBINANT SP-C BASED SURFACTANT CAN IMPROVE LUNG FUNCTION IN SALINE LAVAGED SHEEP.

Ruud Veldhuizen, Lynda McCaig, Li-Juan Yao, Carolyn Kerr, Yushi Ito, Jaret Malloy, Jim Lewis.

Lawson Research Institute,
The University of Western Ontario,
London Ont. Canada.

Clinical trials testing exogenous surfactant administration for the treatment of the Acute Respiratory Distress Syndrome (ARDS) have given variable results. Two factors contributing to the variable responses of this treatment are the specific exogenous surfactant preparation used and the method by which the surfactant is delivered to the injured lung. The current study investigated the potential of a new exogenous surfactant, recombinant SP-C based surfactant (rSP-C, Byk Gulden, Germany), to improve oxygenation in a large animal model of ARDS. The surfactant was delivered to the injured lungs by either tracheal instillation, bronchoscopic instillation or aerosolization. Lung injury was induced in adult sheep by repetitive saline lavage and subsequent mechanical ventilation for 1 hour. Two separate experiments were performed. In the first experiment, animals were then randomized into three different surfactant-instillation groups; 1) tracheal instillation of 100mg/kg; 2) tracheal installation of 25 mg/kg; 3) bronchoscopic installation of 25 mg/kg. The second experiment was performed to evaluate the efficacy of rSP-C surfactant when delivered as an aerosol. The aerosol rSP-C surfactant was delivered using a small catheter placed at the distal end of the endotracheal tube (Trudell Medical, London Ont) and was timed to deliver surfactant during inspiration only. This group of animals was compared to animals receiving air through a similar catheter. In both experiments the r-SP-C product was labeled with $[^{14}C]$- dipalmitoyl phosphatidylcholine to assess the recovery and lobar distribution patterns of the exogenous surfactant. Blood gases were monitored at 30 min intervals for four hours after instillation or after the start of aerosolization. After the four hour monitoring period, animals were euthanized by an overdose of pentobarbitol and the lung were processed to examine the recovery and lobar distribution patterns of the exogenous material.

The results of the first experiment showed that rSP-C based surfactant significantly improved PO2 values in all groups receiving instilled surfactant when compared to pre-treatment values (group 1, 87±6 vs 374±35, group 2, 89±8 vs 258±39 group 3, 73±10 vs 216±48 mmHg, pre- vs 4h post-treatment) . Comparisons among groups revealed that all

Acute Respiratory Distress Syndrome: Cellular and Molecular Mechanisms and Clinical Management
Edited by Matalon and Sznajder, Springer Science+Business Media New York, 1998

133

treatment groups had significantly higher PO2 values than non treated controls at 4 hours post treatment. Tracheal instillation of 100mg/kg resulted in significantly higher PO2 values compared to tracheal instillation of 25 mg/ml. There was no significant difference between the PO2 values obtained after bronchoscopic instillation compared to tracheal instillation. Surfactant analysis after the treatment period revealed that the percentage of the radioactivity recovered were not significantly different among the three instillation groups. Lobar distribution patterns were not significantly different between tracheal and bronchoscopic instillation. It should be noted however that the bronchoscopic instillation technique took significantly longer than the tracheal instillation 12.6min ±1.7 vs 28.3min ± 4.5, tracheal vs bronchoscopic instillation).

The results of the second experiment showed that aerosolized rSP-C surfactant also significant increased PO2 values during the four hour treatment period (80±6 vs 298±69, pre- vs 4h post-treatment). PO2 values in this group after four hours of treatment were significantly higher than control animals (135±20 vs 298±69, control vs aerosol). The total amount of surfactant aerosolized in this group was 28 ± 2 mg phospholipid/kg, surfactant analysis after sacrifice revealed that 58% of the aerosolized material was recovered from the lung. The lobar distribution of the exogenous material delivered by aerosolization was similar to the distribution observed with tracheal and bronchoscopic instillation.

We conclude that utilizing several different treatment strategies, rSP-C based surfactant was effective in improving oxygenation in a large animal model of ARDS. These results warrant further clinical studies to test the use of this surfactant in the treatment of ARDS. In this animal model there was no difference in the efficacy or in the lobar distribution patterns of the exogenous surfactant when comparing tracheal and bronchoscopic instillation. Since bronchoscopic instillation is significantly more time consuming and more invasive we suggest that tracheal instillation should be employed when instilling this surfactant into an injured lung. Aerosolization of rSP-C based surfactant was also effective in this model. With this method of delivery surfactant is administered slowly to the injured lung. The use of this catheter-aerosolization technique may be at early stages of lung injury or when giving surfactant to a specific area of the lung.
(Funded by: Ontario Thoracic Society and Byk Gulden, Germany)

GENE THERAPY FOR THE ACUTE RESPIRATORY DISTRESS SYNDROME

Phillip Factor, D.O.

Pulmonary and Critical Care Medicine,
COLUMBIA/Michael Reese Hospital
Assistant Professor of Medicine
University of Illinois at Chicago
Associate Director, ICU
314 Kundstader
2929 S. Ellis
Chicago, Il 60616

INTRODUCTION

Gene therapy has been the focus of much attention and expectation in recent years. Many conditions have been identified as being likely targets for gene therapy, most of which are either heritable disorders or cancers. Recently acquired disorders, including acute lung diseases, have been identified as potential applications for gene therapy. Gene therapy has evolved from a concept to clinical application via phase I safety studies in humans. Due to current limitations of gene transfer technology little headway has been made in terms of true impact on human disease. Principal limitations of current technology have included lack of prolonged expression of transgenic proteins at physiologically relevant levels in an adequate number of target cells. Despite these stumbling blocks, optimism for gene therapy continues due to the belief that it will create new avenues for therapies that are not currently possible or imaginable. Gene therapy research continues to thrive due to its unrealized potential and with the hope that it will some day be common place therapy for a large number of conditions.

Gene therapy conjures images of replacement of defective genes for treatments of heritable diseases such as adenosine deaminase deficiency, cystic fibrosis or α_1 antitrypsin deficiency. A much broader area of gene therapy research centers on the treatment of cancers. Gene transfer in an effort to reconstitute tumor suppressor gene function or to deliver intracellular cytotoxic agents are but 2 examples of such research (1,2). However, few investigators consider acute illnesses as targets for gene therapy. Acquired diseases may be a much better fit for gene therapy, they may require short term expression, in a limited site, not requiring gene transfer to every cell. Thus, gene therapy for acute, short-term acquired conditions is a unique, little explored application for this future therapy. Only a single conference dedicated to gene therapy for acquired disorders has been held to date[a].

---

[a] Gene Therapy for Acquired Diseases, Nashville, TN. October 19-21, 1995.

The acute respiratory distress syndrome (ARDS) could be an excellent target for the use of gene therapy. It is an acute disorder that requires short-term expression of a gene product and it involves an organ that is very amenable to gene delivery. Little attention has been assigned to this application of gene therapy, to date there have been no human trials of gene therapy for ARDS. Fortunately, the literature is ripe with examples of gene transfer that may be applicable to the treatment of acute lung injury. Table 1 lists potential applications for gene therapy to the treatment of ARDS. This list is not exhaustive, virtually any pharmacologic agent could be delivered via gene transfer.

Table 1: Potential Gene Therapy Applications for the Treatment of ARDS

Therapeutic Goal	Gene Transfer Candidates	Proposed Mechanism
Pulmonary Edema Clearance	Na^+ channel, Na^+,K^+-ATPase	Accelerate edema clearance
Anti-protease therapy	α_1 anti-trypsin	Limit proteolytic & chemoattractant effects of proteases
Anti-oxidants	SOD, Catalase, Glutathione Peroxidase	Limit oxidant induced lung injury
Intracellular Adhesion Molecules	Integrins	Repair alveolar permeability
Anti-coagulants/Fibrinolytics	Urokinase, Thrombomodulin, tPA, Thrombin Receptor, Anti-thrombin III	Treatment of capillary thrombosis
Vasodilators	Prostaglandin G/H Synthetase iNOS, ceNOS, Guanylate Cyclase	Limitation of hypoxemia & pulmonary hypertension
Growth Factors	KGF	Stimulate epithelial cell proliferation
Inhibitors of Fibrosis	IFg, HGF	Limit repairative fibrosis
Anti-endotoxin	Intracellular anti-endotoxin antibodies	Limitation of lung injury/WBC activation

GENE TRANSFER VEHICLES

Gene transfer is the delivery of genetic material intended to alter or augment ordinary cellular function. Although RNA is sometimes used, most gene therapy vehicles transfer double stranded DNA into mammalian cells. This DNA is then transcribed into mRNA and subsequently translated into protein. Gene therapy is the use of gene transfer techniques to introduce genetic sequences into specific cells of a patient in order to achieve some clinical benefit. An ever increasing variety of gene delivery vehicles are available. Most gene transfer is achieved using either DNA-cationic liposome complexes or genetically modified viruses such as retroviruses or human adenoviruses. A recent review of current gene transfer vehicles has been published (3).

Liposomes were among the first available gene transfer vehicles. These cationic molecules complex with DNA and RNA and have the ability to enter cells via several pathways. They are relatively non-toxic to eukaryotic cells however their gene transfer (transduction) efficiency is generally lower than that observed with viral vehicles. Newer formulations are showing promising improvements in transfer efficiency, additionally, liposomes can be complexed with ligands that can specifically bind to cell surface receptors conferring upon them some degree of cell targeting specificity (4).

Many species of human and primate viruses are being tested for use as gene transfer vehicles. Most studies have employed either retroviruses or human adenoviruses that have been genetically modified to prevent in vivo replication. Retroviruses are single stranded RNA viruses that have been used in human studies. They offer the unique advantage of being able to incorporate into the host's genome. This could allow for transgene expression for the entire life of a cell, or if a stem cell were infected, the entire life span of the cell and its progeny. Unfortunately these RNA viruses have several important limitations; they accommodate limited amounts of DNA (generally <6kb), they infect only replicating cells, and they incorporate into the genome of the host raising fears of proto-oncogene activation and malignant transformation (3,5). This final limitation has dampened enthusiasm for these viruses.

Replication incompetent adenoviruses are double-stranded DNA viruses are frequently used gene transfer vehicles (3). They have been used in phase 1 human gene therapy protocols in patients with cystic fibrosis. (6) They can be produced in high and pure quantities, they infect non-replicating cells, they do not insert into the host genome and they can carry more than 8kb of exogenous DNA. Their major limitations are limited duration of transgene expression and substantial inflammatory effects following infection. Post-infection antibody formation also limits the efficacy of repeat administration, thus currently available adenoviruses are less than ideal choices where long term transgene expression is required. Newer, less inflammatory adenoviruses vectors combined with immunomodulation to limit anti-viral host responses continue to fuel interest in these vectors for human gene therapy of heritable/genetic diseases (3,7,8).

Given that no ARDS specific gene therapy trials have been conducted, the intent of this review is to highlight several avenues of ongoing investigation that might play a role in the treatment of ARDS in the future. Below are examples of potential applications of gene therapy to ARDS that exploit the known pathophysiology of ARDS. They demonstrate the uniqueness of gene therapy and may not be applicable without it.

POTENTIAL APPLICATIONS OF GENE THERAPY TO ARDS

Nitric Oxide (NO) Expression

Substantial interest has been directed toward the use of NO for the treatment of hypoxemia and pulmonary hypertension in patients with ARDS (9). Thus far human studies using NO in humans with ARDS have shown promising physiologic responses but no change in outcome. In animal models, NO production is reduced under hypoxic conditions and thus may contribute to pulmonary hypertension, right ventricular dysfunction and abnormal V/Q relationships in patients with ARDS. Janssens and colleagues have postulated that augmentation of NO production might attenuate these observations (10). They have recently published the results of studies that employed a replication deficient adenovirus to deliver the gene for constitutive endothelial nitric oxide synthetase (ceNOS) to the alveoli of rats. Four days after infection they showed increased l-arginine to l-citrulline conversion, an 86% increase in l-NAME inhibitable l-arginine conversion, and a 10 fold increase in cGMP production. These data demonstrate that their virus increased ceNOS expression and function. To test for physiologic impact, they measured changes in mean pulmonary artery pressure and resistance during 25 minutes of ventilation with a hypoxic gas.

Under these acute hypoxemic conditions ceNOS expressing rats had markedly attenuated changes in mPAP and PVR. Thus they were able to demonstrate that adenoviral mediated overexpression of ceNOS could limit pulmonary hypoxic vasoconstriction in these otherwise normal rats. Subsequent unpublished data have been generated following adenoviral mediated gene transfer of the inducible form of NO synthase (iNOS)[b] . This isoform is calcium and calmodulin independent and is capable of producing greater levels of NO than ceNOS. Surprisingly, this vector did not show any further attenuation of vasoreactivity. Additional studies were conducted using a markedly different approach. These investigators engineered a virus that expresses the receptor for NO; soluble guanylate synthase. NO stimulation of this enzyme increases intracellular cGMP levels in smooth muscle cells and decreases free

---

[b] Data presented at the 1997 American Thoracic Society Meeting, San Fancisco, CA., May 17-21, 1997.

intracellular calcium thereby causing smooth muscle cell relaxation. Subsequent studies using this virus were able to show further reductions in hypoxic pulmonary artery vasoreactivity, greater than that seen with ceNOS and iNOS. This is an excellent example of the applicability of gene therapy for ARDS, overexpression of a molecule to modulate an intracellular pathway to achieve a physiologic response. It is unlikely that other delivery mechanisms would be able to deliver high concentrations of a large molecule to the intracellular space.

Surfactant

Ineffective or insufficient levels of surfactant at the alveolar air-epithelium interface contributes to of acute hypoxemic respiratory failure (11). Several studies have tested the utility of delivering surfactant alveolar airspace of humans. Positive impact has been noted in premature babies with respiratory distress syndrome. Two prospective studies have tested surfactant in adults with ARDS (12). Transient improvements in oxygenation and lung compliance were noted. These studies required multiple daily doses of surfactant combined with position changes during endotracheal or bronchoscopic installation of exogenous surfactant. Unfortunately, they did not impact on patient outcome and the cost of treatment for a 70kg adult would be several hundred thousand dollars. For these reasons the enthusiasm for the use surfactant in ARDS has been dampened (12). Gene therapy potentially could overcome the limitations of previous studies where delivery and cost made this potential therapy impractical. Gene transfer could be employed to create a low cost, efficient mechanism to produce excess surfactant at the site where it is needed.

In an effort to achieve surfactant overexpression Korst and colleagues constructed replication deficient adenoviruses that contained the cDNA's for human surfactant proteins A and B (13). Rats infected with these viruses had increased surfactant protein production. While physiologic impact was not measured, this study showed that gene transfer can be used to augment expression of key components of surfactant. similar results have been reported by Yei, et. al. (14). A more challenging task for gene therapy is to find a way to increase synthesis of the phospholipid components of surfactant. Phosphocholine cytidylyltransferase (CTP) is the rate limiting step for synthesis of disaturated phosphatydylcholine (dspc) in rat alveolar type 2 (AT2) cells. Drs. Sprague and Li have recently published in abstract form the results of their studies using gene transfer to enhance surfactant phospholipid production by overexpressing this key component of the surfactant synthetic pathway (15).

Using a replication deficient adenovirus that contains a cDNA for CTP they infected rat alveolar type 2 cells in culture. They were able to demonstrate significant increases in CTP protein expression as well as a 25-fold increase in ^{14}C-phosphocholine incorporation into ^{14}C-cytidine 5'-disphosphate-choline. The rate of ^{3}H-choline incorporation into dspc at 3 and 4 days was increased to 2.27 and 1.76 fold as compared to controls, respectively. These results were achieved using low titers of virus and without signs of cytotoxicity. More recently they have administered their to virus to rats where they have noted a 6-fold increase in dspc levels. This very clever approach to production of a lipid product represents an excellent application for ARDS gene therapy - delivery of an intracellular constituent of a synthetic pathway that can not otherwise be administered - in essence this virus is serving as a drug delivery vehicle.

α_1 antitrypsin

Proteases released from activated neutrophils not only directly injure lung cells but also have chemoattractant, proinflammatory effects. Canonico and associates have previously shown that overexpression of human α_1 antitrypsin can attenuate these chemoattractant effects (16). They have shown that cationic lipid-plasmid mediated expression of α_1 antitrypsin in an immortalized cystic fibrosis airway epithelial cell line results in a reduction in release ofchemotactic factors in the presence of elastase. this inhibition of protease (e.g. elastase) activity by α_1 antitrypsin may diminish chemoattractant activities of activated neutrophils. Consequently, these investigators have proposed that α_1 antitrypsin may have a role in treatment of acute lung injury.

More recent work by Canonico, et. al. has reported the effects of intravenous (iv) administration of a human α_1 antitrypsin expression plasmid/cationic liposome complex into piglets (17). These animals were compared to vector controls and animals given iv doses of α_1 antitrypsin protein (prolastin). Forty-eight hours after gene transfer changes in pulmonary vascular resistance were measured following administration of $25\mu g/kg$ of endotoxin. Piglets transfected with the α_1 antitrypsin expression vector had human α_1 antitrypsin expression in the liver, kidneys as well as the lungs. In contrast to vector and iv α_1 antitrypsin protein controls, these animals had markedly smaller changes in PVR following endotoxin challenge.

Importantly, iv administration of α_1 antitrypsin protein yielded a 500 fold greater serum α_1 antitrypsin levels than did gene transfer. An explanation of this paradox provides the rationale for the use of gene therapy to overexpress α_1 antitrypsin in the face of acute lung injury. Owen and Campbell have described that the release of protease inhibitors combined with adherence of inflammatory cells to the extracellular matrix compartmentalizes proteolytic enzymes to the pericellular microenvironment that exists between adherent inflammatory cells and the extracellular matrix (18). This isolates proteolytic enzymes from their inhibitors, conceivably even very high serum levels of protease inhibitors would not be able to inhibit proteolytic damage. Gene transfer may allow for the expression of a protease inhibitor in the peri-cellular microenvironment where it could not otherwise obtain access. This may account for the disparity between α_1 antitrypsin serum levels and attenuation of lung injury seen in Canonico's studies.

Anti-oxidant Gene Expression

Substantial data suggest that oxygen free radicals play a role in acute lung injury (19). Augmentation of anti-oxidant defense mechanisms in animals has shown variable results in terms of attenuation of lung injury. However, antioxidants have a poor record in modifying the course of human disease. Whether this is due to minor contributions made by free radicals or a reflection of inadequate delivery mechanisms is as yet unanswered. Gene transfer may provide some clues to this question.

Lemarchand, et. al. have recently demonstrated that overexpression of antioxidant genes can prevent lung injury (20). They used adenoviruses that expressed the human antioxidant genes catalase and Cu,Zn SOD. These viruses were simultaneously instilled into the airways of sprague-dawley rats. Three days after infection these animals were placed in a 100% normobaric oxygen chamber and survival was measured at 62 hours. As compared to controls, animals that received the combination of catalase and SOD adenoviruses had markedly improved survival (70% vs. 10% in sham infected controls). Gene therapy in this setting could provide expression for a long enough period of time to tide patients through their acute lung injury. This study also supports the hypothesis that oxidants are significant contribute to acute lung injury.

Pulmonary Edema Clearance

Our laboratory has reported that rats exposed to subacute hyperoxia have increased capacity to clear water from the alveolus and that a significant fraction of this capacity can be inhibited by the specific Na^+,K^+-ATPase inhibitor ouabain (21). AT2 cells isolated from these same rats have increased Na^+,K^+-ATPase function and protein expression. These studies and others suggest that Na^+,K^+-ATPase contributes to alveolar liquid clearance and plays an important role in keeping the airspace dry. We tested the hypothesis that overexpression of Na^+,K^+-ATPase subunits can be used to augment in vitro Na^+,K^+-ATPase activity by engineering cytomegalovirus driven, replication deficient adenoviruses that contained cDNAs for rat α_1 and β_1 Na^+,K^+-ATPase subunits (22). Rat AT2 cells infected with a multiplicity of infection (moi) of 10 of a similarly constructed Ecoli lac z expressing virus demonstrated gene transfer efficiencies in excess of 90% without signs of cytotoxicity. Transgene activation, message processing and translation were evaluated in AT2 cells 24 hours after infection. In contrast to controls, AT2 cells infected with a β_1 expressing adenovirus showed significant expression of β_1 mRNA and protein whereas no change in α_1 message or protein was noted. We assessed Na^+,K^+-ATPase function by measuring ouabain inhibitable $^{86}Rb^+$ uptake 24 hours following infection. In contrast to sham and lac Z infected cells, ouabain inhibitable Na^+,K^+-ATPase function was increased by 2.5 fold in AT2 cells infected with moi's of 5 and 10 of the β_1 expressing adenovirus. We have been able to generate similar results using an α_1 expressing virus in human A549 cells (22).

These data suggest that Na^+,K^+-ATPase expression and function in rat AT2 cells can be increased via adenoviral mediated gene transfer of Na^+,K^+-ATPase subunits. This approach could provide for new supportive treatments for pulmonary edema be it cardiogenic or non-cardiogenic.

CONCLUSIONS

Gene therapy holds promise for ARDS. As gene transfer technology improves we will be able to apply new treatments that allow delivery of medications that can not currently be administered as well as therapy that is not even imaginable at this time. Better delivery and/or improved function of currently used therapies will also be possible. As gene therapy matures, ARDS may be among the first conditions to benefit from this new form of medical treatment.

BIBLIOGRAPHY

1) D.P. Carbone and D.N. Minna, In vivo gene therapy of human lung cancer using wild-type p53 delivered by retrovirus, J Natl Cancer Inst. 86:1437-1438 (1994).
2) R.D. Alvarez and D.T. Curiel, A phase 1 study of recombinant adenovirus vector-mediated intraperitoneal delivery of herpes simplex virus thymidine kinase (HSV-TK) gene and intravenous ganciclovir for previously treated ovarian and extraovarian cancer patients, Hum Gene Ther. 8:597-613 (1997).
3) D.T. Curiel, J.M. Pilewski, and S.M. Albelda, Gene therapy approaches for inherited and acquired lung diseases, Am J Respir Cell Mol Biol. 14:1-18 (1996).
4) F.D. Ledley, Nonviral gene therapy: the promise of genes as pharmaceutical products, Hum Gene Ther. 6:1129-1144 (1995).
5) R.G. Hawley, F.H. Lieu, A.Z.C. Fong, and T.S. Hawley, Versatile retroviral vectors for potential use in gene therapy, Gene Ther 1:136-138 (1994).
6) R.G. Crystal, N.G. McElvaney, M.A. Rosenfeld. C.S. Chu, A. Mastrangeli, J.G. Hay, S.L. Brody, H.A. Jaffe, N.T. Eissa, and C. Danel, Administration of an adenovirus containing the human CFTR cDNA to the respiratory tract of individuals with cytic fibrosis, Nat Genet. 8:42-51 (1994).
7) S.L. Brody, and R.G. Crystal. Adenovirus-mediated in vivo gene transfer, Ann NY Acad Sci. 716:90-101 (1994).
8) B.C. Trapnell and M. Gorziglia, Gene therapy using adenoviral vectors, Curr opin Biotech. 5:617-625 (1994).
9) M.P. Fink and D. Payen, The role of nitric oxide in sepsis and ARDS: synopsis of a roundtable conference held in Brussels on 18-20 March 1995, Int Care Med. 22:158-165 (1996).
10) S.P. Janssens, K.D. Bloch, Z. Nong, R.D. Gerard, P. Zoldhelyi, and D. Collen, Adenoviral-mediated transfer of the human endothelial notric oxide synthase gene reduces acute hypoxic pulmonary vasoconstriction in rats, J Clin Invest. 98:317-324 (1996).
11) A. Hartog, D. Gommers, and B. Lachman, Role of surfactant in the pathophysiology of the acute respiratory distress syndrome(ARDS), Monaldi Arch Chest Dis. 50:327-377 (1995).

12) T.E. Nicholas, I.R. Doyle, and A.D. Bersten, Surfactant replacement therapy in ARDS: white knight or noise in the system, Thorax 52:195-197 (1997).

13) R.J. Korst, B. Bewig, and R.G. Crystal, In vitro and in vivo transfer and expression of human surfactant SP-A and SP-B associated protein cDNAs mediated by replication deficient, recombinant adenoviral vector. Hum Gene Ther. 6:277-87 (1995).

14) S. Yei, C.J. Bachurski, T.E. Weaver, S.E. Wert, B.C. Trapnell, and J.A. Whitsett, Adenoviral-mediated gene transfer of human surfactant protein B to respiratory epithelial ells, Am J Resp Cell Mol Biol. 11:329-336 (1994).

15) Li, J., and R. Spragg, Adenovirus-mediated transfer of phosphocholine cytidylyltransferase gene to adult rat alveolar type II cells, Am J Resp Crit Care Med. 155:A213 (1997).

16) A.E. Canonico, K.L. Brigham, L.C. Carmichael, J.D. Plitman, G.A. King, T.R. Blackwell, and J.W. Christman, Plasmid-liposome transfer of the α_1 antitrypsin gene to cystic fibrosis bronchial epithelial cells prevents elastase-induced cell detachment and cytokine release, Am J Resp Cell Mol Biol. 14:348-355 (1996).

17) A.E. Canonico, R.E. Parker, B.O. Meyrick, X. Gao, K. Lane, D. Cannon, T. Wilson, and K.L. Brigham, Alpha$_1$ antitrypsin (AAT) gene therapy is superior to protein therapy in its protection against endotoxemia in a piglet in situ lung model. Am J Resp Crit Care Med. 155:A265 (1997).

18) C.A. Owen and E.J. Campbell, Proteinases, in: ARDS: Acute Respiratory Distress in Adults, T.W. Evans and C. Haslett, eds., Chapman and Hall, London (1996).

19) A.K. Tanswell and B.A. Freeman, Antioxidant therapy in critical care medicine, New Horizons. 3:330-341 (1995).

20) P. Prayssac, S.C. Erzurum, C. Danel, N.T. Eissa, R.G. Crystal, P. Herve, M. Mazmanian, and P. Lemarchand, Adenovirus-mediated transfer to the lungs of catalase and superoxide dismutase cDNAs prevents hyperoxia toxicity but not ischemia-reperfusion injury in rats, Am J Resp Crit Care Med. 155:A265, (1997).

21) W. Olivera, K. Ridge, L.D.H. Wood, and J.I. Sznajder, Active sodium transport and alveolar epithelial Na-K-ATPase increase during subacute hyperoxia in rats, Am J Physiol. 266:L577-L584 (1994).

22) P. Factor, C. Senne, K. Ridge, H.A. Jaffe, and J.I. Sznajder, Differential effects of adenoviral-mediated transfer of Na^+,K^+-ATPase subunit genes in lung epithelial cells, Chest 111(6 Suppl):110S-111S (1997).

THE USE OF RECOMBINANT ANTIOXIDANTS IN NEONATAL

RESPIRATORY DISTRESS SYNDROME

Jonathan M. Davis, M.D.

Department of Pediatrics and the CardioPulmonary Research Institute
Winthrop University Hospital
SUNY at Stony Brook School of Medicine
259 First Street
Mineola, NY 11501
USA

INTRODUCTION

Bronchopulmonary Dysplasia (BPD) is a chronic lung disease that develops in infants treated with oxygen and mechanical ventilation for a primary lung disorder. BPD has been traditionally defined as oxygen requirement with an abnormal chest radiograph at 28 days of life.[1,2] BPD can affect 20-60% of premature neonates who are born each year with respiratory distress syndrome (RDS).[3] The development of BPD is associated with significant mortality and morbidity such as repeated hospitalizations and developmental handicaps.[4] With the increasing survival of many premature infants, BPD is emerging as the most common sequella of neonatal intensive care.[5]

The etiology of BPD is poorly understood. It was initially postulated that exogenous surfactant replacement therapy (by lowering oxygen and ventilator requirements) would decrease the incidence and severity of BPD in infants with RDS. However, recent surfactant trials have demonstrated that despite an improvement in clinical course and overall survival, the incidence of BPD has not been substantially affected and the prevalence of BPD may actually be increasing.[3] This suggests that the pathogenesis of BPD is multi factorial and associated with a variety of causative factors. Since the lungs of premature infants with RDS are exposed to supraphysiological oxygen concentrations during therapy, it seems logical that an oxidative insult could be an important component of the injury process. BPD has been

Acute Respiratory Distress Syndrome: Cellular and Molecular Mechanisms and Clinical Management
Edited by Matalon and Sznajder, Springer Science+Business Media New York, 1998

143

hypothesized to begin as acute inflammatory changes secondary to toxic free oxygen radicals which then evolve into chronic lung disease.[3,6,7]

The premature neonate may be particularly vulnerable to oxidant damage since endogenous antioxidant enzyme activity may be deficient at birth.[8,9] Preliminary animal and human studies have suggested that acute and chronic lung injury secondary to prolonged hyperoxia may be ameliorated by intratracheal (either alone or encapsulated in liposomes), intravenous or intraperitoneal administration of one of these antioxidants, specifically superoxide dismutase (SOD).[10-17] There are three forms of SOD that have been identified in mammals. The first is a low molecular weight Cu/Zn containing protein (MW = 31 kDa) which is present in the cytoplasm of all mammalian cells. The second form is a Mn containing protein (MW = 45kDa) which is exclusively located in the mitochondria (MnSOD). The third form is also a Cu/Zn containing protein, but it is located in extracellular spaces (EC-SOD). The only known function of SOD is to catalyze the conversion of toxic superoxide anions (oxygen free radicals) to potentially less toxic hydrogen peroxide and water. SOD activity has been detected in natural lung surfactants, but is absent in commercial surfactant preparations.[18]

In genetically engineered mice with disrupted extracellular SOD genes and no functional extracellular Cu/Zn SOD, exposure to 100% oxygen resulted in significantly more lung injury and reduced survival relative to diploid controls.[19] In addition, genetically engineered mice that overexpressed MnSOD in type II pneumocytes were able to survive much longer in 100% oxygen compared to normal controls.[20] These two studies demonstrate that SOD is critically important in protecting the lung from the damaging effects of hyperoxia. It is possible that augmenting antioxidant enzyme activity may mitigate cell damage and inflammatory changes in the lung and prevent the development of acute and chronic injury in premature infants.

As a first step in determining whether a newly developed human recombinant Cu/Zn superoxide dismutase (rhSOD; Bio-Technology General Corporation, Iselin, NJ) could prevent neonatal lung injury from hyperoxia and positive pressure mechanical ventilation, 26 newborn piglets were studied.[10] The term newborn piglet model was used since piglets are relatively inexpensive, are large enough at birth (1.2-1.9 kg) for relevant physiological and biochemical studies and their lungs are similar to that of a premature infant. Ten piglets were hyperventilated (PaCO$_2$ 15-20 torr) with 100% oxygen for 48 hours. Ten additional piglets received identical treatment, but were given 5 mg/kg of rhSOD at time zero. Six piglets were normally ventilated (PaCO$_2$ 40-45 torr) with room air for 48 hours. In piglets treated with hyperoxia and hyperventilation, lung compliance decreased 42% while tracheal aspirates (TAF) showed an increase in neutrophil chemotactic activity (32%), total cell counts (135%), elastase activity (93%) and albumin concentration (339%) over 48 hours (P < 0.05). All variables were significantly lower in rhSOD-treated piglets and comparable to normoxic control levels. Pharmacokinetic studies demonstrated that SOD concentration and activity remained significantly elevated in TAF, bronchoalveolar lavage (BAL), lung tissue and serum at 48 hours. Immunohistochemistry showed that rhSOD was distributed relatively homogeneously along terminal airways, even 48 hours after administration. Adding rhSOD to tracheal aspirates from hyperoxic, hyperventilated piglets did not alter neutrophil chemotaxis suggesting that rhSOD protected the lung by preventing cell injury and the production of inflammatory mediators. These studies suggested that the acute inflammatory changes and lung injury caused by 48 hours of hyperoxia and mechanical ventilation (similar to that seen in preterm neonates in the early stages of BPD) were significantly ameliorated by the prophylactic administration of rhSOD.

Next, we were interested learning more about the localization, activity and metabolism of rhSOD following IT administration. Twenty-six newborn piglets were intubated, mechanically ventilated and given either saline or fluorescently labeled rhSOD (5mg/kg) IT by instillation or nebulization.[21] Animals were sacrificed 1, 6 or 12 hours later. Intact rhSOD (% total fluorescence still associated with macromolecules) and total SOD activity in lung tissue was then determined. Results indicated that after 1 and 6 hours of administration, the majority of rhSOD present in the lung was still associated with the fluorescent label. By 12 hours, most of the rhSOD was no longer fluorescently labeled. At 1 hour, lung SOD activity increased by 100% compared to untreated control values, with activity remaining significantly elevated at 6 and 12 hours. Laser confocal microscopy of lung tissue showed that at 1 hour, labeled rhSOD was found throughout the lung, inside a variety of cell types of airways, respiratory bronchioles and alveoli. Negative controls had minimal background fluorescence. These data indicated that following IT administration, rhSOD was rapidly incorporated into cells in the lung and significantly increased lung SOD activity.

Since we had demonstrated that rhSOD is rapidly incorporated into cells of airways, respiratory bronchioles and alveoli following IT administration, we then wanted to determine whether this cellular uptake was specific for Cu/Zn SOD, or if other proteins are similarly incorporated into lung cells.[22] Eighteen newborn piglets (2 to 3 days old, 1.2 to 2.0 kg) were intubated and mechanically ventilated. Six piglets received fluorescently labeled rhMnSOD and 6 received labeled albumin IT (larger molecular weight proteins). Protein was delivered with the animals' right side down and head elevated to enhance deposition to the right lower lobe. In addition, four piglets were made surfactant deficient by repeated BAL, then given rhSOD IT to determine whether endogenous surfactant was important in the process of intracellular uptake. Finally, two piglets received free fluorophore (unbound to any protein) as an additional control group. All animals were sacrificed 30 - 60 minutes after IT administration of the proteins. The RLL was again examined using laser confocal microscopy in a blinded fashion. Similar to our previous observations with rhSOD, intracellular uptake of rhMnSOD and albumin was noted throughout the lung. The uptake of protein was not affected by surfactant deficiency. The free fluorescent label did not localize intracellularly. Data suggest that the cellular uptake of antioxidants and other proteins delivered IT may occur via a nonspecific mechanism. Although clearance of proteins that accumulate in the lumen of the airways and alveoli occurs primarily through paracellular mechanisms, low concentrations of protein have been shown (by electron microscopy and specific inhibitors of cellular function) to be cleared by intracellular routes.[23,24] Intracellular clearance occurs via endocytosis, transcytosis and restricted diffusion through the epithelium. This may explain how the rhSOD is taken up by cells in the lung and why the half-life of the protein is approximately 24 - 48 hours. These observations have important implications for potential clinical trials in premature infants.

As a first step in the evaluation of rhSOD in the prevention of neonatal lung injury caused by exposure to hyperoxia and positive pressure mechanical ventilation, safety and pharmacokinetics of IT administered rhSOD were studied.[16] Twenty-six preterm infants weighing 750 -1250 g with RDS were studied in three sequential groups (placebo, 0.5 and 5.0 mg/kg). Placebo or rhSOD was administered IT 30 minutes after the first surfactant dose. Serial blood and urine studies, rhSOD levels, TAF markers of acute inflammation, radiographs and ultrasounds were performed over the 28 day study period.

Serum rhSOD concentrations were similar at baseline for all three groups [geometric mean (upper-lower limit): 0.2 (0.1-0.2) μg/ml]. In the 0.5 mg/kg group, levels were highest at 12 hours [0.7 (0.5-0.8) μg/ml] and returned to baseline by day 3. In the 5.0 mg/kg group, levels were highest at 6 hours [3.0 (2.3-4.0) μg/ml] and returned to baseline by day 4. Concentrations of rhSOD in TAF were also similar at baseline for all three groups [0.2 (0.2-0.3) μg/ml]. There were no significant increases in the placebo group, but levels in the 0.5 mg/kg group were highest when first sampled at 24 hours [1.1 (0.8-1.4) μg/ml] and returned to baseline by day 3. In the 5.0 mg/kg group, levels were also highest when sampled at 24 hours [1.4 (0.9-2.1) μg/ml] and returned to baseline by day 4. Urine levels were highest at 12 hours in both the 0.5 mg/kg [1.3 (1.0-1.7) μg/ml] and 5.0 mg/kg infants [6.4 (3.9-10.4) μg/ml] and decreased significantly by day 2-3. rhSOD activity assays (serum, TAF and urine) demonstrated that the enzyme still possessed significant activity. No adverse effects of rhSOD were found. TAF neutrophil chemotactic activity and albumin concentrations, important acute lung injury markers, were significantly lower in the high dose rhSOD group compared to the other groups.

Data suggested that a single IT dose of rhSOD resulted in significant increases in both concentration and activity of the antioxidant in serum, TAF and urine for 2-3 days. The enzyme appeared to be well tolerated and TAF inflammatory markers were reduced following administration. This was interesting preliminary information, but it seemed unlikely that a single dose of rhSOD at birth would be able to completely prevent the development of acute and chronic lung injury caused by prolonged exposure to hyperoxia and mechanical ventilation (i.e. 28 days of life). Multiple doses of rhSOD would be needed to prevent or significantly ameliorate the development of BPD.

The next series of clinical trials examined the safety and pharmacokinetics of multiple IT doses of rhSOD in premature infants with respiratory distress syndrome who were at risk for developing BPD.[17] Thirty-three infants (700-1300 g) were randomized and blindly received saline, 2.5 mg/kg or 5 mg/kg of rhSOD IT within 2 hours of surfactant administration. Infants were treated every 48 hours (while endotracheal intubation was required) up to 7 doses. Serial blood and urine studies, chest radiographs,

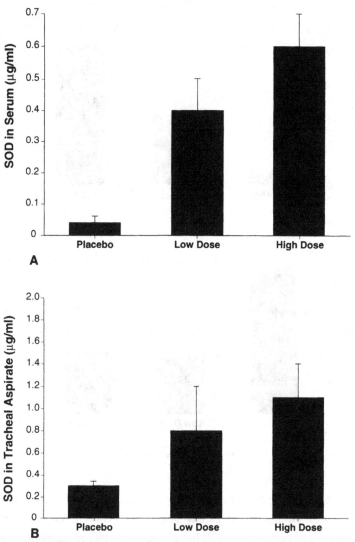

Figure 1: SOD concentrations in serum (**A**), tracheal aspirates (**B**), and urine (**C**) presented as geometric mean ± 1 SEM. No significant changes occurred in the placebo group, but both rhSOD groups were increased when first sampled on day 3 and remained elevated thereafter (*p < 0.05, ANOVA). Data from days 3 to 13 have been pooled and averaged. The activity of SOD correlated with its concentration (p < 0.05). **D.** rhSOD reduced neutrophil chemotactic activity (NCA) in TAF. NCA (which reflects inflammation and lung injury) in TAF is expressed as percent of positive control (zymosan activated serum). Baseline values were similar for all groups, but activity was significantly decreased over the two week dosing period in the rhSOD groups compared to the placebo. Values for each group are pooled and averaged (*p < 0.05). Reprinted with permission, Pediatrics.

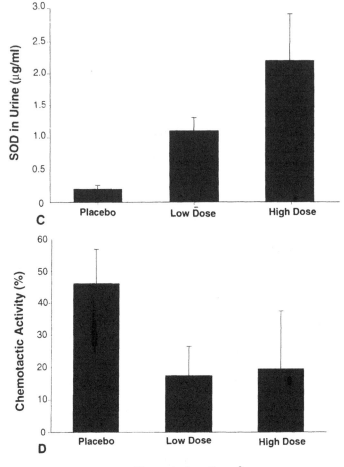

Figure 1: (*continued*)

neurosonograms, SOD concentration and activity measurements and TAF inflammatory markers were assessed throughout the 28 day study. SOD concentrations in serum (0.1 [0.05/0.15] µg/ml - geometric mean with lower/upper confidence intervals), TAF (0.2 [0.1/0.3] µg/ml) and urine (0.3 [0.2/0.4] µg/ml) were similar at baseline in all three groups and did not change significantly in the placebo group. In the rhSOD treatment groups, SOD concentrations were increased on day 3 and did not change significantly thereafter over the 14 day dosing period (also measured on days 5, 7, 13). SOD concentrations averaged 0.4 [0.3/0.5] µg/ml in serum, 0.8 [0.6/1.2] µg/ml in TAF and 1.1 [1.0/1.3] µg/ml) in urine for the low dose group and 0.6 [0.5/0.7] µg/ml in serum, 1.1 [0.9/1.5] µg/ml in TAF, and 2.2 [1.6/2.9] µg/ml in urine for the high dose group (p < 0.05, ANOVA) over the 14 day dosing period (Fig. 1). Enzyme activity directly correlated with SOD concentration and rhSOD was active even when excreted in urine. TAF markers of acute lung injury (neutrophil chemotactic activity, albumin concentration) were lower in the rhSOD groups compared to placebo. No significant differences in any clinical outcome variable were noted between groups.

These data indicated that multiple IT doses of rhSOD increased the concentration and activity of the enzyme in serum, TAF and urine, reduced TAF lung injury markers and was well tolerated. Further clinical trials examining the efficacy of rhSOD in the prevention of BPD were then warranted.

A Phase III trial examining the safety and efficacy of rhSOD in the prevention of BPD in 360 premature infants (600 - 1200g) began at 20 hospitals in the United States in January of 1997. Patient enrollment and statistical analyses will occur over two years. If rhSOD is effective in premature neonates, then further trials are planned in older children and adults with Acute Respiratory Distress Syndrome (ARDS) to determine if the enzyme can prevent lung damage in this patient population.

The pathogenesis and prevention of acute and chronic lung damage in premature infants has become an area of intense scientific interest and investigation. A variety of techniques have been used in an attempt to reduce the incidence and severity of BPD. The early use of exogenous surfactant and high frequency jet ventilation in premature infants with RDS has been recently shown to decrease the development and severity of BPD.[25] Therapeutic trials using agents such as corticosteroids (inhaled and systemic) and vitamin A have been also been performed with variable responses.[26-28] The use of recombinant antioxidants such as rhSOD appears to hold great promise. Advances in gene therapy may allow the delivery and uptake of the genes for SOD to be administered and expressed and this is currently being investigated.

References

1. E. Bancalari, G.E. Abdenour, R. Feller, et al, Bronchopulmonary dysplasia: clinical presentation, J Pediatr. 95:819 (1979).

2. W.H. Tooley, Epidemiology of bronchopulmonary dysplasia, Pediatrics 95:851 (1979).

3. R.F. Soll, M.C. McQueen, Respiratory distress syndrome, in: Effective Care of the Newborn Infant, J.C.Sinclair and M.B. Bracken, eds., Oxford University Press, New York (1992).

4. J. Bregman and E.E. Farrell, Neurodevelopmental outcome in infants with bronchopulmonary dysplasia, Clin. Perinatol. 19:673 (1992).

5. J.M. Davis and W. Rosenfeld, Chronic lung disease, in: Neonatology, G.B.Avery, M.A. Fletcher and M.D. MacDonald, eds., J.B. Lipincott Co., Philadelphia (1994).

6. A. Bagchi, R.M. Viscardi, V. Taciak, et al, Increased activity of interleukin-6 but not tumor necrosis factor α in lung lavage of premature infants is associated with the development of bronchopulmonary dysplasia, Pediatr. Res.36:244 (1994).

7. T.A. Merritt, C.G. Cochrane, K. Holcomb, et al, Elastase and α-1 proteinase inhibitor activity in tracheal aspirates during respiratory distress syndrome. Role of inflammation in the pathogenesis of bronchopulmonary dysplasia, J. Clin. Invest. 72:656 (1983).

8. L. Frank, E.E. Groseclose, Preparation for birth into an O_2 rich environment: The antioxidant enzymes in the developing rabbit lung, Pediatr. Res. 18:240 (1984).

9. L. Frank, Developmental aspects of experimental pulmonary oxygen toxicity, Free Rad. Biol. Med. 11:463 (1991).

10. J.M. Davis, W.N. Rosenfeld, R.J.Sanders, A. Gonenne, Prophylactic effects of recombinant human superoxide dismutase in neonatal lung injury, J. Appl. Physiol. 74:2234 (1993).

11. J.F. Turrens, J.D. Crapo, B.A. Freeman, Protection against oxygen toxicity by intravenous injection of liposome-entrapped catalase and superoxide dismutase, J. Clin. Invest.73:87 (1984).

12. R.V. Padmanabhan, R. Gudapaty, I.E. Liener, B.A. Schwartz, J.R. Hoidal, Protection against pulmonary oxygen toxicity in rats by the intratracheal administration of liposome-encapsulated superoxide dismutase or catalase, Am. Rev. Respir. Dis. 132:164 (1985).

13. F.J. Walther, C.E.M. Gedding, I.M. Kuipers, et al, Prevention of oxygen toxicity with superoxide dismutase and catalase in premature lambs, J. Free Rad. Biol. Med. 2:289 (1986).

14. B.A. Freeman, S.L. Young, J.D. Crapo, Liposome - mediated augmentation of superoxide dismutase in endothelial cells prevents oxygen injury, J. Biol. Chem. 258: 12534 (1983).

15. W. Rosenfeld, H. Evans, L. Concepcion, et al, Prevention of bronchopulmonary dysplasia by administration of bovine superoxide dismutase in preterm infants with respiratory distress syndrome, J. Pediatr. 105:781 (1984).

16. W.N. Rosenfeld, J.M. Davis, L. Parton, S.E. Richter, A. Price, E. Flaster, N. Kassem, Safety and pharmacokinetics of recombinant human superoxide dismutase administered intratracheally to premature neonates with respiratory distress syndrome, Pediatrics 97: 811 (1996).

17. J.M. Davis, W.N. Rosenfeld, S.E. Richter, et al, Safety and pharmacokinetics of multiple doses of recombinant human CuZn superoxide dismutase administered intratracheally to premature neonates with respiratory distress syndrome, Pediatrics 100:24 (1997).

18. S. Matalon, B.A. Holm, R.R. Baker, M.K. Whitefield, B.A. Freeman, Characterization of antioxidant activities of pulmonary surfactant mixtures, Biochem. Biophys. Acta.1035:121 (1990).

19. L.M. Carlsson, J. Jonsson, T. Edlun, S.L. Marklund, Mice lacking extracellular superoxide dismutase are more sensitive to hyperoxia, Proc. Natl. Acad. Sci. 92:6264 (1995)

20. J.R. Wispe, B.B. Warner, J.C. Clark, et al, Human Mn-superoxide dismutase in pulmonary epithelial cells of transgenic mice confers protection from oxygen injury, J. Biol. Chem. 267:23937 (1992).

21. N. Sahgal, J.M. Davis, C.G. Robbins,et al, Localization and activity of recombinant human CuZn superoxide dismutase after intratracheal administration, Am. J. Physiol. 71:L230 (1996).

22. S. Das, S. Horowitz, C. Robbins, et al, Intracellular uptake of recombinant superoxide dismutase after intratracheal administration, Am. J. Physiol., submitted.

23. R.H. Hastings, J.R. Wright, K.H. Albertine, R. Ciriales, M.A.Matthay, Effect of endocytosis inhibitors on alveolar clearance of albumin, immunoglobulin G,and SP - A in rabbits, Am. J. Physiol. 266: L544 (1994).

24. R.H. Hastings, H.G. Folkesson, V. Petersen, R. Ciriales, M.A. Matthay, Cellular uptake of albumin from lungs of anesthetized rabbits, Am. J. Physiol. 269:L453 (1995).

25. M. Keszler, H.D.Modanlou, S. Brudno, et al, Multi-center controlled clinical trial of high frequency jet ventilation in preterm infants with uncomplicated respiratory distress syndrome, Pediatrics, in press.

26. J.M. Davis, R.A. Sinkin, J.V. Aranda. Drug therapy for bronchopulmonary dysplasia. Pediatr. Pulmonol. 8:117 (1990).

27. J.J. Cummings, D.B. D'Eugenio, S.J. Gross, A controlled trial of dexamethasone in preterm infants at high risk for bronchopulmonary dysplasia, N. Engl. J. Med. 320:1505 (1989).

28. J.P. Shenai, M.G. Rush, M.T. Stahlman, F. Chytil, Plasma retinol-binding protein response to vitamin A administration in infants susceptible to bronchopulmonary dysplasia. J. Pediatr. 116:607 (1990).

THE EARLY INFLAMMATORY RESPONSE IN ACUTE RESPIRATORY DISTRESS SYNDROME (ARDS)

Nikhil Hirani and Seamas C Donnelly
Department of Respiratory Medicine
Rayne Laboratory
Teviot Place
Edinburgh.
EH8 9AG

INTRODUCTION

The acute respiratory distress syndrome (ARDS) represents the severe end of a spectrum of acute lung injury[1]. Its pathogenesis classically evolves through a sequence of events: an initiating insult oftentimes distant to the lung, followed by a 'latent period' lasting from a few hours to a few days during which there is no clinical or radiographic evidence of lung injury and finally, in those destined to do so, progression to fulminant lung inflammation characteristic of ARDS. It is now clear however that the latent period is deceptive, for at the cellular level within the lung and indeed throughout the systemic vascular endothelium[2][3], there is an orchestrated cascade of events culminating in sustained overwhelming inflammation. Hence by the time the clinical characteristics by which we recognise ARDS have been established, significant intra-pulmonary inflammation and injury has already occurred. This may in part explain the disappointing results observed with anti-inflammatory therapy initiated in patients with established ARDS. One strategy which may resolve this dilemma is to identify at an early stage specific inflammatory mediators which are associated with ARDS progression and then target high-risk patients early in the diseasee process with specific anti-inflammatory therapy. The role inflammatory mediators in acute lung injury has recently been excellently reviewed[4]. Here we focus on the early at-risk period of ARDS and review possible candidates for a biological mediator which may predict ARDS progression, with particular emphasis on our experience in Edinburgh.

PREDICTING ARDS: THE EARLY INFLAMMATORY RESPONSE

The at-risk period of ARDS provides an ideal 'window of opportunity' in which to dissect out the early inflammatory events, anticipating that the process becomes

Acute Respiratory Distress Syndrome: Cellular and Molecular Mechanisms and Clinical Management
Edited by Matalon and Sznajder, Springer Science+Business Media New York, 1998

increasingly complex thereafter. However there are a number of particular obstacles to this strategy. Ideally patients at-risk for ARDS need to be identified as early as possible. The majority of studies have recruited patients already on the intensive care unit and requiring ventilatory support. Amongst these patients there already exists a spectrum of lung inflammation and injury, requiring mechanical ventilation though not yet satisfying the diagnostic criteria for ARDS. Secondly the risk factors for ARDS are heterogenous, and the pathogenesis of lung injury may reflect this. Differences in case-mix may explain differences in findings between studies. Thirdly for a given mediator to be clinically useful as a predictor of ARDS, it must attain not simply a statistically significant difference between patients who do and do not develop ARDS, but exhibit a sufficiently high sensitivity and specificity to reliably predict disease progression in an individual patient. Despite these difficulties several recent studies have demonstrated that the early identification of patients at-risk of ARDS is an attainable goal.

(I) Chemokines

Regardless of the initiating cause, a hallmark pathological feature of ARDS is neutrophil sequestration within the alveoli and pulmonary interstitium. This infiltration occurs before the development of lung injury[5] and suggests the presence of an intrapulmonary chemotactic signal. Early studies of leukotrienes, in particular LTB4 the potent neutrophil chemoattractant, have reported high levels in bronchalveolar lavage (BAL) fluid of those at risk of ARDS[6] and enhanced neutrophil LTB4 synthesis may have a predictive role in ARDS progression[7]. These findings require further investigation. More recently however the focus has been on the role of chemokines in lung injury. In patients with trauma, acute pancreatitis and bowel perforation, we have demonstrated that elevated levels of interleukin-8 (IL8) in BAL fluid are associated with subsequent progression to ARDS[8]. Indeed retrospective analysis of 75 cases suggests that this single lavage mediator possesses a positive predictive value of 75%. Interestingly BAL fluid levels of the archetypal early-response inflammatory cytokines, namely TNF-α and IL-1β, although present in the airspaces of those at-risk, were not shown to predict progression to ARDS in our study. Similarly serum levels of IL-8, TNF-α and IL-1β had no predictive value for ARDS progression. Other groups have similarly reported that TNF-α levels in BAL or serum do notcorrelate with disease progression[9][10][11][12]. BAL levels of IL-1β have been shown to be higher in at-risk patients that develop ARDS compared to those that do not, but whether this is of clinically useful value is unknown[13][14].

(II) Adhesion molecules

In addition to this rapid chemotactic signal, it is recognised that neutrophil and endothelium activation is an early feature of the generalised inflammatory response. The transition from the quiescent to the activated state results in important phenotypic changes in these cells. These include up-regulation of specific receptor molecules on the cell surface which allow adhesion and migration of neutrophils through the endothelium and into the the tissue space[15]. The selectins are a family of three receptor molecules found on activated endothelium (E-selectin and P-selectin), platelets (P-selectin) and circulating leukocytes (L-selectin). All three are involved in the adhesion of inflammatory cells to endothelium and during this process, are proteolytically cleaved from their parent cell and shed into the circulation in a soluble form. In view of the importance of neutrophil activation and neutrophil-endothelial interaction in early ARDS, we have investigated circulating soluble selectins in patients at-risk of ARDS. We found that soluble L-selectin was significantly lower in patients who progressed to ARDS compared to those who did not and low soluble L-selectin levels were associated with subsequent increased mortality[16]. We have proposed that this finding may be explained by soluble L-selectin binding to its counter - ligand on activated endothelium; thus levels of circulating soluble L-selectin may represent an indirect marker of widespread endothelial activation in these patients. Another potential marker of endothelial dysfunction is von Willebrand factor antigen (vWf-Ag). In patients with non-pulmonary sepsis, a normal chest radiograph and at least one-organ failure, raised plasma vWf-Ag levels have been shown to have a positive predictive value of 80% for the development of ARDS. However in a subsequent study of a group of at-risk patients which included pulmonary sepsis, the sensitivity and specificity of a raised vWf-Ag were considered too low to be of clinical value[17]. Again the heterogeneity of patients involved is an important confounding factor, and indeed endothelial cell activity has been clearly shown to vary between patients with trauma and those with sepsis[18].

(III) Elastase

Activated neutrophils are implicated in actual tissue damage via the release of a variety of potentially histotoxic products, including proteases and reactive oxygen intermediates. Attempts have been made to measure these mediators in patients at-risk of ARDS. Elastase, a serine protease, has potent degradative properties and all structural protein components of the lung matrix are potential targets[19]. In patients with established ARDS, BAL fluid contains significantly raised levels of elastase[20][21][14]. In a study of patients with multiple trauma on initial hospital presentation, plasma elastase levels were significantly higher in the those who progressed to ARDS compared to those who did not[22]. Whilst there was considerable overlap between the two groups of patients such that a given plasma elastase level could not reliably predict ARDS progression, this work clearly implicates a product of recent neutrophil degranulation in the development of ARDS in a homogenous at-risk population.

(IV) Oxidants/anti-oxidants

Oxidant stress as a feature of acute lung injury has been known of for many years[23]. There is considerable evidence for oxidant/antioxidant imbalances in patients with acute lung injury: the expired breath of patients with ARDS has raised levels of hydrogen peroxide[24] while alveolar glutathione levels are reduced[25] and ceruloplasmin and transferrin levels raised[26][27][28]. In patients at-risk of ARDS, both BAL and plasma levels of transferrin and ceruloplasmin have been measured but not shown to be of predictive value[27]. However a number of recent studies have reported more promising results. Lipofuscin, an indicator of lipid peroxidation, was measured sequentially in 66 surgical patients on an ICU. Levels at 24 hours post-admission were found to be significantly higher in the the the ten patients who subsequently developed ARDS or multi-organ failure, though the sensitvity was low at 47%[29]. In a study which included 20 patients at-risk of ARDS, a serum ratio of >1.45 of C18 unsaturated fatty acids to fully saturated palmitate predicted subsequent progression to ARDS in 7 patients with 84% sensitivity and 87% specificity[30]. The same group have reported comparable figures for the predictive value of the antioxidant enzymes manganese superoxide dismutase and catalase in 26 patients with sepsis[31]. Recently serum ferritin was measured in 120 patients, predominantly with sepsis, trauma or hypertransfusion, at-risk of ARDS. In females, a value exceeding 270ng/ml predicted ARDS with a sensitvity of 83%

and specificity of 71%. The corresponding figures in men were 60% and 90% for a value exceeding 680ng[32].

CONCLUSIONS AND FUTURE STUDIES

In the last few years, the possibility of reliably identifying patients at highest risk of developing ARDS has been made into a realistic prospect. Attempting to identify a prognostic biological mediators at the earliest possible point in the development of acute lung injury not only selects high risk patients, but also provides the widest possible therapeutic time-frame in which to deliver therapy prior to the development of severe lung injury. As yet however none of the promising candidates discussed here have been shown to fulfil the required criteria in a sufficiently large **prospective** study. Indeed it is highly conceivable that rather than a single mediator, a combination of mediators may prove more powerful predictors of disease progression. Retrospective analysis of our current data suggests that a combined circulating IL-8, soluble L-selectin and elastase has a specificity of 95% and sensitivity of 86% for predicting subsequent ARDS development in at-risk patients. This is currently being investigated in a prospective study. In the meantime as our understanding of the acute inflammatory response evolves, novel mediators will no doubt emerge and the goal of reliably predicting patients at high-risk of ARDS come ever closer.

REFERENCES

1. Murray JF, Matthay MA, Luce JM, Flick MR: Pulmonary Perspectives. An expanded definition of the adult respiratory distress syndrome. *Am Rev Respir Dis* 138:720 (1988).

2. Mizer LA, Weisbrode SE, Dorinsky PM: Neutrophil accumulation and structural changes in nonpulmonary organs after acute acute lung injury induced by phorbyl myristate acetate. *Am Rev Respir Dis* 139:1017 (1988).

3. Bone RC, Balk RA, Slotman G, Maunder R, Silverman H, Hyers TM, Kerstein MD, Prostaglandin E1 Study Group: Adult respiratory distress syndrome. Sequence and importance of development of multiple organ failure. *Chest* 101:320 (1992).

4. Pittet IF, Mackersie RC, Martin TR, Matthay MA: Biological markers of acute lung injury: Prognostic and pathogenetic significance. *Am J Respir Crit Care Med* 155:1187 (1997)

5. Fowler AA, Hyers TM, Fisher BJ, Bechard DE, Centor RM, Webster RO: The adult respiratory distress syndrome: cell populations and soluble mediators in the air spaces of patients at high risk. *Am Rev Respir Dis* 136:1225 (1987).

6. Stephenson AH, Longigro AJ, Hyers TM, et al: Increased concentrations of leukotrienes in bronchoalveolar lavage fluid of patients with ARDS or at-risk of ARDS. *Am Rev Respir Dis* 138:714 (1988).

7. Davis JM, Meyer JD, Barie PS, Yurt RW, Duhaney R, Dineen P, Shires GT: Elevated production of neutrophil leukotriene B4 precedes pulmonary failure in critically ill surgical patients. *Surg Gyneacol Obstet* 170:495 (1990).

8. Donnelly SC, Strieter RM, Kunkel SL, Walz A, Robertson CR, Carter DC, Grant IS, Pollock AJ, Haslett C: Interleukin-8 and development of adult respiratory distress syndrome in at-risk groups. *Lancet* 341:643 (1993).

9. Hyers TM, Tricomi SM, Dettenmeier PA, Fowler AA: Tumour necrosis factor in serum and bronchoalveolar lavage fluid of patients at risk of the adult respiratory distress syndrome. *Am Rev Respir Dis* 144:268 (1991)

10. Roten R, Markert M, Feihl F, et al: Plasma levels of tumour necrosis factor in the adult respiratory distress syndrome. *Am Rev Respir Dis* 143:590 (1991).

11. Raponi G, Antonelli M, Gaeta A, Bufi M, Blasi RAD, Conti G, D'Errico RR, Mancini C, Filadoro F, Gasperetto A: Tumour necrosis factor in serum and in bronchalveolar lavage of patients at risk for the adult respiratory distress syndrome. *J Crit Care* 7:183 (1992).

12. Parsons PE, Moore F, Moore EE, Ikle DN, Henson PM, Worthen GS: Studies in the role of tumour necrosis factor in adult respiratory distress syndrome. *Am Rev Respir Dis* 146:694 (1992).

13. Siler TM, Swierkosz JE, Hyers TM, et al: Immunoreactive interleukin-1 in bronchoalveolar lavage fluid of high risk patients and patients with ARDS. *Experimental Lung Research* 15:881 (1989).

14. Suter PM, Suter E, Girardin E, Roux-Lombard P, Grau GE, Dayer JM: High bronchoalveolar levels of tumour necrosis factor and its inhibitors, interleukin-1, interferon, and elastase in patients with adult respiratory distress syndrome after trauma, shock or sepsis. *Am Rev Respir Dis* 145:1016 (1992).

15. Nourshargh S, Williams TJ: Molecular and cellular interactions mediating granulocyte accumulation *in vivo*. *Seminars in cell biology* 6:317 (1995).

16. Donnelly SC, Haslett C, Dransfield I, Robertson CE, Carter DC, Ross JA, Grant IS, Tedder TF: Role of selectins in development of adult respiratory distress syndrome. *Lancet* 344:215 (1994).

17. Moss M, Ackerson L, Gillespie MK, et al: Von Willebrand factor antigen levels are not predictive for the adult respiratory distress syndrome. *Am J Respir Crit Care Med* 51:15 (1995).

18. Moss M, Gillespie MK, Ackerson L, Moore F, Moore EE, Parsons PE: Endothelial cell activity varies in patients at risk for the adult respiratory distress syndrome. *Crit Care Med* 24:1782 (1996).

19. Vender RL: Therapeutic potential of neutrophil-elastase inhibition in pulmonary disease. *J Immunol* 44:531 (1996).

20. Lee CT, Fein AM, Lippman M, Hotzmann H, Kimble P, Weinbaum G: Elastolytic activity in pulmonary lavage fluid from patients with adult respiratory distress syndrome. *New Engl J Med* 304:192 (1981).

21. McGuire WW, Spragg RC, Cohen AB, Cochrane CG: Studies in the pathogenesis of the adult respiratory distress syndrome. *J Clin Invest* 69:543 (1982).

22. Donnelly SC, MacGregor I, Zamani A, Gordon MWG, Robertson CE, Steedman DJ, Little K, Haslett C: Plasma elastase levels end the development of the adult respiratory ditress syndrome. *Am J Respir Crit Care Med* 151:1428 (1995).

23. Cochrane CG, Spragg R, Revak SD: Pathogenesis of the adult respiratory distress syndrome: evidence of oxidant activity of bronchoalveolar lavage fluid. *J Clin Invest* 71:754 (1983).

24. Baldwin SR, Grum CM, Boxer LA, Simon RH, Ketai LH, Devall LJ: Oxidant activity in expired breath of patients with adult respiratory distress syndrome. *Lancet* I:11 (1986).

25. Pacht ER, Timerman AP, Lykens MG, Merola AJ: Deficiency of alveolar fluid glutathione in patients with sepsis and the adult respiratory distress syndrome. *Chest* 100:1397 (1991).

26. Lykens MG, Davis WB, Pacht ER: Antioxidant activity of bronchoalveolar lavage fluid in the adult respiratory distress syndrome. *Am J Physiol* 262:L169 (1992).

27. Krsek-Staples JA, Kew RR, Webster RO: Ceruloplasmin and transferrin levels are altered in serum and bronchoalveolar lavage fluid of patients with the adult respiratory distress syndrome. *Am Rev Respir Dis* 145:1009 (1992).

28. Stites SE, Nelson ME, Wesselius LJ: Transferrin concentrations in serum and lower respiratory tract fluid of mechanically ventilated patients with COPD or ARDS. *Chest* 107:1681 (1995).

29. Roumen RMH, Hendriks T, deMan BM, Goris RJA: Serum lipofuscin as a prognostic indicator of adult respiratory distress syndrome and multiple organ failure. *B J Surg* 81:1300 (1994).

30. Bursten SL, Federighi DA, Parsons PE, Harris WE, Abraham E, Moore EE, Moore FA, Bianco JA, Singer JW, Repine JE: An increase in serum C18 unsaturated fatty acids as a predictor of the development of acute respiratory distress syndrome. *Crit Care Med* 24:1129 (1996).

31. Leff JA, Parsons PE, Day CE, Taniguchi N, Jochum M, McCord JM, Repine JE: Serum oxidants as predictors of adult respiratory distress syndrome in patients with sepsis. *Lancet* 341:777 (1993).

32. Connelly KG, Moss M, Parsons PE, Moore EE, Moore FA, Giclas PC, Seligman PA, Repine JE: Serum ferritin as a predictor of the acute respiratory distress syndrome. *Am J Respir Crit Care Med* 155:21 (1997).

THE *N*-NITROSO-*N*-METHYLURETHANE INDUCED ACUTE LUNG INJURY ANIMAL MODEL FOR ARDS: MINIREVIEW

Wilhelm S. Cruz and Michael A. Moxley

Edward A. Doisy Department of Biochemistry and Molecular Biology
St. Louis University School of Medicine
St. Louis, MO 63104

INTRODUCTION

The acute lung injury resulting from subcutaneous injection of *N*-nitroso-*N*-methylurethane (NNMU) exhibits pathological characteristics of acute respiratory distress syndrome (ARDS) such as severe impairment of gas exchange, pulmonary edema, neutrophilic influx into the alveolar space, alterations in surfactant composition, and surfactant dysfunction. This chapter summarizes the published data concerning the use of the NNMU animal model in the investigation of ARDS.

N-NITROSO-*N*-METHYLURETHANE

In the early 1960s, *N*-nitroso-*N*-methylurethane (NNMU) was initially used to investigate the carcinogenic actions of diazomethane, a known alkylating agent. Diazomethane, released from NNMU *in vivo*, has been shown to methylate DNA[1] and thiol groups[2]. Depending on species, route of administration, and dosage, the administration of NNMU can produce a variety of effects. In hamsters, NNMU has been used to produce pulmonary fibrosis[3,4]. In the investigation of the mechanisms of carcinogens, researchers have discovered that NNMU can induce DNA synthesis[5], produce single-strand DNA scission in cells of the pyloric mucosa from intraesophogeal administration of NNMU[6], damage DNA of the central nervous system[7], and induce tumor formation in the rat stomach[8]. Despite these publications on the probable actions of NNMU, the precise molecular mechanisms which lead to the production of acute lung injury are yet to be defined.

Acute Respiratory Distress Syndrome: Cellular and Molecular Mechanisms and Clinical Management
Edited by Matalon and Sznajder, Springer Science+Business Media New York, 1998

159

THE NNMU ANIMAL MODELS FOR ARDS

The species which have been used to study ARDS by NNMU administration include the cat, dog, rabbit, and rat. As seen in Table 1, the effects are similar in each species with respect to a variety of parameters. Ryan *et.al.* characterized the NNMU induced lung injury morphologically in dog showing necrosis of the alveolar epithelium with alveolar type II pneumocytes displaying decreased lamellar body inclusions, interstitial edema, and alveolar collapse[9]. Moreover, using a saline-filled lung and tracer molecules (serum albumin 69,000 MW and dextran 500,000 MW), Glauser *et. al.* demonstrated in the NNMU treated dog an increased alveolar permeability based on the decreased time needed to reach 50% equillibrium of the tracer molecules [10]. A thorough review on the NNMU induced acute lung injury in dogs has been prepared by Ryan *et. al.*[11]. In recent years smaller and less expensive animal models (*e.g.*, rabbit and rat) have been developed to investigate probable mechanisms and treatments of ARDS.

The data summarized in Table 1 clearly show that NNMU produces similar effects *in vivo* regardless of which species is used. All species exhibit impaired gas exchange as evidenced by the decreased PaO_2 or elevated A-a dO_2. Furthermore, the composition of the surfactant recovered from bronchoalveolar lavage appears to be altered in comparison to control animals, most notably in the percentage of phosphatidylcholine (PC). The resultant change in surfactant composition may be the cause for the decrease in surface activity of the crude surfactant isolated from bronchoalveolar lavage fluid. Collectively, the elevation in A-a dO_2 may result from the combination of cellular influx, increased alveolar protein, and decreased PC in the surfactant, and decreased surface activity.

INVESTIGATIONS OF ARDS THERAPIES *IN VIVO*

The NNMU-induced acute lung injury animal model has been used to investigate possible treatments for ARDS. In 1989, Harris, *et. al.* demonstrated in rat that exogenous surfactant instillation can improve several parameters in animals displaying ARDS characteristics[12]. In this report, Survanta®, a bovine derived organic surfactant preparation, was instilled intratracheally into NNMU-treated rats that displayed ARDS symptoms resulting in an improvement in A-a gradient, phospholipid-to-protein ratio of surfactant, minimum surface tension, surfactant composition, and survival of the animal[12].

One drawback of exogenous surfactant instillation for the ARDS patient is the introduction of additional fluid volume into the lung. Therefore, the inhalation of nebulized surfactant preparations was also studied using the NNMU-induced acute lung injury animal model. Lewis *et. al.* reported that in NNMU treated rabbits, nebulized surfactant produced a minimal improvement in lung volume at maximal pressure and ventilation efficiency index[13], despite a lack improvement in PaO_2/FIO_2 compared to control, nebulized saline treated rats[14]. This finding maybe due in part to the small quantity (4.9±1.0 mg lipid/kg) of surfactant deposited in the distal lungs[14].

The application of positive end-expiratory pressure (PEEP) has been one mode of improving gas exchange in the ARDS patient. However, depending on the lung injury model tested, PEEP may have varying effects on lung compliance. Investigations into the mechanisms on how PEEP effects lung compliance in NNMU treated dogs have shown that gas exchange (due to over distention of the alveoli) is improved despite a decrease in lung compliance following PEEP[15].

Recently, Veldhuizen *et. al.* reported that mechanically ventilated NNMU treated rabbits that display acute lung injury have an increased ratio of inactive, small surfactant aggregates

160

(SA) to the surface active, large surfactant aggregates (LA)[16]. It was also reported that ventilation strategies using a high tidal volume can promote a high SA-to-LA ratio in the NNMU treated rabbits[17]. The loss of surface activity of the surfactant system due to the increased ratio of SA-to-LA can further exacerbate the acute lung injury. In contrast, PEEP did not change the conversion of LA-to-SA in NNMU treated rabbits[17]. This finding in the NNMU animal model demonstrates that the mode of mechanical ventilation, a common therapy for ARDS patients, may have detrimental effects on the surfactant system thereby contributing to the lung injury.

Table 1. Characteristics of the NNMU-induced acute lung injury in different animal models.

	Cat[18]	Dog 19,[20]	Rabbit[21]	Rat[12]
Dose	12 mg/kg	7-8 mg/kg	12 mg/kg	7.5 mg/kg
Time	3 days	2-4 days	4 days	4 days
Route	s.c.	s.c.	s.c.	s.c.
Gas exchange				
A-a dO$_2$	ND	ND	increased	increased
PaO$_2$	decreased	decreased	decreased	decreased †
PaCO$_2$	increased*	decreased	decreased*	increased †
BALF Cells				
% Macrophages	ND	ND	decreased	decreased
% PMNs	ND	ND	increased	increased
% Lymphocytes	ND	ND	increased*	increased
BALF				
Phospholipids				
Total	ND	decreased	decreased	decreased
PC	decreased*	decreased	decreased*	decreased
PG	ND	decreased	decreased	decreased*
PS	ND	ND	increased	increased
PE	ND	decreased	increased	increased
BALF protein	ND	ND	increased	increased †
Phospholipid/ Protein ratio (CSP)	ND	ND	decreased	decreased
Minimal surface tension (CSP)	ND	increased	increased	increased

* = not significant, † = personal communication
s.c. = subcutaneous, CSP = crude surfactant pellet, ND = not determined

INVESTIGATIONS OF SURFACTANT *IN VIVO*

Table 1 clearly shows that surfactant composition is altered and surface activity is impaired in NNMU treated animals. The NNMU animal model allows for the investigation of surfactant metabolism in an ARDS-like condition *in vivo*. Utilizing the NNMU animal model, it has been demonstrated that the acute lung injury results in an increased pool of SA and conversion from the LA to the SA[16, 22]. This finding supports the data that shows, *in vitro*, the conversion of LA-to-SA by surface area cycling is enhanced by a variety of factors (*e.g.*, proteolytic enzymes, extravascular seum proteins, and decreased percentage of SP-A and phosphatidylglycerol) which can be found in the ARDS pathology[23]. Using radiolabelled exogenous surfactant (^{3}H-dipalmitoylphosphatidylcholine, DPPC), Lewis *et. al.*, demonstrated that the NNMU injured lung exhibited similar rates of clearance for the surfactant lipids as control animals. However, since the total saturated phosphatidaylcholine pool sizes of the lavage and whole lung homogenate was significantly lower in the NNMU injured lung, it was concluded that the metabolism of the surfactant phospholipids *in vivo* is lower in lung injury[24].

One of the more controversial areas in ARDS research is the role(s) of the free radical nitric oxide (NO•). Despite data published on the potentially harmful effects of NO• inhalation[25], NO• has been administered to patients with ARDS providing temporary benefits[26]. Several laboratories have demonstrated the involvement of nitric oxide, through the production of peroxynitrite, in the progression of lung injury using the paraquat, immune-complex, and ishemia-reperfusion animal models[27-30]. *In vitro*, it has also been reported that peroxynitrite can damage the pulmonary surfactant resulting in a loss of lipid aggregation and surface activity through the nitration of surfactant proteins (SP-A, SP-B, SP-C) and lipid peroxidation[30-32]. In human lung tissue derived from deceased patients with acute pulmonary inflammation, intense nitrotyrosine staining was evident in the alveolar epithelium, inflammatory cells, and the intersitium[33].

Recently, our laboratory has demonstrated that the NNMU induced acute lung injury may, in part, involve the generation of nitric oxide (NO•)[34]. Administration by intraperitoneal injection of a nitric oxide synthase inhibitor (either L-nitro-ω-arginine methyl ester (L-NAME) or aminoguanidine (AG), the latter being a selective inhibitor for the inducible nitric oxide synthase isoform) can attenuate the NNMU induced impairment of blood gas exchange. Furthermore, the administration of the inactive stereoisomer of L-NAME, D-nitro-ω-arginine methyl ester (D-NAME), did not reduce the elevated A-a dO$_2$ in the NNMU treated rats. Both L-NAME and AG were capable of attenuating the NNMU induced increase of polymorphonucleated neutrophils, whereas D-NAME demonstrated no effect on this cellular influx. Both L-NAME and AG were effective in attenuating the decrease in the phospholipid-to-protein ratio of the crude surfactant pellet, and in turn, attenuating the impairment of surfactant function resultant from NNMU injections. Furthermore, Western blots of whole lung homogenates displayed an increased expression of the inducible nitric oxide synthase enzyme in NNMU treated rats with or without inhibitor injections. These findings imply that the activity of nitric oxide synthase in generating NO• is involved in many aspects of the pathology of ARDS.

CONCLUSION

The NNMU induced acute lung injury animal model is a unique system for the investigation of ARDS. As an *in vivo* model it offers the possibility to characterize the mechanisms and players involved in the progression of the pathological characteristics of acute lung injury. Furthermore, it provides a tool for investigations of therapies designed to treat the ARDS patient.

REFERENCES

1. R. Schoental, Methylation of nucleic acids by *N*-^{14}C-methyl-*N*-nitrosourethane *in vitro*

and *in vivo, Biochem. J.* 102: 5c-7c (1967).

2. R. Schoental, and D. J. Rive, Interaction of *N*-alkyl-*N*-nitrosourethanes with thiols, *Biochem. J.* 97: 466-474 (1965).

3. J. O. Cantor, B. A. Bray, S. F. Ryan, I. Mandl, and G. M. Turino, Glycosaminoglycan and collagen synthesis in *N*-nitroso-*N*-methylurethane induced pulmonary fibrosis, *Proceedings of the Society for Experimental Biology & Medicine.* 164: 1-8 (1980).

4. J. O. Cantor, M. Osman, J. M. Cerreta, I. Mandl, and G. M. Turino, Glycosaminoglycan synthesis in explants derived from bleomycin-treated fibrotic hamster lungs, *Proceedings of the Society for Experimental Biology & Medicine.* 173: 362-6 (1983).

5. C. Furihata, S. Yoshida, Y. Sato, and T. Matsushima, Inductions of ornithine decarboxylase and DNA synthesis in rat stomach mucosa by glandular stomach carcinogens, *Jpn. J. Cancer Res.* 78: 1363-9 (1987).

6. C. Furihata, E. Ikui, and T. Matsushima, DNA single-strand scission in the pyloric mucosa of rat stomach induced by four glandular stomach carcinogens and three other chemicals, *Mutat. Res.* 368: 1-6 (1996).

7. L. Robbiano, and M. Brambilla, DNA damage in the central nervous system of rats after *in vivo* exposure to chemical carcinogens: correlation with the induction of brain tumors, *Teratogenesis, Carcinogenesis, & Mutagenesis.* 7: 175-81 (1987).

8. W. Lijinsky, and H. W. Taylor, Carcinogenesis in Sprague-Dawley rats of *N*-nitroso-*N*-alkylcarbamate esters, *Cancer Lett.* 1: 275-9 (1976).

9. S. F. Ryan, A. L. Bell Jr, and C. R. Barrett Jr, Experimental acute alveolar injury in the dog. Morphologic--mechanical correlations, *Am. J. Pathol.* 82: 353-72 (1976).

10. F. L. Glauser, R. K. Falls, J. A. Mathers Jr, and J. E. Millen, Pulmonary microvascular and alveolar epithelial permeability characteristics in *N*-nitroso-*N*-methylurethane-injected dogs, *Chest.* 79: 217-21 (1981).

11. S. F. Ryan, C. R. Barrett, and D. F. Liau, Nitrosourethane-induced lung injury, in: *Handbook of Animal Models of Pulmonary Disease*, J. Cantor, ed., CRC press, Inc., Boca Raton (1989).

12. J. D. Harris, F. Jackson Jr, M. A. Moxley, and W. J. Longmore, Effect of exogenous surfactant instillation on experimental acute lung injury, *J. Appl. Physiol.* 66: 1846-51 (1989).

13. R. H. Notter, E. A. Egan, M. S. Kwong, B. A. Holm, and D. L. Shapiro, Lung surfactant replacement in premature lambs with extracted lipids from bovine lung lavage: effects of dose, dispersion technique, and gestational age, *Pediatr. Res.* 19: 569-577 (1985).

14. J. Lewis, M. Ikegami, R. Higuchi, A. Jobe, and D. Absolom, Nebulized vs. instilled exogenous surfactant in an adult lung injury model, *J. Appl. Physiol.* 71: 1270-6 (1991).

15. C. R. Barrett Jr., A. L. Bell Jr., and S. F. Ryan, Effect of positive end-expiratory pressure on lung compliance in dogs after acute alveolar injury, *Am. Rev. Respir. Dis.* 124: 705-8 (1981).

16. R. A. Veldhuizen, Y. Ito, J. Marcou, L. J. Yao, L. McCaig, and J. F. Lewis, Effects of lung injury on pulmonary surfactant aggregate conversion *in vivo* and *in vitro*, *Am. J. Physiol.* 272: L872-8 (1997).

17. Y. Ito, R. A. Veldhuizen, L. J. Yao, L. A. McCaig, A. J. Bartlett, and J. F. Lewis, Ventilation strategies affect surfactant aggregate conversion in acute lung injury, *Am. J.Respir. Crit. Care Med.* 155: 493-9 (1997).

18. P. Richardson, C. L. Bose, V. Dayton, and J. R. Carlstrom, Cardiopulmonary function

of cats with respiratory distress induced by *N*-nitroso *N*-methylurethane, *Pediatr. Pulmonol.* 2: 296-302 (1986).

19. D. F. Liau, C. R. Barrett, A. L. Bell, and S. F. Ryan, Functional abnormalities of lung surfactant in experimental acute alveolar injury in the dog, *Am. Rev. Respir. Dis.* 136: 395-401 (1987).

20. R. A. Smith, B. Venus, M. Mathru, and Y. Shirakawa, Hemodynamic alterations in canine acute lung injury induced with *N*-nitroso-*N*-methylurethane, *Crit. Care. Med.* 12: 576-8 (1984).

21. D. Pappert, N. Gilliard, G. Heldt, T. A. Merritt, P. D. Wagner, and R. G. Spragg, Effect of *N*-nitroso-*N*-methylurethane on gas exchange, lung compliance, and surfactant function of rabbits, *Intensive Care Med.* 22: 345-52 (1996).

22. J. F. Lewis, M. Ikegami, and A. H. Jobe, Altered surfactant function and metabolism in rabbits with acute lung injury, *J. Appl. Physiol.* 69: 2303-10 (1990).

23. R. Higuchi, J. Lewis, and M. Ikegami, *In vitro* conversion of surfactant subtypes is altered in alveolar surfactant isolated from injured lungs, *Am. Rev. Respir. Dis.* 145: 1416-20 (1992).

24. J. F. Lewis, M. Ikegami, and A. H. Jobe, Metabolism of exogenously administered surfactant in the acutely injured lungs of adult rabbits, *Am. Rev. Respir. Dis.* 145: 19-23 (1992).

25. S. Matalon, V. DeMarco, I. Y. Haddad, C. Myles, J. W. Skimming, S. Schurch, S. Cheng, and S. Cassin, Inhaled nitric oxide injures the pulmonary surfactant system of lambs *in vivo*, *Am. J. Physiol. (Lung Cell. Mol. Physiol.)*. 270: L273-L280 (1996).

26. M. Rossetti, H. Guenard, and C. Gabinski, Effects of nitric oxide inhalation on pulmonary serial vascular resistances in ARDS, *Am. J. Respir. Crit. Care Med.* 154: 1375-1381 (1996).

27. H. I. Berisha, H. Pakbaz, A. Absood, and S. I. Said, Nitric oxide as a mediator of oxidant lung injury due to paraquat, *Proc. Natl. Acad. Sci. USA.* 91: 7445-9 (1994).

28. M. S. Mulligan, J. M. Hevel, M. A. Marletta, and P. A. Ward, Tissue injury caused by deposition of immune complexes is L-arginine dependent, *Proc. Natl. Acad. Sci. USA.* 88: 6338-42 (1991).

29. H. Ischiropoulos, A. B. al-Mehdi, and A. B. Fisher, Reactive species in ischemic rat lung injury: contribution of peroxynitrite, *Am. J. Physiol. (Lung Cell. Mol. Physiol.)*. 269: L158-64 (1995).

30. I. Y. Haddad, H. Ischiropoulos, B. A. Holm, J. S. Beckman, J. R. Baker, and S. Matalon, Mechanisms of peroxynitrite-induced injury to pulmonary surfactants, *Am. J. Physiol. (Lung Cell. Mol. Physiol.)*. 265: L555-L564 (1993).

31. I. Y. Haddad, J. P. Crow, P. Hu, Y. Ye, J. Beckman, and S. Matalon, Concurrent generation of nitric oxide and superoxide damages surfactant protein A, *Am. J. Physiol. (Lung Cell. Mol. Physiol.)*. 267: L242-9 (1994).

32. I. Y. Haddad, S. Zhu, H. Ischiropoulos, and S. Matalon, Nitration of surfactant protein A results in decreased ability to aggregate lipids, *Am. J. Physiol. (Lung Cell. Mol. Physiol.)*. 270: L281-L288 (1996).

33. N. W. Kooy, J. A. Royall, Y. Z. Ye, D. R. Kelly, and J. S. Beckman, Evidence for *in vivo* peroxynitrite production in human acute lung injury, *Am. J. Respir. Crit. Care Med.* 151: 1250-1254 (1995).

34. W. S. Cruz, M. A. Moxley, J. A. Corbett, and W. J. Longmore, Inhibition of nitric oxide synthase attenuates NNMU induced alveolar injury in *in vivo*, *Am. J. Physiol. (Lung Cell. Mol. Physiol.)*. (in press).

HYPEROXIA INDUCED FIBROSIS IN ARDS AND BPD: CURRENT PATHOLOGY AND CYTOKINE PROFILES

Jacqueline J Coalson, Theresa M Siler-Khodr, Vicki T Winter, and Bradley A Yoder

Departments of Pathology and Obstetrics and Gynecology
University of Texas Health Science Center at San Antonio
San Antonio, Texas 78284

Department of Pediatrics
Wilford Hall Medical Center
San Antonio, Texas, 78236

Southwest Foundation for Biomedical Research
San Antonio, Texas, 78227

INTRODUCTION

In the February 16, 1967 issue of the New England Journal of Medicine, the pathologic features of two lung diseases, bronchopulmonary dysplasia (BPD) in infants and adult respiratory distress syndrome (ARDS) in adults, were described in ICU patients[1]. This era marked the time mechanical ventilators capable of delivering high pressures and elevated oxygen tensions were being introduced and utilized in intensive care units. Northway and his associates described the distinctive radiographic pattern of bronchopulmonary dysplasia and accompanying pathologic changes in 32 infants managed in the neonatal intensive care unit[2]. The average gestational age and birth weight of surviving BPD infants who did not receive exogenous surfactant treatment was 34 weeks and 2,224 grams respectively. Northway and his group suggested that the pathogenesis of BPD was a prolongation of the healing phase of severe hyaline membrane disease combined with generalized pulmonary oxygen toxicity, and that endotracheal intubation and mechanical ventilation might contribute to disease development. The pathologic lesions described by this group and others during the 1970's, were alternating zones of overdistended alveoli and fibrosis, squamous metaplasia of the airway epithelium, peribronchial fibrosis, obliterative bronchiolitis, and hypertrophy of peribronchial smooth muscle[3-11].

In adult patients with ARDS, the pathologic hallmark was the presence of fibrinous exudation and fibroblastic proliferation in the heavy, beefy lungs examined at autopsy[1].

Acute Respiratory Distress Syndrome: Cellular and Molecular Mechanisms and Clinical Management
Edited by Matalon and Sznajder, Springer Science+Business Media New York, 1998

165

This lesion of diffuse alveolar damage (DAD) was subsequently separated into exudative and reparative (proliferative) phases[12]. Microscopic findings during the 1- to 6-day exudative phase included pulmonary edema, hyaline membranes, alveolar wall edema, and microatelectasis. Polymorphonuclear leukocytes (PMNs) appeared in the edematous alveolar walls and alveolar spaces within 24 to 36 hours following injury. After 6 days, the reparative phase of diffuse alveolar damage was characterized by hyperplasia of epithelial type II cells, an interstitial mononuclear inflammatory infiltrate, and increased numbers of fibroblasts in thickened alveolar walls and in organizing alveolar exudates. If the reparative phase persisted, widespread fibrosis ensued.

INFLUENCE OF CHANGES IN TREATMENT MODALITIES ON DISEASE PATHOGENESIS AND EVOLUTION

As noted above, original descriptions and early publications regarding both of these ICU associated diseases emphasized that lung fibrosis constituted a major pathologic finding. However, the advent of positive end expiratory pressure (PEEP) and continuous positive airway pressure (CPAP), lower oxygen tensions and better ventilatory strategies in the 1970's influenced the pathologies of bronchopulmonary dysplasia and ARDS, but these were not documented in published pathologic studies. Even before the advent of exogenous surfactant treatment, neonatologists were successfully treating more immature babies with low birth weights[13]. The more recent pathology of BPD of these very low birth weight infants, not treated with exogenous surfactant, and ranging from 520 to 1658g in birth weight and 23.5 to 30w gestation age, showed a persistence of simple, evenly-distributed terminal air spaces lined by undifferentiated cuboidal epithelium, and separated by evenly widened septa with variable amounts of fibrosis and increased amounts of subepithelial elastic tissue[14,15]. An arrested development of alveoli was supported by low Emery counts of the terminal respiratory units[14]. Airway changes, such as squamous metaplasia and peribronchial fibrosis were minimal or absent[14,15] (Figure 1a and b).

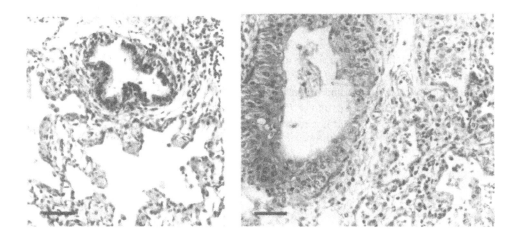

Figure 1. Lung from a 26w gestation human infant with BPD who died at 1 month in the early 1970's (classical BPD) (Fig. 1a) is compared to a 1 month lung specimen from a 25w gestation infant with BPD who died after the use of appropriate oxygenation and ventilatory strategies (Fig. 1b). Differences in the airway epithelial lesions, the inflammatory response and fibroproliferation are evident. Scale bar= 100µm

A lack of severe lung fibrosis has likewise been documented in the majority of patients who died with ARDS. Until the early 1980's it was generally believed that patients with ARDS died with hypoxemic respiratory failure due to underlying proliferative fibrotic lung lesions as a consequence of the DAD process[16]. However, in 1986 a pathologic study of 77 patients autopsied in an ARDS study population documented that over 70% of the patients had evidence of bronchopneumonia at autopsy, and in 21%, it was the only lung finding[17]. Only 9 patients had severe pulmonary fibrosis (reparative diffuse alveolar damage). In this same ARDS study population, it was shown that multiple organ failure, death and infection were highly interrelated, the most common cause of death being multiple organ failure rather than uncorrectable hypoxemia of which only 2 of 141 patients died[18]. It is now widely recognized that most patients with ARDS die with multiple organ failure, not respiratory failure[18-20].

BABOON MODELS OF ARDS AND BPD

Elevated oxygen levels induce DAD lesions and have been used to produce animal models of ARDS. Adult baboon models have been used to study the development of experimental respiratory failure in adult nonhuman primates[21,22]. An intravenous injection of oleic acid followed by ventilation with 100% oxygen or 100% oxygen only for 5-7 days produced acute, diffuse lung injury. Intrapulmonary shunting reduces lung compliance and causes hypoxemia[21]. Histologic studies revealed endothelial injury, pulmonary edema, hyaline membranes, an influx of PMNs after 24 hours, and a reparative response of epithelial type 2 hyperplasia and intraalveolar organization of fibrinous exudate in animals surviving 5 days or more[22].

The pathological lesions known to be associated with experimentally-induced oxygen toxicity and barotrauma are also comparable to those seen in BPD, especially the classical type[23,24]. A moderately severe baboon model of HMD-BPD was elicited with treatment of positive pressure ventilation and 100% oxygen for 10 days or longer in the 140 day prematurely delivered baboon (term gestation = 185 days). This protocol produced severe airway damage and alveolar fibrocellular deposition, comparable to infants with classically described BPD. Recently, a new model of chronic lung injury (CLD, a term used for the current "BPD" disease in human infants) has been developed in which premature baboons of borderline viability (125 days gestation; 67% fullterm) develop a clinical, radiographic, and histologic picture comparable to present day human infants who, despite treatment with exogenous surfactant, develop BPD following only minimal but clinically appropriate amounts of supplemental oxygen and assisted ventilation[25].

THE INFLAMMATORY/INFECTION ETIOLOGIC FACTOR

So at this time, with appropriate oxygen levels and ventilatory strategies being used in intensive care units, the dual injury pattern of hyperoxia and baro/volutrauma has been blunted. Although this has resulted in less severe fibrosis in both diseases, a variable fibroproliferative response does occur. ARDS and BPD still share a common pathogenetic factor that no doubt participates in each one's dysregulated tissue responses - the lung inflammatory/infection autoinjury factor[26,27]. Cell-cell adhesive interactions between neutrophils and endothelium determine how cytokine activated endothelium controls the intensity, temporal pattern and composition of the white blood cells that are recruited into the alveolar spaces. These complex interactions have received extensive study in humans with ARDS and sepsis, and in many adult animal models of lung injury.

Since cytokines appear to participate in the pathophysiology of ARDS and BPD, we have characterized the presence and role of several proinflammatory cytokines in our baboon models of adult induced DAD and infant BPD. Because of the importance of tumor necrosis factor-α (TNF-α) and interleukins-1β and 6 (Il-1 and Il-6) in systemic septic and endotoxic shock responses, plus interleukin-8's (Il-8) role in chemotaxis, these four proinflammatory cytokines have been studied in lavage fluids aspirates from our baboon study groups.

Bronchoalveolar lavage fluid (BALF) was collected using standard techniques at necropsy from normal adults, 6 day 100% O_2-trt adults, 6 day 100% O_2-trt 140d gestation premature baboons , 6 day PRN O_2-trt 140d control infants, 10 day 100% O_2-trt 140d infants, 10 day PRN O_2-trt 140d control infants and 125d gestation PRN O_2-trt CLD infants at 11-14, 21-33, and 41-71 days. A 125d 100% O_2-trt CLD infant group has not been developed. The 125 day PRN group (oxygen as required to maintain appropriate PAO_2 and CO_2 levels) replicates the human disease now seen clinically. The total cell and differential cell counts of the BAL fluids from all study group animals are shown in Figures 2 and 3.

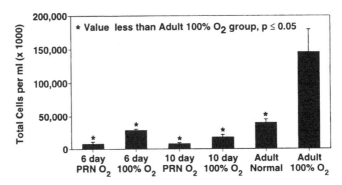

Figure 2. BALF total cell counts of baboon study groups. Anova and Fisher PLSD, significance (p$\leq$0.05).

There was about a threefold increase in the total number of cells in the 6 day 100% O_2-trt 140 d group when compared to the 6 day PRN O_2-trt 140d control infants, a twofold increase in the 10 day 100% O_2-trt 140d infants when compared to the 10 day PRN O_2-trt 140d control infants, and a fourfold increase in the 6 day 100% O_2-trt adult animals when compared to the normal adult value. The adult 6 day 100% O_2-trt group had significantly more cells than all study groups (p$\leq$0.05).

In Figure 3, differential counts of cytospin preparations showed that the relative percent of polymorphonuclear leukocytes was signficantly increased in the 6 day 100% O_2-trt adult animals, establishing the presence of a more extensive alveolitis, when compared to the other study groups (p$\leq$0.05).

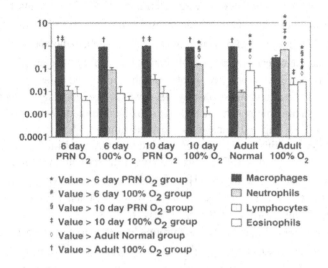

Figure 3. BALF differential cell counts of baboon study groups. Anova and Fisher PLSD, significance (p≤0.05).

Values for the 125 day PRN O_2-trt CLD infant group are shown for comparison to those of the other study groups in Figure 4. This group is a PRN oxygen-treated model and would be treated with appropriate oxgyen as needed comparable to the 6d and 10d PRN study groups.

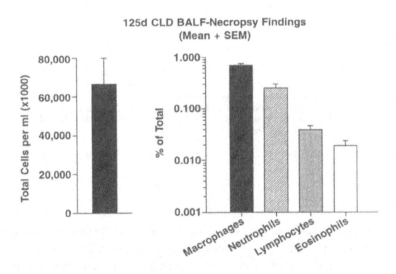

Figure 4. BALF total and differential cell counts of the 125d CLD study group.

IL-6 and TNF-α concentrations were determined in BALF aliquots by specific and sensitive radioimmunoassays. TNF-α was measured using a specific antiserum to human TNF-α (Caltag Laboratories) at a final dilution if 1/100,000, radiolabelled human TNF-α from New England Nuclear and purified human TNF-α for the standard (Collaborative Research). Assay sensitivity was 16 pg/tube and the intra- and interassay coefficients of variation were 5.5% and 6.9% respectively. IL-6 was measured using a specific antiserum to human IL-6 (Sigma Immunochemicals) at a final dilution of 1/100,000, radiolabelled human IL-6 from New England Nuclear and purified human IL-6 from the standard (Austral Biologicals). Assay sensitivity was 0.6 pg/tube and the intra- and interassay coefficients of variation were 6.5% and 11.9%, respectively. Enzyme immunoassays (PerSeptive Diagnostics) were used to measure IL-1β and IL-8. Assay sensitivities were 10pg/ml and the intra- and interassay coefficients of variation were 4.8% and 12%, respectively for IL-1β, and 100 pg/ml and 10% and 24%, respectively for IL-8.

In all adult and infant baboon study groups, whether treated with hyperoxia or not, TNF-α levels at the time of autopsy did not show significant differences (Figure 5). This may reflect that TNF-α is released early in injury, especially in sepsis, so the comparable autopsy levels among the baboon study groups would not negate its presence at higher levels in the hyperoxic injury models at earlier study times.

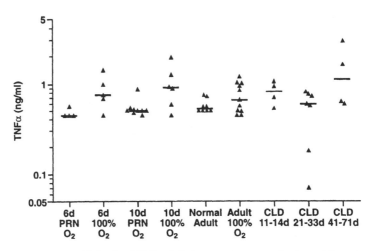

Figure 5. Necropsy BALF levels of TNF-α in baboon study groups. Horizontal lines show median.

IL-1β release usually follows that of TNF-α. Among the baboon study groups, the normal and 100% O$_2$-trt adult baboons both had comparably elevated levels of IL-1β when compared to any of the infant study groups, among which there were no significant differences in IL-1β levels(Figure 6). Whether an age difference exists in Il-β between human adult and infant groups with lung injury has not been documented.

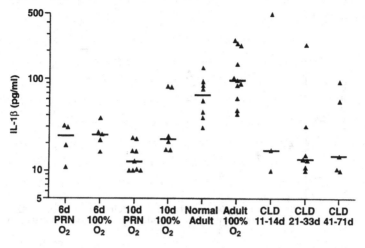

Figure 6. Necropsy BALF levels of IL-1β in baboon study groups. Horizontal lines show median.

Il-6 and Il-8 cytokine levels more clearly separated the injury and control study animals among the baboon study groups. The 100% O_2-trt 10 d infant and adult groups plus the CLD infants (125 d PRN O_2-trt) had significantly elevated Il-8 levels when compared to the 10 day PRN O_2-trt (control) infants and normal adult levels (Figure 7).

A bothersome finding is that the levels of Il-6 and Il-8 in the CLD infants (125 d PRN O_2-trt) are substantively elevated when compared to the two other PRN 140d infant groups, suggesting that there is an ongoing injury and repair response over a number of weeks in these extremely immature infants. Il-6 has many known functions among which

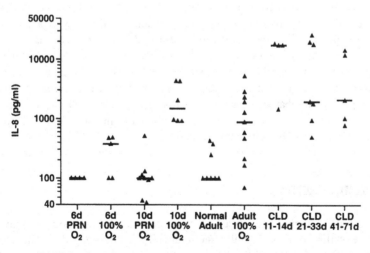

Figure 7. Necropsy BALF levels of Il-8 in baboon study groups. Horizontal lines show median.
Whereas the Il-8 levels were different between the infant PRN and 100% O_2-trt 6 day 140d infants, Il-6 levels at this 6 day study time were similar (Figure 8).

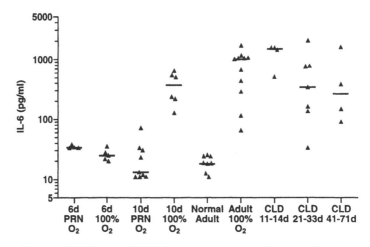

Figure 8. Necropsy BALF levels of Il-6 in baboon study groups. Horizontal lines show median.

are those of regulating the acute-phase response, functioning as a hematopoietic factor, and possessing anti-inflammatory properties[28]. Prominent cell sources are stimulated monocytes/macrophages, cytokine (Il-1 and TNF) stimulated fibroblasts and endothelial cells, and epithelial type 2 cells. Several investigators have shown that Il-6 is increased in the bronchoalveolar lavage samples obtained from adult patients with ARDS[29-31] and human infants with BPD/CLD[32-34].

Recruitment and persistance of neutrophils appear to correlate with the development of ARDS and BPD[35-37]. IL-8 is a major chemotactic factor for neutrophil recruitment. A number of cells, including appropriately stimulated fibroblasts, epithelial cells, endothelial cells, neutrophils and macrophages can produce Il-8[38]. Il-8 levels are elevated in both tracheal aspirate and bronchopulmonary lavage samples from human infants with BPD[34,39-44] and from patients with ARDS[29-31, 45-52]. In several studies, Il-8's persistent elevation in the plasma or BAL fluid of adult patients with acute lung injury predicted poor outcome[29,52], however a more recent study did not substantiate the finding[53].

Many questions remain concerning the significance of elevated cytokine levels in lung injury. A primary one is what constitutes the definition of an increased cytokine concentration. In our ongoing studies with baboons, the use of appropriate control groups that are comparably intubated, ventilated, suctioned, fed, etc., will allow us to separate what may be protective releases of cytokines versus more massive or sustained cytokine releases that may prove to be cell and tissue destructive.

ACKNOWLEDGEMENTS

The writers thank our excellent technical staffs for their support. The administrative skills of Sharyl Reilly are gratefully acknowledged. This work was supported by NIH HL52636 and HL52646 grants.

REFERENCES

1   G. Nash, J.B. Blennerhassett H. Pontoppidan, Pulmonary lesions associated with oxygen therapy and artificial ventilation, *N Engl J Med*. 276:368-74 (1967).

2.   W.H. Northway, R.C. Rosan and D.Y. Porter, Pulmonary disease following respirator therapy of hyaline-membrane disease: bronchopulmonary dysplasia, *N Engl J Med*. 276:357-368 (1967).

3.   M.J. Becker and J.G. Koppe, Pulmonary structural changes in neonatal hyaline membrane disease treated with high pressure artificial respiration, *Thorax*. 24:689-994 (1969).

4.   V.A. Pusey, R. MacPherson, and V. Chernick, Pulmonary fibroplasia following prolonged artificial ventilation of newborn infants, *Can Med Assoc J*. 100:451-457 (1969).

5.   W.R. Anderson and M.B. Strickland, Pulmonary complications of oxygen therapy in the neonate. Postmortem study of bronchopulmonary dysplasia with emphasis on fibroproliferative obliterative bronchitis and bronchiolitis, *Arch Pathol*. 91:506-514 (1971).

6.   C.K. Banerjee, D.J. Girling, and J.S. Wigglesworth, Pulmonary fibroplasia in newborn babies treated with oxygen and artificial ventilation, *Arch Dis Child*. 47:509-518 (1972).

7.   W.R. Anderson, M.B. Strickland, S.H. Tsai, and J.J. Haglin, Light microscopic and ultrastructural study of the adverse effects of oxygen therapy on the neonate lung, *Am J Pathol*. 73:327-348 (1973).

8.   A. Taghizadeh and E.O.R. Reynolds, Pathogenesis of bronchopulmonary dysplasia following hyaline membrane disease, *Am J Pathol*. 82:241-264 (1976).

9.   D.S. Bonikos, K.G. Bensch, W.H. Northway and D.K. Edwards, Bronchopulmonary dysplasia: the pulmonary pathologic sequel of necrotizing bronchiolitis and pulmonary fibrosis, *Hum Pathol*. 7:643-666 (1976).

10. W.R. Anderson and R.R. Engel, Cardiopulmonary sequelae of reparative stages of bronchopulmonary dysplasia, *Arch Pathol Lab Med*. 107:603-608 (1983).

11. J.T. Stocker, Pathologic features of long-standing "healed" bronchopulmonary dysplasia, *Hum Pathol*. 17:943-961 (1986).

12. A.A. Katzenstein and F.B. Askin, Surgical pathology of non-neoplastic lung disease, in: *Major Problems in Pathology*, Volume 13, 2nd edition, WB Saunders Co, Philadelphia (1990).

13. J.T. Wung, A.H. Koons, J.M. Driscoll and L.S. James, Changing incidence of bronchopulmonary dysplasia. *J Pediatr*. 95:845-47 (1979).

14. H.M. Chambers and D. Van Velzen, Ventilator-related pathology in the extremely immature lung, *Pathology*. 21:79-83 (1989).

15. S. Van Lierde, A. Cornelis, H. Devlieger, P. Moerman, J. Lauweryns and E. Eggermont, Different patterns of pulmonary sequelae after hyaline membrane disease: heterogeneity of bronchopulmonary dysplasia?, *Biol Neonate*. 60:152-162 (1991).

16. P. Pratt, Pathology of adult respiratory distress syndrome: implications regarding therapy, Semin Respir Med. 4:79-85 (1982).

17. J.J. Coalson, Pathology of sepsis, septic shock and multiple organ failure, in: *New Horizons: Perspectives on Sepsis and Septic Shock*, W.J. Sibbald and C.L. Sprung, eds, The Society of Critical Care Medicine, Fullerton, CA (1986).

18. R.C. Bell, J.J. Coalson, J.D. Smith, and W.G. Johanson, Jr., Multiple organ failure and infection in adult respiratory distress syndrome, *Ann Intern Med*. 99:293-298 (1983).

19. J.J. Seidenfeld, D.F. Pohl, R.C. Bell, G.D. Harris and W.G. Johanson, Jr., Incidence, site, and outcome of infections in patients with the adult respiratory distress syndrome, *Am Rev Respir Dis*. 134:12-16 (1986).

20. A.B. Montgomery, M.A. Stager, C.J. Carrico and L.D. Hudson, Causes of mortality in patients with the adult respiratory distress syndrome, *Am Rev Respir Dis*. 132:485-489 (1985).

21. R. de los Santos, J.J. Coalson, J.R. Holcomb and W.G. Johanson, Jr., Hyperoxia exposure in mechanically ventilated primates with and without previous lung injury, *Exper Lung Res*. 9:255-275 (1985).

22. J.J. Coalson, R.J. King, V.T. Winter, T.J. Prihoda, A.R. Anzueto, J.I. Peters and W.G. Johanson, Oxygen and pneumonia-induced lung injury in baboons. I. Pathologic and morphologic studies, *J. Appl. Physiol*. 67: 346-356 (1989).

23. J.J. Coalson, T.J. Kuehl, M.B. Escobedo,J.L. Hilliard, F. Smith, K. Meredith, D.M. Null,Jr., W. Walsh, D. Johnson and J.L. Robotham, A baboon model of bronchopulmonary dysplasia. II. Pathologic features, *Exp Mol Pathol*. 37:335-350 (1982).

24. J.J. Coalson, T.J. Kuehl, T.J. Prihod, and R.A. deLemos, Diffuse alveolar damage in the evolution of bronchopulmonary dysplasia in the baboon, *Pediatr Res*. 24:357-366 (1988).

25. J.J. Coalson, V. Winter and B. Yoder, Dysmorphic vascular development in premature baboons with bronchopulmonary dysplasia, *Am J Respir Crit Care Med.* 155:A262, (1997).

26. M.R. Pierce and E. Bancalari, The role of inflammation in the pathogenesis of bronchopulmonary dysplasia, *Pediat Pulmonol.* 19:371-378 (1995).

27. W.W. McGuire, R.G. Spragg, A.B. Cohen and C.G. Cochrane, Studies of the pathogenesis of the adult respiratory distress syndrome, *J Clin Invest.* 69:543-553 (1982).

28. G. Cox and J. Gauldie, Interleukin-6, in *Cytokines In Health and Disease*, 2nd edition, D.G. Remick and J.S. Friedland, eds, Marcel Dekker, New York (1997).

29. G. U. Meduri, S. Healey, G. Kohler, F. Stentz, E. Tolley, R. Umberger and K. Leeper, Persistent elevation of inflammatory cytokines predicts a poor outcome in ARDS. Plasma IL-1β and IL-6 levels are consistent and efficient predictors of outcome over time, *Chest* 107:1062-1073 (1995).

30. S. Chollet-Martin, B. Jourdain, C. Gibert, C. Elbim, J. Chastre and M.A. Gougerot-Pocidalo, Interactions between neutrophils and cytokines in blood and alveolar spaces during ARDS, *Am J Respir Crit Care Med.* 153:594-601 (1996).

31. H. Schutte, J. Lohmeyer, S. Rosseau, S. Ziegler, C. Siebert, H. Kielisch, H. Pralle,l F. Grimminger, H. Morr and W. Seeger, Bronchoalveolar and systemic cytokine profiles in patients with ARDS, severe pneumonia and cardiogenic pulmonary oedema, *Eur Respir J.* 9:1858-1867 (1996).

32. A. Bagchi, R.M. Viscardi, V. Taciak, J.E. Ensor, K.A. McCrea, and J.D. Hasday, Increased activity of interleukin-6 but not tumor necrosis factor-∝ in lung lavage of premature infants is associated with the development of bonchopulmonary dysplasia, *Pediatr Res.* 36: 244-252 (1994).

33. S. Kotecha, L. Wilson, A. Wangoo, M. Silverman, and R.J.Shaw, Increase in interleukin-1 beta and interleukin-6 in bronchoalveolar lavage fluid obtained from infants with chronic lung disease of prematurity, *Pediatr Res.* 40:250-256 (1996).

34. S. Kotecha, Cytokines in chronic lung disease of prematurity, *Eur J Pediatr.* 155(Suppl 2):S14-S17 (1996).

35. T.A. Merritt, C.G. Cochrane, K. Holcomb, B. Bohl, M. Hallman, D. Strayer, D.K. Edwards III and L. Gluck, Elastase and α_1-proteinase inhibitor activity in tracheal aspirates during respiratory distress syndrome, *J Clin Invest.* 72:656-666 (1983).

36. B.E. Ogden, S.A. Murphy, G.C. Saunders, D. Pathak, and J.D. Johnson, Neonatal lung neutrophils and elastase/proteinase inhibitor imbalance, *Am Rev Respir Dis.* 130:817-821 (1984).

37. K.P. Steinberg, J.A. Milberg, T.R. Martin, R.J. Maunder, B.A. Cockrill, and L.D. Hudson, Evolution of bronchoalveolar cell populations in the adult respiratory distress syndrome, *J Respir Crit Care Med.* 150:113-122 (1994).

38. S.L. Kunkel, N.W. Lukacs, S.W. Chensue, and R.M. Strieter, Chemokines and the inflammatory response, in *Cytokines In Health and Disease*, 2nd edition, D.G. Remick and J.S. Friedland, eds, Marcel Dekker, New York (1997).

39. P. Groneck and C.P. Speer, Interleukin-8 in pulmonary effluent fluid of preterm infants, *J Pediatr.* 123:839-840 (1993).

40. J.R. McColm and N. McIntosh, Interleukin-8 in bronchoalveolar lavage samples as a predictor of chronic lung disease in premature infants, *Lancet.* 343:729 (1994).

41. P. Groneck, B. Gotze Speer, M. Oppermann, H. Eiffert, and C.P. Speer, Association of pulmonary inflammation and increased microvascular permeability during the development of bronchopulmonary dysplasia: a sequential analysis of inflammatory mediators in respiratory fluids of high-risk preterm neonates, *Pediatr.* 93:712-718 (1994).

42. S. Kotecha, B. Chan, N. Azam, M. Silverman, and R.J. Shaw, Increase in interleukin-8 and soluble intercellular adhesion molecule-1 in bronchoalveolar lavage of premature infants with chronic lung disease, *Arch Dis Child.* 72:F90-F96 (1995).

43. C.A. Jones, R.G. Cayabyab, K.Y.C. Kwong, C. Stotts, B. Wong, H. Hamdan, P. Minoo, and R.A. deLemos, Undetectable interleukin (Il-10) and persistent Il-8 expression early in hyaline membrane disease: a possible developmental basis for the predisposition to chronic lung inflammation in preterm newborns, *Pediatr Res.* 39:966-975 (1996).

44. K. Tullus, G.W. Noack, L.G. Burman, R.Nilsson, B. Wretlind, and A. Brauner, Elevated cytokine levels in tracheobronchial aspirate fluids from ventilator treated neonates with bronchopulmonary dysplasia, *Eur J Pediatr.* 155:112-116 (1996).

45. E.J. Miller, A.B. Cohen, S. Nagao, D. Griffith, RJ. Maunder T.R. Martin, J.P. Weiner-Kronish, M. Sticherling, E. Christophers and M.A. Matthay, Elevated levels of NAP-1/interleukin-8 are present in the airspaces of patients with the adult respiratory distress syndrome and are associated with increased mortality, *Am Rev Respir Dis.* 146:427-432 (1992).

46. P.G.Jorens, J.V. Damme, W.D. Backer, L. Bossaert, R.F.D. Jongh, A.G. Herman and M. Rampart, Interleukin-8 in the bronchoalveolar lavage fluid from patients with the adult respiratory distress syndrome and in patients at risk for ARDS, *Cytokin.* 4:592-597 (1992).

47. S.C. Donnelly, R.M. Strieter, S.L. Kunkel, A. Walz, C.R. Robertson, D.C. Carter, I.S. Grant, A.J. Pollok and C. Haslett, Interleukin-8 and development of respiratory distress syndrome in at-risk patient groups, *Lancet.* 341:643-647 (1993).

48. S. Chollet-Martin, P. Montravers, C. Gibert, C. Elbim, J.M. Desmonts, J.Y. Fagon, and A. Gougerot-Pocidalo, High levels of interleukin-8 in the blood and alveolar spaces of patients with pneumonia and adult respiratory distress syndrome, *Infect Immun.* 61:4553-4559 (1993).

49. D. Torre, C. Zeroli, M. Giola, G.P. Fiori, G. Minoja, D. Maraggia and M. Chiaranda, Levels of interleukin-8 in patients with adult respiratory distress syndrome, *J Infect Dis.* 167:505-506 (1993).

50. R.P. Baughman, K.L. Gunther, M.C. Rashkin, D.A. Keeton, E.N. Pattishall, Changes in the inflammatory response of the lung during acute respiratory distress syndrome: Prognostic indicators, *Am J Respir Crit Care Med.* 154:76-81 (1996).

51. E.J. Miller, A.B. Cohen and M.A. Matthay, Increased interleukin-8 concentrations in the pulmonary edema fluid of patients with acute respiratory distress syndrome from sepsis, *Crit Care Med.* 24:1448-54 (1996).

52. G.U. Meduri, G. Kohler, S. Headley, E. Tolley, F. Stentz, and A. Postlethwaite, Inflammatory cytokines in the BAL of patients with ARDS; persistent elevation over time predicts poor outcome, *Chest.* 108:1303-1314 (1995).

53. R.B. Goodman, R.M. Strieter, K.P. Steinberg, J.A. Milberg, D.P. Martin, R.J. Mauder, A. Walz, L.D. Hudson, and T.R. Martin, Inflammatory cytokines in patients with persistence of the acute respiratory distress syndrome, *Am J Respir Crit Care Med.* 154:602-611 (1996).

MAP KINASES IN AIRWAY DISEASE: STANDING AT THE CONFLUENCE OF BASIC AND CLINICAL SCIENCE

Paul Vichi and James Posada

University of Vermont
College of Medicine
Department of Molecular Physiology
Burlington, Vermont   05405

THE MAP KINASE FAMILY

Mitogen activated protein (MAP) kinases are a family of serine/threonine protein kinases that currently include three major subgroups; the extracellular regulated kinase 2 (ERK2), the c-Jun NH_2 terminal kinase (JNK), and the p38 kinase. Each of the three major subgroups of MAP kinase has additional isozymes (reviewed in 1). For example, there are three ERK isozymes, and up to 10 JNK isozymes, including differentially processed transcripts[2]. The nomenclature of the JNK MAP kinases is somewhat disjoined owing primarily to the contemporaneous cloning of these kinases. Kyriakis et al. cloned what they termed stress activated protein kinases (SAPK) α, β, γ, which are identical to the JNK MAP kinases cloned by Derijard et al.[3].

BIOLOGICAL FUNCTION OF THE MAP KINASES

The ERK Subgroup

The ERK MAP kinases were cloned in 1990 by Boulton and co-workers[4] and found to be stimulated by many different mitogenic agonists. The flurry of work that followed the cloning of the ERK MAP kinases established a role for ERK in regulating proliferation. A considerable body of evidence now clearly implicates the ERK MAP kinase as playing a central role in controlling proliferation of mammalian cells. Pages et al. demonstrated that microinjection of dominant interfering mutants of ERK MAP kinase blocked growth factor-induced cell proliferation[5]. Cowley et al. expressed constitutively active and dominant negative forms of the MAP kinase kinase, MEK in PC12 cells. In these experiments MEK was made constitutively active by substituting the regulatory serine residues (Ser218, Ser222) with acidic residues to mimic phosphorylation; the dominant

Acute Respiratory Distress Syndrome: Cellular and Molecular Mechanisms and Clinical Management
Edited by Matalon and Sznajder, Springer Science+Business Media New York, 1998

negative form has non-phosphorylatable residues at those sites. Expression of dominant negative MEK inhibited growth factor-induced proliferation, while activated MEK stimulated PC12 cell differentiation and proliferation[6]. Furthermore, the dominant negative MEK mutants were able to revert v-Src- and Ras-transformed cells. Mansour et al. expressed a constitutively active form of MEK in NIH 3T3 cells to demonstrate that constitutive activation of the ERK pathway was sufficient to induce cellular transformation, as well as induce tumors in nude mice[7]. While ERK activation is required for some mitogens to stimulate cell proliferation, it is not sufficient to stimulate cell proliferation, since potent stimulators of ERK activity such as phorbol ester do not appear to stimulate cell proliferation (J. Posada, unpublished observation). Taken together the studies clearly demonstrate the key role played by the ERK MAP kinase in regulating cell proliferation.

The JNK/p38 Subgroup

The JNK and p38 MAP kinases are more recent additions to the MAP kinase family. JNK was purified as a kinase activity that is able to phosphorylate the amino terminus, transactivating portion, of the c-jun transcription factor[2]. The JNK MAP kinase appears to be widely activated in cells treated with agents that induce cell stress (reviewed in 8). For example, JNK is catalytically activated by exposure of cells to ultraviolet light, tumor necrosis factor (TNF), and anisomycin[2]. The p38 MAP kinase is more similar to JNK in this regard and is activated by some of the same agents, but also appears to be activated by cytokines[9,10]. The JNK MAP kinase may regulate cell death, or apoptosis, in a limited number of cell types. For example, in PC-12 cells, withdrawal of nerve growth factor (NGF), resulted in a decrease in ERK activity and an increase in JNK activity, and apoptosis, suggesting that JNK activation is associated with apoptosis. Furthermore, constitutively activating the JNK pathway induced apoptosis[11]. However, in other cell types apoptosis can occur in the absence of JNK activation; JNK activation is also observed without any apoptosis[12]. The biological function of the JNK MAP kinases is therefore less clear than the function of the ERK MAP kinases. The evidence to date implicates the JNKs in the cell stress response, and in a limited set of cell types, mediating apoptosis, although this role may not be generally applicable.

MAP Kinase Activation Mechanism

Prior to the first cloning of the ERK MAP kinase, Ray and Sturgill noticed that the 42 kDa protein that they were studying was a protein kinase whose catalytic activity was itself regulated by threonine and tyrosine phosphorylation[13]. With the cloning of the ERK MAP kinases, it was determined that phosphorylation on a single threonine and a single tyrosine within what we now know as the catalytic cleft of the kinase was required for catalytic activity; removal of phosphate from either residue resulted in a loss of activity[14]. The results of these studies stimulated many investigators to identify and clone the kinase that was responsible for phosphorylating and activating the ERK MAP kinase. These efforts culminated in the cloning of MAP or ERK kinase (MEK), also known functionally as MAP kinase kinase or (MKK)[15]. Serine residues 218 & 222 in MEK were identified as regulatory sites of phosphorylation, revealing that MEK was itself regulated by another kinase[16]. The kinase that activates MEK was identified as Raf, a serine/threonine kinase activated by Ras and cell surface growth factor receptors[17]. Raf therefore is the third kinase in the cascade and is operationally known as the MAP kinase kinase kinase, completing the circuit from the cell surface receptor to ERK activation.

With the cloning of JNK and p38 it became clear that the MAP kinases were actually a family of kinases with the regulatory phosphorylation sites being highly conserved. Alignment of the ERK, JNK, and p38 kinases reveals that the threonine and tyrosine residues whose phosphorylation is required for catalytic activation are conserved in all three kinases. Functionally JNK and p38 are identical to ERK in requiring threonine and tyrosine phosphorylation of the regulatory sites for catalytic activity. The MEK enzyme was found to be specific for ERK and did not phosphorylate JNK or p38, suggesting the existence of additional, specific kinases for JNK and p38. Indeed, at this writing six kinases, termed MAP kinase kinase (MKK) have been cloned that phosphorylate the ERK, JNK and p38 MAP kinases[18]. While the MKKs share a good deal of homology they appear to have substantial substrate specificity. For example, MKK1/MKK2 (also known as MEK) specifically activate ERK1/ERK2, but not JNK or p38. MKK3 specifically activates p38, not ERK or JNK. MKK4 and MKK6 specifically activate JNK, but not p38 or ERK. MKK5 is a specific activator of ERK5/BMK1 (discussed below). Finally, MKK6 phosphorylates p38 but not JNK or ERK[19,20].

While the immediate upstream activators of JNK and p38 are known, the upstream activators of these kinases are not as well worked out as in the ERK pathway. At this point there are a few candidate kinases that may be involved in activating JNK or p38. For example, the MEK kinase-1 (MEKK-1) has been shown to be capable of activating JNK when it is overexpressed, although the physiological significance of this finding has yet to be determined[21]. Ichijo et al. have cloned an upstream activator of the JNK and p38 MAP kinase pathways called ASK1 that is activated during TNF-induced apoptosis, and a kinase-inactive version of Ask1 can inhibit TNF-induced apoptosis, making this kinase an attractive candidate for a physiologically relevant upstream activator of the JNK/p38 pathways[22].

MAP Kinase Substrates

The ERKs are found to phosphorylate a number of important substrates involved in cell proliferation. For example, phospholipase A_2 is an enzyme involved in regulating cell signal transduction pathways, and is an *in vivo* substrate of ERK whose activity is positively regulated by ERK phosphorylation[23]. The ERKs have also been found to phosphorylate transcription factors such as Elk-1 and p62 TCF, both involved in stimulating the transcription of genes that regulate cell proliferation[24,25].

The JNK MAP kinases were initially discovered as an activity that phosphorylated the c-jun transcription factor[26]. JNK is known to phosphorylate serines 63 & 73 in the amino terminus of the c-jun molecule[2]. These residues are in the transactivation domain of c-jun which is distinct from the c-terminal DNA binding domain, and is responsible for stimulating the transcription of numerous genes. JNK also phosphorylates the ATF2 transcription factor[27]. The p38 MAP kinase phosphorylates an IL-1 induced transcription factor called SAP-1[28]. The common theme that arises from a study of various MAP kinase substrates is that many of them are transcription factors that when phosporylated stimulate the transcription of specific genes involved in the biological response to a given agonist. Since activation of ERK can result in very different biological outcomes, additional regulators may be involved in stimulating agonist-specific gene transcription.

MAP KINASE FUNCTION IN AIRWAY BIOLOGY AND PATHOPHYSIOLOGY

Airway Smooth Muscle Cell Proliferation and ERK Activation

There is a large body of evidence implicating increased airway smooth muscle (ASM) cell proliferation in the pathophysiology of asthma (reviewed in 29). Increased ASM proliferation alters the compliance and architecture of the airway and is a common feature of chronic severe asthma. Dissecting the molecular circuitry that is responsible for this pathology is somewhat daunting because there are numerous growth factors in the lung milieu that can stimulate the proliferation of ASM cells in culture. Some of the growth factors that are emerging as important players in regulating the proliferation of ASM cells include growth factors that stimulate receptor tyrosine kinases, such as platelet derived growth factor (PDGF), as well as other mitogens that stimulate G-protein coupled receptors (GPCR), such as thrombin and endothelin[30-32]. One common feature of these mitogens is their ability to stimulate the ERK MAP kinase. While these receptors have distinct upstream activation cascades, they ultimately funnel down to activate the ERK MAP kinase. Activation of ERK in airway smooth muscle is a critical step in ASM cell proliferation. For example, inhibiting ERK activation with the specific MEK inhibitor formulated at Parke-Davis (PD 098059) prevents endothelin-stimulated ERK activation, and ASM cell proliferation, underscoring the requirement for ERK activation in proliferation[33].

MAP Kinases in the Oxidant Stress Response

Many of the mitogens, inflammatory agonists, or reactive oxygen species thought to be important in airway diseases, such as adult respiratory distress syndrome (ARDS) and asthma, involve activation of a MAP kinase pathway. MAP kinase activation is associated with differentiated cell function, cell proliferation, and apoptosis in cultured cells. Determining if a particular MAP kinase isozyme is causally related to the development of a given pathological state has been enigmatic because of the lack of model systems that accurately mimic the disease state. However, many investigators use the increased oxygen model of lung tissue damage to investigate basic questions pertaining to oxidant-induced lung damage. Using cultured intact lung tissue, Shapiro et al. have demonstrated that increased oxygen is associated with an increase in ERK activity but not JNK activity[34]. The data suggest that increased oxygen may trigger a proliferative rather than a stress response.

Reactive oxygen species and their metabolites are thought to play an important role in the pathophysiology of ARDS. For example, nitric oxide (NO) is though to be a key mediator in ARDS. Superoxide, when combined with NO, results in the production of peroxynitrite, another important mediator in ARDS[35-37]. These molecules have extremely complex chemistry that makes it difficult to dissect the molecular systems responsible for mediating their physiological effect. Notwithstanding those difficulties, many labs have demonstrated the activation of various MAP kinase isozymes by reactive oxygen species. Using hydrogen peroxide to model airway remodeling observed in hyperoxic exposed rats, Abe et al. demonstrated the rapid activation of the ERK MAP kinase following exposure to hydrogen peroxide[38]. The authors suggest that ERK activation precedes the increased smooth muscle cell proliferation associated with airways remodeling. In lung, the hydrogen peroxide oxidant stress model appears to specifically involve activation of the ERK pathway. The JNK pathway does not appear to be significantly activated[34] (J. Posada unpublished observation). Therefore, in this tissue, oxidant stress such as hydrogen peroxide is likely to induce a proliferative response consistent with the observed thickening of the airway smooth muscle layer. However, in other cell types, hydrogen peroxide has been reported to induce JNK activation. For example, in chondrocyte cultures treated with hydrogen peroxide investigators have demonstrated activation of the JNK MAP kinases[39]. Activation of these MAP kinases may mediate the response of the cell to stress, which may enable it to survive the insult, although no clear evidence to this effect has been provided. The fission yeast MAP kinase, Sty1, is involved in the yeast stress response to nutrient deprivation and interestingly, oxidants such as hydrogen peroxide, indicating that the MAP kinase pathway has been evolutionarily conserved from yeast to mammals, possibly to enable cells to cope with oxidant injuries[40].

The oxidant stress response activates several kinases in addition to the well known ERK MAP kinases. For example, a distant member of the MAP kinase family, ERK5, (also known as Big MAP Kinase, BMK1) is a 95 kDa kinase that contains the canonical threonine and tyrosine regulatory phosphorylation sites[41]. ERK5/BMK1 is potently activated by hydrogen peroxide, but unlike ERK it is not activated by growth factors. Therefore, ERK5/BMK1 is a MAP kinase that appears to be **specifically** involved in the oxidant stress pathway. At this point little is known regarding the biological consequences of oxidant-induced BMK activation. Another kinase that appears to be specifically induced by oxidant stress is the SOK-1 (Ste20/Oxidant stress response kinase) kinase recently cloned by Pombo et al. SOK-1 is distantly related to the yeast Ste20 kinase, which is an upstream activator of a yeast MAP kinase pathway[42]. The oxidant stress pathway appears to involve the activation of the ERK MAP kinases as well as other kinase pathways that are poorly defined at this point. While the ERKs are activated by growth factors and oxidant stress, BMK and SOK-1 are specifically activated by hydrogen peroxide and not mitogenic agonists, suggesting they are confined to the oxidant stress pathway. The substrates of these kinases are poorly defined.

Nitric oxide has complex and poorly understood biological effects. One of the primary effector molecules thought to be activated by NO is soluble guanylate cyclase, which activates, the cyclic GMP-dependent kinase[43]. Additional signal transduction pathways appear to be involved as well. Recently, investigators have described the activation of the ERK, JNK and p38 MAP kinase isozymes by nitric oxide in chrondrocytes[44]. In Jurkat T cells nitric oxide potently induced JNK activity and weakly activated ERK and p38[39]. The JNK pathway appears to be utilized by cells stimulated by nitric oxide, perhaps to help overcome the stress induced by the agent. The MAP kinases have in common the phosphorylation and stimulation of transcription factors, perhaps these kinases activated by oxidant stress induce the transcription of genes related to disarming the oxidant molecules. On the other hand, JNK activation is associated with apoptosis in some systems, it may be possible that the observed JNK activation in response to nitric oxide is causally related to the induction of apoptosis by this molecule. Studies connecting a biological endpoint such as apoptosis and JNK activation by nitric oxide need to be conducted to determine what the biological consequences of JNK activation are in the context of oxidant stress.

SIGNALING MOLECULES AS THERPEUTIC DRUG TARGETS

The diversity of responses mediated by the MAP kinases requires a complete dissection of a given pathway before intervention is feasible. For example, JNK activation is associated with apoptosis in PC12 cells following withdrawal of nerve growth factor[11], and JNK activation is also associated with Fas activation in 293 and HeLa cells[45]. However, in MCF7 cells TNF causes apoptosis without JNK activation[12]. Therefore, future work will include determining if there is a causal relationship between a MAP kinase activation and a given lung pathology. There is evidence to suggest that in airway smooth muscle cells and epithelial cells oxidant stress results in MAP kinase activation. The critical questions include determining if the injury induced by reactive oxygen species can be prevented by interfering with MAP kinase activation.

Inhibiting MAP Kinase Function in the Lung

Many bioactive compounds in the lung activate the ERK MAP kinase via a cascade of phosphorylation events including Raf-1 -> MEK -> ERK. While this pathway appears to hold for many different agonists there are agonists such as PDGF that may activate ERK by Raf-independent mechanisms. Furthermore, the activation mechanism of Raf is not completely understood, making it possible that there are different activation mechanisms for Raf depending on the agonist used. Therefore, tracing the pathways that are stimulated by individual agonists to understand the upstream activators may be useful in terms of future drug design efforts. For example, if mitogens that stimulate airway smooth muscle cell proliferation such as PDGF, thrombin or endothelin, utilize Raf to activate the ERK MAP kinase. Then the pathway could be more specifically inhibited by targeting this upstream kinase, rather than the ERK molecule itself whose function may serve other functions that may be adversely affected. Understanding the intermediate kinases and activating molecules in the MAP kinase cascades will form the foundation for future drug design.

There are currently no specific inhibitors of the ERK molecules themselves, but there is an inhibitor of MEK, the MAP kinase kinase. Parke-Davis identified the first highly specific inhibitor of MEK. Its mechanism of action may involve inhibiting the activation of MEK by Raf. The compound, PD098059, now commercially available, appears to bind to the regulatory region of MEK and may prevent Raf from phosphorylating serines 218 & 222 which control the catalytic activity of the molecule[46]. Thus, the MEK inhibitor interferes with activation of MEK, thereby preventing it from subsequently activating ERK. The MEK inhibitor appears to be highly specific for MEK and does not inhibit other protein kinases even at higher doses, which is consistent with its presumed mechanisms of action being specific for the MEK-Raf interaction. When used in airway smooth muscle cells, the MEK inhibitor prevented endothelin-induced mitogenesis. However the drug had much less effect on PDGF-induced mitogenesis of the same cell type, suggesting that PDGF may employ additional mechanisms for the activation of ERK[33]. The data indicate PD098059 may not be clinically useful because it does not inhibit cell proliferation by several important agonists that are present in the lung milieu. However, the drug **is** useful for teasing apart various MAP kinase signaling pathways.

There is an effective inhibitor of the p38 kinase produced by SmithKline Beecham, SB 203580, that is also commercially available[47]. This drug appears to bind to the catalytic cleft of the kinase and competes for ATP binding. Competitive ATP inhibitors tend to be somewhat non-specific, but the p38 inhibitor is specific for p38 and does not inhibit JNK or ERK. SB 203580 binds to the catalytic cleft of the p38 kinase but not the ERK or JNK kinase. This finding infers that although the MAP kinases share considerable homology in this region these protein kinases have distinct three dimensional structure in the catalytic cleft region. The p38 MAP kinase does not appear to be one of the primary oxidant stress response MAP kinases in lung tissue, although it is activated by hydrogen peroxide in human Jurkat T cells[39].

Evidence to date indicates that oxidants such as hydrogen peroxide and nitric oxide activate the ERK MAP kinases, and in some cell types the JNK MAP kinases. The cellular damage of oxidant stress is pleiotropic and includes DNA damage, lipid damage and in some cases apoptosis. The role of various MAP kinase isozymes in mediating these effects is not well understood. Using the available inhibitors may provide additional information regarding the role of the MAP kinases in oxidant-induced cell damage. If the MAP kinases do mediate the biological effects of oxidant stress, then inhibiting the enzymes may be a potentially clinically useful treatment modality. Again, further basic research will be required to work out the oxidant signaling pathways.

Gene Therapy: Manipulating MAP Kinase Pathways in the Lung

In addition to pharmacological intervention in the lung, it is feasible to inhibit the ERK or JNK pathways by genetic means. In cultured cells investigators have been able to interfere with MAP kinase pathways by making dominant negative versions of the kinases involved in the pathway. For example, through site directed mutagenesis, Cowley et al. constructed a MEK protein that was kinase-inactive. Expression of the kinase-inactive MEK inhibited the ability of various growth factors to stimulate proliferation[6]. This approach may be useful in the lung to prevent proliferation of epithelial and possibly smooth muscle cells. Since these cell usually have a low proliferation index this treatment may not be associated with significant toxicity to non-proliferating cells. In the JNK MAP kinase pathway, investigators have changed the regulatory phosphorylation sites from threonine and tyrosine, to alanine to prevent phosphorylation and activation, rendering the molecule kinase-inactive. The kinase-inactive JNK had a dominant negative phenotype. For example, expression of the kinase-inactive JNK molecule prevented the cytokine- and UV-induced JNK activation and ATF2 transcription factor phosphorylation[48]. This dominant negative approach may be a useful one to inhibit these pathways *in vivo* in the lung airways and parenchyma.

Future molecular therapy for lung diseases such as asthma or ARDS may include interfering with various MAP kinase pathway in the lung. In the near future this will be feasible by using recombinant, replication deficient adenovirus that harbors the dominant negative MEK or JNK kinase. Inhalation of recombinant adenovirus may allow the airway epithelial cells, and possibly smooth muscle cells, to be infected with the adenovirus containing the dominant negative MAP kinases. Expression of these dominant negative molecules may interrupt the MAP kinase pathways *in vivo* at the molecular level, and ameliorate some of the pathophysiology of the disease. Although gene therapy involving MAP kinases seems futuristic today, the field is rapidly progressing in that direction!

*The authors wish to acknowledge the important contributions of investigators whose work may not have been cited due to space limitations.

LITERATURE CITED

1. Ferrell, J.E. Jr. Tripping the switch fantastic: how a protein kinase cascade can convert graded inputs into switch-like outputs. *Trends in Biochemical Sciences* 21: 460 (1996).
2. Derijard B, Hibi M, Wu I-H, Barrett T, Su B, Deng T, Karin M, and Davis RJ. JNK1: a protein kinase stimulated by UV light and Ha-Ras that binds and phosphorylates the c-Jun activation domain. *Cell* 76: 1025 (1994).
3. Kyriakis, J.M., Banerjee, P., Nikolakaki, E., Dai, T., Rubie, E.A., Ahmad, M.F., Avruch, J., Woodgett, J.R. The stress-activated protein kinase subfamily of c-Jun kinases. *Nature* 369: 156 (1994).
4. Boulton TG, Yancopoulos C, Slaughter C, Moomaw J, Hsu Gregory JS, and Cobb M. An insulin-stimulated protein kinase similar to yeast kinases involved in cell cycle control. *Science* 249: 64 (1990).

5. Pages, G., Lenormand, P., L'Allemain, G., Chambard, J.C., Meloche, S., and Pouyssegur, J. Mitogen-activated protein kinases p42^{mapk} p44^{mapk} are required for fibroblast proliferation. *Proc. Natl. Acad. Sci. USA* 90: 8319 (1993).

6. Cowley, S., Paterson, H., Kemp, P., and Marshall, C. J. Activation of MAP Kinase Kinase is Necessary and Sufficient for PC12 Differentiation and for Transformation of NIH 3T3 cells. *Cell* 77: 841 (1994).

7. Mansour, S. J., Matten, W. T., Hermann, A. S., Candia, J. M., Rong, S., Fukasawa, K., Vande Woude, G. F., and Ahn, N. G. 1994. Transformation of mammalian cells by constitutively active MAP kinase kinase. *Science* 265: 966 (1994).

8. Fanger, G.R., Gerwins, P., Widman, C., Jarpe, M.B., Johnson, G.L. MEKKs, GCKs, MLKs, PAKs, TAKs, and tpls: upstream regulators of the c-jun amino terminal kinases? *Curr. Opin. Gen. & Develop.* 7: 67 (1997).

9. Han, J., Lee, J.D., Bibbs, L., Ulevitch, R.J. 1994. A MAP kinase targeted by endotoxin and hyperosmolarity in mammalian cells. *Science* 265: 808 (1994).

10. Raingeaud, J., Gupta, S., Roger, J.S., Dickens, M., Han, J., Ulevitch, R.J., and Davis, R.J. Pro-inflammatory cytokines and environmental stress cause p38 mitogen-activated protein kinase activation by dual phosphorylation on tyrosine and threonine. *J. Biol. Chem.* 270: 7420 (1995).

11. Xia, Z., Dickens, M., Raingeaud, J., Davis, R. J., and Greenberg, M. E. Opposing Effects of ERK and JNK -p38 MAP Kinases on Apoptosis. *Science* 270, 1326 (1995).

12. Liu Z., Hsu, H., Goeddel, D. V., Karin, M. 1996. Dissection of TNF receptor 1 effector functions: JNK activation is not linked to apoptosis while NF kappa B activation prevents cell death. *Cell* 87: 565 (1996).

13. Ray, L.B., and Sturgill, T.W. Rapid stimulation by insulin of a serine/threonine kinase in 3T3-L1 adipocytes that phosphorylates microtubule-associated protein 2 in vitro. *Proc. Natl. Acad. Sci USA.* 84: 1502 (1987).

14. Posada, J.A. and Cooper, J.A. Requirements for phosphorylation of MAP kinase during meiosis in Xenopus oocytes. *Science*, 255: 212 (1992).

15. Crews, C. M., Alessandrini, A., and Erikson, R. L. The primary structure of MEK, a protein kinase that phosphorylates the ERK gene product. Science 258: 478 (1992).

16. Zheng, C.-F., and Guan, K.-L. Activation of MEK family kinases requires phosphorylation of two conserved Ser/Thr residues. *EMBO J.* 13: 1123 (1994).

17. Dent, P., Haser, W., Haystead, T.A., Vincent, L.A., Roberts, T.M., and Sturgill, T.W. Activation of mitogen-activated protein kinase kinase by v-Raf in NIH 3T3 cells and in vitro. *Nature* 257: 1404 (1992).

18. Derijard, B., Raingeaud, J., Barrett, T., Wu, I-H, Han, J., Ulevitch, J., and Davis, R.J. Independent human MAP kinase signal transduction pathways defined by MEK and MKK isoforms. Science, 267: 682 (1995).

19. Raingeaud, J., Whitmarsh, A.J., Barrett, T., Derijard, B., and Davis, R.J. MKK3- and MKK6-regulated gene expression is mediated by the p38 mitogen-activated protein kinase signal transduction pathway. *Mol. Cell. Biol.* 16: 1247 (1996).

20. Han, J., Lee, J.-D., Jiang, Y., Li, Z., Feng, L., and Ulevitch, R.J. Characterization of the structure and function of a novel MAP kinase kinase (MKK6). *J. Biol. Chem.* 271: 2886 (1996).

21. Minden A, Lin A, McMahon M, Lange-Carter C, Derijard B, Davis RJ, Johnson GL, and Karin M. Differential activation of ERK and JNK mitogen-activated protein kinases by Raf-1 and MEKK. *Science* 266: 1719 (1994).

22. Ichijo, H., Nishida, E., Irie, K., ten Dijke, P., Saitoh, M., Moriguchi, T., Takagi, M., Matsumoto, K., Miyazono, K., Gotoh, Y. Induction of apoptosis by ASK1, a mammalian MAPKKK that activates SAPK/JNK and p38 signaling pathways. *Science.* 275: 90 (1997).

23. Lin, L.-L., Wartmann, M., Lin, A. Y., Knopf, J. L., Seth, A., and Davis, R.J. cPLA$_2$ is phosphorylated and activated by MAP kinase. *Cell* 72: 269 (1993).

24. Gille, H., Kortenjann, M., Thomae, O., Moomaw, C., Slaughter, C., Cobb, M.H., and Shaw, P. E. ERK phosphorylation potentiates Elk-1-mediated ternary complex formation and transactivation. *EMBO J.* 14: 951 (1995).

25. Whitmarsh, A.J., Shore, P., Sharrocks, A. D., and Davis, R.J. Integration of MAP kinase signal transduction pathways at the serum response element. *Science* 269: 403 (1995).

26. Hibi, M., Lin, A., Smeal, T., Minden, A., and Karin, M. Identification of an oncoprotein-and UV-responsive protein kinase that binds and potentiates the c-jun activation domain. *Genes and Development*, 7: 2135 (1993).

27. Gupta, S., Cambell, D., Derijard, B., Davis, R.J. Transcription factor ATF2 regulation by the JNK signal transduction pathway. *Science*, 267: 389 (1996).

28. Whitmarsh, A.J., Yang, S.H., Su, M.S., Sharrocks, A.D., and Davis, R.J. Role of p38 and JNK mitogen-activated protein kinases in the activation of ternary complex formation. *Mol. Cell. Biol.* 17: 2360 (1997).

29. Panettieri R.A., Jr. Airways smooth muscle cell growth and proliferation. *In Airway Smooth Muscle: Development and Regulation of Contractility.* D. Raeburn and M. A. Giambycz, editors. Birchauser Verlag, Basel, Switzerland, (1994).

30. Kellerher, M.D., Abe, M.K., Chao, T-S. O., Soloway, J., Rosner, M.R., and Hershenson, M.B. Role of MAP kinase in bovine airway smooth muscle proliferation. *Am. J. Physiolo.* 268: L894 (1995).

31. Shapiro, P.S., Evans, J.N., Davis, R.J., and Posada, J.A. The seven-transmembrane-spanning receptors for endothelin and thrombin cause proliferation of airway smooth muscle cells and activation of the extracellular regulated kinase and c-Jun NH$_2$-terminal kinase groups of mitogen activated protein kinases. *J. Biol. Chem.* 271: 5750 (1996).

32. Panettieri, R.A., Hall, I.P., Maki, C., and Murray, R.K. α-thrombin increases cytosolic calcium and induces human airway smooth muscle cell proliferation. *Am. J. Resp. Cell Mo. Biol.* 13: 205 (1995).

33. Whelchel, A., Evans, J., and Posada, J., Inhibition of ERK activation attenuates endothelin-stimulated airway smooth muscle cell proliferation. *Am. J. Respir. Cell Mol. Biol.* 16: 589 (1997).

34. Shapiro, P., Absher, M.P., Posada, J.A., and Evans, J.N. Activation of ERK and JNK1 MAP kinases in cultured lung tissue. *Am. J. Physiol.* In press (1997)

35. Matalon, S., DeMarco, V., Haddad, I.Y., Myles, C., Skimming, J.W. Schurch, S., Cheng, S., Cassin, S. Inhaled nitric oxide injures the pulmonary surfactant of lambs in vivo. *Am. J. Physiol.* 270: L273 (1996).

36. Haddad, I.Y., Pataki, G., Hu, P., Galliani, C., Beckman, J.S., Matalon, S. Quantitation of nitrotyrosine levels in lung sections of patients and animals with acute lung injury. *J. Clin. Inves.* 94: 2407 (1994).

37. Ischiropoulos, H., al-Mehdi, Fisher, A.B. Reactive oxygen species in ischemic rat lung injury: contribution of peroxynitrite. *Am. J. Physiol.* 269: L158 (1995).

38. Abe, M.K., Chao, T.O., Solway, J., Rosner, M.R., Hershenson, M.B. 1994. Hydrogen Peroxide Stimulates Mitogen-activated Protein Kinase in Bovine Tracheal Myocytes: Implications for Human Airway Disease. *Am. J. Respir. Cell and Mol. Biol.* 11: 577 (1994).

39. Lo, Y.Y.C., Wong, J.M.S., Cruz, T.F. Reactive oxygen species mediate cytokine activation of c-Jun NH2—terminal kinases. *J. Biol.Chem.* 271: 15703 (1996).

40. Shieh, J.C., Wilkinson, M.G., Buck, V., Morgan, B.A., Makino, K., Millar, J.B. The Mcs4 response regulator coordinately controls the stress-activated Wak1-Wis-Sty1 MAP kinase pathway and fission yeast cell cycle. *Genes & Development* 11: 1008 (1997).

41. Abe, J.-I., Kusuhara, M., Ulevitch, R.J., Berk, B.C., and Lee, J.-D. Big mitogen-activated protein kinase 1 (BMK) is a redox-sensitive kinase. *J. Biol. Chem.* 271: 16586 (1996).

42. Pombo, C.M., Bonventure, J.V., Molnar, A., Kyriakis, J., Force, T. Activation of a human Ste20-like kinase by oxidant stress defines a novel stress response pathway. *EMBO J.* 15: 4537 (1996).

43. Roczniak, A., Burns, K.D. Nitric oxide stimulates guanylate cyclase and regulates sodium transport in rabbit proximal tubule. *Am. J. Physiol.* 270: F106 (1996).

44. Lander, H.M., Jacovina, A.T., Davis, R.J., and Tauras, J.M. Differntial activation of mitogen-activated protein kinases by nitric oxide. *J. Biol. Chem.* 271: 19705 (1996).

45. Yang, X., Khosravi-Far, R., Chang, H., and Baltimore, D. Daxx, a novel Fas-binding protein that activates JNK and apoptosis. *Cell.* 89: 1067 (1997).

46. Alessi, D.R., Cuenda, A., Cohen, P., Dudley, D.T., and Saltiel, A.R. PD 098059 is a specific inhibitor of the activation of mitogen-activated protein kinase kinase in vitro and in vivo. *J. Biol. Chem.* 270: 27489 (1995).

47. Cuenda, A., Rouse, J., Doza, Y.N., Meier, R., Cohen, P., Gallagher, T.F., Young, P.R., Lee, J.C. SB 203580 is a specific inhibitior of a MAP kinase homologue which is stimulated by cellular stress and interleukin-1. *FEBS Letters* 364: 229 (1995).

48. Gupta, S., Campbell, D., Derijard, B., Davis, R.J. Transcription factor ATF2 regulation by the JNK signal transduction pathway. *Science,* 267: 389 (1995).

OXIDANT-REGULATED GENE EXPRESSION IN INFLAMMATORY LUNG DISEASE

Linda D. Martin[1], Thomas M. Krunkosky[1], Judith A. Voynow[2], and
Kenneth B. Adler[1]

[1]North Carolina State University
College of Veterinary Medicine
4700 Hillsborough Street
Raleigh, NC 27606

[2]Pediatric Pulmonary Diseases Dept.
Duke University Medical Center
Durham, NC 27710

INTRODUCTION

Acute respiratory distress syndrome (ARDS) is characterized by severe hypoxemia, diffuse pulmonary infiltrates and poor lung compliance in the absence of left heart failure[1,2]. Early pathogenic changes include pulmonary neutrophil sequestration and intravascular fibrin-platelet aggregates[3-5]. Subsequent injury to the alveolar-capillary barrier leads to increased pulmonary vascular permeability causing progressive lung inflammation and pulmonary edema[4]. Persistent inflammation frequently leads to fibrosis.

The presence of activated neutrophils appears to be a major contributor to the lung injury observed in ARDS. Normally present in very small numbers in the alveoli, the transendothelial migration of large numbers of neutrophils is observed during ARDS. This migration may be due in part to inappropriate adhesion, as the expression of intercellular adhesion molecule 1 (ICAM-1) is increased[6-8]. This overwhelming presence of neutrophils results in the release of substantial amounts of reactive oxygen species and destructive enzymes such as elastase[9].

Activated neutrophils also release cytokines such as tumor necrosis factor alpha (TNFα), interleukin 6 (IL-6) and interleukin 8 (IL-8), which can increase the inflammatory response observed in ARDS[10]. Interestingly, some studies have noted a correlation between the degree of neutrophil activation and the level of TNFα in plasma[11]. Similar correlations have been observed for IL-6 and IL-8[12]. These results suggest cytokines may play a role in priming neutrophils, exacerbating the destructive effects of neutrophils and cytokines during ARDS.

Further injury in ARDS may also be due to additional reactive oxygen and nitrogen species (ROS/RNS) generated during treatment. Since there is no effective therapy for altering the pulmonary vascular permeability associated with ARDS, therapy is relegated to treating the underlying causes of the increased permeability and attending to adequate perfusion of the organs[2]. Prolonged exposure to high concentrations of oxygen can produce further acute lung injury. In addition, use of nitric oxide (NO) to decrease the oxygen requirement may have toxic effects, reacting with oxygen and superoxide present in the ventilated patient to form additional toxic products[1].

Acute Respiratory Distress Syndrome: Cellular and Molecular Mechanisms and Clinical Management
Edited by Matalon and Sznajder, Springer Science+Business Media New York, 1998

Stress-induced genes and the roles played by their products in inflammatory or fibrotic respiratory diseases such as ARDS have been the focus of much recent research. Work in our laboratory has focused on inflammatory cell-released substances such as TNFα and ROS/RNS, and their role in the regulation of expression of genes whose products affect further injury during ARDS. Recent topics of interest include molecular regulation of ICAM-1 and IL-6 gene expression in response to exogenous oxidative stress, TNFα stimulation, and signal transduction pathways involving ROS/RNS.

INTERACTION OF OXIDANTS AND CELLS

Sources of Oxidants

Airway epithelial cells are continuously exposed to exogenous and endogenous sources of ROS/RNS. Environmental pollutants are a major source of exogenous oxidative species. Such pollutants include particles and fibers such as industrially-derived ground silica and asbestos which may themselves contain free radicals[13,14]. In addition, particles such as residual oil fly ash and ambient air dusts may contain ionizable concentrations of transition metals which are capable of generating ROS in human airways[15]. Inhaled gaseous pollutants such as ozone, nitrogen dioxide, automobile exhaust, and cigarette smoke also contain numerous oxidants, adding to the oxidant burden in the airway upon inhalation[16].

Therapies used to increase oxygenation of the body's organ systems during congestive lung diseases such as ARDS can also be a source of exogenous ROS/RNS[1]. Supplemental oxygen and positive pressure mechanical ventilation are essential in the treatment of ARDS in order to maintain adequate gas exchange and provide ventilatory muscle rest. However, based on observed pathological changes, this prolonged exposure to high oxygen concentrations can cause acute lung injury[17]. While no prospective randomized clinical trials have been conducted on the use of NO as a therapeutic agent for the treatment of ARDS, its use suggests some positive benefits such as increased oxygenation and reduced pulmonary arterial pressure. However, the potential for NO toxicity during treatment is very real, with NO capable of further reaction with oxygen and superoxide in the ventilated airway to form toxic products such as peroxynitrite[1].

Probably the greatest source of reactive species within the airway, however, derives from the endogenous and infiltrating inflammatory cells present following exposure to exogenous pollutants, antigens, bacteria or viruses. These reactive species result mainly from the respiratory burst of stimulated macrophages and from stimulated polymorphonuclear leukocytes (neutrophils)[9].

Endogenous reactive species can also be generated within airway epithelial cells in response to proinflammatory cytokines such as TNFα[18]. This relationship is evidenced by the conversion of cellular glutathione to its oxidized form, presumably as a protection against increased intracellular oxidants, following exposure of human epithelial type II cells (A549) to TNFα[19]. At least some of the reactive species generated in response to TNFα appear to be side products derived from electron transport reactions taking place in the mitochondria of stimulated cells[20]. TNFα can also upregulate gene expression and activity of xanthine oxidase, a very abundant enzyme in epithelial cells, which can lead to generation of superoxide anion and hydrogen peroxide from oxygen present in airways. The presence of additional cytokines such as interferon-γ (IFN-γ), interleukin-1 (IL-1), and IL-6 causes additive effects on xanthine oxidase up regulation, adding further to the oxidant burden within epithelial cells[21,22]. It has been proposed that such xanthine oxidase-mediated events contribute to the morbidity and mortality associated with ARDS[23].

Cellular Defenses Against Oxidants

As exposure to potentially damaging oxidative stress is unavoidable in airway epithelial cells, these cells utilize numerous antioxidant defenses evolved by cells in aerobic organisms to counter the cytotoxic effects of ROS/RNS. These mechanisms include both enzymatic and non-enzymatic defenses. The most important antioxidant enzymes within airway epithelial cells include catalase (CAT), superoxide dismutase (SOD), and glutathione cycle enzymes such as glutathione peroxidase (GPO). The ability of CAT and GPO to consume hydrogen peroxide (H_2O_2) in airway epithelial cells has been suggested using inhibitory experiments. While inhibition of GPO potentiates the damage observed in epithelial cells following exposure to exogenous H_2O_2, only inhibition of CAT significantly inhibits the ability of tracheal and alveolar epithelial cells to consume H_2O_2[24-26]. In addition, the glutathione oxidation-reduction cycle aids in the reduction of intracellular hydroperoxides, and plays an important role in degrading lipid peroxides and products of lipoxygenase-catalyzed reactions[27].

The Cu, Zn form of SOD is constitutively expressed in the cytosol and nucleus of epithelial cells, while the inducible manganese form (MnSOD) is located in the mitochondria[28]. Evidence suggests MnSOD can be upregulated by reactive oxygen species or cytokines such as TNF, IL-1 and IL-6 or inflammatory minerals[29-32]. This upregulation may provide epithelial cells with protection in inflammatory situations where increases in reactive species are likely[33].

The prevalence of antioxidant enzymes in airway epithelial cells likely plays a role in maintaining a healthy airway. For example, greater concentrations of total glutathione are observed in bronchial and alveolar fluid from asthmatics compared to concentrations in normal control subjects[34]. This finding, coupled with an observed reduction in GPO levels in asthmatic adults and children[35], suggests ROS are inappropriately defended against in inflammatory respiratory diseases, leading to increased morbidity.

Nonenzymatic epithelial antioxidants include vitamins E and C, β-carotene, uric acid, and thiols. Vitamin E acts by stopping the chain reaction involved in lipid peroxidation. Carotenoids and vitamin C act to quench radicals in nonlipid cellular compartments[36]. Uric acid, the major product of oxidative reactions catalyzed by xanthine dehydrogenase, protects against oxidative damage of lipids, proteins, and nucleobases by directly scavenging hydroxyl radical, singlet oxygen, hypochlorous acid, oxoheme oxidants and hydroperoxyl radicals[37]. In addition, uric acid is a potent iron chelator, and can immobilize iron which could contribute to production of intracellular oxidant species via the Fenton reaction[38]. Recent investigation has suggested that S-thiolation-dethiolation may be a reversible process whereby protein sulfhydryls are protected against irreversible oxidation. In these reactions, sulfhydryl groups on proteins become covalently bound to glutathione, γ-glutamylcysteine or cysteine following a large oxidative insult such as the monocyte respiratory burst. The glutathione/thioredoxin reductase systems are implicated in the dethiolation process[39]. Finally, some proteins produced in response to cellular stress may also play a role in protecting the cell from damage by oxidative species. One such protein is heme oxygenase 1 (HO-1) which catalyzes the rate-limiting step in the conversion of heme to bilirubin. Thus, with the production of additional bilirubin, an efficient free radical scavenger, the intracellular oxidant burden can be decreased[40,41].

The airway epithelium is also protected from ROS by mucus overlying the epithelium. The mucus, with its many sugar moieties, may act similarly to mannitol and glucose by scavenging hydroxyl radical and H_2O_2[42,43]. In cultured guinea pig epithelial (GPTE) cells, removal of the apical mucus layer reduces the efficiency of H_2O_2 scavenging[44]. Thus, in diseases such as cystic fibrosis, asthma, or ARDS, where the epithelial layer is injured via denuding, fibrosis, pressure injury, or increased mucus viscosity, further injury may occur due to the decreased ability of mucus to serve as an antioxidant.

MECHANISMS ALTERING GENE EXPRESSION IN RESPONSE TO OXIDANTS

The airway epithelium responds functionally to exposure to ROS/RNS via increases in secretion of mucus, surface expression of adhesion molecules, and release of cytokines. Most of these changes are regulated by changes in the expression of genes within the epithelial cells. Thus, it is important to understand the numerous molecular mechanisms by which oxidant species can alter gene expression. In this way, it may be possible to understand how functional reactions to oxidative species can at times serve to prevent disease and at other times exacerbate airway inflammation.

Oxidant-Sensitive Transcription Factors

Stimulation with exogenous ROS/RNS may change the redox state of a cell. Inflammatory mediators such as TNFα also appear to have this capability. In order for cellular function to be modulated, the cell must respond to these shifts in oxidation/reduction equilibria or the production of new reactive signaling molecules through the modulation of regulatory proteins that have the potential to initiate gene transcription[45,46]. Two transcription factors shown to be regulated by the redox state of the cell are nuclear factor kappa beta (NFκB) and activator protein 1 (AP-1)[47]. Oxidants such as H_2O_2 and hypochlorous acid as well as multiple cytokines have been shown to alter the NFκB complex, increasing its nuclear translocation and ability to bind DNA[48,49]. In addition, activation of NFκB has also been demonstrated in inflammatory cells from patients with airway diseases such as ARDS, suggesting an in vivo role for the complex in regulating the expression of inflammatory genes[50]. The activation of AP-1 appears to oppose that of NFκB. For example, expression of AP-1 and its DNA binding ability in response to TNFα can be increased by exposure to antioxidants such as pyrrolidine dithiocarbamate (PDTC)[51].

NFκB defines a family of transcription factors that share a common structural motif for DNA binding and dimerization. Once in the nucleus, these dimers bind a consensus sequence found in the promoter region of many genes including those coding for cytokines such as IL-6, IL-8, IL-1 and TNFα. NFκB activity and nuclear translocation is controlled by a family of inhibitory subunits called IκB. When IκB is bound to NFκB, an inactivated NFκB:IκB complex is formed which cannot translocate to the nucleus, thus inhibiting the ability of NFκB to bind DNA. Phosphorylation of IκB serves as a tag for the addition of ubiquitin. Following ubiquitination, IκB can be recognized by a proteosome and degraded, allowing NFκB to translocate to the nucleus[52]. Intracellular ROS appear to play a key role in regulating phosphorylation of IκB. For example, cells overexpressing GPO exhibit decreased translocation of NFκB and decreased phosphorylation of IκB in response to oxidative stress or TNFα exposure. A correlation also exists between the rise in ROS levels in TNFα-stimulated cells and the degradation of IκB[53].

Effects of Oxidants on Signal Transduction Pathways

While some of the mechanisms that directly alter the function of oxidant-sensitive transcription factors are known, far less is understood about how the signal transduction pathways which trigger these mechanisms are affected by oxidative stress. Oxidant stress is known to cause cellular injury via lipid peroxidation. This effect can then trigger the activation of membrane-associated enzymes such as phospholipases and protein kinase C (PKC). For example, t-butyl hydroperoxide activates phospholipase A_2 (PLA_2), ultimately stimulating cyclooxygenase metabolism[54]. Exposure to ozone results in release of arachidonic acid, eicosanoids and platelet-activating factor (PAF) from a variety of pulmonary cell types[55,56]. PKC translocation and activation can be stimulated by ROS produced from purine + xanthine oxidase (PXO) in GPTE and BEAS-2B cells (B. Fischer, unpublished observations). Ciliary activity in sheep airways is also inhibited by H_2O_2 via activation of PKC second messenger pathways[57].

Other enzymes associated with intracellular signaling pathways may also be redox sensitive. At least two members of the src family of protein tyrosine kinases have been found to be activated by H_2O_2. In addition, the protein tyrosine kinase syk is responsive to treatment with H_2O_2[58]. Protein-tyrosine phosphatases have reactive cysteine residues in their active sites which may make these enzymes oxidant-sensitive[59]. Studies using inhibitors have also suggested that IL-8 gene expression in response to asbestos exposure may be governed by redox-induced changes leading to phosphorylation events, mediated by tyrosine kinases. In addition, this signaling pathway may involve PKC. Ultimately, the phosphorylation events set in motion by this redox-sensitive pathway appear to activate nuclear proteins which recognize the NFκB/NF-IL-6 sites of the IL-8 promoter, contributing to IL-8 gene expression[60].

Other steps in intracellular signaling pathways may also be oxidant regulated. For example, hydroperoxides have been shown to increase the concentration of intracellular calcium[61]. Some studies suggest that H_2O_2 treatment induces inositol-1,4,5-triphosphate (IP_3) production, which allows IP_3 to bind to intracellular receptors and release stored calcium[58]. It may be this increase in calcium which then stimulates phospholipase activity resulting in proteolytic events which have direct effects on the activation of oxidant-sensitive transcription factors such as NFκB[45,61].

ROLE OF ROS/RNS IN THE EXPRESSION OF INFLAMMATION-ASSOCIATED GENES

Regulation of ICAM-1 Expression

Much of the epithelial destruction observed in ARDS is due to the large number of infiltrating neutrophils releasing substantial amounts of ROS and elastase within the airways. One protein which may play a role in the recruitment of these neutrophils is ICAM-1, a transmembrane glycoprotein that promotes immunological and inflammatory reactions such as leukocyte diapedesis within the lung. Distinct cis-regulatory elements within the ICAM-1 promoter which respond to oxidants as well as TNFα have been functionally identified using endothelial cells[62]. Recent experiments in our laboratory demonstrate that ICAM-1 is constitutively expressed on the surface of normal human bronchial epithelial (NHBE) cells grown in primary culture. Both ICAM-1 gene and surface expression can be stimulated by TNFα. Studies utilizing northern analysis demonstrate that TNFα increases ICAM-1 steady-state mRNA by one hour. This TNFα-stimulated increase in ICAM-1 mRNA above control levels can be inhibited by actinomycin D, an inhibitor of RNA polymerase II. Investigations of the effects of antioxidants dimethylthiourea (DMTU) and PDTC on TNFα-induced ICAM-1 expression reveal the ability of these antioxidants to inhibit ICAM-1 expression at the surface and mRNA level. These data suggest a role for intracellular ROS/RNS in the signal transduction pathway regulating TNFα-stimulated ICAM-1 gene expression in airway epithelium.

Since TNFα-induced ICAM-1 expression appears to be regulated mainly at the level of transcription in NHBE cells, it is possible that intracellular reactive species are acting to alter the activation of oxidant-sensitive transcription factors such as NFκB. One consensus NFκB-binding site located in the human ICAM-1 promoter has been functionally identified as important for expression of ICAM-1 in response to TNFα in endothelial cells[62]. Utilizing this NFκB binding site in an electrophoretic mobility shift assay (EMSA), we have demonstrated that proteins in nuclear extracts from cells stimulated with TNFα bind to this site in vitro. Examination of the bound complex with appropriate antibodies indicates binding of NFκB protein subunits.

Regulation of IL-6 Expression

Inflammatory cell-derived substances such as TNFα likely play a role in regulating expression of other genes which may increase the inflammation observed during ARDS. In addition, gene expression of some of these secondary inflammatory mediators may also be regulated by ROS/RNS released from infiltrating macrophages and neutrophils. One such gene which is regulated in this manner is IL-6. An increase in secreted IL-6 is observed in response to treatment of NHBE cells with oxidants, mainly superoxide generated by PXO, or with TNFα. Interestingly, a time course of mRNA increase during these treatments reveals that the IL-6 steady-state message rapidly increases in response to PXO (in 30 minutes) and returns to near control levels by one hour. In contrast, treatment of NHBE cells with TNFα results in an increase in steady-state message by 1 hour which remains elevated up to eight hours post-treatment. Thus, it appears that these two types of stimuli may exert their effects on IL-6 expression via different molecular mechanisms or signal transduction pathways.

The cell permeable oxidant scavenger DMTU can block the increase in secreted IL-6 observed following exposure of cells to PXO or TNFα. These data suggest that intracellular reactive species may play a role in signaling an increase in IL-6 expression in response to these stimuli. Species that may be involved in this signaling include hydroxyl radical, H_2O_2, hypochlorous acid or peroxynitrite, since all of these species have been shown to react with DMTU[63].

Efforts are currently underway in our laboratory to determine the level at which IL-6 gene expression is regulated in response to PXO or TNFα. Work to date suggests that newly synthesized proteins are not needed for IL-6 expression in response to PXO, as cycloheximide cannot block the PXO-induced increase in IL-6 mRNA. By contrast, when NHBE cells are incubated with actinomycin D plus PXO, the level of steady-state mRNA falls below that observed in untreated control cells. While this suggests some role for transcription in the regulation of IL-6 gene expression in response to oxidative stress, it also suggests a possible mRNA stability component to this regulation. A modulation in mRNA stability would also help account for the rapid rise and fall observed in the steady-state IL-6 mRNA in response to PXO.

Since there appears to be some component of transcriptional regulation to IL-6 expression, we have begun to examine the ability of nuclear proteins to bind to sites present in the IL-6 promoter using EMSAs. Nuclear extracts from cells exposed to PXO or TNFα were used in shift assays with either the NFκB or NF-IL-6 site found in the IL-6 promoter. Interestingly, extracts treated with PXO or TNFα had different patterns of binding to these sequences. TNFα-treated extracts appeared to have more proteins capable of binding the NFκB site when compared with binding of proteins from control extracts. However, no changes in the ability of the treated extracts to bind the NF-IL-6 site were observed. By contrast, nuclear extracts from PXO-treated cells caused an increase in binding to the NF-IL-6 site, but not the NFκB site. These data suggest different modes of transcriptional regulation of IL-6 gene expression in response to PXO or TNFα. In addition, they suggest that while NF-IL-6 has not been traditionally thought of as an oxidant-sensitive transcription factor, it does appear to respond to exogenous oxidant stress. It now remains to determine whether these transcription factor sites play a role in differential expression of IL-6 in response to these two stimuli in vivo.

CONCLUSION

ROS/RNS can play a variety of roles in signaling pathways which lead to changes in the expression of proteins important to the development of inflammatory airway and lung disease. ROS/RNS can serve directly as stimuli, setting in motion signaling pathways which ultimately alter gene expression. Additional stimuli, such as the cytokine TNFα, can also trigger these pathways. Both types of stimuli may, under certain circumstances, lead to production of intracellular reactive species which can serve directly as signaling molecules, or indirectly alter the redox state of the cell. This change ultimately results in activation or production of transcription factors which may or may not be traditionally thought to be oxidatively-regulated. Such activation of transcription factors may occur via numerous mechanisms such as phosphorylation/dephosphorylation, increased transcription, nuclear localization or conformational change. In turn, these transcription factors can cause expression of transcriptionally-controlled genes, ultimately affecting changes in the expression of proteins like ICAM-1 and IL-6 which are important in inflammatory airway and lung disease.

REFERENCES

1. W. Fulkerson, N. MacIntyre, J. Stamler, and J. Crapo, Pathogenesis and treatment of the adult respiratory distress syndrome, *Arch Intern Med.* 156:29 (1996).
2. S. Watling and J. Yanos, Acute respiratory distress syndrome, *Ann Pharmacother.* 29:1002 (1995).
3. M. Lamy, R. Fallat, E. Koeniger, H.P. Dietrich, J.L. Ratliff, R.C. Eberhart, H.J. Tucker, and J.D. Hill, Pathologic features and mechanisms of hypoxemia in adult respiratory distress syndrome, *Am Rev Respir Dis.* 114:267 (1976).
4. M. Bachofen and E. Weibel, Alterations of the gas exchange apparatus in adult respiratory insufficiency associated with septicemia, *Am Rev Respir Dis.* 116:589 (1977).

5. N. Ratliff, J. Wilson, F. Mikat, D. Hack, and T. Graham, The lung in hemorrhagic shock. IV: the role of the polymorphonuclear leukocyte, *Am J Pathol.* 65:325 (1971).

6. M. Mulligan, J. Varani, M.K. Dame, C.L. Lane, C.W. Smith, D.C. Anderson, and P.A. Ward, Role of endothelial-leukocyte adhesion molecule (ELAM-1) in neutrophil-mediated lung injury in rats, *J Clin Invest.* 88:1396 (1991).

7. P. Eichacker, A. Farese, W.D. Hoffman, S.M. Banks, T. Monginis, S. Richmond, G.C. Kuo, T.J. Macvittie, and C. Natanson, Leukocyte CD11B/18 antigen-directed monoclonal antibody improves early survival and decreases hypoxemia in dogs challenged with tumor necrosis factor, *Am Rev Respir Dis.* 145:1023 (1992).

8. I. Engelberts, S. Samyo, J. Leeuwenberg, C. van der Linden, and W. Buurman, A role for ELAM-1 in the pathogenesis of MOF during septic shock, *J Surg Res.* 53:136 (1992).

9. Y. Sibille and H. Reynolds, Macrophages and polymorphonuclear neutrophils in lung defense and injury, *Am Rev Respir Dis.* 141:471 (1990).

10. C. Dinarello, J. Gelfand, and S. Wolff, Anticytokine strategies in the treatment of the systemic inflammatory response syndrome, *JAMA.* 269:1829 (1993).

11. S. Chollet-Martin, P. Montravers, C. Gilbert, C. Elbim, J.M. Desmonts, J.V. Fagon, and M.A. Gougerot-Pocidalo, Subpopulation of hyperresponsive polymorphonuclear neutrophils in patients with adult respiratory distress syndrome, *Am Rev Respir Dis.* 146:990 (1992).

12. S. Chollet-Martin, B. Jourdain, C. Gibert, C. Elbim, J. Chastre, and M. Gougerot-Pocidalo, Interactions between neutrophils and cytokines in blood and alveolar spaces during ARDS, *Am J Respir Crit Care Med.* 154:594 (1996).

13. V. Vallyathan, X. Shi, N. Dalal, W. Irr, and V. Castranova, Generation of free radicals from freshly fractured silica dust, *Annu Rev Respir Dis.* 18:1213 (1988).

14. A. Brody, Asbestos-induced lung disease, *Environ Health Perspect.* 100:21 (1993).

15. R. Pritchard, A. Ghio, J. Lehmann, D.W. Winsett, J.S. Tepper, P. Park, M.I. Gilmour, K.L. Dreher, and D.L. Costa, Oxidant generation and lung injury after particulate air pollutant exposure increase with the concentration of associated metals, *Inhalation Toxicol.* 8:457 (1996).

16. C. Cross and B. Halliwell. Biological consequences of general environmental contaminants, in: *The Lung. Scientific Foundations*, R. Crystal and J. West, eds., Raven Press, New York (1991).

17. R. Bell, J. Coalson, J. Smith, and W. Johanson, Multiple organ system failure and infection in the adult respiratory distress syndrome, *Am Intern Med.* 99:293 (1983).

18. Y. Shoji, Y. Uedono, H. Ishikura, N. Takeyama, and T. Tanaka, DNA damage induced by tumour necrosis factor-alpha in L929 cells is mediated by mitochondrial oxygen radical formation, *Immunology.* 84:543(1995).

19. I. Rahman, X. Li, K. Donaldson, D. Harrison, and W. MacNee, Glutathione homeostasis in alveolar epithelial cells in vitro and lung in vivo under oxidative stress, *Am J Physiol.* 269:L285 (1995).

20. K. Schulze-Osthoff, R. Beyaert, V. Vandevoorde, G. Haegeman, and W. Fiers, Depletion of the mitochondrial electron transport abrogates the cytotoxic and gene-inductive effects of TNF, *EMBO J.* 12:3095 (1993).

21. A. Kooji, K. Bosch, W. Frederks, and C.V. Noorden, High levels of xanthine oxidoreductase in rat endothelial, epithelial and connective tissue cells, *Virchows Arch B Cell Pathol.* 62:143 (1992).

22. K. Pfeffer, T. Huecksteadt, and J. Hoidal, Xanthine dehydrogenase and xanthine oxidase activity and gene expression in renal epithelial cells: cytokine and steroid regulation, *J Immunol.* 153:1789 (1994).

23. C. Grum, R. Ragsdale, L. Ketai, and R. Simon, Plasma xanthine oxidase activity in patients with adult respiratory distress syndrome, *J Crit Care.* 2:22 (1987).

24. R. Simon, P. DeHart, and D. Nadeau, Resistance of rat pulmonary alveolar epithelial cells to neutrophil- and oxidant-induced injury, *Am J Respir Cell Mol Biol.* 1:221 (1989).

25. P. Engstrom, L. Easterling, R. Baker, and S. Matalon, Mechanisms of extracellular hydrogen peroxide clearance by alveolar type II pneumocytes, *J Appl Physiol.* 69:2078 (1990).

26. J. Heffner, S. Katz, P. Halushka, and J. Cook, Human platelets attenuate oxidant injury in isolated rabbit lungs, *J Appl Physiol.* 65:1258 (1988).

27. B.A. Freeman and J.D. Crapo, Biology of disease: free radical and tissue injury, *Lab Invest.* 47:412 (1982).
28. H. Hassan and J. Scandolios, Superoxide dismutases in aerobic organisms, in: *Stress Responses in Plants: Adaptation and Acclimation Mechanisms*, Wiley-Liss, New York (1990).
29. G. Wong and D. Goeddel, Induction of manganese superoxide dismutase by tumor necrosis factor: possible protective mechanisms, *Science.* 242:941 (1988).
30. G. Visner, W. Dougall, J. Wilson, I. Burr, and H. Nick, Regulation of manganese superoxide dismutase by lipopolysaccharide, interleukin-1 and tumor necrosis factor. Role in the acute inflammatory responses, *J Biol Chem.* 265:2856 (1990).
31. M. Ono, H. Kohda, T. Kawaguchi, M. Ohhira, C. Sekiya, M. Naomiki, A. Takeyasu, and N. Taniguchi, Induction of Mn-superoxide dismutase by tumor necrosis factor, interleukin-1 and interleukin-6 in human heptoma cells, *Biochem Biophys Res Commun.* 182:1100 (1992).
32. Y. Janssen, J. Marsh, K. Driscoll, P. Borm, G. Oberdorster, and B. Mossman, Increased expression of manganese-containing superoxide dismutase in rat lungs after inhalation of inflammatory and fibrogenic minerals, *Free Radical Biology & Medicine.* 16:315 (1994).
33. G. Wong, J. Elwell, L. Oberley, and D. Goeddel, Manganese superoxide dismutase is essential for cellular resistance to cytotoxicity of tumor necrosis factor, *Cell.* 58:923 (1989).
34. L. Smith, M. Houston, and J. Anderson, Increased levels of glutathione in bronchoalveolar lavage fluid from patients with asthma, *Am Rev Respir Dis.* 147:1461 (1993).
35. C. Powell, A. Nash, H. Powers, and R. Primhak, Antioxidant status in asthma, *Pediatric Pulmonology.* 18:34 (1994).
36. G. Burton and K. Ingold, Mechanisms of antioxidant action: preventive and chain-breaking antioxidants, in: *CRC Handbook of Free Radicals and Antioxidants in Biomedicine, II*, A, Quintanilha, ed., CRC Press, Boca Raton, FL (1989).
37. B. Ames, R. Cathcart, E. Schwiers, and P. Hochstein, Uric acid provides antioxidant defense in humans against oxidant and radical caused aging and cancer: a hypothesis, *Proc Natl Acad Sci, USA.* 78:6842 (1981).
38. K. Davies and A. Seranian, Uric acid-iron ion complexes, *Biochem J.* 235:747 (1986).
39. T. Seres, V. Ravichandran, T. Moriguchi, K. Rokutan, and J. Thomas, R.B. Johnston, Jr., Protein S-thiolation and dethiolation during the respiratory burst in human monocytes: a reversible post-translational modification with potential for buffering the effects of oxidant stress, *J Immunol.* 156:1973 (1996).
40. R. Stocker, Y. Yamamoto, A. McDonagh, A. Glazer, and B. Ames, Bilirubin is an antioxidant of possible physiological importance, *Science.* 235:1043 (1987).
41. S. Keyse and R. Tyrrell, Heme oxygenase is the major 32-KDa stress protein induced in human skin fibroblasts by UVA radiation, hydrogen peroxide, and sodium arsenite, *Proc Natl Acad Sci, USA.* 11:787 (1989).
42. C. Cross, B. Halliwell, and A. Allen, Antioxidant protection: a function of tracheobronchial and gastrointestinal mucus, *Lancet.* 1:1328 (1984).
43. M. Grisham, C, VonRitter, B. Smith, J. Lamont, and D. Granger, Interaction between oxygen radicals and gastric mucin, *Am J Physiol.* 253:G93 (1987).
44. L. Cohn, V. Kinnula, and K. Adler, Antioxidant properties of guinea pig tracheal epithelial cells in vitro, *Am J Physiol.* 266:L397 (1994).
45. C. Sen and L. Packer, Antioxidant and redox regulation of gene transcription, *FASEB J.* 10:709 (1996).
46. J. Remacle, M. Raes, O. Toussaint, P. Renard, and G. Rao, Low levels of reactive oxygen species as modulators of cell function, *Mutat Res.* 316:103 (1995).
47. J.M. Muller, R.A. Rupec, and P.A. Baeuerle, Study of gene regulation by NF-kappa B and AP-1 in response to reactive oxygen intermediates, *Methods.* 11:301 (1997).
48. A.S. Baldwin, Jr., The NF-kappa B and I kappa B proteins: new discoveries and insights, *Annu Rev Immunol.* 14:649 (1996).
49. S. Schoonbroodt, S. Legrand-Poels, M. Best-Belpomme, and J. Piette, Activation of NF-kappaB transcription factor in a T-lymphocytic cell line by hypochlorous acid, *Biochemical J.* 321:777 (1997).

50. M. Schwartz, J. Repine, and E. Abraham, Xanthine oxidase-derived oxygen radicals increase lung cytokine expression in mice subjected to hemorrhagic shock, *Am J Respir Cell Mol Biol.* 12:434 (1995).

51. C. Munoz, D. Pascual-Salcedo, M. Castellanos, A. Alfranca, J. Aragones, A. Vara, M.J. Redondo, and M.O. de Landazuri, Pyrrolidine dithiocarbamate inhibits the production of interleukin-6, interleukin-8, and granulocyte-macrophage colony-stimulating factor by human endothelial cells in response to inflammatory mediators: modulation of NF-kappa B and AP-1 transcription factors activity, *Blood.* 88:3482 (1996).

52. T. Blackwell and J. Christman, The role of nuclear factor-κB in cytokine gene regulation, *Am J Respir Cell Mol Biol.* 17:3 (1997).

53. C. Kretz-Remy, P. Mehlen, M. Mirault, and A. Arrigo, Inhibition of I kappa-B alpha phosphorylation and degradation and subsequent NF-kappa B activation by glutathione peroxidase overexpression, *J Cell Biol.* 133:1083 (1996).

54. S. Chakraborti, G. Gurtner, and J. Michael, Oxidant-mediated activation of phospholipase A_2 in pulmonary endothelium, *Am J Physiol.* 257:L430 (1989).

55. J. Samet, T. Noah, R. Devlin, J. Yankaskas, K. McKinnon, L. Dailey, and M. Friedman, Effect of ozone on platelet activating factor production in phorbol-differentiated HL60 cells, a human bronchial epithelial cell line (BEAS S6), and primary human bronchial epithelial cells, *Am J Respir Cell Mol Biol.* 7:514 (1992).

56. K.P. McKinnon, M.C. Madden, T.L. Noah, and R.B. Devlin, *In vitro* ozone exposure increases release of arachidonic acid products from a human bronchial epithelial cell line, *Tox Appl Pharm.* 118:215 (1993).

57. K. Kobayashi, M. Salathe, M. Pratt, N.J. Cartagena, F. Soloni, Z.V. Seybold, and A. Wanner, Mechanism of hydrogen peroxide-induced inhibition of sheep airway cilia, *Am J Respir Cell Mol Biol.* 6:667 (1992).

58. G. Schieven, J. Kirihara, D. Myers, J. Ledbetter, and F. Uckun, Reactive oxygen intermediates activate NF-κB in a tyrosine kinase dependent mechanism and in combination with vanadate activate the p56[lck] and p59[fyn] tyrosine kinases in human lymphocytes, *Blood.* 82:1212 (1993).

59. A. Weiss and D. Littman, Signal transduction by lymphocyte antigen receptors, *Cell.* 76:263 (1994).

60. P. Simeonova and M. Luster, Asbestos induction of nuclear transcription factors and interleukin 8 gene regulation, *Am J Respir Cell Mol Biol.* 15:787 (1996).

61. C. Hoyal, E. Gozal, H. Zhou, K. Foldenauer, and H. Forman, Modulation of the rat alveolar macrophage respiratory burst by hydroperoxides is calcium dependent, *Arch Biochem Biophys.* 326:166 (1996).

62. K.A. Roebuck, A. Rahman, V. Lakshminarayanan, K. Janakidevi, and A.B. Malik, H_2O_2 and tumor necrosis factor-alpha activate intercellular adhesion molecule 1 (ICAM-1) gene transcription through distinct cis-regulatory elements within the ICAM-1 promoter, *J Biol Chem.* 270:18966 (1995).

63. M. Whiteman and B. Halliwell, Thiourea and dimethylthiourea inhibit peroxynitrite-dependent damage: nonspecificity as hydroxyl radical scavengers, *Free Radical Biol & Med.* 22:1309 (1997).

PULMONARY EDEMA OF MIXED OR INCOMPLETELY UNDERSTOOD PATHOGENESIS

George J.Baltopoulos[1], Anastasios J.Damianos[1]

[1]Agioi Anargyroi Cancer Hospital of Kifisia
Intensive Care Unit, 145 64 Athens Greece

INTRODUCTION

Although the categorization of pulmonary edema into high (hydrostatic) and low (permeability -ARDS) pressure is convenient in teaching sessions, it should be understood that these are not two different entities. There is a significant overlap in the pathophysiology , clinical presentation, and treatment of both forms. It is conceivable for example that a hole on a rubber tube (the equivalent of permeability) increases as long as the intratubular pressure increases (the equivalent of hydrostatic pressure). The pathophysiology of the entities that will be presented in the following could not be explained either with the high hydrostatic pressure or the high permeability approach. Both approaches could partially explain the underlying pathophysiology.

PULMONARY EDEMA OF MIXED OR INCOMPLETELY UNDERSTOOD PATHOGENESIS I

High Altitude Pulmonary Edema

High Altitude Pulmonary Edema (HAPE) is one of the manifestations of high altitude disease which is defined as acute respiratory failure post high altitude hypoxia exposure, developing in normal subjects with no pre-existing pulmonary or cardiac disease (1). It is still a hazard to people ascending to a high altitude with approximately 20 HAPE deaths per year globally (2). From a historical aspect of view, HAPE has been described in 1913 by Ravenhill in the Chilean Andes, but the Peruvian doctors of the Chulec General Hospital (altitude 3782 m) are the people who first publish in Spanish, on 1954, their experience from treating HAPE (2). The first short English publication on this issue in 1956 (3), has been followed by several more detailed ones, and by now many reports on various HAPE aspects have been published(1,2).

Acute Respiratory Distress Syndrome: Cellular and Molecular Mechanisms and Clinical Management
Edited by Matalon and Sznajder, Springer Science+Business Media New York, 1998

197

HAPE is developed after a rapid ascent in altitudes higher than 2450 m especially when it is combined with cold exposure and moderate to heavy physical exertion(1,2). The higher the altitude and the faster the ascent the more likely it is to occur. Symptoms which are worst during the night, begin within 1-9 days (mean 2.9 days) post high altitude arrival.(4)

The symptoms of HAPE are dyspnea, weakness, cough (dry initially, which progressively becomes productive) , chest congestion and pain, nausea, vomiting and headache. It is conceivable that the last three symptoms represent the CNS involvement. Clinical and laboratory examination reveals tachycardia, tachypnea, (generally more than 120 bpm and 20 bpm respectively), rales, cyanosis, hypoxemia (SaO2 ranges from 40% to 70%) & hypocapnia, leycocytosis (rarely more than $14000/mm^3$), and commonly the findings from the nervous system involvement, confusion, mental obtundation and coma. Electocardiography shows tachycardia and signs of acute right ventricular overload and echocardiography may show tricuspid regurgitation or insufficiency. In terms of spirometry the vital capacity and peak expiratory flow are decreased. The radiologic picture of HAPE could be characterized by vascular congestion, and patchy or homogenous diffuse infiltrates which spared the apices and the supradiaphragmatic areas (1,2).The common denominator in all hemodynamically studied cases was the pulmonary hypertension which, in most cases, was absolutely reversible when victims were treated with O2.The pulmonary capillary wedge pressure was normal or low, cardiac output slightly decreased, and the diastolic right ventricle pressures within normal limits (2). Autopsy from 25 cases revealed diffuse pulmonary edema, thrombi, hemorrhages and hyaline membranes, leucocytic infiltration, and normal heart/coronaries (2,5).

The proposed diagnostic criteria of HAPE, as they have been decided by the Consensus Committee of the International Hypoxia Symposium of 1991 are: history of a recent altitude gain and at least two of the following symptoms: dyspnea at rest, cough, weakness or decreased exercise performance, chest tightness or congestion and two of the following signs: rales or wheezing in one at least lung field, central cyanosis, tachycardia, tachypnea (1,2).

The HAPE altitude of occurrence is extremely rare less than 2000 m. and it seems that the altitude of 2500m is the altitude above which the HAPE is likely to happen. Among 56 patients with HAPE evacuated from the Swiss Alps , no one develop it below 2500m (6). It is assumed that the sleeping altitude is important rather the altitude attained during the day (2). People living in altitudes higher than 2500m and are moving to lower ones for 10-14 days -less days in children- may develop HAPE during re-ascent, which is not different compared the one which is developed the first time(2). Although the incidence of sub-clinical (rales in auscultation) HAPE may approach 30% or even more, the one of the clinically evident ranges between 3-5.2%, depending mainly on the altitude attained, the speed of ascent and the physical activity post arrival. The gender, the age group and the individual susceptibility also play a role. Among 229 cases of HAPE it was revealed a male preponderance of 87% (2).Children are more susceptible to HAPE than adults. In individuals aged 13-20 years and >21 years, the HAPE incidence when they ascent to the altitude of 3782m was 17% and 3% respectively(7). People with isolated absence of the right pulmonary artery as well as people with a history of HAPE are more susceptible to HAPE. The first group people may develop HAPE in lower altitudes.

Mortality is ranging between 3.9-49% (8,9). Prompt access to medical care or the ability to descent rapidly can influence the mortality dramatically. An analysis of 166 cases of HAPE worldwide revealed 19 deaths (11.4%). Among 61 patients treated with O2 and/or descended 4 died (6%), whereas among 23 patients not treated or descended 10 died (49%).(9)

Although the HAPE has been described meticulously, its pathogenesis remains obscure. It has been proposed that the hypoxia constricts the pulmonary vascular bed in a not uniformly distributed way and therefore there are areas overperfused and others underperfused (2). This focal overperfusion hypothesis explains the majority of findings we have in HAPE. It has been shown in rabbits that ink injected to rabbit vein was found in autopsy trapped inhomogeneously in the lung when the animal was hypoxic and not trapped at all when it was normoxic (10).

If in a man there are areas of the vascular bed severely constricted then the blood flow is detoured to vessels with less constriction , dilate the precapillary arterioles and subjects the distal capillary bed to increased pressure and flow. The increased pressure results in fluid leakage (hydrostatic edema) and the increased flow may result in shear forces, which may damage the capillary wall and result in leakage of protein and red cells (increased permeability).

Mechanically ventilated dogs were submitted to right atrium/left upper lobe pulmonary branch bypass. The blood was moving forward with a roller pump. The right pulmonary artery and the middle and lower left pulmonary artery branches were ligated. In that situation the left upper lobe was perfused with 0.5-1.5 l/min (sixfold the normal flow for left upper lobe), which represents the normal dog CO. During perfusion, the lobe became engorged , distended and stiff, the pulmonary artery pressure elevated (=38-68mmHg - as in HAPE-) and the pulmonary capillary wedge pressure (PCWP) and left atrium pressure remained within normal limits. The protein content of the hemorrhagic pulmonary edema which soon after filled the lobe and the airways, was compatible with an increased permeability pulmonary edema.

The idea of high blood flow has been challenged recently (11,12). It has been shown that even the high pressure alone can result in high permeability edema by disrupting the capillary endothelial layer and/or the alveolar epithelial layer and/or all the layers of the respiratory membrane.

Human data are supporting the idea of high permeability. The bronchoalveolar lavage fluid from 3 patients with HAPE was found rich in high molecular weight proteins, erythrocytes and leukocytes, compared to the controls (13).

Although this experimental approach explains how HAPE might happen, the increased pulmonary vein pressure could also be followed by such results. A higher than a fourfold increase in the pulmonary blood flow is followed by a diminished precapillary vessels pressure drop and a resistance shift to the pulmonary veins with an increase in the capillary pressure(14). The hypoxia also, when it is prolonged, results in pulmonary venous constriction as well (15).

At sea level, there are conditions such as pulmonary embolism, successful balloon dilation of peripheral pulmonary artery, surgical removal of pulmonary emboli and congenital absence of right pulmonary artery, which can illustrate that pulmonary edema occurs if portion of the pulmonary vascular bed is occluded and the remaining is subjected to an increased blood flow perfusion (2). In the case of pulmonary embolism and congenital absence of right pulmonary artery, the blood flow is deviated to the not occluded pulmonary vasculature. The blood flow therefore is inappropriately high for the available pulmonary vessels. In the case of successful balloon dilatation, the lung perfused by the stenotic vessel, develops edema post stenosis repair because the blood flow is again inappropriately high for the post-stenotic vascular bed. The same happens in the case of post-pulmonary embolectomy edema; although other mechanism have been proposed (ischemia- reperfusion injury).

Several different treatments are now available for HAPE, depending on the location and the availability of medical facilities. Bed rest which is also the patient's choice, is not always successful. The clinical improvement should be ascertain with frequent observations. The method is less successful than the oxygen therapy, which if it is available should be started upon the symptoms are evident. Descent is the therapy of choice. All other therapies should be administered to save time for the descent. The hyperbaric bag therapy is consisted of putting the patient in a specially designed textile bag in which the inside pressure can be increased to 2 psi, which mimics an elevation at 2662 m, when it is applied at an altitude of 4246m. The method has equivalent results to the oxygen therapy (16). Nifedipine given in HAPE patients decrease the pulmonary artery pressure, improves the symptoms score, the exercise tolerance and the arterio- alveolar oxygen difference. When it is given in a dose 20 mg tid upon the ascent it might prevent HAPE (17). The rest of the drugs used, are not proven of any significant effect. Acetazolamide given to children having a sojourn at sea level has greatly reduced the previously high incidence of HAPE. The NO administration could be beneficial in HAPE, improving the arterial oxygenation (18).

Neurogenic Pulmonary Edema

Pulmonary edema may occur in conjunction with various CNS insults, such as head trauma, cerebral hemorrhage, seizures and neurosurgery (Neurogenic Pulmonary Edema-NPE). It may appear acutely within minutes or hours post predisposing factor exposure or insidiously over several days (19,31).

Its pathogenesis is unclear. Theodore and Robin (20), in order to explain the pathophysiology of the disease, develop the blast theory, which incorporates findings suitable for both the hydrostatic and high permeability pulmonary edema. According to this theory immediately after the CNS insult a great quantity of catecholamines is secreted (CNS sympathetic discharge) which flood the circulation and may result in systemic and pulmonary vasoconstriction which in sequence increases LV afterload and preload (increase in venous return and transient increase in PCWP), and capillary permeability (capillary endothelium disruption), resulting in pulmonary edema. Although the increased PCWP has been accused for the edema formation (hydrostatic edema), the increased edema fluid to serum albumine ratio(=0.86 without any increase in pressures) (23), is suggesting for the coexistence of an increased permeability factor (increased permeability edema) (21,22, 23).

The hydrostatic abnormalities involved to the pathogenesis of NPE could be attributed to LV failure and to pulmonary venoconstriction. The left heart failure (24) can be the result of the increased venous return, the altered ventricular compliance caused by shortened diastolic relaxation time, the increased afterload and of the well documented direct, but reversible (by dobutamine administration), depressant effect on myocardium (25,26) of the CNS insult (trauma or subarachoid hemorrhage). Pulmonary venoconstriction can also increase the pulmonary capillary pressure as much as the increased LA and LV pressure do. Increase of the intracranial pressure in dogs, is followed by pulmonary veins constriction, an effect attributed to nor-epinephrine (27).

It is also well known that increased pulmonary permeability follows intracranial hypertension in sheep (28) and in rabbit as well (29). This increased permeability seems to be mediated through a pressure independent pathway, since it has been demonstrated a decrease in lung lymph flow in a CNS insult model, after alpha blockade (30).

At this point it is worth mention that diphenylhydantoin which is a drug extensively administered in head injuries, is preventing from the high permeability edema formation (31).

The incidence of NPE is difficult to say. In a recent study the incidence of NPE in isolated head injury patients, dying at the scene was 32% and within 96 hours post trauma

50%. Surviving head injury patients manifest a significantly decreased PaO2/FiO2 ratio in the presence of a normal chest x-ray film, which seems to be related to a decrease in cerebral perfusion pressure and may be caused, in part, by NPE (32).

In terms of therapy, in all NPE cases the therapy is symptomatic. On the other hand the therapy of the head injury itself is necessary.

Pulmonary Edema Associated with Narcotic Overdose

The parenteral drug abuse results frequently in pulmonary edema. Most opioids have been accused for its formation. Both the high permeability and high pressure pathogeneses have been proposed (19,34). However there are not strong experimental or clinical data to support any of these hypotheses (33,34,35,36,37,38). The hypotheses of hypersensitivity, direct toxic effect, vasoactive substance release, cardiac or neurologic defects (38), are still working.

Although most of the cases are responding easily to therapy, which is symptomatic, there are and some cases not responding as expected. In a recent report (39) the delay in resolution of a cocaine / heroin abuse related pulmonary edema, has been attributed in a cocaine induced impairment of sodium and thus fluid transport across alveolar epithelium.

Pulmonary Edema Associated with Tocolytic Therapy

The causes of pregnancy associated pulmonary edema are the tocolytic therapy with sympathomimetic amines, the eclampsia and other hypertensive disorders and the acid aspiration. The eclampsia and /or hypertension-related pulmonary edema and the aspiration one, are assumed to be cardiogenic and high permeability ones respectively.

The tocolytic therapy associated pulmonary edema is the most common form of hypoxemic respiratory failure (40,41) seen during pregnancy (0.0-4.4% of the patients receiving tocolytic therapy will develop hypoxemic respiratory failure). Most described cases have resulted from intravenous (iv) use of beta-mimetics, particularly ritodrine, terbutaline, isoxuprine, salbutamol, fenoterol and magnesium sulfate (40,41,42). Co-existing maternal infection increases the incidence of pulmonary edema (43). The incidence of the syndrome is higher in twin pregnancies. Most of the patients have intact membranes at presentation. The iv beta2-mimetic therapy related pulmonary edema, develops acutely within 30-72 hours (44) of the initiation of therapy. Any one else developed 24 hour post discontinuation should be considered as pulmonary edema of another etiology. When it develops postpartum the vast majority are encountered within 12 h of delivery. In asthma patients treated with high doses of beta-mimetics it is never observed pulmonary edema. It seems that some pregnancy related factors predispose the pregnant women to such a complication. Also a beta receptor down regulation has been proposed for peripartum heart failure (and pulmonary edema) in patients in prolonged per os tocolytic therapy (45) .

From an echocardiographic and hemodynamic aspect of view, in all studied cases, no heart dysfunction was confirmed. Chest roentgenograms usually show bilateral alveolar infiltrates and a normal-sized heart (40).

The speculation that the beta2-mimetics increase the pulmonary capillary leak, never confirmed and in contrary they may decrease it, enhancing the lung liquid clearance, as it has been shown in animal models. It is assumed that volume overload may play a role in such patients since: Large crystalloid volumes are often given to combat the sympathomimetic induced tachycardia in addition to the expected intravascular volume expansion because of the pregnancy. The colloid oncotic pressure is reduced peripartum (increase in the susceptibility of hydrostatic edema formation). The water and sodium excretion may be impaired in the supine pregnant patient (sodium retention in ritodrine treated patients is the primary cause of plasma volume expansion) . The beta2-mimetic may increase the ADH secretion and finally. The response to the diuretic therapy is prompt (40,42).

All patients manifest tachycardia, tachypnea and crackles in lung auscultation; and 25% of them precordial pain. Hypoxemia (PaO2 = 50.0 +- 3.0 mmHg) and hypocapnea (PaCO2 = 28.0 +- 2 mmHg) accompany the clinical symptoms. Positive fluid balance does often exist at the symptom onset. History and clinical findings will differentiate this disorder from other entities (40).

Although all patients should be admitted to the ICU the invasive monitoring is not warranted, since the disease is usually benign. Discontinuation of the tocolytic therapy, oxygen therapy and diuresis is the appropriate therapy. Response to therapy is expected within the first 12 hours. The maternal mortality in Pisani and Rosenow series (40) was 3% (2 deaths) and the fetal survival 95%.

PULMONARY EDEMA OF MIXED OR INCOMPLETELY UNDERSTOOD PATHOGENESIS II

Re-expansion Pulmonary Edema (REPE)

The rapid expansion of a lung made chronically atelectatic secondary to a pleural effusion or pneumothorax may lead to the development of pulmonary edema. Acute re-expansion pulmonary edema following re-expansion of a collapsed lung after pleurocentesis or pneumothorax is a well described entity (19). Most cases are ipsilateral, but a few cases (four to my knowledge) of contralateral or bilateral pulmonary edema have been reported (46).

Although case reports abound, the true incidence of it, is unknown. Old reviews with big series of pneumothoraces treated with thoracentesis / thoracostomy do not describe any REPE (47,49). In a relatively recent retrospective study (47) of 146 spontaneous pneumothoraces treated with thoracentesis and /or continuous low negative pressure suction drainage, 21 (14.4 %) patients developed re-expansion pulmonary edema.

The etiology of REPE is unclear but it is related to the amount of air or pleural fluid evacuated, the rapidity of evacuation and the chronicity of lung collapse. Chronicity of lung collapse has been shown experimentally as a predisposing factor when the duration of pneumothorax is more than three days (48,49). Although there are case reports with REPE in patients with shorter duration of pneumothorax (50).

The pathophysiology of REPE is not clear and its pathogenesis is not completely understood. Several mechanisms have been proposed: The high negative intrathoracic pressure which can be higher when suctioning is used. The more negative the intrapleurar pressure is the more negative the interstitial pressure is and therefore the more the filtration pressure is. (48,50). The relatively more negative intrapleural pressure following the re-expansion post air or pleural fluid evacuation, with or without suction, may decrease the hydrostatic perivascular pressure leading to extrvascular fluid accumulation (increase fluid transudation to the interstitium). This process is facilitated when there is a bronchial obstruction. The more the negative pressure the more the fluid extravasation (51). Although the increased capillary pressure and flow have also been accused for its pathogenesis, there are not enough experimental or clinical data to support such an idea. The increased microvascular permeability is the most widely accepted theory for the REPE. Marland & Glauser (52), in a case of REPE, catheterized the patient and found normal PCWP and high edema to serum protein ratio (as high as 0.85), indicating high permeability. Also in rabbits Pavlin et al (53) showed increased concentration of the radio-labeled albumin in the re-expanded lung. Decreased surfactant concentration has also been proposed, a hypothesis not experimentally confirmed (54). In the possible mechanisms of the increased microvascular permeability are included the mechanical damage of endothelium (stretch forces increase the endothelial pores) and the local prior to the lung expansion hypoxia which facilitates the development of re-perfusion injuries. Animal studies indicate that re-expansion pulmonary edema is the result of alveolar damage mediated by the re-perfusion of collapsed alveoli, activated inflammatory cells (51). However oxygen radical scavenger administration and / or neutrophil depletion prevents partially the REPE (51).

The REPE incidence is higher in the 20-39 year age group and it is rarely seen unless at least 2 L of pleural fluid are rapidly removed. The onset of symptoms is usually rapid within hours post re-expansion, although they may be delayed up to 24 hours.

The treatment is consisted of oxygen therapy, diuretics and mechanical ventilation if needed. Most cases are responding very well to therapy, although deaths have been described in a series by Mahfood and associates (54) approaching a mortality of 21 %.

Postobstructive Pulmonary Edema (POPE)

The postobstructive pulmonary edema represents a small minority among the pulmonary edemas of other etiologies. Its causes have in common some acute or chronic upper airway obstruction. In the group of acute causes are included: laryngospasm, epiglottitis, laryngotracheobronchitis, croup, foreign body aspiration, tumor, upper airway trauma, strangulation, interrupted hanging, retropharyngeal or peritonsilar abscess, Ludwig;s angina, angioedema, near drowning, asthma; whereas in the group of chronic causes are included: obstructive sleep apnea syndrome, adenoidal or tonsilar hypertrophy, nasopharyngeal mass, thyroid goiter, acromegaly. Among those causes, the most frequently reported one, in adults, is the post anesthesia laryngospasm, and in children and infants the one following epiglottitis, spasmodic croup and laryngotracheobronchitis. (51)

The incidence of the post obstructive pulmonary edema among the different causes, varies and has been reported as much as 11-12% (51). In terms of pathogenesis the markedly negative intrapleural pressure generated by the forced inspiratory effort against the obstructed upper airway, which is fully transmitted to the perivascular interstitium, has been accused. This negative pressure facilitates the fluid extravasation into the interstitium and therefore the edema formation. During quiet respiration the intrapleural inspiratory pressure is -2 to -5 and during expiration up to +10 cmH2O. In acute upper airways obstruction, the peak negative pressure may reach to -50 cmH2O or more and in obstructive sleep apnea cases to -100 cmH2O. At this high level pressures the filtration pressure is also high because the increased venous return, the increased pulmonary blood volume and PCWP and the decreased peri-vascular interstitial pressure which is lower than the normal of -5 to -10 (51,55).

The clinical characteristics of the POPE, are those of the underlying upper airway obstructive disease, and the ones of pulmonary edema. Treatment should be focused on the underlying disease and on the pulmonary edema itself.

Overdistention Pulmonary Edema (OPE)

The clinically evident primary OPE is a rare entity, but the not clinically evident injuries of the alveolo- capillary membrane, which can be caused by the ventilator are common (51). It is assumed that the OPE is an ''Intensive Care Unit '' induced one, it has to do with mechanical damage of the alveolo-capillary membrane, the volume is the critical factor and not the pressure; and PEEP may act protectively to edema formation. It has been shown that cytocines participate also to the extravascular lung water volume increase, in experimental models (51,56,57,58).

In general, the mechanical ventilation can cause: shear stress , capillary stress fracture, stretch pore phenomenon and changes in Kf, and surfactant dysfunction, which directly or indirectly, can affect the alveolo-capillary membrane permeability, the lung anatomical and functional integrity and why not, the other organs' integrity (56,57,58).

REFERENCES

1. J. P. Richalet , High altitude pulmonary oedema: still a place for controversy? , Thorax. 50:923 (1995).
2. H.N. Hultgren, High altitude pulmonary edema : Current concepts, Annu Rev Med. 47: 267 (1996).
3. B.Vega, Edema of the lung in mountain sickness, JAMA. 160:698 (1956)
4. H. Hultgren, B.Honigman, K.Theis,D.Nicholas, High altitude pulmonary edema in a ski resort, West J Med. 164:222 (1996).
5. H. Hultgren,W.Spickard, D.Lopez, Further studies of high altitude pulmonary edema. Br Heart J. 24:95 (1962).
6. J.Hochstrasser, A.Nazer, C.Oetz, Altitude edema in the Swiss Alps:Observations on the incidence and clinical course in 50 patients, Schweitz Med Wochenschr. 116:866 (1986).
7. H. Hultgren, E.Marticorena, High altitude pulmonary edema: epidemiologic observations in Peru, Chest. 74:372 (1978).
8. N.Menon, High altitude pulmonary edema, N Engl J Med 273:66 (1965).
9. H.Lobenhoffer, R.Zink,W.Brendel, High altitude pulmonary edema: analysis of 166 cases. in: High Altitude Physiology and Medicine, W.Brendel,R.Kink, ed., Springer Verlag, New York (1982).
10. D. Lehr, M.Triller, L.Fisher, et al, Induced changes in the pattern of pulmonary blood flow in the rabbit, Circ Res. 13:119 (1963).
11. J.B. West, K.Tsukimoto, O. Mathieu-Costello, R.Prediletto, Stress failure in pulmonary capillaries, J Appl Physiol. 70:1731 (1991).
12. K.Tsukimoto, O. Mathieu-Costello, R.Prediletto,A.R. Elliot, J.B. West, Ultrastractural appearances of pulmonary capillaries at high transmural pressures. J Appl Physiol. 71:573 (1991).
13. R.B. Shoene,P.H. Hackett, W.R.Henderson, E.H. Sage, M. Chow, R.C. Roach, W.J.Mills, T.R. Martin, High altitude pulmonary edema: characteristics of lung lavage fluid, JAMA. 56:63 (1986).
14. A.Hyman, Effects of large increases in pulmonary blood flow on pulmonary venous pressure, J Appl Physiol. 217:1177 (1961).
15. K.Welling, R. Sanchez, J.Raun et al, Effect of prolonged alveolar hypoxia on pulmonary arterial pressure and segmental vascular resistance, J Appl Physiol. 75:1194 (1993).
16. J. Kasic, M. Yaron, R.Nicholas et al, Treatment of acute mountain sickness: hyperbaric versus oxygen therapy, Ann Emerg Med. 20:1109 (1991).
17. P. Bartsch, M. Maggiorini, M. Ritter et al, Prevention of high altitude pulmonary edema by nifedipine, N Engl J Med. 325:1284 (1991).
18. U. Scherrer, L.Volleneider, A.Delabay et al. Inhaled nitric oxide for high altitude pulmonary edema, N Engl J Med. 334:624 (1996).

19. A. Kides, R. McCaffree, Pulmonary edema. in: Textbook of Critical Care. S.M. Ayres, A. Grenvik, P.R. Holbrook, W.C. Shoemaker, ed., W.B.Saunders Company. A division of Harcourt Brace & Company, Philadelphia, London, Toronto, Montreal, Sydney, Tokyo (1995).

20. J.Theodore,E.D. Robin, Speculations on neurogenic pulmonary edema (NPE), Am Ren Respir Dis. 113:405 (1976).

21. R.W. Carlson , R.C. Shaeffer Jr., S.G.Michaels, M.H.Weil, Pulmonary edema following intracranial hemorrhage, Chest. 75:731 (1979).

22. N.P. Wray, M.B.Nicotra, Pathogenesis of neurogenic pulmonary edema. Am Ren Respir Dis. 118:783 (1978).

23. I.A. Fein, E.C. Rackow, Neurogenic pulmonary edema, Chest. 81:318 (1982).

24. A.A. Luisada, Mechanisms of neurogenic pulmonary edema, Am J Cardiol 20:66 (1967)

25. S.C. Deehan, I.S.Grant, Haemodynamic changes in neurogenic pulmonary oedema: effect of dobutamine, Intensive Care Med. 22:672 (1996).

26. S.A. Mayer, M.E. Fink, S. Homma et al, Cardiac injury associated with neurogenic pulmonary edema following subarachnoid hemorrhage, Neurology. 44:815 (1994).

27. B.T. Peterson, S.E. Grauer, R.W. Hyde, C. Ortiz, H. Moosavi, M.J.Utell, Response of pulmonary veins to increased intracranial pressure and pulmonary air embolism. J Appl Physiol. 48:957 (1980).

28. R.E. Bowers, C.R. McKeen, B.E. Park, K.L. Brigham,.Increased pulmonary permeability follows intracranial hypertension in sheep, Am Rev Respir Dis. 119:637(1979).

29. F.L. Minnear, C. Kite, L.A. Hill, H. Van der Zee, Endothelial injury and pulmonary congestion characterize neurogenic pulmonary edema in rabbits, J Appl Physiol 63:335 (1987).

30. H. Van der Zee, A.B. Malik, B.C. Lee, T.S.Hakim, Lung fluid and protein exchange during intracranial hypertension and role of sympathetic mechanisms, J Appl Physiol 48:273 (1980).

31. G.L. Colice, M.A. Matthay, E. Bass, R.A. Matthay, Neurogenic pulmonary edema, Am Rev Respir Dis. 130:941 (1984).

32. F.B. Rogers, S.R. Shackford, G.T. Trevisani, J.W. Davis, R.C. Mackersie, D.B. Hoyt, Neurogenic pulmonary edema in fatal and nonfatal head injuries, The Journal of Trauma: Injury, Infection and Critical Care. 39:860 (1995).

33. S.Katz, A. Aberman, U.I.Frand, I.M.Stein, M.Fulop, Heroin pulmonary edema. Evidence for increased pulmonary capillary permeability, Am Rev Respir Dis. 108:472 (1972).

34. A.D.Steinberg, J.S.Karliner, The clinical spectrum of heroin pulmonary edema, Arch Intern Med. 122:122 (1968).

35. U.I.Frand, C.S. Chim,M.H.Williams Jr, Methadone-induced pulmonary edema, Ann. Intern Med. 76:975 (1972).

36. D.S.Prough, R.Roy,J.Bumgarner, G.Shannon, Acute pulmonary edema in healthy teenagers following conservative doses of intravenous naloxone, Anaesthisiololgy. 60:485 (1984).

37. J.Glassroth,G.D.Adams,S.Schnoll, The impact of substance abuse on the respiratory system, Chest. 91:596 (1987).

38. S.B.Wilen,S.Ulreich,J.G.Rabinowitz, Roentgenographic manifestations of methadone induced pulmonary edema, Radiology. 114:51 (1975).

39. P.G.H.M. Raijmakers, A.B.J.Groeneveld, M.C.M. deGroot, G.J.J.Teule,L.Thijs, Delayed resolution of pulmonary oedema after cocaine /heroin abuse, Thorax. 49:1038 (1994).

40. R.J. Pisani, E.C. Rosenow, Pulmonary edema associated with tocolytic therapy, Ann Int Med. 110 : 714 (1989).

41. T.J. Benedetti, Life-threatening complications of betamimetic therapy for preterm labor inhibition, Clin in Perinatol. 13:843 (1986).

42. B.Armson, P.Samuels, F.Miller,J.Verbalis, E. Main, Evaluation of maternal fluid dynamics during tocolytic therapy with ritodrine hydrochloride and magnesium sulfate, Am J Obstet Gynecol. 167: 758 (1992).

43. C.G. Hatjis, M.Swain, Systemic tocolysis for premature labor is associated with increased incidence of pulmonary edema in the presence of maternal infection. Am J Obstet Gynecol 159: 723 (1988).

44. H.M. Hollingsworth, M.R. Pratter, R.S. Irwin, Acute respiratory failure in pregnancy, Intensive Care Med. 4:11 (1989).

45. M.B. Lambert, J.Hibbard, L.Weinert, J. Briller, M. Lindheimer, R.M. Lang, Peripartum heart failure associated with prolonged tocolytic therapy. Am J Obstet Gynecol. 168:493 (1993).

46. M.W. Ragozzino, R. Greene, Bilateral reexpansion pulmonary edema following unilateral pleurocentesis, Chest. 99:506 (1991).

47. Y. Matsuura,T. Nomimura , H. Murakami,T. Matsusima,M. Kakehashi,H.Kajihara, Clinical analysis of re-expansion pulmonary edema, Chest, 100:1562 (1991).

48. W.C.Miller,R.Toon,H. Palat, J.Lacroix, Experimental pulmonary edema, following re-expansion of pneumothorax, Am Ren Respir Dis.108:664 (1973).

49. G.Baltopoulos,J.Floros,V.Kadas, G.Ladas, A. Hatzimichalis, N.Exarchos, Unilateral pulmonary edema post chest intubation for pneumothorax. Nosokomiaka Chronika. 50: 71 (1988).

50. S.Sherman K.P.Ravikrishnan, Unilateral pulmonary edema following re-expansion of pneumothorax of brief duration, Chest. 77:714 (1980).

51. J.Timby,C. Reed,S. Zeilender, F.L.Glouser, "Mechanical " causes of pulmonary edema.Chest. 96:973 (1990)

52. AS.Marland,F.L.Glauser, Hemodynamic and pulmonary edema protein measuremaent in a case of re-expansion pulmonary edema, Chest. 81:250 (1982).

53. D.J. Pavlin, M.L.Nessly,F.W. Cheney, Increased pulmonary vascular permeability as a cause of re-expansion edemain rabbits, Am Rev Respir Dis. 124:422 (1981).

54. S.Mahfood, W.R. Hix,B.L. Aaron,P. Blaes,D.C. Watson, Re-expansion pulmonary edema, Ann Thorac Surg. 45:340 (1988).

55. B.A. Chaudhary,N.Manouchehr,T.K.Chaudhary, W.A.Speir, Pulmonary edema as a presenting feature of sleep apnea, South Med J. 77:499 (1984).

56. J.C.Parker, L.A.Hernandez, K.J. Peevy, Mechanisms of ventilator-induced lung injury, Crit Care Med. 21:131 (1993).

57. J.B.West, O. Mathieu-Costello, Stress failure of pulmonary capillaries: role in lung and heart disease, Lancet. 340:956 (1992).

58. D.Dreyfuss, G.Saumon, Should the lung be rested or recruited ? The Charybdis and Scylla of ventilator management, Am Rev Respir Dis. 149:1066 (1994).

BIOLOGIC MARKERS OF ACUTE LUNG INJURY

Michael A. Matthay,[1] Edward J. Conner,[1] Lorraine Ware,[1] and
George Verghese[1]

[1] Cardiovascular Research Institute
University of California, San Francisco
San Francisco, CA 94143-0130

INTRODUCTION

The pathogenetic mechanisms of acute lung injury have been examined in many clinical studies. There are at least three good reasons to search for sensitive and specific biologic markers of clinical acute lung injury (Pittet et al., 1997). First, these markers may improve the prediction of acute lung injury in high-risk clinical conditions such as sepsis, pneumonia, severe trauma, and following aspiration of gastric contents. Secondly, these biologic markers may provide new insights into the pathogenesis of clinical lung injury. Third, the biologic markers may help to predict the outcome of patients once clinical acute lung injury has developed.

This article will briefly review some of the important developments in measurements of biologic markers of lung injury. The discussion will focus on early biologic markers of lung injury that may be involved in the pathogenesis of acute lung injury as well as discuss other markers that may have prognostic significance, particularly relative to the development of fibrosing alveolitis.

INFLAMMATION IN ACUTE LUNG INJURY

Several studies have demonstrated that the early phase of clinical lung injury is usually associated with the influx of large numbers of neutrophils into the distal air spaces of the lung along with protein-rich edema fluid. The protein-rich edema fluid results from an increase in the protein permeability of both the lung endothelial and alveolar epithelial barriers. Experimentally, it is possible to measure accurately the total excess quantity of edema fluid (extravascular lung water) and plasma protein that has entered the interstitial and alveolar spaces of the lung in acute lung injury (Folkesson et al., 1995). Clinically, it is more difficult to be as precise, but it is possible to estimate the degree of inflammation and injury in the lung. For example, Table 1 shows the data from several studies in which bronchoalveolar lavage (BAL) was done early in the clinical course of acute lung injury. As shown in the table, there was a high concentration of protein in the BAL in lung injury patients compared to hydrostatic edema or saline instillation alone. Interestingly, in one study (Clark et al., 1995), the quantity of protein in the BAL was higher in patients who died than in those who survived, suggesting that the magnitude of the permeability defect was of prognostic significance. The ability to remove some of the protein-rich edema fluid in the first 24 hours after acute lung injury has developed may have a favorable prognostic significance (Matthay and Wiener-Kronish, 1990).

Acute Respiratory Distress Syndrome: Cellular and Molecular Mechanisms and Clinical Management
Edited by Matalon and Sznajder, Springer Science+Business Media New York, 1998

207

Table 1. Total Protein Concentration and Percent Neutrophils in the BAL Fluid from Control Patients and Patients with ARDS

Clinical Condition and Year of Study	Total Protein Concentration in BAL (mg/100 ml)	Neutrophils in BAL (%)
Saline — 1989	16	12
Hydrostatic edema — 1986	6	5
LTB$_4$ instillation — 1989	25	56
ARDS — 1982	40	76
ARDS — 1986	92	68
ARDS — 1986	101	73
ARDS — 1994	147	78
ARDS — 1995	180*	64
ARDS — 1995	472[†]	79

* Survivors
[†] Nonsurvivors

Reprinted with permission of the American Thoracic Society from Pittet et al., 1997.

BIOLOGIC MARKERS OF EARLY ACUTE LUNG INJURY

A variety of pro-inflammatory molecules is present in the air spaces of the lung early in the course of acute lung injury. For example, several studies have reported the presence of prostaglandin metabolites, products of neutrophil degradation, cytokines, complement fragments, platelet activating factor, and coagulation factors (Pittet et al., 1997). One of the critical questions in trying to establish the importance of these inflammatory markers has been to determine if they are capable of causing structural damage to the lung. One novel approach that has been adopted recently has been to measure the ability of broncho-alveolar lavage fluid (BAL) or pulmonary edema fluid collected from patients with acute lung injury to release intercellular adhesion molecule-1 (ICAM-1) from cultured alveolar epithelial cells (A549 cells) (Pugin et al., 1996). This approach is of particular value because the release of ICAM-1 indicates that BAL has intrinsic pro-inflammatory activity. In this study, the pro-inflammatory activity of BAL could be attributed primarily to interleukin-1 (IL-1). Although there was immunologic evidence supporting the presence of tumor necrosis factor-α (TNF-α), neutralization of TNF-α activity in the BAL fluid only minimally reduced the inflammatory activity. Interestingly, similar results have now been demonstrated in a study at our institution of patients in whom pulmonary edema fluid was collected within the first 24 hours of the development of acute lung injury. Control patients with hydrostatic pulmonary edema were included. The results confirm that the pro-inflammatory activity could be primarily attributed to IL-1. These results do not mean that TNF-α does not exert a pro-inflammatory activity at an even earlier time point, but they do emphasize the value of a biologic assay for assessing the primary mediator of inflammatory activity of fluid sampled from the lung (or the circulating plasma) in patients with early acute lung injury.

Several lines of evidence indicate that the early phase of acute lung injury in patients is characterized also by the presence of large quantities of interleukin-8 (IL-8), a molecule that is important for chemoattracting neutrophils to the distal air space of the lung. In several studies from our institution as well as other investigators, high levels of IL-8 have been detected in patients with acute lung injury with very low levels in patients with hydrostatic pulmonary edema (Miller et al., 1996; Donnelly et al., 1993). *In vitro* studies indicate that inhibition of IL-8 activity in the edema fluid removes most of the neutrophil chemotactic properties of the fluid (Miller et al., 1992). Since several experimental studies support the importance of IL-8 as a neutrophil chemotactic agent in experimental acute lung injury (Folkesson et al., 1995), it is likely that generation of IL-8 from alveolar macrophage and other pulmonary cells is an important mechanism for the acute inflammatory response in many patients with acute lung injury.

In addition, there is new evidence from several studies that the presence of neutrophils in the distal air spaces is associated with the release of metalloproteinases

(Delclaux et al., 1997; Ricou et al., 1996). These metalloproteinases may well be important in producing structural injury to the interstitium of the lung as well as to the alveolar epithelial barrier.

It is difficult to identify one particular mediator or pathway that may always be critical for the development of lung injury. It is clear, for example, that acute lung injury can occur in some patients without circulating neutrophils. Thus, neutrophils are not always required for the development of acute lung injury. On the other hand, there is considerable circumstantial evidence that neutrophils contribute to lung injury in many patients.

Table 2 summarizes several studies of cell-specific markers that have some predictive value for the development of acute lung injury or its outcome. This list is not comprehensive, but provides an overview of several important pathways.

Table 2. Studies of Cell-Specific Markers with a Predictive Value for Development of ARDS and/or Outcome*

Biologic Marker	Cell Type	Biologic Fluid	Predictive Value for ARDS	Predictive Value for Outcome
vWf-Ag	Endothelium	Plasma	+	+
E/P-selectins	Endothelium	Plasma	-	+
SP-A	Epithelium	BAL	+	+
Macrophage/ neutrophil count	Macrophage Neutrophil	BAL	-	+
PCPIII	Fibroblasts	BAL	-	+
Elastase	Neutrophil	Plasma	+	-
L-selectin	Neutrophil	Plasma	+	-
$CD_{11}b/CD_{18}$	Neutrophil	Plasma	+	-

* The sign + indicates that the particular marker has a predictive value for development of ARDS and/or outcome; the sign - indicates that the particular marker does not predict development of ARDS and/or outcome, or that this marker has not been studied for this purpose.

Reprinted with permission of the American Thoracic Society (Pittet et al., 1997).

There are, as mentioned above, several soluble biochemical markers of acute lung injury that have demonstrated inflammatory activity *in vitro* as well as in experimental studies of lung injury. Many of these markers have also been measured in the air spaces or plasma of patients with acute lung injury. Some of the markers have some predictive value for the development of ARDS or for outcome from ARDS, although in some of the studies, the data to support these conclusions is only moderately persuasive. For example, TNF-α is elevated in the plasma of some patients at risk of ARDS with sepsis, but the biological activity of TNF-α has not been well studied in these patients. Most of the data is based on immunologic assays of TNF-α. The predictive value is only fair for the development of ARDS. Elevated levels of complement have been predictive of ARDS in some patients with sepsis, but not in all studies. IL-8 has been measured in the plasma, BAL, and pulmonary edema fluid. The plasma levels are the least predictive of ARDS, while the BAL levels of IL-8 have some predictive value in high risk patients. Elevated levels of IL-8 in pulmonary edema fluid or BAL from patients with acute lung injury seem to predict nonsurvival in some studies, but the predictive value has not been confirmed in all the published work. More studies are needed in which patients are divided into the clinical disorder that is responsible for the lung injury (i.e., sepsis, trauma, aspiration) to test the predictive value for outcome early in the clinical course of acute lung injury. PCP III does have some predictive value for nonsurvival; this issue will be discussed in the last section of this article. Table 3 summarizes some of this information.

Table 3. Studies of Markers of Acute Inflammation with a Predictive Value for Development of ARDS and/or Outcome*

Biologic Marker	Biologic Fluid	Predictive Value for ARDS	Predictive Value for Outcome
TNF-α	Plasma	+	-
IL-1ß	BAL	+/-	+/-
IL-8	Plasma	+	-
	BAL/pulmonary edema fluid	+	+
C_3a	BAL	+	-
LTB_4	Plasma	+	-
Sulfopeptide leukotrienes	BAL	+	-
Hydrogen peroxide	Urine	-	+
Lipofuscin	Plasma	+	-
Catalase MnSOD	Plasma	+	-

Definition of abbreviation: MnSOD = manganese superoxide dismutase.

* The sign + indicates that the particular marker has a predictive value for development of ARDS and/or outcome; the sign - indicates that the particular marker does not predict development of ARDS and/or outcome, or that this marker has not been studied for this purpose.

Reprinted with permission of the American Thoracic Society (Pittet et al, 1997).

In summary, markers of acute inflammation that have been measured in the plasma and the distal air spaces of the lung early in acute lung injury have provided considerable new information about the mechanisms of clinical acute lung injury. The pro-inflammatory environment in the distal air spaces of the lung is complex and interpretation of the results of clinical studies will depend on integrating the mechanistic studies of the early inflammatory response from controlled experimental studies.

BIOLOGIC MARKERS OF FIBROSING ALVEOLITIS IN LUNG INJURY

It is well recognized that clinical acute lung injury is frequently complicated by fibrosing alveolitis (Zapol et al., 1979; Crouch, 1990). Clinically, this problem becomes apparent usually after several days (5-7 days) when the patient has a progressive reduction in lung compliance and often a persistent and evolving ground glass appearance to the chest radiograph. In addition, patients may develop a further increase in dead space ventilation, thus indicating a loss of capillary blood flow to several lung units. Several pathologic studies have documented that the fibroproliferative phase of acute lung injury can be appreciated within 4-5 days of the development of acute lung injury. Recently, there has been a growing interest in measuring biologic markers in the distal air spaces of the lung that may identify which patients are destined to develop fibrosing alveolitis. The first study to evaluate this hypothesis in detail in patients with ARDS was published in 1995 (Clark et al., 1995). In that study, the investigators measured the concentration of the N-terminal peptide of type III procollagen. In the BAL of patients at 3, 7, and 14 days after the development of acute lung injury. This molecule (type III procollagen peptide) is cleaved from the precursor procollagen molecule by specific proteinases in the extracellular space; it has been used as a biological marker of collagen synthesis in several experimental and clinical studies. In the lung injury study, type III procollagen peptide was significantly more elevated in patients who eventually died at both 3 and 7 days after the development of acute lung injury (Clark et al., 1995).

This initial study was followed recently by another clinical study at our institution in which measurements were made of pulmonary edema fluid that had been collected from patients with acute lung injury within 24 hours of the development of the syndrome as well as from control patients with hydrostatic edema (Chesnutt et al., 1997). These patients were well matched since they had similar severity of oxygenation defect and even had a similar four-point lung injury score. However, the patients with acute lung injury had higher edema fluid to plasma protein concentrations. This data is summarized in Table 4.

Table 4. Characteristics of 44 Patients with Pulmonary Edema

Characteristic	Acute Lung Injury (n=33)	Hydrostatic Edema (n=11)
Age (y)*	53 (21-93)	66 (22-91)
Men/women (n/n)	16/17	5/6
PaO_2/FiO_2[+]	127±66	132±68
Lung injury score (day 1)[+∂]	2.76±0.57	2.53±0.49
Edema fluid/plasma protein ratio	0.93±0.29[†]	0.54±0.12

* Values are mean, range
+ Values are mean ± SD
∂ Values were obtained on day 1 at the time of edema fluid collection
† p<0.05 compared to hydrostatic edema group

Reprinted with permission of the American Thoracic Society (Chesnutt et al., 1997).

Interestingly, type III procollagen peptide was already markedly elevated in patients with acute lung injury on day 1, suggesting that this molecule was released very early in the course of acute lung injury. Perhaps the magnitude of early acute lung injury determines which patients will develop fibrosing alveolitis. Type III procollagen peptide levels were much higher in patients with acute lung injury than in patients with hydrostatic pulmonary edema as controls **(Figure 1)**.

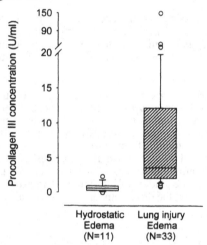

Figure 1. Pulmonary edema fluid procollagen III concentrations in patients with lung injury edema and hydrostatic edema. Median procollagen III level was higher in patients with lung injury edema (hatched box) than in patients with hydrostatic edema (open box) within 24 hours of intubation (p = 0.0001). Horizontal lines indicate the median value. Boxes represent the 25th to 75th percentile. Vertical lines represent the 10th to 90th percentile. Individual values below the 10th percentile and above the 90th percentile are shown as solid circles.

Reprinted with permission of the American Thoracic Society (Chesnutt et al., 1997).

Further analysis was made regarding the potential prognostic significance of elevated procollagen peptide levels. Interestingly, procollagen peptide III levels were significantly higher in the 21 patients who died from acute lung injury than in the 12 patients who lived **(Figure 2)**.

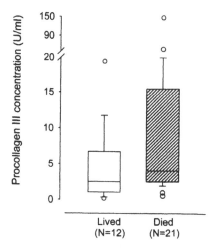

Figure 2. Pulmonary edema fluid procollagen III concentrations in patients with acute lung injury who lived or died. Median procollagen III level was higher in patients with lung injury who died (hatched box) than in those who lived (open box) ($p = 0.05$). Horizontal lines indicate that median value. Boxes represent the 25th to 75th percentile. Vertical lines represent the 10th to 90th percentile. Individual values below the 10th percentile and above the 90th percentile are shown as solid circles.

Reprinted with permission of the American Thoracic Society (Chesnutt et al., 1997).

Furthermore, if the level of procollagen peptide III was > 1.75 units/ml, a previously established cutoff from the earlier clinical study by Clark et al., then the positive predictive value for death was 75% and the negative predictive value for death was 83%. Thus, this marker may have prognostic value as well as provide insight into the pathogenesis of fibrosing alveolitis.

Further work is needed to examine the correlation of the release of type III procollagen peptide with morphologic evidence for the development of fibrosis. Clearly, mortality in patients in this study was not strictly attributable to pulmonary failure from irreversible lung fibrosis. However, the release of type III procollagen peptide may be a good marker of the severity of acute inflammatory lung injury. There is evidence, for example, that interleukin-1 and other proximal cytokines may be an important pro-inflammatory molecule in activating fibroblasts. Thus, the data in this study may provide an interesting link to the results of the studies of interleukin-1 in providing a mechanistic link between early acute lung injury and the development of fibrosing alveolitis.

SUMMARY

Considerable progress has been made in the last 10-15 years in determining the mechanisms of acute inflammatory lung injury in patients with clinical lung injury. Interpretation of the results of some of the studies has been complicated by heterogeneous patient populations, differences in criteria for diagnosis of acute lung injury, and measurement of pro-inflammatory molecules at different time points in the course of lung injury. Nevertheless, in conjunction with experimental models of acute lung injury, it appears that a cohesive pattern is emerging to explain some of the early mechanisms of acute lung injury. Release of proximal cytokines such as interleukin-1 appear to be important for priming the endothelium and perhaps the epithelium for acute lung injury. The influx of neutrophils, as chemoattracted by IL-8, may be an important mechanism for mediating neutrophil-induced lung injury in many patients. The release of metalloproteinases by neutrophils themselves may be very important in actually mediating structural damage to the lung. Several investigators are examining the important apoptosis of neutrophils in the response to acute lung injury as well. Finally, there does appear to be a significant new insight into the potential role of early inflammation in stimulating fibrosing alveolitis. Since type III procollagen peptide is present in high quantities in the edema fluid in patients with acute lung injury within 24 hours after intubation and initiation of positive pressure ventilation, it is likely that early inflammatory phase of lung injury determines in part the development of fibrosing alveolitis over the next several days. If this hypothesis is correct, then anti-inflammatory strategies early in the course of acute lung injury may attenuate the severity of fibrosing alveolitis.

Finally, future directions will need to focus on mechanisms that can enhance the capacity of the lung to recover from acute lung injury. In this regard, the release of growth factors that can stimulate alveolar epithelial repair without stimulating fibroblast proliferation may be important. Recently, it has been discovered that hepatocyte growth factor (HGF) and keratinocyte growth factor (KGF) are the most important mitogens for alveolar epithelial type II cells. Interestingly, experimental studies indicate that both of these growth factors may have the potential for attenuating lung injury. Recently, we have measured the concentration of both of these growth factors in the pulmonary edema fluid of patients with acute lung injury. KGF is present in negligible quantities that are not different between patients with hydrostatic pulmonary edema compared to those with acute lung injury. In contrast, HGF is present in high quantities in patients with acute lung injury, and in fact those patients with the highest levels seem to have a worse outcome.

Paradoxically, this result seems to be counterintuitive if HGF is potentially protective. However, HGF may be released from both pulmonary and non-pulmonary sources as a response to severity of acute lung injury. It is conceivable that treatment of patients with KGF might prove to be beneficial if it were given soon enough to stimulate alveolar epithelial repair. There is also emerging evidence that KGF may have other beneficial properties other than simply restoring the epithelial barrier. For example, there is preliminary evidence that KGF can accelerate alveolar epithelial fluid clearance by increasing the number of alveolar type II cells. Furthermore, there is data that KGF may also have some benefit in terms of increase in surfactant secretion as well as some anti-inflammatory properties. Although work is just beginning to evaluate these possibilities, it is important to consider mechanisms that might hasten recovery from lung injury as well as strategies that might limit the severity of lung injury.

REFERENCES

Chesnutt, A.N., Matthay, M.A., Tibayan, F.A., and Clark, J.G., 1997, Early detection of type III procollagen peptide in acute lung injury: pathogenetic and prognostic significance, Am. J. Resp. Crit. Care Med. 156: 840-845.

Clark, J.G., Milberg, J.A., Steinberg, K.P., and Hudson, L.D., 1995, Type III procollagen peptide in the adult respiratory distress syndrome: association of increased peptide levels in bronchoalveolar lavage fluid with increased risk for death, Ann. Intern. Med. 122: 17-23.

Crouch, E., 1990, Pathobiology of pulmonary fibrosis, Am. J. Physiol. 259:L159-L184.

Delclaux, C., d'Ortho, M.P., Delacourt, C., Lebargy, F., Brun-Buisson, C., Brochard, L., Lemaire, F., Lafuna, C., and Harf, A., 1997, Gelatinoses in epithelial lining fluid of patients with adult respiratory distress syndrome, Am. J. Physiol. (Lung) (In press).

Donnelly, S.C., Strieter, R.M., and Kunkel, S.L., Waltz, A., Robertson, C.R., Carter, D.C., Grant, I.S., Pollok, A.J., and Haslett, C., 1993, Interleukin-8 and development of adult respiratory distress syndrome in at-risk patient groups, Lancet 341:643-647.

Folkesson, H.G., Matthay, M.A., Hébert, C.A., and Broaddus, V.C., 1995, Acid aspiration–induced lung injury in rabbits is mediated by interleukin-8–dependent mechanisms, J. Clin. Invest. 96:107-116.

Matthay, M.A., and Wiener-Kronish, J.P., 1990, Intact epithelial barrier function is critical for the resolution of alveolar edema in humans, Am. Rev. Respir. Dis. 142: 1250-1257.

Miller, E.J., Cohen, A.B., and Matthay, M.A., 1996, Increased interleukin-8 concentrations in the pulmonary edema fluid of patients with acute respiratory distress syndrome from sepsis, Crit. Care Med. 24:1448-1454.

Miller, E.J., Cohen, A.B., Nagao, S., Griffith, D., Maudner, R.J., Martin, T.R., Wiener-Kronish, J.P., Sticherling, M., Christophers, E., and Matthay, M.A., 1992, Elevated levels of NAP-1/interleukin-8 are present in the airspaces of patients with adult respiratory distress syndrome and are associated with increased mortality, Am. Rev. Respir. Dis. 146:427-432.

Pittet, J.F., Mackersie, R.C., Martin, T.R., and Matthay, M.A., 1997, Biological markers of acute lung injury: prognostic and pathogenetic significance, Am. J. Respir. Crit. Care Med. 155:1187-1205.

Pugin, J., Ricou, B., Steinberg, K.P., Suter, P.M., and Martin, T.R., 1996, Proinflammatory activity in bronchoalveolar lavage fluids from patients with ARDS: a prominent role for interleukin-1ß, Am. J. Respir. Crit. Care Med. 153:1850-1856.

Ricou, B., Nicod, L., Lacaraz, S., Welgus, H.G., Suter, P.M., and Dayer, J.-M., 1996, Matrix metalloproteinases and TIMP in acute respiratory distress syndrome, Am. J. Respir. Crit. Care Med. 154:346-352.

Zapol, W.M., Trelstad, R.L., Coffey, J.W., Salvador, R.A., 1979, Pulmonary fibrosis in severe acute respiratory failure, Am. Rev. Respir. Dis. 119:547-554.

PATHWAYS TO OXIDATIVE CELL DEATH

Stuart Horowitz

CardioPulmonary Research Institute
SUNY Stony Brook School of Medicine
Winthrop-University Hospital
222 Station Plaza North
Mineola, NY 11501

INTRODUCTION

Therapy with supraphysiological concentrations of oxygen (O_2) is required in a number of clinical situations, but this use of O_2 may be accompanied by tissue damage (Northway et al, 1967; Horowitz and Davis, 1996). For example, hyperoxia plays a role in the etiology of bronchopulmonary dysplasia (Pappas et al, 1983; Roberts and Frank, 1984) a disorder resulting from ventilatory O_2 therapy for the respiratory distress syndrome of premature infants. The general problem of O_2 toxicity involves several organs, most notably the lungs, which receive direct exposure (Wispe and Roberts, 1987). The pathology of O_2-induced lung injury includes inflammation and permeability changes of the alveolar-capillary membrane, causing extensive pulmonary edema and severe decreases in respiratory function (Freeman et al, 1986). Lung pathology resulting from acute O_2 toxicity is accompanied by, and may result from biochemical alterations in lung cells.

OVERALL PATHOPHYSIOLOGY OF HYPEROXIC LUNG INJURY

Studies in a variety of animal models have shown that the first visible signs of lung injury after hyperoxic exposure are a swelling of endothelial cells and an altered appearance of mitochondria and microsomes (Crapo et al, 1978). With time, gross morphological changes of individual cells become apparent. There are shifts of lung cell populations as a result of proliferation and hypertrophy of type II pneumocytes, hyperplasia of interstitial cells and loss of capillary endothelial cells. An inflammatory response also occurs (Crapo et al, 1980). These alterations make it difficult to distinguish between biochemical changes that may be occurring in a single cell type from overall changes as a result of shifts in lung cell populations (Freeman et al, 1986). Often, both pulmonary endothelium and epithelium are damaged, resulting in the leak of fluid and macromolecules into the air spaces (Freeman et al, 1986), attenuating respiratory function, and causing death in the worst cases.

Not surprisingly, these complex events have roots in alterations in gene expression at multiple levels. For example, the overall rate of protein synthesis decreases by 30% in hyperoxic lung (Kelly, 1988). Technological advances have helped improve our understanding of genes, mRNAs, and proteins that are involved in hyperoxic lung injury

Acute Respiratory Distress Syndrome: Cellular and Molecular Mechanisms and Clinical Management
Edited by Matalon and Sznajder, Springer Science+Business Media New York, 1998

215

(see Horowitz and Davis, 1997). Studies of lung gene expression usually involve one of two approaches. Most typically, cDNA probes encoding proteins of known function or suspected importance have been used to measure changes in specific mRNA abundance. By a second approach, subtractive hybridization cloning or differential screening of cDNA libraries have been used to isolate and identify cDNA clones corresponding to mRNAs that increase in lung injury. Our laboratory has taken both these approaches (Horowitz et al, 1989, Piedboeuf et al, 1996). Despite these advances, relatively little is known about the biochemical pathways to cell death from hyperoxia.

OXYGEN TOXICITY IN CELL CULTURE

The enormous complexity of hyperoxic lung injury *in vivo* precludes a straightforward assessment of the mechanisms underlying O_2 toxicity, giving rise to scientific investigation of this problem in cell culture. In mammalian cell lines the toxic effects of normobaric hyperoxia have been observed to occur only after several days of exposure. The typical symptoms are progressive growth inhibition, loss of reproductive capacity, and chromosomal instability. In cultured transformed alveolar epithelial (A549) cells, lethal exposure to hyperoxia (95% O_2) is associated with cell cycle arrest at G2 (Clement et al, 1992). It is generally thought that O_2 toxicity results from increased steady-state levels of highly reactive ocygen-derived free radicals, also called reactive oxygen species (ROS), which overwhelm the cellular antioxidant defenses (Gille et al, 1994; Wagner et al, 1994; Schacter et al, 1994). Hyperoxia not only stimulates ROS production, but may also increase the production of oxidizable substrates from reactions that are normally limited at physiological levels of O_2. Hyperoxic substrate conversion rates could create a metabolic balancing problem which may eventually jeopardize cell survival. It should thus be kept in mind that O_2 toxicity might not only result from direct ROS-induced damage, but could also be an indirect consequence of other metabolic imbalances, which may result in the inactivation of oxygen-sensitive target enzymes. For example, respiratory failure — resulting in an 80% inhibition of O_2 consumption within 3 days of hyperoxia in Chinese Hamster Ovary (CHO) cells and HeLa cells — is associated with the selective inactivation of 3 mitochondrial key enzymes: NADH dehydrogenase, succinate dehydrogenase, and alpha-ketoglutarate dehydrogenase (Schoonen et al, 1990a, 1990b, 1991; Gille et al, 1991; Joenje et al, 1997). This last enzyme controls the influx of glutamate into the Krebs cycle and is particularly critical for oxidative ATP generation in most cultured cells. These observations indicate that cellular O_2 toxicity is correlated with an impairment of mitochondrial energy metabolism resulting from inactivation of SH-group-containing flavoprotein enzymes localized at or near the inner mitochondrial membrane. In A549 cells, hyeroxia causes the inactivation of aconitase (Gardner et al, 1994) In addition, poly-ADP ribose polymerase (PARP) is inactivated by hyperoxia in HeLa cells (Gille et al, 1989). Thus, hyperoxia is a pleiotropic insult, and to date, it seems likely that only a portion of the biochemical changes leading to cell death by hyperoxia has been identified

OXIDANT-INDUCED CELL INJURY AND DEATH

Because hyperoxia causes excess ROS production, O_2 toxicity is often considered to be similar to other oxidant stresses. Not surprisingly, cells can also be injured directly by ROS. For example, exposure of cells to peroxide or superoxide radicals can result in cell death (Janssen et al, 1993; Hollan, 1995). In most of the cases in which it has been examined, cell death caused by what are thought to be physiologically-relevant levels of these oxidants appears to occur via apoptosis, also known as programmed cell death. However, cytotoxic effects of oxidants are dose and cell type-dependent (Gille and Joenje, 1992; Spitz et al, 1992), and at very high oxidant levels, most cells die within a few short hours. In addition, at sufficiently low levels, oxidants may no longer be cytotoxic, but still can wreak havoc upon the cell due to genotoxic activity, as demonstrated by clonogenic survival assays (Gille and Joenje, 1992). Thus, oxidants can damage cells and organs over both short and long time periods. Importantly, an increased oxidant burden is a critical component of inflammatory diseases, especially in the lungs of premature infants or adult patients with sepsis, trauma, or ARDS (Horowitz and Mantell, 1997). Oxidants probably also have a pivotal role in the lung's response to asbestos fiber inhalation (Bérubé et al, 1996). For these reasons, the question of how oxidants trigger pulmonary cell death is of vital importance.

APOPTOSIS AND NECROSIS

Cell death is currently thought to occur via one of two distinct modes: apoptosis or necrosis. Apoptosis, also called programmed cell death, is the process of cellular self destruction and involves a specific series of events (Steller 1995). Apoptosis is often a scheduled and physiologically-regulated event, occurring during normal cell turnover, development, and in response to viral infection. In contrast, necrosis is the result of unscheduled, acute injury, and *in vivo* can result in or from an inflammatory response, which can be local or systemic (Thompson 1995). On the other hand, when cells in tissue die via apoptosis, they are removed through phagocytosis by neighboring healthy cells (Steller 1995). Thus, the observation that inflammation can be circumvented has given rise to the popular notion that apoptosis is a protective mode of cell death (Steller 1995). Apoptosis is characterized by cell shrinkage, and in nucleated cells, by chromosome condensation together with internucleosomal cleavage of DNA. However, these are relatively late manifestations of apoptosis, and a great deal of attention has been focused on the biochemical and molecular events that constitute intermediate steps toward along the apoptotic pathway. In addition, the early signal transduction events leading to apoptosis are also subjects of intense study in a variety of nonpulmonary systems (although very little is known about apoptosis in lung cells). Suffice it to say that apoptosis can be regulated at a growing number of checkpoints, depending on the cell type and the initial trigger.

APOPTOSIS AND OXIDANTS

As mentioned above, apoptosis appears to be the major mode of cell death when cells experience lethal oxidative insult from exposure to oxidants, including H_2O_2 and superoxide (Albrecht et al, 1994; Sandstrom et al, 1993; Forrest et al, 1994; Cappelletti et al, 1994; Slater et al, 1995). Interestingly, even cells that undergo apoptosis following non-oxidative insults, such as steroid treatment or viral infection, have been shown to accumulate lipid peroxides, which is evidence of oxidative damage (Jones, 1992; Favier et al, 1994; Zamzami et al, 1995). Moreover, apoptosis can be prevented in such cells by the overexpression of cellular antioxidant enzymes or the oncogene Bcl-2, which has been suggested to be an antioxidant or to impinge on an antioxidant pathway (Hockenberry et al, 1993). Taken together, there is a tight correlation between oxidative damage to cells and apoptosis, and lipid peroxidation may be a key step leading to apoptosis in some cases, and resulting from it in others. For these reasons, we anticipated that another form of oxidant injury, O_2 toxicity, would also result in apoptosis.

HYPEROXIA DOES NOT KILL EPITHELIAL CELLS BY APOPTOSIS

A549 cells, derived from human alveolar type II pneumocytes, have been extensively studied with respect to their responses to hyperoxia and other oxidant injuries (Gardner et al, 1994; Cappelletti et al, 1994). Cells were cultured either in 95% room air or 95% O_2, and assayed for apoptosis by 3 complimentary techniques (Kazzaz et al, 1997). Cells cultured in hyperoxia showed overt signs of death (Trypan blue inclusion) by day 4,

217

and nearly 70% of the cells were dead by day 7. The remainder of the cells died over the next 3 days. These kinetics of A549 cell death in hyperoxia were similar to those reported for other cell types (Christie et al, 1994; Bowden et al, 1994; Yuan et al, 1995). Because cell death from other oxidants occurs via apoptosis (Albrecht et al, 1994; Sandstrom et al, 1993; Forrest et al, 1994), we tested whether hyperoxia also induces apoptosis.

In the presence of the DNA-binding dye Hoechst 33258, nuclei that are condensed during apoptosis are much smaller and more intensely fluorescent when excited by UV than nuclei in non-apoptotic cells. However, the nuclei of cells exposed to hyperoxia were larger than controls with no apparent increase in fluorescence. In contrast, cells exposed to the oxidants H_2O_2 or paraquat (which generates intracellular superoxide) underwent apoptosis, as shown by their typical shrunken and intensely-fluorescent nuclei.

We also utilized the TUNEL (Terminal deoxynucleotidyl transferase dUTP Nick End Labeling) assay to study apoptosis in these cultures. This is an in situ assay that exploits the capacity of terminal transferase to add nucleotides to 3'-OH ends of DNA in chromatin, which result from endonucleolytic cleavage occurring during apoptosis (Tornusciolo et al 1995; Gavrieli et al, 1992). The vast majority of cells cultured in hyperoxia were TUNEL negative. In contrast, a large population of cells exposed to the other oxidants were clearly TUNEL positive. Similar observations were made in two other epithelial cell lines, HeLa cells, and MLE 12 cells.

An unassailable means of assessing whether cells are or are not apoptotic is to study their morphology by electron microscopy. An EM analysis showed that when cells were exposed to 95% O_2 for 6 d they became swollen, with enlarged nuclei and mitochondria. By contrast, cells that were exposed to paraquat or H_2O_2 had condensed chromatin, which is a hallmark of apoptosis. These data confirm the results obtained using DNA fluorescence microscopy and TUNEL assays, and further indicate that hyperoxia did not result in apoptosis in these epithelial cells.

If the kinetics of cell death is correlated with the extent of apoptosis at each time point, it would be reasonable to conclude that the cells had died via apoptosis. Conversely, if there is no correlation, cells would have died via necrosis, not apoptosis. To quantify the extent of apoptosis, we used computer-aided image analysis. When apoptosis was induced by exposure to H_2O_2 or paraquat, a large population of cells with shrunken and brightly-fluorescent nuclei were clearly distinguished from untreated cells. Using micrographic CCD video images analyzed by a Universal Imaging Image-1 computer workstation, we established an objective threshold for the area of apoptotic nuclei. Cells were scored as apoptotic only when their nuclear area was beneath that threshold. Similar quantitative analyses have been achieved by fluorescence-activated cell sorting of non-adherent cells, although in our hands computer-aided image analysis is better-suited to adherent cells. This quantitative analysis showed that there was no correlation between cell death and the extent of apoptosis at any time during hyperoxic exposure, as determined either by image analysis after Hoechst staining or by counting TUNEL-positive nuclei. In contrast, cell death and apoptosis were tightly correlated in cultures exposed either to 5 mM H_2O_2 or 20 mM paraquat.

LOW DOSE OXIDANTS DO NOT KILL EPITHELIAL CELLS BY APOPTOSIS

Hyperoxic cell death is different than death from other oxidants several respects, including the kinetics of cell killing. At the relatively high oxidant doses typically used to study apoptosis (Ueda and Shah, 1992; Sandstrom et al, 1993; Slater et al, 1995; Iwata et al, 1994; Pierce et al, 1991) and used here, cells are killed in a matter of hours. However, even at the highest doses of hyperoxia ($\geq$95% O_2), it takes days before appreciable cell death is evident. To determine if oxidant-induced apoptosis occurs when the rate of cell death is substantially reduced, experiments were performed at much lower concentrations of the oxidants H_2O_2 and paraquat. However, unlike hyperoxia, these low oxidant doses were not 100% lethal, and a sub-population of cells began to adapt and divide.

218

Nevertheless, similar to cell death by hyperoxia, we observed virtually no apoptosis at these lower oxidant concentrations. This may be important, because the high levels of oxidants required to induce apoptosis might not be achieved in vivo. Thus, it seems unlikely that apoptosis is the major mode of oxidant-induced cell death, except perhaps in the culture dish.

APOPTOTIC PATHWAYS IN LUNG EPITHELIAL CELLS

Almost nothing is known about the pathways to apoptotic cell death in lung epithelium. In preliminary experiments, we addressed the question of whether any apoptosis-related genes are regulated in lung epithelial cells that undergo either oxidant-induced apoptosis or necrosis. Nedd-2 is a member of the IL-1b converting enzyme (ICE) family of proteases, and other members of the ICE family have been shown to be activated in apoptosis. To determine if Nedd-2/Ich-1 activation (proteolytic cleavage) occurs in oxidant-induced lung epithelial cell death, A549 cells were exposed either to hyperoxia or 5 mM H_2O_2. Western blota show that Nedd-2 is cleaved when A549 cells are driven into apoptosis by H_2O_2. This is similar to many published accounts of Nedd-2 regulation during apoptosis of other cell types. Immunofluorescence assays indicated that not only Nedd-2, but also ICE itself was activated by H_2O_2. The enzyme poly-ADP-ribose polymerase (PARP) is proteolytically cleaved during apoptosis, and is then translocated to the nucleus. Immunofluorescence assays show that oxidant-induced apoptosis of A549 cells causes the nuclear translocation of PARP. These are among the first data implicating these apoptotic pathways in lung epithelium. Importantly, neither PARP nor ICE are induced during cell death from hyperoxia, suggesting that the molecular pathways to oxidant-induced cell death are distinct, depending on the oxidant and mode of death. Preliminary data also suggest that PARP is downregulated in hyperoxia, consistent with a previous report of its functional inactivation (Gille et al, 1989).

TRANSCRIPTION FACTORS AND SIGNALING OF KINASES

Recent reports from several laboratories show that apoptotic cell death can be prevented by the expression of NFκB (Beg and Baltimore, 1996; Antwerp et al, 1996; Wang et al, 1996; Liu et al, 1996), a multisubunit transcription factor that can rapidly activate the expression of genes involved in inflammation, infection, and stress (Schreck et al, 1991). These recent reports suggest that the induction of NFκB may be part of a survival mechanism used to escape cell death (Barinaga, 1996). We examined the expression of NFκB in cells exposed to lethal concentrations of hyperoxia or H_2O_2. Our observations (summarized below) show that despite the induction of NFκB by molecular O_2, the cells do not escape death (Li et al, 1997).

It is known that following release from the inhibitory binding protein IκB, NFκB translocates from cytosol to the nucleus, where it regulates transcription. We therefore studied NFκB activation during hyperoxia by examining nuclear translocation, using an antibody to the p65 subunit of NFκB. Control A549 cells grown in room air had weak NFκB immunofluorescence. The signal was evident primarily in the cytoplasm, although in a small number of cells there was weak fluorescence. Within a half hour of hyperoxia nuclear fluorescence was brighter, and it increased over the course of 1 d. By 24 h, the cells already showed signs of swelling, and fluorescence was more intense not only in the nuclei but also in the cytoplasm of many cells. Interestingly, there was no significant nuclear translocation of NFκB when cells were exposed to concentrations of H_2O_2 that caused apoptosis. By 4 h of H_2O_2 treatment, the majority of cells were undergoing apoptosis. However, in a small number of cells that had not yet undergone apoptosis, nuclear fluorescence for NFκB was observed.

The immunofluorescence data suggested that in addition to nuclear translocation, an increase in NFκB protein levels might have occurred during hyperoxia. To test this, western blots were performed. NFκB levels were found to increase as soon as 30 min after exposure to 95% O_2, and peak levels were achieved by 24 h. Levels remained elevated for 3 d. In sharp contrast, apoptotic levels of H_2O_2 caused no increased NFκB protein, and there was even a slight decrease after 2 h.

To determine if elevated NFκB protein was correlated with increased mRNA abundance, northern blots were performed. There was a slight increase in steady state levels of NFκB mRNA within 30 min of culture in hyperoxia. Message levels increased over the course of 1 d and remained elevated for 2 d. The abundance of the message encoding the housekeeping enzyme, glyceraldehyde-3-phosphate dehydrogenase (GAPDH) was unchanged in this experiment. Unlike exposure to hyperoxia, treatment of the culture with a dose of H_2O_2 that caused apoptosis failed to induce NFκB expression. Rather, a notable decrease was observed after 2 h.

The differential activation and expression of NFκB during different modes of cell death imply that different signals are transduced by hyperoxia and H_2O_2. We therefore undertook a study aimed at deciphering the transcriptional regulatory and signaling events that occur in the early phases of oxidant-induced death of these cells. In many cell types, one of the earliest responses to stress involves the transient expression of Fos and Jun. Moreover, the Fos-Jun dimer constitutes the transcription complex called AP-1, which is reported to be redox sensitive, and is activated in lungs of hyperoxic rats (Choi et al, 1995). Cells (MLE-12, transformed mouse lung epithelial cells) were exposed to 95% O_2 for up to 24 h. At various time points, cells were harvested and RNA assayed by northern blots for Fos and Jun expression. Relative to control cells (that were plated at the same time as hyperoxic cells, and also transferred to fresh medium), both Fos and Jun transcript levels were elevated transiently (at the 30 min time point) and then returned rapidly to baseline. However, both mRNAs were again increased in abundance, at 16 to 24. These data suggest that there may be (at least) two signals that activate Fos/Jun expression. The first may be related to oxygen tension or initial oxidant stress. Perhaps the second signal is a consequence of cell-injury.

The activation of the c-Fos promoter is known to be under the control of the p42 and p44 Mitogen Activated Protein kinase, or MAP kinase cascade in some cases (Treisman, 1994).We therefore investigated whether p42 and p44 MAP kinases were activated by hyperoxia. To study MAP kinase we used antibodies that are specific to the phosphorylated, or activated form of the proteins. There were no changes detected in the levels of phosphorylated p42 or p44 at any time during exposure to hyperoxia. In contrast, as soon as 10 min after incubation in an apoptosis-inducing concentration of H_2O_2, there was a significant increase in phosphorylated p42 and p44. Interestingly, this activation was transient, and the phosphorylated protein levels decreased rapidly. This experiment shows that the p42 and p44 pathway is activated by an apoptotic dose of H_2O_2, but not by hyperoxia. These observations further support the notion that signaling events are different in these two modes of lung epithelial cell death.

SUMMARY

The mode of epithelial cell death from oxidant injury depends largely on the oxidant and the dose. There is some controversy over the question of what oxidant levels are physiologically meaningful. At extremely high oxidant doses (which vary according to cell type), cells are rapidly oxidized, killing them — *not* by apopoptosis — within a couple of hours or sooner. At reduced doses (again, which vary according to cell type) cells die via apoptosis, requiring a few hours more and sometimes as long as one day. Perhaps because it is now such a popular notion that apoptosis is the only physiologically-relevant mode of cell death, some feel that H_2O_2 doses resulting in apoptosis are likely to be relevant in biology. However, we have observed that at still lower oxidant doses, epithelial cells do not undergo apoptosis, yet they still die. This is a slower death (requiring days), morphologically similar to hyperoxic cell death, and with similar kinetics. Unfortunately, the dichotomous nature of prevailing thoughts on cell death give us a limited vocabulary;

death is either necrotic or apoptotic. Thus, not only is the very rapid oxidation of cells at high oxidant doses labeled as necrosis, but also the slower cell death that occurs below oxidant levels that induce apoptosis. Importantly, contrary to the notion that the only mode of cell death involving a program is apoptisis, the nonapoptotic cell death that occurs in hyperoxia and lower levels of oxidants appears to involve specific signallingevents and alternate pathways.

REFERENCES

Albrecht, H., Tschopp, J., and Jongeneel, C. V. 1994. Bcl-2 protects from oxidative damage and apoptotic cell death without interfering with activation of NF-kappa b by TNF. FEBS Lett 351, 45-48.

Antwerp, D. J., Martin, S. J., Kafri, T., Green, D. R., and Verma, I. M. 1996. Suppression of TNF-alpha-induced apoptosis by NF-kB. Science 274.

Barinaga, M. (1996). Life-death balance within the cell. Science 274.

Beg, A. A., and Baltimore, D. 1996. An essential role for NF-kB in preventing TNF-alpha-induced cell death. Science 274, 782-784.

BeruBe KA, Quinlan TF, Fung H, Magae J, Vacek P, Taatjes DJ, Mossman BT., 1996. Apoptosis is observed in mesothelial cells after exposure to crocidolite asbestos. Am J Resp Cell Mol Biol 15, 141-147.

Bowden, D. H., Young, L., and Adamson, I. Y. R. 1994. Fibroblast inhibition does not promote normal lung repair after hyperoxia. Exp Lung Res 20, 251-262.

Cappelletti, G., Incani, C., and Maci, R. 1994. Paraquat induces irreversible actin cytoskeleton disruption in cultured human lung cells. Cell Biol Toxicol 10, 255-263.

Choi AM, Sylvester S, Otterbein L, Holbrook NJ. 1995. Molecular responses to hyperoxia in vivo: relationship to increased tolerance in aged rats. Am. J. Resp. Cell & Mol. Biol. 13(1): p. 74-82.

Christie, N. A., Slutsky, A. S., Freeman, B. A., and Tanswell, A. K. 1994. A critical role for thiol, but not ATP, depletion in 95% O-2-mediated injury of preterm pneumocytes in vitro. Arch Biochem Biophys 313, 131-138.

Clement, A., Edeas, M., Chadelat, K., and Brody, J. S. 1992. Inhibition of lung epithelial cell proliferation by hyperoxia. Posttranscriptional regulation of proliferation-related genes. Journal of Clinical Investigation 90, 1812-8.

Crapo, J. D., Peters-Golden, M., Marsh-Salin, J., and Shelburne, J. S. 1978. Pathologic changes in the lungs of oxygen-adapted rats: a morphometric analysis. Lab. Invest. 39, 640-653.

Crapo, J. D.; Barry, B. E.; Foscue, H. A.; Shelburne, J. 1980 Structural and biochemical changes in rat lungs occurring during exposure to lethal and adaptive doses of oxygen. Am Rev Respir Dis; 122: 123-143.

Favier A, Sappey C, Leclerc P, Faure P, and Micoud M. 1994. Antioxidant status and lipid peroxidation in patients infected with HIV. Chemico-Biological Interactions; 91(2-3): p. 165-80.

Forrest, V. J., Kang, Y. H., McClain, D. E., Robinson, D. H., and Ramakrishnan, N. 1994. Oxidative stress-induced apoptosis prevented by Trolox. Free Radical Biology & Medicine 16, 675-84.

Freeman BA, Mason, RJ, Williams MC, Crapo JD. 1986. Antioxidant enzyme activity in alveolar type II cells after exposure of rats to hyperoxia. Exp Lung Res 10(2): p. 203-22.

Gavrieli Y, Sherman Y, and Ben-Sasson SA. 1992. Identification of programmed cell death in situ via specific labeling of nuclear DNA fragmentation. J. Cell Biol. 119(3): p. 493-501.

Gille, J. J., and Joenje, H. 1989. Chromosomal instability and progressive loss of chromosomes in HeLa cells during adaptation to hyperoxic growth conditions. Mutation Research 219, 225-30.

Gille, J. J., and Joenje, H. 1992. Cell culture models for oxidative stress: superoxide and hydrogen peroxide versus normobaric hyperoxia. [Review]. Mutation Research 275, 405-14.

Gille, J. J., Pasman, P., van-Berkel, C. G. and Joenje, H. 1991. Effect of antioxidants on hyperoxia-induced chromosomal breakage in Chinese hamster ovary cells: protection by carnosine. Mutagenesis 6, 313-318.

Hockenbery, D. M., Oltvai, Z. N., Yin, X. M., Milliman, C. L., and Korsmeyer, S. J. (1993). Bcl-2 functions in an antioxidant pathway to prevent apoptosis. Cell 75, 241-51.

Hollan, S. 1995. Free radicals in health and disease. [Review]. Haematologia 26, 177-89.

Horowitz S and Davis JM. 1997. Lung injury when development is interrupted by premature birth. pp. 577-610. In: Growth and development of the lung. J.A. McDonald, Ed. Marcel Dekker, NY.

Horowitz S, Dafni N, Shapiro DL, Holm BA, Notter RH, Quible DJ. 1989. Hyperoxic exposure alters gene expression in the lung. Induction of the tissue inhibitor of metalloproteinases mRNA and other mRNAs. J. Biol. Chem; 264(13): p. 7092-7095.

Horowitz S, and Mantell L. 1997. Oxygen radicals in sepsis and multiorgan failure. In: Sepsis and Multiorgan Failure. AM Fein, E Abraham, R Balk, G Bernard, D Dantzker M Fink, eds. Williams and Wilkins, Media.

Iwata, M., Myerson, D., Torok-Storb, B., and Zager, R. A. 1994. An evaluation of renal tubular DNA laddering in response to oxygen deprivation and oxidant injury. Journal of the American Society of Nephrology 5, 1307-13.

Janssen, Y. M., Van, H. B., Borm, P. J., and Mossman, B. T. 1993. Cell and tissue responses to oxidative damage. [Review]. Laboratory Investigation 69, 261-74.

Joenje H, Gille JJP, Horowitz S, Li ZG, Whyzmuzis CA. 1997, Metabolic effects of hyperoxia: lessons from permanent cell lines. pp. 67-73. In: Oxygen, Gene Expression, and Cellular Function LB Clerch & D Massaro, eds. Marcel Dekker, Inc, New York.

Jones, G. R. 1992. Cancer destruction in vivo through disrupted energy metabolism. Part I. The endogenous mechanism of self-destruction within the malignant cell, and the roles of endotoxin, certain hormones and drugs, and active oxygen in causing cellular injury and death. [Review]. Physiological Chemistry & Physics & Medical NMR 24, 169-79.

Kazzaz, J. A., Xu, J., Palaia, T. A., Mantell, L., Fein, A. M., and Horowitz, S. 1996. Cellular oxygen toxicity - oxidant injury without apoptosis. J Biol Chem 271, 15182-15186.

Kelly, F. J. 1988. Effect of hyperoxic exposure on protein synthesis in the rat [published erratum appears in Biochem J 1988 Jun 15;252(3):935 Biochemical Journal 249(2) :609-12.

Li Y, Kazzaz JA, Mantell LL, Fein AM, Horowitz S. 1997. NFκB is activated by hyperoxia but does not protect from cell death. J. Biol. Chem. 272, 20646-2064

Liu, Z. G., Hsu, H., Goeddel, D. V., and Karin, M. 1996. Dissection of TNF receptor 1 effector functions: JNK activation is not linked to apoptosis while NF-kB activation prevents cell death. Cell 87, 565-576.

Mantell, L. Horowitz S.1997. Oxygen radicals in sepsis and organ failure. In: Sepsis in multiorgan failure: Mechanisms and treatment strategies. Alan Fein, Edward Abraham, Robert Balk, Gordon Bernard, Roger Bone, David Dantzker, Mitchell Fink, Eds. Williams & Wilkins, PA.

Northway, W. H. Jr, Rosan, R. C., Porter, D. Y. 1967. Pulmonary disease following respirator therapy of hyaline-membrane disease. Bronchopulmonary dysplasia New England Journal of Medicine 276(7) :357-68.

Pappas CT, Obara H, Bensch KG, Northway WH Jr. 1983. Effect of prolonged exposure to 80% O_2 on the lung of the newborn mouse. Laboratory Investigation 48(6): p. 735-48.

Piedboeuf, B., Frenette, J., Petrov, P., Welty, S. E., Kazzaz, J. A., and Horowitz, S. 1996. In vivo expression of intercellular adhesion molecule 1 in type II pneumocytes during hyperoxia. Am. J. Respir. Cell Mol. Biol. 15, 71-77.

Pierce, G. B., Parchment, R. E., and Lewellyn, A. L. 1991. Hydrogen peroxide as a mediator of programmed cell death in the blastocyst. Differentiation 46, 181-6.

Roberts, R. J.; Frank, L. 1984 In: Toxicology and the Newborn. Eds: S. Kacew and M.J. Reasor; 141-171.

Sandstrom, P. A., Roberts, B., Folks, T. M., and Buttke, T. M. 1993. HIV gene expression enhances T cell susceptibility to hydrogen peroxide-induced apoptosis. AIDS Research & Human Retroviruses 9, 1107-13.

Schacter, E., Williams, J. A., Lim, M., and Levine, R. L. 1994. Differential susceptibility of plasma proteins to oxidative modification: examination by western blot immunoassay. Free Radical Biology & Medicine 17, 429-37.

Schoonen, W. G., Wanamarta, A. H. van-der-Klei-van-Moorsel, J. M., Jakobs, C. and Joenje, H. 1990a. Hyperoxia-induced clonogenic killing of HeLa cells associated with respiratory failure and selective inactivation of Krebs cycle enzymes. Mutat. Res. 237, 173-181.

Schoonen, W. G., Wanamarta, A. H. van-der-Klei-van-Moorsel, J. M., Jakobs, C. and Joenje, H. 1990b Respiratory failure and stimulation of glycolysis in Chinese hamster ovary cells exposed to normobaric hyperoxia. J. Biol. Chem. 265:1118-1124.

Schoonen, W. G., Wanamarta, A. H., van, d. K.-v. M. J. M., Jakobs, C., and Joenje, H. 1991. Characterization of oxygen-resistant Chinese hamster ovary cells. III. Relative resistance of succinate and alpha-ketoglutarate dehydrogenases to hyperoxic inactivation. Free Radical Biology & Medicine 10, 111-8.

Schreck, R., Rieber, P., and Baeuerle, P. A. 1991. Reactive oxygen intermediates as apparently widely used messengers in the activation of the NF-kappa B transcription factor and HIV-1. EMBO J. 10, 2247-2258.

Slater AFG, Nobel CSI, and Orrenius S. 1995. The role of intracellular oxidants in apoptosis. Biochim. Biophys. Acta-Molecular Basis of Disease 1271(1): p. 59-62.

Spitz, D. R., Adams, D. T., Sherman, C. M., and Roberts, R. J. 1992. Mechanisms of cellular resistance to hydrogen peroxide, hyperoxia, and 4-hydroxy-2-nonenal toxicity: the significance of increased catalase activity in H2O2-resistant fibroblasts. Arch. Biochem. & Biophys. 292, 221-7.

Steller, H. 1995. Mechanisms and genes of cellular suicide. Science 267, 1445-1449.

Stone, K. C., Mercer, R. R., Gehr, P., Stockstill, B., and Crapo, J. D. 1992. Allometric relationships of cell numbers and size in the mammalian lung. Am. J. Respir. Cell. Mol. Biol. 6, 235-243.

Thompson, C. B. 1995. Apoptosis in the pathogenesis and treatment of disease. Science 267, 1456-1462.

Tornusciolo, D. R. Z., Schmidt, R. E., and Roth, K. A. 1995. Simultaneous detection of TdT-mediated dUTP-Biotin nick end labeling (TUNEL)-positive cells and multiple immunohistochemical markeres in single tissue sections. Biotechniques 19, 800-805.

Treisman, R. 1994. Ternary complex factors: growth factor regulated transcriptional activators. Curr. Opin. Genet. Dev. 4, 96.

Ueda, N., and Shah, S. V. 1992. Endonuclease-induced DNA damage and cell death in oxidant injury to renal tubular epithelial cells. Journal of Clinical Investigation 90, 2593-7.

Wagner, B. A., Buettner, G. R., and Burns, C. P. 1994. Free radical-mediated lipid peroxidation in cells: oxidizability is a function of cell lipid bis-allylic hydrogen content. Biochemistry 33, 4449-53.

Wang XD, Deng XM, Haraldsen P, Andersson R, and Ihse I. 1995. Antioxidant and calcium channel blockers counteract endothelial barrier injury induced by acute pancreatitis in rats. Scandinavian Journal of Gastroenterology 30(11): p. 1129-36.

Wispe, J. R.; Roberts, R. J. 1987 Molecular basis of pulmonary oxygen toxicityClin Perinatol; 14(3): 651-666.

Yuan, H., Kaneko, T., Kaji, K., Kondo, H., and Matsuo, M. 1995. Species difference in the resistibility of embryonic fibroblasts against oxygen-induced growth inhibition. Comp Biochem Physiol [B] 110, 145-154.

Zamzami, N., Marchetti, P., Castedo, M., Decaudin, D., Macho, A., Hirsch, T., Susin, S. A., Petit, P. X., Mignotte, B., and Kroemer, G. 1995. Sequential reduction of mitochondrial transmembrane potential and generation of reactive oxygen species in early programmed cell death. Journal of Experimental Medicine 182, 367-377.

224

DIVERGENT EFFECTS OF HYPOXIA AND OXIDANTS ON MITOCHONDRIAL SUPEROXIDE DISMUTASE (MnSOD) GENE EXPRESSION

Robert M. Jackson[1,2], Gregory Parish[1,2] and Eric S. Helton[1,2]

Birmingham VA Medical Center[1], Birmingham, AL 35233 and The University of Alabama at Birmingham[2], Birmingham, AL 35294

INTRODUCTION

We reported previously that lung cellular hypoxia downregulated superoxide dismutase expression in alveolar type II (ATII) epithelial cells (Jackson *et al.* 1996). ATII cells were exposed *in vitro* to air (controls) or 2.5% O_2 (hypoxia) for 24 hours. MnSOD mRNA and CuZnSOD mRNA expression were measured by quantitative RT-PCR. MnSOD and CuZnSOD mRNA expression in ATII cells both decreased significantly after one day in hypoxia. The decrease in MnSOD mRNA (-69%) was greater than that in CuZnSOD mRNA (-48%). ATII cell SP-A transcript expression remained constant. MnSOD (-52%) and CuZnSOD (-54%) mRNA expression decreased similarly in lung fibroblasts cultured in hypoxia. The half-life of the MnSOD mRNA measured in lung fibroblasts exposed to air or hypoxia for 24h decreased significantly from 4.8 ± 0.4 to 3.8 ± 0.7 hours (-21%). The half-life for the CuZnSOD decreased significantly from 4.0 ± 0.9 to 2.3 ± 0.2 hours (-43%). MnSOD protein expression (assessed on western blots) also decreased after exposure of the cells to hypoxia (1% O_2 for 24 hours). The present studies were designed to learn whether oxidant stress upregulated expression of MnSOD gene expression in epithelial cells

Nitric oxide (˙NO) synthases generate ˙NO from arginine (Moncada and Higgs 1993). Lung epithelial type II (ATII) cells, endothelial cells and macrophages contain isoforms of the ˙NO synthase. Inflammatory cells express an inducible, calcium-independent enzyme (iNOS). ATII cells apparently contain both an inducible and calcium-dependent constitutive form (cNOS) of the NOS (Miles *et al.* 1996).

Production of ˙NO during inflammation leads to formation of peroxynitrite (ONOO⁻) by rapid (7×10^9 $M^{-1}s^{-1}$) reaction of ˙NO and superoxide (O_2^{-}) (Beckman *et al.* 1990). This reaction has important implications in tissue injury because protonated ONOO⁻ (peroxynitrous acid, ONOOH) has apparent hydroxyl radical (˙OH) reactivity and nitrates tyrosine residues in structural proteins (Ischiropoulos *et al.* 1992). ONOO⁻ reaction products, immunoreactive 3-nitrotyrosine residues, occurs in lung tissue sections from ARDS patients and in atherosclerotic blood vessels (Kooy *et al.* 1995 and White *et al.* 1994).

The rat mitochondrial, manganese-containing superoxide dismutase (MnSOD) gene contains promoter elements, which may permit regulation by oxidant-sensitive transcription

Acute Respiratory Distress Syndrome: Cellular and Molecular Mechanisms and Clinical Management
Edited by Matalon and Sznajder, Springer Science+Business Media New York, 1998

225

factors (Ho *et al.* 1991). Coordinate induction of MnSOD gene expression and nuclear factor kappa B (NFκB) activation occurs in human lung epithelial cells in response to tumor necrosis factor-alpha (TNF) (Warner *et al.* 1996).

Peroxynitrite has an important proinflammatory role in acute lung injury. MnSOD is a key cellular defense against oxidant stress, and it is induced in many models of acute lung injury. Peroxynitrite inactivates the MnSOD protein's enzymatic activity by tyrosine nitration in rejected kidney tissue, although the quantity of MnSOD protein present actually increases (MacMillan-Crow *et al.* 1996). Therefore, we investigated a possible role for ONOO⁻ in signal transduction during inflammation. We postulated that ONOO⁻, like some other oxidants, induces MnSOD gene expression.

MATERIALS AND METHODS

Materials

We used human lung epithelial cells (A549, CCL 185 from ATCC, Rockville, MD) in these studies. 1-propanamine, 3-(2-hydroxy-2-nitroso-1-propylhydrazine (PAPA NONOate) was obtained from Cayman Chemical Company (Ann Arbor, MI). 3-morpholinosydnonimine·HCl (SIN-1) was obtained from Alexis Corporation (San Diego, CA). Dr. Joseph S. Beckman, University of Alabama at Birmingham, generously provided stock solutions (175 mM) of authentic peroxynitrite in 0.1 N NaOH. ONOO⁻ remained stable at -80° C until use.

Experimental Design

Human lung epithelial cell monolayers (A549 cells) were exposed to an ˙NO generator (PAPA NONOate), a peroxynitrite generator (SIN-1), or authentic peroxynitrite in various concentrations. We assessed MnSOD and HPRT mRNA expression by RT-PCR and Northern blots to establish concentration-response relationships between the compounds and MnSOD gene expression. Inhibitors (N-acetyl cysteine [N-Ac], pyrrole dithiocarbamate [PDTC], L-cysteine [L-cys], and bovine erythrocyte superoxide dismutase [SOD]) were used in some experiments to find whether antioxidants affected the peroxynitrite-induced expression of MnSOD mRNA.

In other experiments, epithelial cells were transfected with a reporter gene construct (herein designated pMnSODpr-luc[iferase]) consisting of the rat MnSOD gene 5' region (2505 bp) (Ho *et al.* 1990) luciferase reporter in pGL2-Basic® (Promega). We tested transiently transfected epithelial cells' luciferase expression after exposure to authentic peroxynitrite.

Cell Cultures

We cultured A549 cells in DMEM/F-12 containing 10% fetal bovine serum and 1X penicillin-streptomycin at 37° C in air/5% CO_2. Cells were cultured in 100 mm Primaria® dishes for the mRNA induction studies and in 6-well Primaria® plates for the luciferase transfection assays. Cells were consistently 80 to 90% confluent when used in experiments.

Nitrate/nitrite Assays

We used the Griess reaction to assay total nitrate plus nitrite produced by SIN-1 or peroxynitrite after enzymatic reduction by nitrate reductase (Green *et al.* 1982).

226

SIN-1 Exposures

SIN-1 was diluted (10 mM final stock concentration) in 100 mM KPi buffer at pH 5 immediately before addition to the plates. Complete medium was removed and monolayers rinsed with PBS. Ten ml of serum-free DMEM/F-12 were added to each plate and SIN-1 or vehicle added to the desired concentration (10 or 1000 μM). Cells were incubated with SIN-1 or vehicle at 37° C in 5% CO_2 with frequent agitation (SIN-1 consumes oxygen) for 2 hours. At the end of the 2-hour incubation period, medium was removed and aliquots stored for later nitrate/nitrite assay. Monolayers were washed with PBS. Total RNA was prepared from monolayers. RNA was stored at -80° C until used in PCR or Northern blots.

PAPA NONOate Exposures

We exposed A549 cells to ˙NO by incubating with an ˙NO generator, PAPA NONOate, in serum-free DMEM/F-12. PAPA NONOate was diluted (10 mM stock) in DMEM/F-12 immediately before experiments. Monolayers were washed in PBS as above, and the ˙NO generator (100 or 1000 μM) added to serum-free DMEM/F-12. Cells were incubated (37° C, 5% CO_2) with the ˙NO generator for two hours. Medium was then collected for later nitrate/nitrite assay. Monolayers were washed with PBS and total RNA prepared. In some experiments, cells were incubated with inhibitors in addition to the ˙NO generator. Inhibitors included N-Ac, L-cysteine and SOD as described above.

Authentic ONOO⁻ Exposures

ONOO⁻ in stock solution (175 mM in 0.1 N NaOH) was kept frozen at -80° C until used in experiments. Monolayers were rinsed with PBS, and 10 ml of PBS was then added to the culture dish. ONOO⁻ (100 or 500 μM) was added and swirled rapidly over the entire monolayer surface. We allowed the PBS containing the decomposing ONOO⁻ to remain in contact with the cell monolayers for 10 minutes. PBS was then removed and an aliquot stored frozen at -20° C for later nitrate/nitrite assay. Complete DMEM/F-12 with FBS was then replaced for 6 to 8 hours.

Cells in 6-well plates were treated similarly. The ONOO⁻ incubations took place in serum free buffer, but otherwise conditions were comparable. Cells were incubated with ONOO⁻ for 10 minutes, after which buffer was collected for nitrate/nitrite assay. Complete DMEM/F-12 with FBS was then replaced for 6 to 8 hours. Monolayers were rinsed with PBS and lysed in luciferase assay reagent (Passive Lysis Buffer, Promega, Madison, WI).

Control Experiments

We also investigated whether sodium nitrite (NO_2^-) or hydrogen peroxide (H_2O_2) induced MnSOD gene expression. The rationale for these controls is that peroxynitrite breaks down to nitrate and nitrite, and H_2O_2 spontaneously or enzymatically dismutes from O_2^- produced by SIN-1. We exposed cells to H_2O_2 (100 μM) or sodium nitrite (40 μM) for two hours in PBS. We then replaced complete medium for 4 - 6 hours and prepared total RNA from the cells as above.

Northern Blots

Epithelial cells were lysed in RNAzol® according to the manufacturer's (Biotecx Laboratories, Houston, TX) protocol. Total RNA was prepared (Chomczynski and Sacchi 1987). Total cellular RNA was denatured with glyoxal and 25 - 35 μg loaded at each lane.

RNA was electrophoresed in 1.2% agarose gels containing 10 mM sodium phosphate and then transferred to nylon membranes. RNA on the membrane was fixed by UV light before prehybridization. Blots were hybridized subsequently with a ^{32}P-labelled cDNA probes for MnSOD (HMS from Y.-S. Ho, Ph.D.) (Ho and Crapo 1987) and 18s RNA (Mrockzka *et al.* 1984).

RT-PCR Estimation of MnSOD and HPRT mRNA

PCR using *Taq* polymerase from Gibco-BRL was then done in the Stratagene RoboCycler. We used the following primers for MnSOD: sense 5'-AAGTTCAAGGGTGGAGGTCA-3'; and, antisense 5'-TGCAGGTAGTAAGCGTGTTC-3' (Jackson *et al.* 1996). Cycle conditions for the MnSOD reactions were: 95° C, 5 min.; 95° C, 1 min.; 51° C, 1 min; 72° C, 1 min. x 35 cycles; 72° C, 7 min. These conditions produced a single 311 bp product.

We also did RT-PCR also to estimate abundance of HPRT mRNA transcripts. HPRT is a constitutive gene, which does not change response to SIN-1, ONOO⁻, or ·NO (see below). We used the following primers for human HPRT: sense 5'-ATTTATGGACAGGACTGAACGTC-3'; antisense 5'-CGTGGGGTCCTTTTCACCAGCAAG-3' (Jiralerspong and Patel 1996). The cycle conditions were: 95° C, 5 min.; 95° C, 1 min.; 60° C, 1 min; 72° C, 1 min. x 35 cycles; 72° C, 7 min. These conditions produced a single 386 bp product.

PCR products in 1% agarose gels were visualized by ethidium bromide staining. Products were quantified by densitometry (BioRad model GS-670 imaging densitometer) of Polaroid negatives.

MnSOD Promoter-Luciferase Construct

To assess possible direct effects of ONOO⁻ on MnSOD promoter activity directly, we constructed a luciferase reporter vector driven by the rat MnSOD gene promoter. We obtained a 2.5 kb fragment of the rat MnSOD 5' flanking region by HindIII digestion of the full length (3.3 kb) rat MnSOD promoter (Ho *et al.* 1991). The 2.5 kb fragment extended from the transcription initiation site to a HindIII site at -2,505. The fragment was ligated in sense and antisense orientations into the pGL2-Basic® vector (GibcoBRL).

A549 cells were transfected using Lipofectamine (GibcoBRL) consisting of a 3 to 1 formulation of 2, 3-dioleyloxy-N-[2-(sperminecarboxamido)ethyl]-N,N-dimethyl-1-propanaminium trifluoroacetate (DOSPA) and dioleylphosphatidylethanolamine (DOPE) and water. A549 cells were 80 to 90% confluent in 6-well Primaria plates at the time of transfection. Each well was transfected with 1 μg DNA in 2.5 μg Lipofectamine. In some experiments, we co-transfected cells with a cytomegalovirus (cmv) promoter-driven Renilla luciferase (Lorenz et al. 1996) reporter vector (pRL-cmv® from Promega) to test transfection efficiency and possible effects of ONOO⁻ on the unrelated cmv promoter. The ratio of pMnSODpr-luc to pRL-cmv® in the cotransfection experiments was 100: 1.

RESULTS

SIN-1 Induction of MnSOD mRNA

We exposed cells to SIN-1, a compound that releases superoxide and nitric oxide at physiologic pH and generates peroxynitrite (Henry *et al.* 1989). We then assayed MnSOD mRNA expression by RT-PCR using specific primers and densitometry to estimate the product yield. A549 cells incubated with SIN-1 at 10 or 1000 μM expressed increased steady state levels of MnSOD mRNA. The increase was concentration-dependent, as the MnSOD mRNA appeared greater than control after exposure to 10 μM SIN-1, although the increase was not statistically significant. Nitrite levels measured in buffer at the end of the SIN-1 exposures showed a concentration-dependent increase, confirming the predicted ONOO⁻ release by SIN-1.

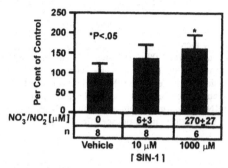

Figure 1. <u>ONOO⁻ generated by SIN-1 increases MnSOD gene expression in human lung epithelial cells</u>. MnSOD gene expression (steady state mRNA level) is indicated by the height of the bars. Gene expression was determined by densitometry of ethidium bromide stained agarose gels after electrophoresis of the PCR products. Control expression is normalized to 100%. The concentration of nitrate/nitrite (μM) after each SIN-1 incubation is shown in boxes below the bar graph. Data are means ± SE for n replicates indicated below the bars. HPRT gene expression (mRNA) in these samples did not change significantly. These results show that SIN-1, an ONOO⁻ generator, induces MnSOD mRNA in human lung epithelial cells. SIN-1 does not change expression of HPRT mRNA in these cells, so this result is not a nonspecific stimulatory effect of SIN-1 on mRNA expression.

Expression of the HPRT message did not change in response to SIN-1, suggesting that the increase in MnSOD mRNA did not simply represent induction of all mRNA's. We also found that N-acetyl cysteine (50 mM), L-cysteine (10 mM), and SOD (300 units/ml) all inhibited the SIN-1 induced increase in MnSOD mRNA steady state level.

ONOO⁻ Induction of MnSOD mRNA

We tested the effects of authentic ONOO⁻ *per se* on MnSOD and HPRT message expression in A549 cells. These experiments were similar to those conducted with SIN-1, but ONOO⁻ exposures were limited to 10 minutes because of the extremely short half-life (<2 seconds) of the compound (Beckman *et al*. 1990). Authentic ONOO⁻ (100 or 500 μM) had an apparent dose-dependent, stimulatory effect on MnSOD mRNA induction. We found that ONOO⁻ (100 or 500 μM) significantly increased MnSOD mRNA expression, but did not increase HPRT mRNA expression. The increase due to ONOO⁻ (500 μM) was similar in magnitude to that due to TNF-α (10 ng/ml).

Antioxidants including N-Ac (50 mM), L-cysteine (10 mM), and PDTC (10 mM) inhibited the induction of MnSOD mRNA by ONOO⁻. Northern blots show that both the 1 and 4 kb MnSOD messages are induced by 500 μM ONOO⁻.

Effects of Hydrogen Peroxide and Nitrite on MnSOD Expression

Neither H_2O_2 (100 μM) nor sodium nitrite (40 μM) had any effect on MnSOD gene induction.

229

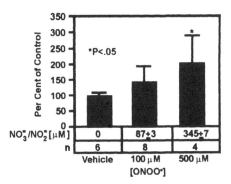

Figure 2. <u>Peroxynitrite significantly increases MnSOD mRNA expression in human lung epithelial cells.</u>
MnSOD gene expression relative to control (100%) is shown by height of the bars as in Figure 1. Gene
expression was determined by densitometry of ethidium bromide stained agarose gels after electrophoresis of
the PCR products. The concentration of nitrate/nitrite (μM) at the end of the 10 minute ONOO⁻ incubation
is shown in boxes below the bar graph. Data are means ± SE for the n indicated below the bars. These results
show ONOO⁻ induces MnSOD mRNA in human lung epithelial cells. ONOO⁻ does not change expression of
HPRT mRNA in these cells, so this result is not a nonspecific stimulatory effect of ONOO⁻ on mRNA
expression.

Effects of ONOO⁻ on MnSOD Promoter-Luciferase Reporter Constructs

To directly assess the effects of ONOO⁻ on the rat MnSOD gene promoter region, we
transiently transfected lung epithelial cells with a 2.5 kb HindIII fragment (Ho *et al.* 1991)
of the MnSOD gene 5' prime region ligated into a luciferase vector, pGL2-Basic
(GibcoBRL). Authentic ONOO⁻ (100 or 500 μM) induced expression of the luciferase
reporter, indicating that ONOO⁻ had direct, concentration-dependent, stimulatory effects on
the MnSOD gene promoter. We also did experiments using a cmv-driven Renilla luciferase
vector, to detect whether ONOO⁻ affected the activity of the unrelated cmv promoter.
ONOO⁻ exposure decreased cmv-driven Renilla luciferase expression slightly. Transfection
efficiency (assessed as the ratio of firefly luciferase activity to Renilla luciferase activity) did
not differ significantly among groups.

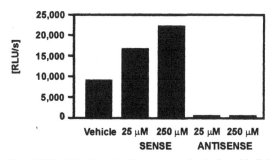

Figure 3. <u>Peroxynitrite (25 and 250 μM) induced luciferase expression in the epithelial cells.</u> We exposed the
cells to peroxynitrite (0, 25, 250 μM) in PBS for 10 minutes, and then replaced complete DMEM/F-12 for 6
to 8 hours. Cells were lysed and promoter activity (luciferase) assayed. Promoter activity, measured as light
is indicated by the height of the bars. Data are mean relative light units (RLU) for n ≥ 3 in each group.
Peroxynitrite did not affect the MnSOD promoter in the antisense orientation, and induction was concentration-
dependent. These results are not due to differences in transfection efficiency, because cotransfection with
Renilla luciferase and normalization revealed similar results.

DISCUSSION

These experiments revealed a potentially important and novel function of peroxynitrite in oxidant-stressed cells. Classical interpretations of peroxynitrite's effects in inflammation assign it a pure cytodestructive function (by necrosis and apoptosis) by virtue of its strong oxidizing potential (1.74 V) (Beckman and Crow 1993). In the system described here, peroxynitrite induces MnSOD gene expression. MnSOD is a key component of the acute inflammatory response (Visner *et al.* 1996). Induction of MnSOD expression would lower steady-state superoxide concentrations, thus decreasing generation of the superoxide-nitric oxide reaction product, peroxynitrite. We show that peroxynitrite may participate in a negative feedback loop that inhibits its own production. The physiological significance of these observations is not addressed by the present studies.

Induction of MnSOD required micromolar concentrations of ONOO⁻. These results do not imply physiological significance, because the maximum physiological concentration of nitric oxide is around 0.1 μM. These results do show for the first time that ONOO⁻ *per se* is a strong inducer of MnSOD transcript expression. The biological effect of ONOO⁻ depends on both concentration and time of exposure. Peroxynitrite decomposes completely in a few seconds, so that the actual dose applied to the cell monolayer is the integral of the concentration versus time (for discussion see Beckman and Crow 1993). The 500 μM bolus ONOO⁻ represents a far smaller oxidant exposure than does 500 μM hydrogen peroxide that remains stable in buffer. In this regard, similar induction of the MnSOD message in H441 cells required exposure to 500 μM hydrogen peroxide for 6 - 12 hours (Warner *et al.* 1996). This comparison shows that ONOO⁻ is a much more potent stimulus to MnSOD mRNA induction than is transient exposure to hydrogen peroxide.

Peroxynitrite produced continuously at localized sites of inflammation may be present at sufficient concentration to induce MnSOD gene expression. Evidence in favor of this mechanism includes the observation by MacMillan-Crow *et al.* (1996) that the MnSOD protein is present in increased concentration in the tubular epithelium of rejected kidneys, where the enzyme protein is both nitrated and enzymatically inactivated.

SIN-1, a chemical donor of superoxide and nitric oxide (Henry *et al.* 1989), generates peroxynitrite in aqueous solution. Incubation of SIN-1 with lung epithelial cells also increased MnSOD gene expression. Thiol antioxidants (N-Ac, PDTC) and SOD inhibited the increase in MnSOD gene expression due to SIN-1. In contrast, an •NO donor, PAPA NONOate, did not increase MnSOD gene expression. These results suggest that peroxynitrite produced by SIN-1, but not •NO alone, was responsible for increased MnSOD gene induction. The results did not reveal the specific mechanism of induction of MnSOD gene expression by SIN-1, but taken together are consistent with an oxidant mechanism that depends on superoxide, since both thiol antioxidants and SOD inhibit the response (Warner *et al.* 1996).

We obtained similar results with authentic peroxynitrite, further suggesting that SIN-1 induction of MnSOD was due to generation of peroxynitrite *per se*. Despite its extremely rapid decomposition in solution at pH 7.4 - 7.6 (Beckman *et al.* 1990), peroxynitrite significantly increased MnSOD gene expression. The increase appeared concentration related, although as noted above, this technique tends to underestimate the degree of MnSOD mRNA induction.

Peroxynitrite also appears to stimulate the rat MnSOD promoter region in transiently transfected epithelial cells. Peroxynitrite increased MnSOD promoter activity, as shown by the MnSOD promoter-luciferase reporter assay. Peroxynitrite did not affect the MnSOD mRNA promoter subcloned in the antisense orientation, confirming that the observed increase in luciferase activity was due to peroxynitrite's action on the promoter region *per se*. The actual increase in MnSOD promoter activity may be greater than that detected by the luciferase reporter, because peroxynitrite may decrease luciferase enzyme activity. The mechanism of peroxynitrite's stimulation of the rat MnSOD 5' region may involve a direct effect of peroxynitrite on the promoter (Ho et el. 1991), activation of an intracellular signaling pathways (Kong et al. 1996), or indirectly by activation of an oxidant-sensitive nuclear factor (Meyer *et al.* 1994).

In these experiments, we have made the novel observation that peroxynitrite increases MnSOD mRNA expression in lung epithelial cells. This suggests that lung inflammation, e.g. due to activated macrophages or hyperoxia, could increase expression of antioxidant defenses in the alveolar epithelium. Shull *et al.* (1991) reported that superoxide from xanthine (50 μM) and xanthine oxidase (0.2 units/ml) increased MnSOD message level in human tracheal epithelial cells about 2 - 3 fold. Satriano and Schlondorff (1994) also reported that generation of superoxide anion by xanthine oxidase activated NFκB in mouse glomerular mesangial cells. The activation was inhibited by PDTC, suggesting a role for ROS as mediators for NFκB activation.

REFERENCES

Baeuerle, P. 1991. The inducible transcription activator NFκB: regulation by distinct protein subunits. Biochim. Biophys. Acta **1072**: 63-80.

Beckman, J. and Crow, J. 1993. Pathological implications of nitric oxide, superoxide and peroxynitrite formation. Biochem. Society Trans. **21**: 330-334.

Beckman, J., Beckman, T., Chen, J., Marshall, P. and Freeman, B. 1990. Apparent hydroxyl radical production by peroxynitrite: Implications for endothelial injury from nitric oxide and superoxide. PNAS USA **87**: 1620-1624.

Chomczynski, P. and Sacchi, N. 1987. Single-step method of RNA isolation by acid guanidinium-thiocyanate-phenol-chloroform extraction. Anal. Biochem. **162**: 156-159.

Green, C., Wagner, D., Glogowski, J., et al. 1982. Analysis of nitrate, nitrite and [^{15}N] nitrite in biological fluids. Anal. Biochem. **126**: 131-138.

Henry, P., Horowitz, J. and Louis, W. 1989. Nitroglycerin-induced tolerance affects multiple sites in the organic nitrate bioconversion cascade. J. Pharmacol. Exper. Therap. **248**: 762-768.

Ho, Y. and Crapo, J. 1987. Nucleotide sequences of cDNA's coding for the rat MnSOD. Nuc. Acids Res. **15**(23): 10070.

Ho, Y., Howard, A. and Crapo, J. 1991. Molecular structure of a functional rat gene for manganese-containing superoxide dismutase. Am. J. Respir. Cell. Mol. Biol. **4**: 278-286.

Ischiropoulos, H., Zhu, L., Chen, J., Martin, J., Smith, C. and Beckman, J. 1992. Peroxynitrite-mediated tyrosine nitration by superoxide dismutase. Arch. Biochem. Biophys. **298**(2): 431-437.

Jackson, R., Parish, G. and Ho, Y-S. 1996. Effects of hypoxia on expression of superoxide dismutases in cultured ATII cells and lung fibroblasts. Am. J. Physiol. **271** (Lung Cell. and Mol. Physiol. 15): L955-L962.

Jiralerspong, S. and Patel, P. 1996. Regulation of the HPRT gene: *in vitro* and *in vivo* Approaches. Proc. Soc. Exper. Biol. Med. **212**(2): 116-127.

Kong, S., Yim, M., Stadtman, E. and Chock, P. 1996. Peroxynitrite disables the tyrosine phosphorylation regulatory mechanism: Lymphocyte specific tyrosine kinase fails to phosphorylate nitrated cdc2(6-20)NH2 peptide. PNAS USA **93**(8): 3377-3382.

Kooy, N., Royall, J., Ye, Y., Kelly, D. and Beckman, J. 1995. Evidence for *in vivo* peroxynitrite production in human acute lung injury. Am. J. Resp. Crit.Care Med. **151**(4): 1250-1254.

Lorenz, W.W., Cormier, M.J., O'Kane, D.J., Hua, D., Escher, A.A. and Szalay, A.A. 1996. Expression of the Renilla reniformis luciferase gene in mammalian cells. J. Bioluminescence Chemiluminescence **11**(1): 31-37.

MacMillan-Crow, L., Crow, J., Kerby, J., Beckman, J., Thompson, J. 1996. Nitration and inactivation of manganese superoxide dismutase in chronic rejection of human renal allografts. PNAS USA **93**(21): 11853-11858.

Meyer, M., Pahl, H.L. and Baeuerle, P.A. 1994. Regulation of the transcription factors NF-κB and AP-1 by redox changes. Chemico-Biol. Inter. **91**: 91-100.

Miles, P., Bowman, L. and Huffman, L. 1996. Nitric oxide alters metabolism in isolated alveolar type II cells. Am. J. Physiol. **271**(Lung Cell. Mol. Physiol. 15): L23-L30.

Moncada, S. and Higgs, A. 1993. The L-arginine-nitric oxide pathway. New Engl. J. Med. **329**(27): 2002-2011.

Mrockzka, D., Casidy, B., Busch, H. and Rothblum, L. 1984. Characterization of rat ribosomal DNA. The highly repetitive sequences that flank the ribosomal RNA transcription unit are homologous and contain RNA polymerase III transcription initiation sites. J. Molec. Biol. **174**: 141-162.

Satriano, J. and Schlondorff, D. 1994. Activation and attenuation of transcription factor NF-kB in mouse glomerular mesangial cells in response to tumor necrosis factor-alpha, immunoglobulin G, and adenosine 3':5'-cyclic monophosphate. Evidence for involvement of reactive oxygen species. J. Clin. Invest. **94**(4): 1629-36.

Shull, S., Heintz, N.H., Periasamy, M., Manohar, M., Janssen, Y. and Marsh, J. 1991. Differential regulation of antioxidant enzymes in response to oxidants. J. Biol. Chem. **266**(36): 24398-403.

Visner, G., Dougall, W., Wilson, J., Burr, I. and Nick, H. 1990. Regulation of MnSOD by LPS, IL-1, and TNF. J. Biol. Chem. **265**(5): 2856-2864.

Warner, B.B., Stuart, L., Gebb, S. and Wispe', J.R. 1996. Redox regulation of manganese superoxide dismutase. Am. J. Physiol. **271** (Lung Cell. Mol. Physiol. 15): L150-L158.

White, C., Brock, T., Chang, L., Crapo, J., Briscoe, P., Ku, D., Bradley, W., Gianturco, S., Gore, J., Freeman, B., et al. 1994. Superoxide and peroxynitrite in atherosclerosis. PNAS USA **91**(3): 1044-1048.

CELLULAR SOURCES OF CYTOKINES IN PULMONARY FIBROSIS

Sem H. Phan, Mehrnaz Gharaee-Kermani, and Hong-yu Zhang

Department of Pathology
University of Michigan Medical School
Ann Arbor, MI 48109-0602, U. S. A.

INTRODUCTION

One of the graver consequences of any insult to the lung is pulmonary fibrosis. The adult respiratory distress syndrome (ARDS) is no different in this regard, since it may progress to end-stage fibrotic lung disease. Hence understanding the pathogenesis of this response to lung injury and inflammation would be of benefit in planning future strategies for determining prognosis and management of ARDS patients.

Current understanding of mechanisms underlying pulmonary fibrosis suggests that this process is a result of a complex interaction or internetworking between several cell types and their mediator products, such as cytokines, which results in the abnormal deposition of extracellular matrix. Given the potential importance of these cytokines in regulating the fibrotic process, identification of their cellular sources would provide important clues as to the key cellular participants driving the process. Despite abundant published data concerning regulation of cytokine production by isolated and cultured cells in vitro, there remains considerable uncertainty as to the actual cellular sources of cytokines in vivo during the different stages of acute lung injury and fibrosis. Recent studies have yielded novel concepts concerning the role of the eosinophils and myofibroblasts as important sources of these cytokines, and hence their participation in pulmonary fibrosis. This chapter will focus on these new findings.

THE MODEL: BLEOMYCIN-INDUCED PULMONARY FIBROSIS

Several animal models of pulmonary fibrosis have been described (1) with somewhat differing properties designed to model the findings found in the human diseases with their specific etiologies. Bleomycin-induced pulmonary fibrosis in rodents has been extensively used to model lung injury and fibrosis, in part due to its ease of induction and reproducibility, and its many similarities to the histopathology of human interstitial pulmonary fibrosis (1-4). Hence it has been one of the most extensively characterized

Acute Respiratory Distress Syndrome: Cellular and Molecular Mechanisms and Clinical Management
Edited by Matalon and Sznajder, Springer Science+Business Media New York, 1998

models, and most of the findings described in this chapter are based on studies using this model. A brief review of the model follows to introduce the topic of this chapter.

Induction of lung injury by endotracheal injection bleomycin results in acute inflammation with rapid recruitment of neutrophils peaking by about 24 hours. This subsides over the next few days, and is replaced by a mononuclear infiltrate consisting of both monocytes/macrophages and lymphocytes, which become maximal after the first week. Increases in the numbers of fibroblasts are first noticeable as early as a week after injury, and with the subsequent severe distortion of lung architecture, large cellular lesions comprised of these cells become prominent. With the associated abnormal deposition of extracellular matrix, the fibrotic areas subsequently become less cellular after the third to fourth week after induction of lung injury. The fibrosis subsides after this time point, and while the scarred areas remained, there are adequate spared areas of lung to allow the animals to survive. Hence this model is not a true model of progressive human pulmonary fibrosis with end-stage lung disease and a fatal outcome. However the natural history of this model can be divided into stages which may mimic closely the comparable events in the human disease.

Based on these sequence of events, the model can be subdivided into four major phases. The first is the acute inflammatory phase (days 1-2) with rapid onset and subsidence, and associated with the injury to the alveolar epithelium and endothelium (3, 4). Subsequent mononuclear infiltrate is associated with the induction phase (days 3-7) of fibrosis characterized by peak expression of cytokines with well-known fibrosis promoting activities, such as tumor necrosis factor α (TNFα) and transforming growth factor β (TGFβ) (1, 4-8). This phase also sees the initiation of heightened extracellular matrix gene expression (8, 9). The active fibrosis phase (days 7-21) follows which is characterized by the increase in fibroblast numbers, maximal elevation in extracellular matrix gene expression, and organization of the lesions to more densely cellular areas with total obliteration of the normal alveolar architecture (1, 4). These areas of active fibrosis are typically populated by a myriad of cells, from mononuclear to fibroblast-like cells, with the latter predominating in the later stages. Beginning after about the third week, fibrosis begins to subside with the lesions becoming less cellular, and characterized by a decline in extracellular and cytokine gene expression, thus marking the onset of the fibrosis subsidence phase (1, 4, 8, 9). These phases have different major cellular constituents whose role(s) are not yet fully understood, but recent studies have provided important clues which suggest time-dependent changes in the roles of these cells.

CELLS

Inflammation and deposition of extracellular matrix are two key characteristics of pulmonary fibrosis (1). Attempts to control fibrosis have focussed on these two processes with variable success. Hence improved understanding of the factors regulating them would be of value in advancing such therapeutic strategies. The potential roles of several key cell types in both processes have been examined with respect to their ability to elaborate mediators driving the fibroblast recruitment, proliferative and extracellular matrix biosynthetic activities in vitro, but much of these remained to be confirmed in vivo. Based on recent in situ studies and in conjunction with in vitro data, novel roles for two cell types have been uncovered. On the one hand the eosinophil appears to be important during the early inflammatory and inductive stages of fibrosis, while the myofibroblast appears to take over during the subsequent stages, being the major source of cytokines and collagen during the active fibrotic phase. Preliminary evidence further suggests that it is the disappearance of these myofibroblasts, perhaps via apoptotic pathways, which results in termination of fibrosis in this model. In this brief overview, only these two cell types will be discussed, although clearly there are other cells, which potentially could play similar and/or complementary roles.

Eosinophils

The earliest cell to appear after this and most other causes of lung injury is the neutrophil (4). This member of the polymorphonuclear series is capable of removing debris, but also to cause further tissue injury in its own right by virtue of its ability to secrete potent oxidants, proteases and other toxic substances. They can also regulate the local tissue hemodynamics by secretion of these same substances plus eicosanoids and other mediators. However, despite these properties, the impact of its recruitment into the lung on fibrosis appears to be minimal (10-12), which may be related to its temporal separation from the phase of fibrosis induction or active fibrosis.

In contrast to the neutrophil,, the eosinophil appears later in the lungs of injured animals. The eosinophil first becomes noticeable after day 3 with peak influx just after day 7 (13). This cell is known to be present in certain fibrotic lesions, with most attention being focussed on lesions with parasitic infestation, with the assumption being that it is somehow important in the immune and inflammatory response to the parasite. Studies of pulmonary fibrosis have noted the presence of this cell as well, but have not systematically determine its role in the fibrotic process itself. Recent studies to determine cellular sources of cytokines known to be expressed during the active fibrosis phase, have localized them to polymorphonuclear leukocytes (13, 14). Upon further investigation using dual staining methods for identification of the various members of this class of leukocytes and expression for either the C-C chemokine, monocyte chemotactic protein-1 (MCP-1), or TGFβ1, the eosinophil appears to be the predominant source of these cytokines at the peak of their expression during induction of fibrosis (13, 14). The finding that the eosinophil can express these cytokines is not entirely unexpected since other studies have shown it to be capable of expressing a myriad of cytokines in other tissues. However its identification as a key source for these cytokines during the induction phase of fibrosis is surprising given the abundant data concerning the ability of mononuclear, endothelial and other cells to express such cytokines, at least in vitro (15-27). Confirmation of the eosinophil is a major source for these mediators is provided by results of studies using anti-TNFα antibodies. Treatment with such antibodies is known to inhibit fibrosis in this and other models of fibrosis (5). Suppression of fibrosis by such treatment is found to be associated with diminished lung expression of interleukin-5 (IL-5), TGFβ and MCP-1; and, more importantly, significant reduction in eosinophil influx (28). There is considerable evidence to support the critical role of IL-5 in lung eosinophil recruitment, and perhaps, activation as well. Furthermore, lung IL-5 expression is elevated in this model at the same time that eosinophils are being recruited into the lung (29). These observations taken together point to a major role for the eosinophil in the induction of fibrosis by virtue of its ability to elaborate cytokines known to be important in promoting inflammation and extracellular matrix synthesis, two important hallmarks of pulmonary fibrosis (1). According to this scenario, TNFα would play a role upstream to the recruitment and activation of eosinophils, and thus its neutralization with antibodies could abrogate fibrosis by preventing eosinophil recruitment and its expression of such pro-fibrotic cytokines as TGFβ. TNFα may regulate eosinophil recruitment and activation via its effects on IL-5 expression.

Why does suppression of eosinophil influx and selected cytokine expression inhibit fibrosis? There is evidence at least that neutralization of TNFα, TGFβ or MIP-1α could inhibit fibrosis in this model (5, 28, 30, 31). TNFα as mentioned above appears to work upstream to eosinophil influx (via IL-5) and hence, TGFβ and MCP-1 expression (28). Other evidence suggests that TNFα may mediate its effects via its ability to regulate adhesion molecule expression related to platelet aggregation in the lung vasculature (32). MIP-1α appears to be important in recruitment of mononuclear cells, and hence inflammation and the immune response associated with fibrosis (31, 33, 34) TGFβ is thought to be important in fibrotic responses, primarily because of its potent effects in the upregulation of extracellular matrix accumulation (1, 6, 7). Since a major source of this matrix is the fibroblast, the assumption is that this cytokine is important because of its ability to upregulate extracellular matrix in these cells. What is the evidence that such a cell is the primary source of interstitial collagen, which characterizes fibrosis?

Myofibroblasts

The answer to this question is made more complicated by extensive evidence that fibroblasts are phenotypically heterogeneous (35). However, exploiting the observations that a new fibroblast phenotype is present at sites of active wound healing, studies in fibrotic lung tissues reveal the presence of a similar cell phenotype. This cells have fibroblastic features morphologically and biochemically, but distinct from other fibroblasts, they express α-smooth muscle actin (ASMA). Because of these properties, they are referred to as myofibroblasts. Their presence in active fibrotic lesions suggests that they may be the source of interstitial collagen and other matrix components. Direct in situ combined analysis of collagen and ASMA expression unequivocally demonstrated that the primary cell expressing collagen is the myofibroblast in this model (36, 37). Furthermore, kinetic studies show that these cells emerge de novo beginning sometime after day 3, with maximal ASMA expression occurring by day 14. This increase in ASMA protein is associated with increase in ASMA mRNA expression, thus suggesting that these cells emerge de novo by induction of this cytoskeletal protein. The de novo appearance of myofibroblasts argues against pre-existing ASMA expressing cells (e.g. airway or vascular smooth muscle cells) as the cells from which they are derived. The first cells to emerge and express ASMA are recognizable morphologically as fibroblasts in the adventitia of the distal airways and associated vascular structures, areas which are distinctly devoid of pre-existing ASMA positive cells (36, 37). Myofibroblasts are present only in areas undergoing active fibrosis, and represent the primary source of interstitial collagen in the affected lung tissue. Their role however appears to extend beyond the mere production of structural matrix components since they represent a significant source of cytokines as well during the later stages of fibrosis when their numbers peak at about day 14 (13, 14). At this time, the eosinophil and other inflammatory cells are beginning to decline, thus suggesting a transition in the cellular sources of important pro-fibrotic cytokines as the lesion progresses.

During the subsequent subsidence phase of fibrosis, the gradual disappearance of the myofibroblast is associated with decline in ASMA mRNA, collagen and cytokine expression. This correlation confirms the relative importance of myofibroblasts as the cellular source of extracellular matrix and cytokines. Hence an understanding of the mechanisms governing their emergence and disappearance may provide insight into the self-limiting nature of the fibrosis seen in this model, with potential implications toward understanding the basis for continuing or progressive disease in human pulmonary fibrosis. A directly pertinent question is whether persistence of these cells in the fibrotic lesions is the cause for progression to end-stage fibrotic lung disease. There is evidence that the presence of myofibroblasts in renal disease is an indicator of progressive disease (38). Thus given the potential importance of the myofibroblast, recent studies have focussed on the possible mechanisms governing its emergence and disappearance.

Since the de novo appearance of this cell appears to occur via phenotypic alteration or differentiation of perhaps pre-existing fibroblasts, much of the attention has been placed on understanding the factors responsible for regulating this process, namely the expression of

ASMA. Depending on the tissue source and the culture conditions, fibroblasts in vitro contain variable amounts of cells expressing ASMA, or myofibroblasts. Cells obtained from normal rat lung tissue at low in vitro passage number usually contained less than 20% myofibroblasts, although this number goes up with increasing time and passage in culture in the presence of serum (39). The proportion of myofibroblasts is dramatically reduced if the cells were isolated and cultured in the absence of serum, using plasma-derived serum instead, and supplemented with purified growth factors. This suggests that something in whole serum is promoting the emergence of myofibroblasts, or the differentiation of fibroblasts to myofibroblasts. The interpretation that this is due to a differentiation event is supported by evidence that cloning fibroblast populations results eventually in a similar mix of ASMA-positive and ASMA-negative cells as the parent population (40). Given its presence in whole serum, and that its expression temporally precedes the emergence of myofibroblasts in vivo, a logical candidate for the stimulator of ASMA expression (i.e. myofibroblast differentiation) is TGFβ. This has indeed received support from in vitro studies (39, 41). Furthermore, cells isolated from fibrotic lungs contain significantly more myofibroblasts than those from normal uninjured lungs, even after controlling for in vitro factors alluded to above (39). These myofibroblasts from fibrotic lungs have similar phenotypic characteristics as their counterparts in vivo, both in terms of heightened collagen and cytokine gene expression, as well as ASMA protein and mRNA expression (39, 42, 43). Additionally they have enhanced contractile properties measured by their ability to contract collagen gels in which they are cultured, consistent with the increased contractility of fibrotic lung tissue (39). These phenotypic characteristics are associated with heightened endogenous TGFβ expression, whose neutralization with anti-TGFβ antibodies results in suppression of the heightened contractility of these cells (39). In toto these findings are consistent with the concept that TGFβ elaborated initially by eosinophils and other cells may be the initiating stimulus for the emergence of myofibroblasts via its ability to induce ASMA expression in fibroblasts and activate them as manifested by increased contractility, extracellular matrix and cytokine expression. Thus the ability of antiTGFβ antibodies to inhibit fibrosis (30) may be mediated by suppression of myofibroblast emergence. The existence of factors other than TGFβ as promoters of myofibroblast differentiation has not been ruled out. GM-CSF is known to also promote the appearance of myofibroblasts, but its effect appears to be mediated by TGFβ (44).

Resolution of wound healing and self-limited fibrosis are accompanied by the gradual disappearance of myofibroblasts. This presumably can occur via death of the myofibroblasts after they have fulfilled their functions of laying down extracellular matrix and promotion of wound contraction, and/or de-differentiation back to the quiescent fibroblast phenotype with associated loss of ASMA expression. As pointed out previously, understanding the mechanism of this disappearance is important for providing insight as to why certain forms of lung injury resolves with recovery of normal architecture and function, and others progress to end-stage fibrotic disease. Presumably disappearance of myofibroblasts would result in termination of the active fibrosis phase and thus be beneficial in situations where the fibrosis is undesirable, such as when it occurs in vital organs such as the lung. What then is the mechanism by which the myofibroblast mysteriously disappears from resolving sites of injury?

It has been known for some time that cells which disappear during normal differentiation undergo programmed cell death or apoptosis. Subsequent studies show similar apoptotic mechanisms mediating the disappearance of leukocytes from inflammatory sites, which may account for the mechanism by which the inflammation resolves. The possibility that a similar mechanism may be responsible for the disappearance of myofibroblasts is suggested by a recent study showing the presence of apoptotic cells in resolving wound healing sites (45). Additional support for such a possibility is provided by an in vitro study showing that isolated rat lung myofibroblasts are

239

preferentially susceptible to undergo apoptosis upon treatment with IL-1β when compared to fibroblasts not expressing ASMA (46). In that study, IL-1β treatment also causes significant suppression of ASMA expression, thus suggesting that selective apoptosis of the myofibroblast may be responsible for the loss in ASMA expression. The mechanism by which apoptosis is induced in myofibroblasts is unclear. There is evidence however that NO may be important in mediating this process. IL-1β is a potent inducer of NO synthetase (iNOS) in rat lung fibroblasts (46). When these cells are treated with an arginine analog iNOS inhibitor, the effects of IL-1β on ASMA expression and apoptosis are significantly inhibited. The inhibitor has no effect on iNOS protein expression, but appears to exert its effect by inhibition of enzyme activity and production of NO (46). Thus from this standpoint at least, NO is beneficial by virtue of its ability to mediate the induction of myofibroblast apoptosis, and presumably the subsidence of active fibrosis.

Since IL-1β expression is not known to occur during the termination phase of fibrosis in this model, it is unlikely to be the inducer of myofibroblast apoptosis in vivo, at least in this model of pulmonary fibrosis. IL-1 activity is expressed very rapidly and early (within the first day) after the induction of lung injury, and not detectable during the later period of myofibroblast disappearance (beginning after the second week) (47). Although its use in vitro has been helpful in confirming apoptosis in myofibroblasts, and their preferential susceptibility compared to ASMA-negative fibroblasts, other factors may be responsible for the actual induction of apoptosis in myofibroblasts in this model in vivo. As yet no candidate has been implicated in this phenomenon in vivo. However, there are clues from other terminally differentiated cells, which also undergo apoptosis. The eosinophil undergoes apoptosis when they disappear from inflammatory sites, but can be prevented from doing so if a source of growth factor (e.g. GM-CSF) is present (48). These studies suggest that it is the withdrawal from growth or differentiation factors, which trigger apoptosis in certain cells. Transposing such a possibility to myofibroblast apoptosis in the bleomycin model indicates that the period of fibrosis subsidence is associated with a decline in TGFβ expression (6-8). Carrying the eosinophil analogy further, since TGFβ appears to be a potent inducer of fibroblast differentiation to myofibroblast, it is tempting to speculate that withdrawal of such a stimulus may be the trigger in inducing the myofibroblast to undergo apoptosis. The possibility of other apoptosis inducing factors requires further study.

CONCLUSION

This brief review highlights the importance of the eosinophil and myofibroblast in pulmonary fibrosis. These two cells may not be the exclusive cells fulfilling the roles of mediator/cytokine source and structural cell target elaborating extracellular matrix in this complex intercellular networking that comprises fibrosis, but certainly they are representative of the large numbers of cells that are involved. The previously unsuspected role of the eosinophil has been brought out and highlighted by recent advances documenting cytokine expression by this cell type. Extension of this observation to other fibrotic diseases has received some support from reports that many of these have eosinophil components to varying extents, but confirming similar roles for this cell requires further investigation. The novel concept that the myofibroblast is the key cell in elaboration of extracellular matrix and propagation of fibrosis by secretion of both inflammatory and pro-fibrotic cytokines, extends to its key role as a determinant of whether lung injury/inflammation results in resolution (or at least limited fibrosis) or progressive fibrotic lung disease. Hence efforts directed at suppressing its emergence and/or promoting its apoptosis may be productive in designing novel therapeutic strategies in the future.

REFERENCES

1. Phan, S.H.: Diffuse interstitial fibrosis, in, Massaro, D. (Ed.) Lung Cell Biology, Marcel Dekker, New York, 1989; Chapter 19, pp. 907-979.
2. Snider GL, Hayes JA, and Korthy AL: Chronic interstitial pulmonary fibrosis produced in hamsters by endotracheal bleomycin: pathology and stereology. Am Rev Respir Dis 1978; 117: 1099-108.
3. Adamson IY, and Bowden DH: The pathogenesis of bloemycin-induced pulmonary fibrosis in mice. Am J Pathol 1974; 77: 185-97.
4. Thrall RS, McCormick JR, Jack RM, McReynolds RA, and Ward PA: Bleomycin-induced pulmonary fibrosis in the rat: inhibition by indomethacin. Am J Pathol 1979; 95: 117-30.
5. Piguet PF, Collart MA, Grau GE, Kapanci Y, and Vassalli P: Tumor necrosis factor/cachectin plays a key role in bleomycin-induced pneumopathy and fibrosis. J Exp Med 1989; 170: 655-63.
6. Khalil N, Bereznay O, Sporn M, and Greenberg AH: Macrophage production of transforming growth factor beta and fibroblast collagen synthesis in chronic pulmonary inflammation. J Exp Med 1989; 170: 727-37.
7. Hoyt DG, Lazo JS: Alterations in pulmonary mRNA encoding procollagens, fibronectin and transforming growth factor-beta precede bleomycin-induced pulmonary fibrosis in mice. J Pharmacol Exp Ther 1988; 246: 765-71.
8. Phan, S.H., and Kunkel, S.L.: Lung cytokine production in bleomycin-induced pulmonary fibrosis. Exp. Lung Res. 1992; 18:29-43.
9. Phan, S.H., Thrall, R.S. and Ward, P.A.: Bleomycin induced pulmonary fibrosis in rats: Demonstration of increased rate of collagen synthesis. Am. Rev. Resp. Dis. 1980; 121: 501-506.
10. Thrall, R.S., Phan, S.H., McCormick, J.R. and Ward, P.A.: The development of bleomycin-induced pulmonary fibrosis in neutrophil-depleted and complement-depleted rats. Am. J. Pathol. 1981; 105: 76-81.
11. Phan, S.H., Schrier, D., McGarry, B. and Duque, R.: The effect of the beige mutation on bleomycin-induced pulmonary fibrosis in mice. Am. Rev. Resp. Dis., 1983; 127: 456-459.
12. Clark JG, and Kuhn C 3d: Bleomycin-induced pulmonary fibrosis in hamsters: effect of neutrophil depletion on lung collagen synthesis. Am Rev Respir Dis 1982; 126:737-9.
13. Zhang, K., Gharaee-Kermani, M., Jones, M.L., Warren, J.S., and Phan, S.H.: Monocyte chemoattractant protein-1 gene expression in bleomycin-induced pulmonary fibrosis. J. Immunol. 1994; 153:4733-4741.
14. Zhang, K., Flanders, K.C., and Phan, S.H.: Cellular localization of transforming growth factor β expression in bleomycin-induced pulmonary fibrosis. Am. J. Pathol. 1995; 147:352-361.
15. Karmiol, S. and Phan, S.H.: Fibroblasts and cytokines, in, Kunkel, S.L. and Remick, D.G. (Eds.) Cytokines in Health and Disease, Marcel Dekker, New York, 1992; Chapter 16, pp. 271-296.
16. Zhang, K., and Phan, S.H.: Cytokines and pulmonary fibrosis. J. Biol. Signals, 1996; 5:232-239.
17. Phan, S.H.: Endothelial cells in pulmonary fibrosis, in Phan, S.H., and Thrall, R.S. (Eds.) Pulmonary Fibrosis, Marcel Dekker, New York, 1995; Chapter 14, pp 481-510.
18. Strieter, R.M., Kunkel, S.L., Showell, H.J., Remick, D.G., Phan, S.H., Ward, P.A., and Marks, R.M.: Human endothelial cell gene expression of a neutrophil chemotactic factor by TNFα, LPS, and IL-1β. Science, 1989; 243: 1467-1469.

19. Strieter, R.M., Phan, S.H., Showell, H.J., Remick, D.G., Lynch, J.P. III, Genord, M., Raiford, C., Eskandari, M., Marks, R.M., and Kunkel, S.L.: Monokine-induced neutrophil chemotactic factor gene expression in human fibroblasts. J. Biol. Chem. 1989; 264: 10621-10626

20. Strieter, R.M., Wiggins, R., Phan, S.H., Wharram, B.L., Showell, H.J., Remick, D.G., Chensue, S.W. and Kunkel, S.L.: Monocyte chemotactic Protein Gene Expression by cytokine-treated human fibroblasts and endothelial cells. Biochem. Biophys. Res. Commun. 1989; 162: 694-700.

21. Phan, S.H., Gharaee-Kermani, M., Wolber, F., and Ryan, U. S.: Stimulation of rat endothelial cell transforming growth factor-β production by bleomycin. J. Clin. Invest. 1991; 87: 148-154.

22. Standiford, T.J., Kunkel, S.L., Phan, S.H., Rollins, B.J. and Strieter, R.M.: Alveolar macrophage-derived cytokines induce monocyte chemoattractant protein-1 expression from human pulmonary type II-like epithelial cells. J. Biol. Chem. 1991; 266: 9912-9918.

23. Rolfe, M.W., Kunkel, S.L., Standiford, T.J., Chensue, S.W., Allen, R.M., Evanoff, H.L., Phan, S.H., and Strieter, R.M.: Pulmonary fibroblast expression of interleukin-8: A model for alveolar macrophage-derived cytokine networking. Am. J. Resp. Cell Molec. Biol. 1991; 5:493-501.

24. Phan, S.H., Gharaee-Kermani, M, McGarry, B., Kunkel, S.L., and Wolber, F.W.: Regulation of endothelial cell transforming growth factor β production by IL-1β and TNFα . J. Immunol. 1992; 149:103-106

25. Rolfe, M.W., Kunkel, S.L., Standiford, T.J., Orringer, M.B., Phan, S.H., Evanoff, H.L., Burdick, M.D., and Strieter, R.M.: Expression and regulation of human pulmonary fibroblast-derived monocyte chemotactic peptide (MCP-1). Am. J. Physiol. 1992; 263 (*Lung Cell Mol. Physiol. 7*): L536-545.

26. Brieland, J.K., Jones, M.L., Flory, C.M., Miller, G.R., Warren J.S., Phan, S.H., and Fantone, J.C.: Expression of monocyte chemoattractant protein-1 by rat alveolar macrophages during chronic lung injury. Am. J. Resp. Cell Molec. Biol. 1993; 9:300-305.

27. Gharaee-Kermani, M., Denholm, E.M., and Phan, S.H.: Co-stimulation of fibroblast collagen and transforming growth factor β₁ gene expression by monocyte chemoattractant protein-1 via specific receptors. J. Biol. Chem. 1996; 271:17779-17784.

28. Zhang, K., Gharaee-Kermani, M., McGarry, B., Remick, D., and Phan, S.H.: TNFα mediated lung cytokine networking and eosinophil recruitment in pulmonary fibrosis. J. Immunol. 1997; 158:954-959.

29. Gharaee-Kermani, M., and Phan, S.H.: Lung interleukin-5 expression in murine bleomycin-induced pulmonary fibrosis. Am. J. Resp. Cell Molec. Biol. 1997; 16:438-447.

30. Giri SN, Hyde DM, and Hollinger MA: Effect of antibody to transforming growth factor beta on bleomycin induced accumulation of lung collagen in mice. Thorax 1993; 48: 959-66.

31. Smith, R.E., Strieter, R.M., Phan, S.H., Lukacs, N.W., Huffnagle, G.B., Wilkie, C.A., Burdick, M.D., Lincoln, P., Evanoff, H., and Kunkel, S.L.: Production and function of murine macrophage inflammatory protein-1α in bleomycin-induced lung injury. J. Immunol. 1994; 153:4704-4712.

32. Piguet PF, Tacchini-Cottier F, and Vesin C: Administration of anti-TNF-alpha or anti-CD11a antibodies to normal adult mice decreases lung and bone collagen content: evidence for an effect on platelet consumption. Am J Respir Cell Mol Biol 1995; 12: 227-31.

33. Smith, R.E., Strieter, R.M., Zhang, K., Phan, S.H., Standiford, T.J., Lukacs, N.W., Kunkel, S.L.: A role for C-C chemokines in fibrotic lung disease. J. Leukocyte Biol. 1995; 57:782-787.

34. Smith, R.E., Strieter, R.M., Phan, S.H., Kunkel, S.L.: C-C chemokines: Novel mediators of the profibrotic inflammatory response to bleomycin challenge. Am. J. Resp. Cell Mol. Biol. 1996; 15:693-702.

35. Karmiol, S. and Phan, S.H.: Phenotypic changes in lung fibroblast populations in pulmonary fibrosis, in, Phipps, P. (Ed.) Fibroblast heterogeneity in pulmonary fibrosis. CRC Press, Boca Raton, Florida, 1992; Chapter 1, pp. 1-26.

36. Zhang, K., Gharaee-Kermani, M., McGarry, B., and Phan, S.H.: *In situ* hybridization analysis of lung $\alpha_1(I)$ and $\alpha_2(I)$ collagen gene expression in bleomycin-induced pulmonary fibrosis in the rat. Lab. Invest. 1994; 70:192-202.

37. Zhang, K., Rekhter, M.D., Gordon, D., and Phan, S.H.: Co-Expression of α-smooth muscle actin and type I collagen in fibroblast-like cells of rat lungs with bleomycin-induced pulmonary fibrosis: a combined immuno-histochemical and *in situ* hybridization study. Am. J. Pathol. 1994; 145:114-125.

38. Roberts I, Burrows C, Shanks JH, Venning M, and McWilliam LJ: Interstitial myofibroblasts; predictors of progression in membranous nephropathy. J Clin Pathol 1997; 50: 123-7.

39. Zhang, H., Gharaee-Kermani, M., Zhang, K., and Phan, S.H.: Lung fibroblast contractile and α-smooth muscle actin phenotypic alterations in bleomycin-induced pulmonary fibrosis. Am. J. Pathol. 1996; 148:527-537.

40. Desmouliere A, Rubbia-Brandt L, Abdiu A, Walz T, Maciera-Coelho, and Gabbiani G: α-Smooth muscle actin is expressed in a subpopulation of cultured and cloned fibroblasts and is modulated by γ-interferon. Exp Cell Res 1992; 201: 64-73.

41. Desmouliere A, Geinoz A, Gabbiani F, and Gabbiani G: Transforming growth factor-beta 1 induces alpha-smooth muscle actin expression in granulation tissue myofibroblasts and in quiescent and growing cultured fibroblasts. J Cell Biol 1993; 122:103-11.

42. Phan, S.H., Varani J. and Smith, D.: Rat lung fibroblast collagen metabolism in bleomycin-induced pulmonary fibrosis. J. Clin. Invest. 1985; 76: 241-247.

43. Breen, E., Shull, S., Burne, S., Absher, M., Kelley, J., Phan, S.H., Cutroneo, K.: Bleomycin regulation of TGF-β mRNA in rat lung fibroblasts. Am. J. Resp. Cell Molec. Biol. 1992; 6:146-152.

44. Vyalov S, Desmouliere A, and Gabbiani G: GM-CSF-induced granulation tissue formation: relationships between macrophage and myofibroblast accumulation. Virchows Arch B Cell Pathol Incl Mol Pathol 1993; 63: 231-9.

45. Desmouliere A, Redard M, Darby I, and Gabbiani G: Apoptosis mediates the decrease in cellularity during the transition between granulation tissue and scar. Am J Pathol 1995; 146: 56-66.

46. Zhang, H., Gharaee-Kermani, M., and Phan, S.H.: Regulation of lung fibroblast α-smooth muscle actin expression, contractile phenotype and apoptosis by IL-1β. J. Immunol. 1997; 158:1392-1399.

47. Jordana M, Richards C, Irving LB, and Gauldie J: Spontaneous in vitro release of alveolar-macrophage cytokines after the intratracheal instillation of bleomycin in rats. Characterization and kinetic studies. Am Rev Respir Dis 1988; 137: 1135-40.

48. Vancheri C, Gauldie J, Bienenstock J, Cox G, Scicchitano R, Stanisz A, and Jordana M: Human lung fibroblast-derived granulocyte-macrophage colony stimulating factor (GM-CSF) mediates eosinophil survival in vitro Am J Respir Cell Mol Biol 1989; 1: 289-95.

STRATEGIES TO SPEED LUNG HEALING IN ARDS

David H. Ingbar, Craig A. Henke, Vitaly Polunovsky, Hyun J. Kim, Joseph Lasnier, Christine H. Wendt and Peter B. Bitterman

Pulmonary, Allergy, & Critical Care Medicine Division
Department of Medicine
University of Minnesota School of Medicine
Minneapolis, MN 55455

INTRODUCTION

A major problem in the treatment of patients with the acute respiratory distress syndrome (ARDS) is that most of the clinical care consists of treatment of the underlying predispositions and advanced supportive measures. While the prognosis of ARDS has improved somewhat over the last decade, trials of early intervention have yet to demonstrate major improvements in the outcomes from ARDS. Specifically, therapeutic trials that have shown no or little clinical benefit have included: glucocorticoids given either early or for established ARDS or sepsis syndrome, early positive end expiratory pressure for at-risk intubated patients, intravenous prostaglandin E2 (PGE2), non-steroidal anti-inflammatory agents, and aerosolized surfactant lipid therapy with Exosurf. The likely explanation for this is that by the time the syndrome is recognized clinically multiple parallel pathophysiologic abnormalities have been initiated that are largely self-perpetuating; intervening against one component then may not be sufficient to have major clinical benefit. An alternative strategy for treatment of ARDS is the promotion of the repair process. This paper reviews some of the approaches on the horizon for testing in preclinical models and hopefully soon in clinical trials. Table 1 lists these strategies.

MINIMIZE ONGOING INJURY

In addition to treating the precipitating causes of ARDS, it is critical to avoid amplifying the injury if at all possible. Mechanical ventilation-related factors that may worsen the injury include oxygen toxicity and barotrauma. These problems have been reviewed in detail, including in this volume, and hence will not be presented in detail.

Oxygen toxicity can lead to lung injury that is virtually identical in morphology to acute lung injury or diffuse alveolar damage. The precise safe level of oxygen is still somewhat uncertain, especially in the setting of an inflamed and injured lung. In multiple different animals models of lung injury, fractions of inspired oxygen (FiO2)

Acute Respiratory Distress Syndrome: Cellular and Molecular Mechanisms and Clinical Management
Edited by Matalon and Sznajder, Springer Science+Business Media New York, 1998

245

between 0.4 - 0.7 can synergize the degree of injury and lead to more severe and persistent fibrosis. In general clinicians feel that an FiO2 of 0.4-0.5 probably is relatively safe for long periods of time, but that a FiO2 of 0.7 or greater for 3 days or a FiO2 of 1.0 for 24 hours should not be exceeded if possible. The potential for subtle oxygen toxicity is of greater importanc, as low minute ventilation strategies for mechanical ventilation become more widely used, since these strategies often require increasing the FiO2.

Barotrauma can take the clinically recognizable forms of pneumothorax, pneumomediastinum, or pulmonary interstitial emphysema. Alternatively, a plethora of animal studies have demonstrated histologic changes of diffuse alveolar damage when the lung is ventilated with high tidal volumes, high airway pressures or with insufficient positive end expiratory pressure. Conceptually this "microbarotrauma" can be thought of as resulting from either overdistention of residual compliant alveoli ("volutrauma") or from shear stress on alveoli that repetitively open with high pressure and then collapse during exhalation. With surfactant that is abnormal and unable to reduce alveolar surface tension to the usual extent, tremendous force likely is exerted on these alveoli with each opening cycle. At present the optimal ventilatory strategy to avoid contributions of barotrauma to the underlying lung injury has not been determined in randomized clinical trials in humans. Conceptually a strategy of adequate PEEP to avoic expiratory collapse accompanied by moderate tidal volumes seems logical. Multiple human clinical trials of low tidal volume strategies, when accompanied by adequate level of PEEP, have not yet shown any benefit over more conventional tidal volumes of approximately 10 mls/kg ideal body weight when used with adequate PEEP.

CYTOPROTECTION

An interesting strategy based on animal trials is to augment cellular host defenses against injury. The most studied agent to date is keratinocyte growth factor (KGF). This is a member of the fibroblast growth factor (FGF) family that binds to a specific FGF receptor. It is a mitogen for many types of epithelial cells and when given to animals, it induces epithelial hyperplasia in the lung, mammary gland, liver and intestine. Transgenic mice with dominant-negative mutations of the KGF receptor have deficient skin wound healing. If this defect is expressed in the lung, branching morphogenesis and epithelial differentiation are aberrant and the mice die at birth.

For many years investigators searched for growth factors that induced alveolar epithelial proliferation. Many standard growth factors induced DNA synthesis without major increases in cell number. Even combinations of many growth factors led to relatively small degrees of proliferation. Recently KGF has been recognized as a potent type II cell mitogen in vitro and in vivo (Panos, Ulich). Of even greater interest KGF markedly attenuates the degree of lung injury from hyperoxia, bleomycin, radiation and acid aspiration (Panos, Yi) and can be given just prior to or after injury. The mechanism by which KGF is protective is uncertain. It stimulates alveolar type II cell proliferation and surfactant production. It also may augment alveolar epithelial fluid transport or diminish cell loss from apoptosis. In airway epithelial cells in vitro, it protects against oxidant injury, with diminished paracellular leak. Trials of cytoprotective agents in humans have not yet been reported.

AUGMENTATION OF ALVEOLAR FLUID RESORPTION

In patients with pulmonary edema of either cardiogenic or non-cardiogenic origin, increasing sodium and fluid transport (as indicated by rising tracheal edema fluid:plasma protein concentration ratio) correlates with clinical improvement. In ARDS patients, thi transport also correlated with prognosis. This suggests that speeding alveolar fluid resorption in ARDS patients might improve their clinical outcome. The mechanisms for this improvement might include: lesser need for mechanical ventilatory support with high FiO2; better alveolar defenses against infection; or alterations in the fibroproliferative response.

Alveolar fluid resorption in the normal lung by active sodium transport can be upregulated by a variety of agents, as discussed in detail in the chapters by Matthay and Sznajder. In multiple forms of lung injury, alveolar fluid resorption is upregulated. In some types of injury this occurs through increased endogenous catecholamines, whereas in other types the increase is not catecholamine dependent. Several investigators have demonstrated in animal models of lung injury, such as hyperoxia (Lasnier), that terbutaline can augment fluid resorption even in the face of increased paracellular permeability. An analogy that supports the importance of active sodium transport, even in the face of net edema fluid formation, is being in a small boat with holes in the hull; the rate of bailing water out of the boat is an important determinant of how long the boat can remain afloat - provided the holes are not giant.

PROMOTE RE-EPITHELIALIZATION

The alveolar epithelial barrier serves multiple functions. In addition to being a tighter barrier against fluid movement than the endothelium, it also prevents serum proteins from moving into the alveolar space. This prevents fibronectin and fibrinogen from contributing to formation of hyaline membranes or an intra-alveolar meshwork of provisional matrix that will become the site of granulation tissue formation by incoming fibroblasts and endothelial cells. The serum proteins also inhibit the function of surfactant apoprotein A (SP-A) and thereby may contribute to ongoing difficulties with ventilation and oxygenation.

Many years ago, classic studies by Vracko in dogs with oleic acid-induced lung injury led to the belief that retained alveolar basement membrane structure was a requisite for lung repair to recreate normal architecture. Recent electron micrographs of lungs from humans dying of ARDS demonstrate alveolar type II cells on the surface of hyaline membranes or intra-alveolar fibrin meshworks (Anderson). This suggests that alveolar type II cells also may migrate over the surface to reform an epithelial barrier.

Recent studies from 3 different laboratories demonstrated the ability of cultured alveolar epithelial cells to close a surface wound and to migrate in a regulated manner. Using a Boyden chamber assay system and filters coated with denatured type I collagen, alveolar epithelial cell chemotaxis was stimulated by soluble factors, including: epidermal growth factor (EGF); hepatocyte growth factor (HGF); basic FGF; transforming growth factor alpha (TGFa); laminin; or fibronectin (Lesur). However, soluble tumor necrosis factor alpha (TNFa), keratinocyte growth factor (KGF), interleukin-1 beta, and TGF beta had no effect. When filters were coated with fibronectin in solid phase, we found that type II cell migration increased several fold compared with gelatin coating used by Lesur (Kim 1997). Freshly isolated type II cells did not migrate, but this capacity was present after two days in culture. Consequently it is not clear whether the type II cells are transiently injured during isolation and lose this migratory capacity or whether migration represents a transitional (type II to type I) or type I-like cell function. Migration was dependent on specific integrins that differed on fibronectin and type I collagen substrates. Interestingly, although both freshly isolated and cultured type II alveolar epithelial cells adhered to fibrinogen (Kim 1996), they would not migrate on fibrinogen. In summary haptotaxis - directed crawling on surfaces - seems to be a more effective mode of type II cell migration than solution-based chemotaxis and specific integrins and their receptors are involved in this process.

247

Matthay and colleagues also examined type II cell migration in a wound closure model. TGFa, EGF and fibronectin stimulated wound closure through a combination of cell spreading and motility, without cell proliferation (Kheradmand). As in the Boyden chamber, insoluble fibronectin promoted closure more than type I or type IV collagen (Garat). Thus both these models support the ability of alveolar cells to migrate in order to re-epithelialize wounds and suggest that the fibronectin-rich hyaline membrane may promote this process.

As discussed above, KGF is cytoprotective against a variety of experimental forms of lung injury. KGF and HGF also are likely to stimulate alveolar epithelial cell migration in the lung - as they do for multiple other types of epithelial cells. Whether KGF's protective effect is specifically due to the increased numbers of type II cells that recreate the epithelial barrier or to KGF's effects on specific type II cell functions - such as surfactant production, migration or ion transport - is unknown.

In summary, intra-alveolar administration of pro-migratory and/or proliferation-inducing substances might lead to earlier reformation of the alveolar epithelial barrier. This could significantly impact the subsequent course of lung injury by decreasing leakage of the proteins that form provisional matrix and affecting migration of cells into the alveolar space.

INHIBIT FIBROSIS

Early in ARDS there are increases in the lung content of collagen, especially type I collagen - which is quite inflexible. Many agents that inhibit collagen synthesis and/or secretion have been tested in experimental models of lung injury, but no human clinical trials have been performed. A logical contender for trial is colchicine, since it is relatively non-toxic. Since proteases that can destroy collagen, such as elastase, are present and active in the alveolar space during lung injury, the structural scaffolding of the lung can be disrupted. A potential difficulty with the strategy of inhibiting collagen synthesis is that some of the collagen synthesis may be beneficial when remodelling is occurring. However, a relatively acute insult, such as ARDS, may be a situation in which this type of intervention may be beneficial since short-term treatment might be sufficient without interrupting the slower turnover of matrix molecules in blood vessels and other vital structures. Protease inhibitors are now undergoing early clinical trials in humans.

An alternative to direct prevention of collagen, is to inhibit the function of some of the cytokines that promote fibrosis. For example basic FGF would be a logical target since it's production is augmented in alveolar macrophages from ARDS patients (Henke 1993). Obviously there are a number of other potential target molecules for intervention and we do not know which one or ones would be most beneficial to attack.

This strategy of decreasing collagen synthesis or matrix formation has yet to be tested in humans with ARDS.

INHIBIT CHRONIC INFLAMMATION & ALVEOLITIS

Recent studies of the bronchoalveolar lavage fluid (BALF) from patients with ARDS have demonstrated elevations in a variety of pro-inflammatory cytokines, while many patients have lower BALF levels of anti-inflammatory cytokines. Patients with persistently high BALF levels of the procollagen type III propeptide or neutrophils have a worse long term prognosis. These findings led to the concept, promoted by Dr. Meduri (1994), that persistent lung inflammation leads to ongoing fibroproliferation and a worse prognosis. This concept supports the practice of giving high doses of glucocorticoids to patients in "chronic phase" ARDS with evidence of ongoing lung inflammation, but no lung infection. Given the variable clinical course of ARDS in this phase, it has been hard to interpret positive results reported from non-randomized case series. A recent preliminary report of a small randomized trial suggests benefit from glucocorticoid therapy, however, there was considerably more pneumonia in the treated group (Meduri 1997). A definitive recommendation about this therapy awaits publication of the results from large randomized, controlled trials.

PREVENT SECONDARY INFECTION & MULTIORGAN FAILURE SYNDROME (MOFS)

This general approach has been considered for many years. However, practical therapies to accomplish these ends have been elusive. In surgical patients some data supports augmenting oxygen transport as a method of decreasing the likelihood of MOFS, but this approach has not been beneficial in most medical patients. Strategies to prevent infection have been lacking. Selective decontamination of the digestive tract, treating colonizing organisms in the lung with antibiotics and changing endotracheal tubes have not been tested specifically in this patient population.

REMOVE GRANULATION TISSUE & PROMOTE APOPTOSIS

An appealing strategy is to accelerate the removal of the excess intra-alveolar and interstitial granulation tissue, at least in part by promoting apoptosis (Solary). Apoptosis, or programmed cell death, is a non-inflammatory, selective type of cell death that deletes cells without induction of inflammation. Some cell types, such as neurons and lymphocytes, undergo apoptosis when growth factors are withdrawn. Other cell types can be stimulated to undergo apoptosis by TNF, Fas ligand, myc activation or other cytokines. The mechanisms for induction of apoptosis are very complex and only partially defined (Kerr, Kroemer). Usually a trigger signal is required and then several checkpoints must be passed before the program for this type of cell death actually is executed. Checkpoints in some cell types involve the Bcl-2 proto-oncogene family (bcl-2, bax, and bad), p53 and the interleukin-1beta converting enzyme (ICE). In simpler organisms, such as the roundworm, C. elegans, specific genes or proteins can activate or block apoptosis. Not surprisingly, apoptosis in many cells is intimately linked to cell cycle status and cell proliferation. Recent data also indicate that there are linkages between apoptosis and proteins that regulate translation, such as the elongation initiation factors (eIFs). The regulation of apoptosis induction is influenced by components in both the nucleus and cytoplasm (Polunovsky 1996). Recent data suggest that transmitochondrial calcium fluxes also may be critically important for apoptosis. Thus much remains to be defined about the biochemical mechanisms of apoptosis, their universality in different cell types and their interrelationship with other essential cell functions.

In lung injury, two desirable targets for induction of apoptosis are endothelial cells and fibroblasts, since they make up the bulk of cells in intra-alveolar granulation tissue. In the early 1990s, it was recognized that BALF from patients in chronic phase ARDS who subsequently recovered had pro-apoptotic activity for these cell types, whereas this was not present in BALF from early phase ARDS (Polunovsky 1993). The precise components responsible for this activity are still being characterized.

A new conceptual approach is to activate apoptosis using its linkage to translational control of cell proliferation and death. Using an in vivo wound chamber model in which fibrin gels are implanted subcutaneously, several agents have been tested for prevention of granulation tissue formation. Agents that modify translation, such as lovastatin, inhibited this response and ultrastructural morphology demonstrated that this occurred by induction of apoptosis (Tan). In addition to its inhibition of HMG CoA

Table 1. Strategies to Speed Lung Healing in ARDS

Decrease severity of early injury
Decrease ongoing injury
Cytoprotection
Augment alveolar fluid resorption
Inhibit fibrosis
Augment re-epithelialization
Remove granulation tissue/Promote apoptosis

reductase, lovastatin deactivates Ras and increases p27, thereby inducing cell cycle blockade at the R point. Another clinically utilized anti-transplant rejection drug, rapamycin (FK506), also induces apoptosis in experimental models. Presumably this occurs because it inhibits a kinase that is involved in phosphorylation of proteins important for translation (the inhibitory eIF4E binding protein). Since the safety profile of these drugs is relatively well established in humans, their use in clinical trials may be feasible if they demonstrate benefit in animal models of lung injury.

Investigations of endothelial cell apoptosis suggest that ligation of cell surface receptors may be another strategy for inducing apoptosis of cells within the alveolar space. Activation of matrix-cell signaling pathways by inhaled compounds is attractive since it could be targeted specifically to granulation tissue that is aberrantly located in the alveolar space. One example utilizes the CD44 cell surface adhesion receptor. This receptor binds provisional matrix proteins such as fibronectin, fibrinogen and hyaluronic acid, but not laminin. It mediates endothelial and fibroblast adhesion, migration and invasion into fibrin gels. CD44 is expressed intensely in mesenchymal cells of fibrotic regions of ALI lungs and is particularly dense on the motile structural components of fibroblasts (filopodia and lamellipodia) (Svee, Henke 1996a). A cross-linking anti-CD44 antibody induces fibroblast detachment accompanied by apoptosis of attached and detached fibroblasts (Henke 1996b). It also prevents invasion of fibrin gels and apoptosis of fibroblasts within gels. Thus inhalation of drugs that bind CD44 and activate matrix signaling pathways might reduce the endothelial cell component of intra-alveolar granulation tissue.

Finally, it seems logical that the alveolar epithelium would influence events occurring within the alveolar lumen. Teleologically it would be ideal if they activated granulation tissue apoptosis. However, cultured alveolar epithelial cells instead release a bioactivity that inhibits the induction of mesenchymal cell apoptosis by TNF or chronic phase ARDS BALF (Wendt). This may prevent apoptotic destruction of remaining capillaries during the early phases of ARDS when TNF is present in BALF. It also may help protect capillary remodelling during the later repair phases of ARDS. In addition, some of the hyperplastic type II cells that line the repairing alveolus need to be eliminated in order to create space for some cells to differentiate into type I cells. Recently it was demonstrated that apoptosis is involved in this portion of the repair process (Bardales). These cells demonstrate foci with increased p53 and WAF1 staining consistent with the involvement of these proteins in apoptosis (Guinee).

CONCLUSION

Multiple strategies to speed healing are close to trial in preclinical models of lung injury. Some of these trials are hindered by the current lack of good animal models of the chronic phase of lung injury. It is likely that acceleration of the healing process will reduce the need for prolonged mechanical ventilation and ICU care, along with decreasing other complications - such as nosocomial pneumonia. Thus pursuing these strategies is likely to be worthwhile and clinically applicable; at the same time, we should continue efforts to diminish the degree of early injury and the complications of MOFS and nosocomial infection.

ACKNOWLEDGEMENTS

Supported in part by NIH HL50152 SCOR in Acute Lung Injury (DHI, CAH, VP, PBB), NIHKO8-HL03114 (CHW), an American Lung Association Career Investigator Award (DHI) and Research Fellowships (HJK), and American Heart Association Research Grant-In-Aids (DHI, CHW).

REFERENCES

Anderson WR and K Thielen. Correlative study of adult respiratory distress syndrome by light, scanning and transmission electron microscopy. Ultrastruct Pathol 16:615-628, 1992.

Bardales RH, SS Xie, RF Schaefer, SM Hsu. Apoptosis is a major pathway responsible for the resolution of type II pneumocytes in acute lung injury. Am J Pathol 149:845-852, 1996.

Fukuda Y, M Ishizake, Y Masuda, G Kimura, O Kawanami, Y Masugi. The role of intraalveolar fibrosis in the process of pulmonary structural remodeling in patients with diffuse alveolar damage. Am J Pathol 126:171-182, 1987.

Garat C, F Kheradmand, KH Albertine, HG Folkesson and MA Matthay. Soluble and insoluble fibronectin increases alveolar epithelial wound healing in vitro. Am J Physiol (Lung) 271:L844-853, 1996.

Guinee D, M Fleming, T Hayashi, M Woodward, J Zhang, J Walls, M Koss, V Ferrans, W Travis. Association of p53 and WAF1 expression with apoptosis in diffuse alveolar damage. Am J Pathol 149:531-538, 1996.

Henke C, W Marinelli, J Jessurun, J Fox, D Harms, M Peterson, L Chiang, P Doran. Macrophage production of basic fibroblast growth factor in the fibroproliferative disorder of alveolar fibrosis after lung injury. Am J Pathol 143:1189-1199, 1993.

Henke CA, U Roongta, DJ Mickelson, JR Knutson, JB McCarthy. CD44-related chondroitin sulfate proteoglycan, a cell surface receptor implicated with tumor invasion, mediates endothelial cell migration on fibrinogen and invasion into a fibrin matrix. J Clin Invest 97:2541-2552, 1996a.

Henke C, P Bitterman, U Roongta, D Ingbar, V Polunovsky. Induction of fibroblast apoptosis by anti-CD44 antibody. Implications for the treatment of fibroproliferative lung disease. Am J Pathol 149:1639-1650, 1996b.

Kerr JFR. Neglected opportunities in apoptosis research. Trends Cell Biol 5:55-57, 1995.

Kheradmand F, Folkesson HG, Shum L, Derynk R, Pytella R, and Matthay MA. Transforming growth factor alpha enhances AECs repair in a new in vitro model. Am J Physiol (Lung) 267:L728-738, 1994.

Kim HJ, DH Ingbar, CA Henke. Integrin mediation of type II cell adherence to provisional matrix proteins. Am J Physiol (Lung) 271:L277-L286, 1996.

Kim HJ, CA Henke, DH Ingbar. Integrin mediation of alveolar epithelial cell migration on fibronectin and type I collagen. Am. J. Physiol (Lung) 273:L134-L141, 1997.

Kroemer G, P Petit, N Zamzami, JL Vayssiere, B Mignotte. The biochemistry of programmed cell death. FASEB J 9:1277-1287, 1995.

Lasnier J, Wangensteen OD, Ingbar DH. Terbutaline stimulates alveolar fluid resorption in hyperoxic lung injury. J Appl Physiol 81:1723-1729, 1996.

Lesur O, K Arsalane, D Lane. Lung AEC migration in vitro:modulators and regulation processes. Am J Physiol (Lung) 270:L311-319, 1996

Matthay MA, and JP Wiener Kronish. Intaaft epithelial barrier function is critical for the resolution of alveolar edema in humans. Am Rev Respir Dis 142:1250-1257, 1990

Meduri GU, AJ Chinn, KV Leeper, RG Wunderink, E Tolley, HT Winer-Muran, V Khare, M El Torky. Corticosteroid rescue treatment of progressive fibroproliferation in late ARDS. Chest 105:1516-1527, 1994.

Meduri GU, S Headley, E Golden, S Carson, R Umberger, T Kelso, E Tolley. Methylprednisolone tratment of late ARDS. Am J Respir Crit Care Med 155:A391, 1997

Panos RJ, Bak PM, WS Simonet, JS Rubin, LJ Smith. Intratracheal instillaiton of keratinocyte growth factor decreases hyperoxia-induced mortality in rats. J Clin Invest 96:2026-2033, 1995.

Polunovsky VA, B Chen, C Henke, D Snover, C Wendt, DH Ingbar, PB Bitterman. Role of mesenchymal cell death in lung remodelling after injury. J Clin Invest 92:388-397, 1993.

Polunovsky VA, CH Wendt, DH Ingbar, MS Peterson, PB Bitterman. Induction of endothelial cell apoptosis by TNFalpha: modulation by inhibitors of protein synthesis. Exp Cell Res 214:584-594, 1994.

Polunovsky VA, DH Ingbar, M Peterson, PB Bitterman. Cell fusion to study nuclear-cytoplasmic interactions in endothelial cell apoptosis. Am J Pathol 149:115-128, 1996.

Solary E, L Dubrez, B Eymin. The role of apoptosis in the pathogenesis and treatment of diseases. Eur Respir J 9:1293-1305, 1996.

Svee K, J White, P Valliant, J Jessurun, U Roongta, M Krumwiede, D Johnson, C Henke. Acute lung injury fibroblast migration and invasion of a fibrin matrix is mediated by CD44. J Clin Invest 98:1713-1727, 1996.

Tan A, H Levrey, C Dahm, V Polunovsky, PB Bitterman. Lovastatin: a possible new therapy for fibroprooliferative diseases. Am J Respir Crit Care Med 155:A448, 1997.

Ulich TR, ES Yi, K Longmuir, S Yin, R Blitz, CF Morris, RM Housley, GF Pierce. Keratinocyte growth factor is a growth factor for type II pneumocytes in vivo. J Clin Invest 93:1298-1236, 1994.

Wendt CH, VA Polunovsky, MS Peterson, PB Bitterman, DH Ingbar. Alveolar epithelial cells regulate the induction of endothelial cell apoptosis. Am J Physiol (Cell) 267:C893-C900, 1995.

Yi ES, ST Williams, H Lee, DM Malicki, EM Chin, S Yin, J Tarpley, TR Ulich. Keratinocyte growth factor ameliorates radiation- and bleomycin-induced lung injury and mortality. Am J Pathol 149: 1963-1970, 1996.

Address correspondence to: Dr. David H. Ingbar
                               FUMC Box 276
                               Telephone 612-624-0999
                               Fax 612-625-2174
                               Email: ingba001@maroon.tc.umn.edu

PULMONARY MICROVASCULAR OCCLUSION IN AGGREGATE ANAPHYLAXIS

James W. Ryan,[1] R. Sandrapaty,[2] M. Sequeira,[2] F. Valido,[2] A. Chung,[2] and P. Berryer[2]

[1]Vascular Biology Center and the Department of Anesthesiology
Medical College of Georgia
Augusta, GA 30912
[2]Department of Medicine, University of Miami
Miami, FL 33101

The acute respiratory distress syndrome (ARDS) is a common late stage manifestation of several different kinds of severe lung injury (1). A priori, there is little reason to believe that the early pathogenesis of, e.g., sepsis-related lung injury is identical, or even similar, to the early pathogenesis of lung injury caused by aspiration of gastric contents, pneumonia or microthromboembolism even though all causes of serious diffuse lung injury may lead to a common clinical presentation. Thus, it is timely to examine separately the multiple causes of lung injury with the goal of developing appropriate means of treatment of each so as to prevent progression of the lung injuries to frank ARDS.

In the present study, we focus on the diffuse microvascular injury that occurs in the early minutes of aggregate anaphylaxis in guinea pigs. In aggregate anaphylaxis, fibrinogen consumption is brisk and large numbers of lung microvessels are occluded with platelet, leukocyte and erythrocyte plugs; all typically enmeshed in fibrin (2). Here, we attempt to quantify the degree of microvascular occlusion by estimating microvascular surface area during a control period and five min after induction of aggregate anaphylaxis.

MATERIALS AND METHODS

Guinea pigs, ~0.4 kg b.w., were immunized against ovalbumen and anaphylaxis was induced with i.v. ovalbumen (2).

Angiotensin converting enzyme (ACE) is disposed uniformly on the luminal surface of pulmonary endothelial cells; thus, pulmonary ACE in moles is a direct function of microvascular surface area. An index of ACE in moles can be measured by multiple indicator dilution (MID) methods in which tracer radiolabeled ACE substrate or inhibitor (or both) are included in the injectate. Substrate hydrolysis obeys first order enzyme kinetics,

Acute Respiratory Distress Syndrome: Cellular and Molecular Mechanisms and Clinical Management
Edited by Matalon and Sznajder, Springer Science+Business Media New York, 1998

255

and inhibitor uptake obeys pseudo-first enzyme kinetics (3). If an anatomical shunt is not present, pulmonary plasma flow must equal $Q_{cp}/\bar{t}_c$, where Q_{cp} is capillary plasma volume, and $\bar{t}_c$ is mean capillary transit time. Thus, using the integrated form of the first order rate equation and multiplying the natural log of fractional substrate utilization $(\ln(S_o/S))$ or inhibitor uptake $(\ln(I_o/I))$ by plasma flow $(\dot{Q}_p)$, the indices of microvascular surface area A_{max}/K_m and B_{max} can be obtained

$$A_{max}/K_m = \dot{Q}_p * \ln(S_o/S) = (k_{cat}/K_m)E_o \tag{1}$$

$$B_{max} = \dot{Q}_p * \ln(I_o/I) = k_1 E_o \tag{2}$$

where k_{cat}/K_m is the second order rate constant for substrate hydrolysis, k_1 is the second order rate constant for inhibitor binding to ACE and E_o is ACE in moles (4). We used [^{14}C]benzoyl-Ala-Gly-Pro (^{14}C-BAGP) as substrate and ^{3}H-RAC-X-65 (^{3}H-RAC) as inhibitor (3). For convenient notation, we use the following symbols:

$$v = \ln(S_o/S) = (k_{cat}/K_m)[E]_o \bar{t}_c \tag{3}$$

$$\theta = \ln(I_o/I) = k_1[E]_o \bar{t}_c \tag{4}$$

where $[E]_o$ is ACE concentration in moles/liter.

RESULTS

Control studies. The parameters v and θ vary insignificantly for any given control animal studied longitudinally. From definitions above, the relatively small changes of v and θ over time suggest that $[E]_o$ and $\bar{t}_c$ change little under control conditions. However, when endothelium is damaged (as in anaphylaxis), $\bar{t}_c$ could change and ACE could be lost from the endothelial surface. To examine for the latter possibility, we performed control studies and then simulated loss of ACE by treating guinea pigs with a dose of the inhibitor RAC-X-65 sufficient to reduce pulmonary ACE activity by 50 - 70%. As is evident in Fig. 1, we can readily detect loss of ACE activity ($\downarrow[E]_o$).

Effects of anaphylaxis. The most striking change in the indicator-dilution curve obtained 5 min after induction of anaphylaxis is the enormous increase in the area of the curve (Fig. 2). The increase in area is attributable to increases in both fractional concentrations of the indicators (evidence of a marked decrease in central blood volume) and in the time required for the indicator bolus to complete its first pass through the pulmonary vascular bed. Area is the reciprocal of plasma flow; thus increasing area bespeaks decreasing plasma flow. Strikingly, neither v nor θ is much changed, a finding that suggests that neither capillary ACE concentration, $[E]_o$, nor $\bar{t}_c$ is much changed in those few capillaries that remain patent.

Table 1 summarizes findings obtained using 10 sensitized guinea pigs studied 30 min before and 5 min after i.v. challenge with antigen. Central blood volume (CBV) (volume of dilution of the intravascular indicator) was decreased by about 42%, and plasma flow ($\dot{Q}_p$), A_{max}/K_m and B_{max} were decreased by ~87% (range of 44 to 92%). A_{max}/K_m and B_{max} measured simultaneously should differ only in terms of differences in their respective second order rate constants, k_{cat}/K_m and k_1 (see eq. (1) & (2)) and were in fact closely correlated (B_{max} (y) versus A_{max}/K_m: $y = 0.36(x) + 0.04$, $r = 0.94$). Average lung mean transit time ($\bar{t}$) increased by 3.27-fold. In longitudinal control studies using sensitized guinea pigs challenged with i.v. saline instead of ovalbumen in saline, none of $\bar{t}$, $\dot{Q}_p$, CBV, v, θ, A_{max}/K_m or B_{max} varied by more than 8% (intra-animal comparison) (data not shown).

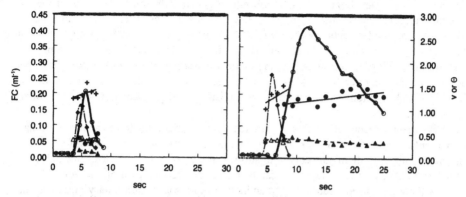

Fig. 1. A mixture of ^{14}C-BAGP, 0.5 μCi, and ^{3}H-RAC, 5 μCi, was injected as a bolus into the superior vena cava 25 min before (♦) and 5 min after (○) treatment of the test animal with a partially-saturating dose of unlabeled RAC-X-65 (15 nmol/kg i.v.). FC: fractional concentration in ml^{-1}; symbols for v are + & ● and for θ are Δ & ▲ (studies 1 and 2 respectively).

Fig. 2. A mixture of ^{14}C-BAGP and ^{3}H-RAC was injected as a bolus 30 min before (♦) and 5 min after (○) induction of anaphylaxis with i.v. ovalbumen. For other symbols, see Fig. 1 legend.

Table 1. Early effects of aggregate anaphylaxis on pulmonary hemodynamics and reactions of pulmonary vascular ACE with ^{14}C-BAGP and ^{3}H-RAC.

Time	$\bar{t}$ (sec)	$\dot{Q}_p$ (ml/sec)	CBV (ml)	v	θ	A_{max}/K_m (ml/sec)	B_{max} (ml/sec)
-30'	3.23	1.82	7.35	1.309	0.373	2.381	0.679
+5'	10.55	0.23	4.27	1.238	0.413	0.284	0.095

DISCUSSION

Our data indicate that aggregate anaphylaxis is characterized in part by a fulminant microthromboembolic process in which upwards of 90% of the pulmonary microvascular bed is occluded. Undoubtedly, humoral mediators also seriously compromise pulmonary hemodynamics; indeed, complement-derived anaphylatoxins likely play a major role in that inhibitors of carboxypeptidases M and N (enzymes also known as "anaphylatoxin inhibitors") greatly worsen aggregate anaphylaxis (5 min mortality rate >80%) (2). Nonetheless, the mechanical occlusion alone is sufficient to raise pulmonary artery blood pressure (PAP). Although we did not measure PAP directly, other data suggest the onset of severe acute pulmonary hypertension and right heart dilation and strain: All test animals in this study developed 2:1 or 3:1 heart block and one died ~30 min after antigenic challenge in ventricular fibrillation. Further, the relatively small decreases in CBV (~42%) compared with greater decreases in $\dot{Q}_p$ (~88%) suggest precapillary blood pooling.

Aggregate anaphylaxis is probably an uncommon cause of ARDS. Still, it appears to be a useful model for more common causes in which the initial damage to lungs is at the level of the microvascular endothelium. Such causes include fractures of large bones (accidental and surgical) that initiate showers of fat, bone marrow and bone spicule microemboli. Complete hip replacement procedures generate the same microemboli and add to the mix surgical "glues" such as methylmethoxyacrylate. The latter compound, an α, β-unsaturated carbonyl, reacts spontaneously, at physiologic pH, with mercapto-bearing substances (e.g. most proteins, glutathione, etc) to form thio ether products and with primary amines to form Schiff bases. Thus, pulmonary microvascular endothelium can be damaged mechanically and chemically. Sickle cell acute chest syndrome can be viewed as a form of ARDS in which pulmonary microthromboembolism is a major contributor to pulmonary insufficiency.

By means described in this report, it is now feasible to detect acute pulmonary microvascular occlusion. Our approach may be most informative in monitoring for episodes of microthromboembolism consequent to surgical fractures of large bones; in which cases pulmonary microvascular surface area can be measured pre- and post-operatively.

Although we used a substrate and an inhibitor to measure pulmonary microvascular ACE (and therefore surface area) and did so via an invasive multiple indicator dilution protocol, our data make evident that an inhibitor can be used alone to obtain information of the same quality (see Table 1). Carboxyalkyl-dipeptide ACE inhibitors such as RAC-X-65 can be labeled with gamma-emitters such as 123-I. Thus, it should soon be feasible to assess pulmonary microvascularity, and decreases consequent to vascular obstruction, by non-invasive gamma scintigraphy, including first pass radionuclide angiography and lung scans.

REFERENCES

1. J.F. Pittet, R.C. Mackersie, T.R. Martin, and M.A. Matthay, Biological markers of acute lung injury: prognostic and pathogenetic significance, *Am J Respir Crit Care Med.* 155:1187-1205 (1997).
2. J.W. Ryan, P. Berryer, M.A. Hart, and U.S. Ryan, Aggregate anaphylaxis and carboxypeptidase N, *Adv Exp Med Biol.* 198A:435-443 (1986).
3. J.W. Ryan, F.A. Valido, M.J. Sequeira, A.Y.K. Chung, P. Berryer, X-L. Chen, and J.D. Catravas, Estimation of rate constants for reactions of pulmonary microvascular angiotensin converting enzyme with an inhibitor and a substrate in vivo, *J Pharmacol Exp Ther.* 270:260-268 (1994).
4. J.D. Catravas, and R.E. White, Kinetics of pulmonary angiotensin converting enzyme and 5'-nucleotidase in vivo, *J Appl Physiol.* 57:1173-1181, (1984).

The Expression of ICAM-1 and P-Selectin Are Increased Following Hemorrhage and Resuscitation in Mice.

Andrea J. Cohen [1,3,4], Robert Shenkar[1], Dietmar Vestweber[5], Rubin Tuder[1], York E. Miller[1],[3] and Edward Abraham[1]

Division of Pulmonary Sciences and Critical Care Medicine and Department of Pathology, University of Colorado Health Sciences Center,[1] Denver Veterans Affairs Medical Center[3], National Jewish   Center for Research[4], Spemann Laboratories, Max-Plank-Institut fur Immunobiologie[5], Freiburg- Zahringer, Germany.

INTRODUCTION

Acute inflammatory lung injury is a common clinical occurrence following trauma and blood loss and is characterized by massive neutrophil infiltration into the lungs. The mechanisms responsible for the increased susceptibility to acute lung injury remain incompletely understood. P-selectin and ICAM-1, two types of adhesion molecules, have been shown to participate in several forms of neutrophil-dependent lung injury. In order to examine the mechanisms which contribute to trafficking of neutrophils in the lungs and development of pulmonary injury following blood loss, we investigated in-vivo mRNA and immunohistochemical expression of P-selectin and ICAM-1 following hemorrhage and resuscitation.

MATERIALS AND METHODS

Hemorrhage and Resuscitation

The murine hemorrhage and resuscitation model used in these experiments was done as previously described[1,2]. Within this model, 30% of the calculated blood volume was withdrawn over a 60-second period. Control mice were subjected to anesthesia and cardiac puncture, but no blood withdrawal. The left upper lobe of the lungs was removed and cDNA was synthesized. Semi-quantitative RT-PCR for P-selectin and ICAM was performed[4].   We synthesized competitor DNA for each set of primers using a MIMIC construction kit[3] (Clontech, Palo Alto, CA). To detect amplified cDNA, the PCR product was analyzed by agarose gel electrophoresis. The gels were then scanned with gel documentation system and densitometry[3].

Immunohistochemistry

Immunohistochemisty was performed on OCT embedded frozen tissue that had been fixed with 1% paraformaldehyde. The consecutive tissue sections were incubated with rabbit anti-mouse P-selectin at a dilution of 1:100 [4].Other consecutive sections were incubated with biotinylated hamster anti-mouse ICAM-1 at a dilution of 0.05 µg/ml (Pharmingen, San Diego, CA).   Controls included omission of the primary antibody, and either the substitution of the primary antibody antibody with normal mouse serum (Biogenex) or biotinylated hamster IgG isotype control (Accurate Chemical, Westbury, NY). Intensity was graded 0-4 by a blinded observer. Five mice were used for each control and experimental group.

RESULTS

The amounts of ICAM-1 mRNA levels were significantly elevated compared to normal in lungs obtained 6 and 72 hr after hemorrhage and resuscitation, (P<0.5), but not 24 hr post hemorrhage. Significant increases in mRNA levels for P-selectin compared to control mice were observed in lungs obtained 72 hr (P<0.001) after hemorrhage and resuscitation.

Immunohistochemistry

In lungs from normal mice, staining for P-selectin was present on the endothelial cells of all visible vessels. The staining was clearly discontinuous; there were prominent areas of staining on the endothelial cells intermixed with areas of faint or absent staining. The most intense staining was in the perinuclear areas. There also were occasional punctate collections of of P-selectin immunoreactivity in the alveolar walls. This staining was presumed to represent decoration of alveolar capillary endothelial cells, but may have represented alveolar epithelial cell staining. As soon as 6 hr after hemorrhage, and as late as 3 days post-injury, the murine lungs exhibited intense P-selectin immunoreactivity on the entire luminal surface of the blood vessels. This intense staining involved all visible blood vessels. Grading of P-selectin endothelium membrane sections revealed significant increases from mean values of 0.7 in controls to 3.7 (P<0.001 vs normal) at 6 hr and 2.0 (P< 0.05 vs control) at 24 and 72 hr after hemorrhage. Alveolar wall (presumably capillary) P-selectin decreased from mean values of 2.3 in normals to 0.4 (P<0.001 vs normal) at 6 hr, 0.0 (P<0.001 vs normal) at 24 hr and 1.0 (P<0.01 vs normal) at 72 hr after hemorrhage. ICAM-1 immunoreactivity was constitutively expressed in most the alveolar epithelium with faint staining on bronchiolar epithelium and endothelium. Six hr after hemorrhage, there was a significant increase in the ICAM-1 staining in alveolar epithelium as compared to normal animals, involving both Type 1 and Type 2 cells. The strongest ICAM-1 staining was present 6 hr and 3 days following hemorrhage. Scoring of alveolar ICAM-1 showed increases in mean values from 1.3 in normals to 2.6 at 6 hr, 1.3 at 24 hr and 2.7 (P<0.05 vs normal) at 72 hr after hemorrhage. Neutrophil accumulation was graded at 72 hr at 2.9 and was significantly different than control.

DISCUSSION

There are several mechanisms operative in the post-hemorrhage period which may be responsible for the upregulation of P-selectin expression. P-selectin is stored intracellularly and following stimuli, presynthesized protein is directed to the cell surface. The rapid increase of membrane-associated P-selectin in the immediate post-hemorrhage period is probably due to the release of preformed adhesion molecules, rather than new protein synthesis given the limited time available to complete transcription and translation, as well as the lack of increase in P-selectin mRNA levels at this early post-hemorrhage time point. In contrast, at when mRNA for P-selectin is increased, newly synthesized protein is more likely to contribute to the enhanced P-selection expression. The mechanisms for increased adhesion molecules include cytokine stimulation[3], histamine release[5], and reactive oxygen[6] generation following hemorrhage and resuscitation. In addition, hemorrhage-induced oxygen free radicals appear to increase proinflammatory cytokine expression in the lungs, as early as 1 hr. Increased expression of adhesion molecules is not sufficient to produce a leukocyte influx early following hemorrhage. Additional factors capable of activating neutrophils to move from the circulation into the lung appear to be required, such as tumor necrosis[7]. Agents that block the interaction of neutrophil with ICAM-1 and P-selectin may have a clinical utility in treatment of acute lung injury.

REFERENCES

1. R. Shenkar, W.F. Coulson, E. Abraham, Hemorrhage and resuscitation induce alterations in cytokine expression and the development of acute lung injury. *Am J Respir Cell Mol Biol.* 10:290, (1994).
2. R. Shenkar, E.Abraham, Effects of hemorrhage on cytokine gene transcription. Lymphokine Cytokine Res. 12:237, (1993).
3. E. Abraham, S. Bursten, R. Shenkar, J. Allbee, R.Tuder, P. Woodson, D.M. Guidot, G. Rice, J.W. Singer, J.E. Repine, Phosphatidic acid signaling mediates lung cytokine expression and lung inflammatory injury after hemorrhage in mice. *J Exp Med.* 181:569, (1995).
4. A.J. Cohen, P.A. Bunn, W. Franklin, L. Gilman, C. Magill Solc, Karen Helm, B. Helfrich, J. Folkvord, Y.E. Miller, Neutral endopeptidase: variability in normal lung, low expression in lung cancer, and

modulation of peptide induced calcium flux. *Cancer Res.* Vol *56:* 831-839, (1996).

5. B..M. Altura, S. Halevy, Beneficial and detrimental actions of histamines H1-and H2-receptor antagonists in circulatory shock. *Proc Natl Acad Sci USA.* 75:2941, (1978).

6. M. Schwartz, J.E. Repine, E. Abraham, Xanthine oxidase-derived oxygen radicals increase lung cytokine expression in mice subjected to hemorrhagic shock. *Am J Respir Cell Mol Biol.* 12:434, (1995).

7. R. Shenkar, W.F. Coulson, E. Abraham, Anti-transforming growth factor-ß monoclonal antibodies prevent lung injury in hemorrhaged mice. *Am J Respir Cell Mol Biol.* 11:351, (1994).

INCREASED LEVELS OF MARKERS OF PROTEIN OXIDATION IN BRONCHOALVEOLAR LAVAGE FLUID FROM PATIENTS WITH ARDS

Gregory J. Quinlan, Nicholas J. Lamb, Timothy W. Evans, John MC. Gutteridge

Unit of Critical Care, NHLI,
Imperial College of Science, Technology and Medicine,
London, United Kingdom.

INTRODUCTION

The formation of reactive oxygen species (ROS), leading to increased oxidative stress has been implicated as a contributing factor to the development and progression of ARDS. Evidence of increased oxidative damage to lipids and proteins has been found in the plasma of patients with ARDS, together with low or compromised antioxidant protection. Oxidative damage to proteins and glutathione has also been observed in bronchoalveolar lavage fluid (BALF) from these patients. Plasma and BALF from ARDS patients can often have increased pro-oxidant levels. Thus, the presence of redox active iron (a catalyst for the production of reactive radicals) has been detected. Hypoxanthine (a substrate for the ROS producing enzyme xanthine oxidase) is also elevated in these patients, particularly non-survivors. ROS can arise *in vivo* in patients with ARDS by inflammatory cell activation, ischemia-reperfusion, and treatment with high FiO_2. These processes can ultimately result in the formation of damaging ROS, such as peroxynitrite, hypochlorous acid, and the hydroxyl radical. Direct measurement of highly reactive species is not possible because of their transient nature. However, measurement of characteristic patterns of damage to biological molecules can provide evidence of their formation. We have therefore measured markers of ROS formation as:-(chlorotyrosine for hypochlorous acid, nitrotyrosine for peroxynitrite, and ortho-tyrosine for hydroxyl radicals) in BALF proteins from patients with ARDS, as well as in ventilated post operative intensive care patients, and normal healthy controls. Neutrophil activation in BALF was determined by measuring myleoperoxidase (MPO) levels. Our aim was to establish whether ROS neutrophil activation was the dominant source of ROS in patients with ARDS.

METHODS AND RESULTS

BALF was collected from ARDS patients and control groups. Samples were spun to remove debris, and proteins concentrated. Free amino acids were obtained by acid hydrolysis.

Acute Respiratory Distress Syndrome: Cellular and Molecular Mechanisms and Clinical Management
Edited by Matalon and Sznajder, Springer Science+Business Media New York, 1998

Samples were purified by ion exchange, and separated by hplc with uv detection. MPO levels in BALF were determined by use of an ELISA kit method .

Comparisons between ARDS patients (n = 28), intensive care (IC) controls (n = 6), and normal healthy (NH) controls (n = 11) for markers of oxidative protein modification, showed significant increases in the ARDS group in all cases. Results are summarized as follows; ortho-tyrosine ARDS (7.98 ± 3.78 nmol/ mg protein) versus IC (0.67 ± 0.67) and NH (0.71 ± 0.22), $p < 0.05$ compared with ARDS. Chlorotyrosine ARDS (4.82 ± 1.07 nmol/mg protein) versus IC (1.55 ± 1.34) and NH (0.33 ± 0.12), $p < 0.05$ compared with ARDS. Nitrotyrosine ARDS (2.21 ± 0.65 nmol/mg protein) versus IC (0.29 ± 0.29) and NH (0.006 ± 0.03), $p < 0.05$ compared with ARDS. BALF MPO levels were also significantly elevated in the ARDS group (84.72 ± 6.46 ng/ml) compared to IC (31.72 ± 13.47, $p < 0.05$) and NH (13.42 ± 3.03, $p < 0.01$) controls.

DISCUSSION

Results show evidence for the formation of oxidatively modified amino acids in BALF proteins from patients with ARDS. When compared to ventilated IC patients and normal healthy controls, the markers of oxidation (ortho-tyrosine, chlorotyrosine and nitrotyrosine) are significantly increased in the ARDS group, thereby indicating that the lungs of these patients are under considerable oxidative stress. Neutrophil activation in the lungs of the ARDS group was significantly increased, as indicated by the increased levels of MPO in the BALF of these patients. Comparisons of MPO levels with the markers of oxidative damage in the ARDS group showed significant correlations for chlorotyrosine ($r = 0.67$, $p = 0.0002$) and nitrotyrosine ($r = 0.58$, $p = 0.0018$). Ortho-tyrosine showed only a weak correlation, possibly reflecting the extreme reactivity of the hydroxyl radical.

In conclusion, our results indicate the the hydroxyl radical, hypochlorous acid, peroxynitrite and possibly another chlorinating and nitrating species, are formed in the lungs of patients with established ARDS. These ROS clearly implicate neutrophil activation as a major source of lung damage in patients with ARDS.

THE USE OF P.E.E.P. IN THE TREATMENT OF A.R.D.S.

Jesus Villar, MD, PhD, FCCM

Director, Research Institute, Hospital Universitario de la Candelaria,
Tenerife, Canary Islands, Spain;
Clinical Professor, Department of Anesthesia & Critical Care Medicine,
Mercer University, Macon, Georgia, U.S.A.

INTRODUCTION

Positive end-expiratory pressure (PEEP) has become an essential component of the care of many critically ill patients who require ventilatory support. With the application of PEEP, the baseline end-expiratory pressure in mechanically ventilated patients is elevated above atmospheric pressure. The use of PEEP was first reported by Barach et al[1] in 1938, and it has been widespread since Ashbaugh et al[2], in their classic description of the acute respiratory distress syndrome (ARDS), reported that PEEP improved oxygenation and allowed ventilation with gas of lower inspired oxygen concentration. The rationale for the use of PEEP in acute lung injury coincides with the theoretical basis for loss of lung compliance in patients with ARDS. PEEP would prevent complete alveolar collapse and improve oxygenation by increasing functional residual capacity (FRC), probably by preventing airway closure and recruiting previously unventilated alveoli[3]. The increase in FRC may also increase lung compliance. However, although there is no question that PEEP can improve oxygenation in selected patients, its beneficial effects on morbidity and mortality have not been conclusively demonstrated[4,5]. It is not well known whether the application of PEEP contributes to lung damage or helps to ameliorate it.

Research on the effects of PEEP in patients with acute respiratory failure and in animal models of acute lung injury has produced a plethora of information during the last two decades. For consistency, clarity and simplicity, in this paper we will focus only on the use of PEEP in conjunction with a positive pressure ventilator in the setting of acute lung injury. To evaluate the effects of PEEP in patients with acute lung injury, it is necessary to be aware of the physiologic manifestations, types and effects of interventional therapy, and the desired outcome.

GOALS OF PEEP: BASIC MECHANISMS AND PULMONARY EFFECTS

PEEP is applied generally to improve oxygenation. Improvement in oxygenation is usually not observed unless there is a concomitant increase in FRC. The application of

Acute Respiratory Distress Syndrome: Cellular and Molecular Mechanisms and Clinical Management
Edited by Matalon and Sznajder, Springer Science+Business Media New York, 1998

265

PEEP is expected to increase PaO_2 and decrease intrapulmonary shunt, alveolar-arterial O_2 pressure difference, and arteriovenous O_2 content difference as it recruits lung volume. Four mechanisms have been proposed to explain the improved pulmonary function and gas exchange with PEEP[6]: (i) increased FRC; (ii) alveolar recruitment; (iii) redistribution of extravascular lung water; and (iv) improved ventilation-perfusion matching.

Increased Functional Residual Capacity and Alveolar Recruitment

Regardless of the status of the pulmonary system, the FRC increases with the application of PEEP. This increase of the lung volume is the result of three separate effects[6,7]. First, PEEP increases lung volume as a result of distension of already patent airways and alveoli by an amount dependent on system compliance. Second, application of PEEP prevents alveolar collapse during expiration. Dependent small airways tend to collapse at low lung volumes. The lung volume at which small airway closure occurs can increase above the passive FRC in anesthetized man[8]. Third, PEEP levels above 10 cmH$_2$O recruit collapsed alveoli in acute lung injury. Alveolar recruitment describes reinflation of previously collapsed alveoli. Kumar et al[9] observed in a series of 8 patients with ARDS that PEEP withdrawal produced an immediate decline in oxygenation within the first minute and a further decline over the next 30 min (Figure 1). Improvements of oxygenation after PEEP was re-established were more gradual, a response that has been attributed to hysteresis[10]. During acute lung injury, and depending on the severity of lung disease, PEEP can markedly alter the compliance of the lung by alveolar recruitment[11]. The greater the alveolar collapse and pulmonary edema, the more the compliance curve shifts downward and to the right. As PEEP is applied and alveoli recruited, the pressure-volume curve shifts upward and to the left.

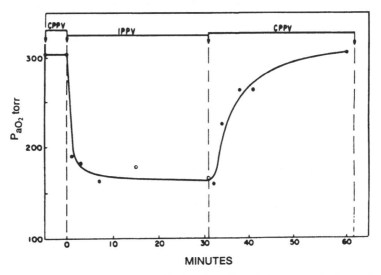

Figure 1. Time sequence of PaO_2 change with alteration of ventilation pattern in 8 patients with acute lung injury. PEEP withdrawal produced an immediate, precipitous decline in oxygenation within 1 minute. The PaO_2 rose gradually to its initial value after reapplication of PEEP (from ref. 9, with permission).

266

Redistribution of Extravascular Lung Water

The effects of PEEP on extravascular lung water are complex and depend on the vascular surface area perfused and the relations among alveolar, arterial, interstitial, and venous pressures. Selected studies in animals looking at the effects of PEEP on extravascular lung water in both cardiogenic and noncardiogenic pulmonary edema, have shown that the application of 5 to 20 cmH_2O is able to recruit flooded alveoli and improved oxygenation without diminishing lung water content[12]. A study by Malo et al[12] in an animal model of oleic acid-induced lung injury supports the conclusion that PEEP redistributes lung water from alveoli to the perivascular space. They examined alveoli after the application of 13 cmH_2O of PEEP and found that although extravascular lung water was unchanged, fewer alveoli were flooded. In general, by increasing intraalveolar pressure, PEEP moves fluid from the interstitial space of alveolar vessels to the interstitial space around extraalveolar vessels.

Ventilation-Perfusion Relationships

In certain patients with ARDS, PEEP produces marked improvement in gas exchange. However, in other patients, PEEP produces equivocal and even detrimental changes, often combined with reductions in cardiac output. In order to understand these conflicting responses, it is useful to understand the effects of PEEP on the distribution of ventilation-perfusion ratios (V/Q)[7].

In normal lungs, the lowest V/Q is found in the gravity-dependent areas. In acute lung injury, low levels of PEEP (5-10 cmH_2O) reduce shunt and abolish low ventilation-perfusion ratio regions at the expense of creating areas of dead space ventilation (very high ventilation-perfusion ratio)[13]. PEEP has no direct effect on $PaCO_2$, but significant alterations in $PaCO_2$ may occur secondary to improved ventilation-perfusion matching resulting from recruitment or a deterioration in the ventilation-perfusion relationships due to overdistension and decreased pulmonary perfusion[14]. Dueck et al[13] assessed the effects of four levels of PEEP (5, 10, 15, and 20 cmH_2O) on the distributions of ventilation-perfusion ratios in normal and oleic acid-induced edematous lungs in dogs. They found that high levels of PEEP (10-20 cmH_2O) increased deadspace units while the number of acceptable V/Q units decreased. These alterations appeared to be related to reduction in cardiac output. In animals with severe pulmonary edema, regions of intrapulmonary shunt were eliminated by PEEP, but dead space emerged only with higher levels of PEEP.

In humans studies, Dantzker and associates[15] examined the distribution of ventilation and perfusion and found that the application of PEEP is accompanied by decreased blood flow to poorly ventilated regions. In addition, PEEP abolished regions of shunt and redistributed blood flow from regions with high shunt to regions of very low V/Q. Manzano et al[16] were able to decrease shunt and increase oxygenation when cardiac output was not decreased during the application of PEEP. After the administration of fluids and dopamine to patients on high levels of PEEP (10-30 cmH_2O), these investigators found that oxygen transport increased and shunt decreased when cardiac output was preserved. This observation supports the hypothesis that the reduction of shunt may be independent of decreased cardiac output induced by PEEP. Available data in acute lung injury can be best interpreted to show that PEEP reduces shunt mainly by alveolar recruitment and redistribution of pulmonary perfusion[6].

There is no question that PEEP can improve oxygenation in selected patients; however, its beneficial effects on morbidity and mortality have not been conclusively demonstrated. The benefits of PEEP may be counterbalanced by such complications as a reduction in cardiac output and the risk of further damage to the lungs.

Patients ventilated with positive pressures may be predisposed to pulmonary barotrauma. It has been suggested that three factors must be present to produce barotrauma: lung disease, overdistension, and pressure[17]. Since ARDS is a non-homogeneous process, overdistension of a given lung unit may be achieved at any PEEP level. However, does PEEP really rupture alveoli? Pulmonary barotrauma typically referes to any disorder that produces extraalveolar gas. Pneumothorax, pneumomediastinum, and subcutaneous emphysema are the most common forms of barotrauma, and they are the result of spontaneous alveolar rupture. Increased end-expiratory pressure is not enough by itself to produce alveolar rupture. The prevalence of barotrauma ranges from 10 to 20 percent in most series whether PEEP is used or not.

Animal studies have shown that an increase in alveolar volume, and not just pressure, is the key element in producing disruption of alveolar walls[18,19]. However, there is evidence that PEEP may actually be protective. The first clear study of the problem was a classic paper by Webb and Tierney[18]. Rats with normal lungs were mechanically ventilated for one hour. Those animals ventilated at peak pressures of 45 cmH_2O and no PEEP developed severe hypoxemia, decreased compliance, and died with extensive alveolar edema. Those ventilated with no PEEP and a peak airway pressure of 14 cmH_2O showed no abnormalities. The most interesting observation was that the application of 10 cmH_2O of PEEP, even when the same peak inspiratory pressure was used (45 cmH_2O), dramatically reduced edema formation and no animals died. Dreyfuss and colleagues[19] ventilated rats with normal lungs at 45 cmH_2O and no PEEP for 20 min after which there was a widespread alveolar flooding and disruption of the epithelium. If high positive pressures were used but the volume expansion limited by thoraco-abdominal strapping, there was no protein leak. Similarly to Webb and Tierney, Dreyfuss and colleagues[19] reported a marked reduction in both edema and protein leak when 10 cmH_2O of PEEP was applied. More recently, Dreyfuss' group have conducted a series of experiments to evaluate the respective influences of PEEP, tidal volume and overall lung distension on ocurrence and amount of high peak inspiratory pressure-induced edema[20]. They found that although permeability alterations were similar, edema was less marked in animals ventilated with PEEP that in those ventilated with zero PEEP.

Thus, our understanding of the mechanism of high peak inspiratory pressure edema has evolved from the concept of barotrauma to that of "volutrauma". To date, no study has looked prospectively at the association of barotrauma with the use of PEEP. Lacking prospective, controlled studies in patients who developed alveolar rupture, clinicians have relied on anecdotal observations and retrospective reviews of available bedside data in the charts of patients identified through their X-rays reports as having any form of extralveolar gas[17]. Although most of those observations have shown that patients who are treated with PEEP have a high incidence of alveolar rupture, they do not demonstrate that high PEEP is the cause, since those patients who require the highest levels of PEEP are likely the sickest patients and thus would be more prone to develop barotrauma.

EFFECTS OF PEEP IN ACUTE LUNG INJURY

In the initial phases of ARDS, the lungs are edematous and show a reduced compliance with an inflection point in the ascending limb of the pressure-volume curve. In late stages of ARDS, fibrosis develops, the lungs become stiff, compliance is very low, and the pressure-volume curve does not show an inflection point. This inflection point represents the recruitment of collapsed alveoli and was first documented by Cook et al[21]. In an oleic acid-induced pulmonary edema, Slutsky et al[22] found that the inflection was explained on the basis of reopening of the units closed during deflation (Figure 2). Analysis of the pressure-volume curves in patients with acute lung injury allowed Matamis et al[23] to separate them into four groups: (i) patients with normal compliance and no inflection point; (ii) patients with normal compliance and the presence of an inflection point; (iii) patients with reduced compliance and no inflection; and (iv) patients with reduced compliance and the presence of an inflection point. These patterns were correlated with the different stages of human ARDS.

Several studies performed by Gattinoni et al[24] provide evidence that the lung disease in patients with ARDS is heterogeneous with collapsed and consolidated units mainly in the dependent regions, and more healthy units in the non-dependent regions. Using computerized tomography, they demonstrated that the healthy zone can represent as little as 20% of the normal lung volume. This *baby* lung must accomplish the entire gas exchange during mechanical ventilation. This finding is clinically relevant, since patients with ARDS are often ventilated with tidal volumes derived from estimates of body size despite having greater

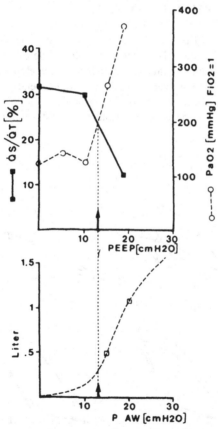

Figure 2. Effect of PEEP on pressure-volume (PV) curve in a typical patient with ARDS. *Lower panel:* PV curve during inflation shows inflection near 12 cmH$_2$O. *Upper panel:* Gas exchange as a function of PEEP. When PEEP is less than 12 cmH$_2$O, PaO$_2$ and intrapulmonary shunt (Qs/Qt) are not influenced. When PEEP is above the inflection point, shunt drops sharply (from ref. 23, with permission)

than 50% air space flooding. Overinflation of the most compliant zones (which are the less damaged) may occur because these zones receive the bulk of ventilation.

The main effect of augmenting PEEP is maintain recruitment of alveolar units that were previously collapsed. Thus, since tidal volume is distributed to more alveoli, peak airway pressure is reduced and compliance is increased. There are a number of animal studies comparing high PEEP and low PEEP strategies in acute lung injury. Corbridge et al[25], in an acid aspiration model, showed a significantly lower shunt fraction and less pulmonary edema with high PEEP compared to low PEEP with the same peak pressure. In a rat model, Muscedere et al[26] ventilated isolated, nonperfused, lavaged rat lungs with tidal volumes between 5 to 6 mL/kg at different PEEP levels (below and above inflection point). Lungs ventilated with no PEEP or PEEP below the inflection point had a marked decrease in compliance, whereas those ventilated above the inflection point had a significant increase in compliance (Figure 3). Histological examination of the lungs revealed that both the no PEEP and the PEEP below the inflection point groups had severe hyaline membrane formation. In contrast, those lungs ventilated with PEEP above the inflection point had no more damage than the control group. From these studies, it appears that maintenance of an end-expiratory lung volume at or above the inflection point is of more importance than other factors in minimizing lung damage[27-29].

PEEP AND CLINICAL APPROACH IN ARDS

It is unlikely that PEEP truly reverses any of the underlying pulmonary derangements once ARDS is established. However, the mode of ventilation may alter the location of densities in the lung fields and the further development of lung injury. A ventilatory mode which fails to prevent partial or complete end-expiratory collapse may worsen lung damage. Sjöstrand's group, using an animal model of acute lung injury induced by surfactant depletion, applied five ventilatory modes, adjusting either volume or

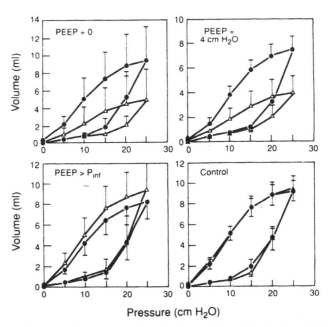

Figure 3. Composite pressure-volume curves before (circles) and after (triangles) ventilation with different levels of PEEP (from ref. 26, with permission).

pressure to keep PEEP at about 16 cmH$_2$O and PaCO$_2$ constant[30]. They found that: a) to open collapsed areas of the lung, a pressure amplitude of 40 cmH$_2$O (peak pressure minus PEEP) had to be applied for 5-10 min; b) after opening the lungs, they could then be adequately ventilated with pressure amplitudes of only about 20 cmH$_2$O (peak pressures of 35 cmH$_2$O). Therefore, in order to prevent lung damage only ventilation modes which result in the smallest possible pressure amplitude should be used[31,32]. In another words, open up the whole lung and keep it totally open[31].

Although the optimal method of applying PEEP is still quite controversial, it is generally agreed that simply using increased PaO$_2$ as the end-point is inappropriate. PEEP is only one of the variables that can be adjusted to improve oxygenation. The initial approach to the application of PEEP attempted to optimize tissue oxygen utilization. To accomplish this goal, investigators have suggested changing the level of PEEP until oxygen delivery and compliance are maximized[11]. Others have recommended the titration of PEEP until intrapulmonary shunt is reduced to 15 percent[33], mixed venous oxygen saturation is increased[34], oxygen consumption is maximized[35], or until adequate oxygenation is achieved with the lowest non-toxic FiO$_2$[36] (Table 1).

Table 1. Approaches used to achieve the optimum level of PEEP.

Investigators (Ref.)	Date	End points	Terminology
Suter et al (11)	1975	Maximum compliance	Optimal PEEP
Kirby et al (33)	1975	Reduction of shunt to 15%	Super-PEEP
Demers et al (34)	1977	Highest SvO$_2$	Best PEEP
Carroll et al (36)	1988	PaO$_2 \geq$60 with FiO$_2$<0.6	Minimal PEEP

Suter et al[11] studied 15 patients with acute respiratory failure supported by volume controlled ventilation. FiO$_2$ ranged between 21 to 75% and the tidal volume between 13-15 mL/kg. PEEP was increased in 3 cmH$_2$O steps until cardiac output fell. "Best" PEEP defined as the maximal oxygen transport corresponded with the maximal point of improvement in static compliance and mixed venous oxygen tension, and the lowest physiological dead space fraction. PEEP above that "optimum" level of PEEP caused diminished oxygen transport. Although PaO$_2$ continued to increase beyond that point, oxygen transport did not increase significantly. Suter et al[11] did not investigate what might have happened to the compliance-transport relationship had cardiac output been manipulated so that higher levels of PEEP would not have had this effect upon oxygen transport. PEEP was not raised above 15 cmH$_2$O in any of their patients.

An alternative method of optimizing PEEP was proposed by Kirby et al[33]. For these investigators, the end-point for optimizing PEEP was the return of the intrapulmonary shunt fraction to 15 percent. This point could almost always be reached with a combination of PEEP and appropiate interventions to maintain cardiac output and oxygen consumption. The mean PEEP value in their series of 28 patients was 25 cmH$_2$O. Carroll et al[36] reported that using the minimum PEEP required to maintain a PaO$_2$ $\geq$60 mmHg on a FiO$_2$ of 50% was associated with fewer complications than using "recruitive PEEP". In practice, PEEP has often been used in the way advocated by Albert[37]; that is, the lowest level of PEEP that maintains an adequate PaO$_2$ on a FiO$_2$ less than 60%. Benito and Lemaire[38] analyzed the pressure-volume curves in patients with ARDS at three levels of PEEP (0, 10, 20 cmH$_2$O). Only the application of 20 cmH$_2$O was able to suppress the inflection point of the ascending limb, reduced hysteresis, and shifted the pressure-volume trace upward and to the left. Gattinoni et al[39] took an interesting approach to determine how much PEEP is required in each patient. They performed transverse computed tomography scans in ARDS patients at the base of the lungs while the patient was supine. The height of each lung was measured from its lowest (posterior) to its highest (anterior) point and the lung was divided into 10 equal horizontal slices. Each of these sections was analyzed to determine its gas and tissue content, and they were compared to study the impact of PEEP. It is easy to understand that in edematous ARDS lungs, the pressure within the dependent sections of the lung will be higher because of the accumulated weight of the water column above. The resultant increase in pressure, if sufficiently high, can produce alveolar collapse. Gattinoni's group hypothesized that is this pressure that mechanical ventilation must offset if the alveoli are to reopen. They confirmed that for recruitment to take place, PEEP levels must equal the superimposed hydrostatic pressure.

EFFECT OF PEEP ON OUTCOME FROM ARDS

Since the original paper by Ashbaugh et al[2], PEEP was quickly adopted as standard practice for the treatment of severe acute respiratory failure. Survival rate in patients treated with 5-10 cmH2O of PEEP was 60%, compared with 30% in those treated with standard mechanical ventilation. Because of this intuitive benefit, no prospective randomised trial of PEEP has ever been carried out in established ARDS patients to evaluate its efficacy and there are very few controlled studies of the effects of PEEP on ARDS outcome[32].

Early application of PEEP appears to alter the degree of pulmonary deterioration, but does not prevent its occurrence. One well controlled study by Pepe et al[40] examining the prophylactic use of PEEP showed that there was no clinical utility to this approach. These investigators randomly assigned 92 patients at risk for ARDS to receive mechanical ventilation either without PEEP or with early PEEP of 8 cmH$_2$O. This protocol continued for 3 days unless respiratory distress developed at 24 hours or later. ARDS developed in 25% of patients with early PEEP and in 27% of patients in the control group. Although the mortality was a little lower in the treated group (30% vs. 38%), the difference was not statistically significant. However, the absence of benefits observed by Pepe et al could be misleading to the practicing intensivist. First, in their study, the level of PEEP chosen was 8 cmH$_2$O, below the inflection point as determined in most studies (between 12-16 cmH$_2$O). Second, the tidal volume used was 12 mL/kg, a relatively large tidal volume which likely produced high peak pressures, overdistension, and possible lung injury. Third, PEEP was intermittently removed for 8 minutes before blood samples were obtained.

Although the data from animal studies cited earlier are very convincing, there are insufficient data in humans to propose an universal approach for the use of PEEP in patients with ARDS. Recently, Slutsky[28] have suggested a randomized clinical trial comparing conventional mechanical ventilation to a strategy that includes continuous volume recruitment. The amount of experimental evidence suggests that we know how to apply PEEP early in the disease, how to measure the alveolar recruitment, and how to maintain a low peak inspiratory pressure.

Supported in part by Fondo de Investigación Sanitaria of Spain (#95/1769)

REFERENCES

1. AL Barach, J Martin, M Eckman. Positive pressure respiration and its application to the treatment of acute pulmonary edema. *Ann Intern Med* 12:754-795 (1938).
2. DG Ashbaugh, DB Bigelow, TL Petty, BE Levine. Acute respiratory distress in adults. *Lancet* 2:319-323 (1967).
3. KJ Falke, H Pontoppidan, A Kumar, DE Leith, B Geffin, MB Laver. Ventilation with end-expiratory pressure in acute lung disease. *J Clin Invest* 51:2315-2323 (1972).
4. RR Springer, PM Stevens. The influence of PEEP on survival of patients in respiratory failure: A retrospective analysis. *Am J Med* 66:196-200 (1979).
5. JA Weigelt, RA Mitchell, WH Snyder. Early positive end-expiratory pressure in the adult respiratory distress syndrome. *Arch Surg* 114:497-501 (1979).
6. BA Shapiro, RD Cane, RA Harrison. Positive end-expiratory pressure therapy in adults with special reference to acute lung injury: A review of the literature and suggested clinical correlations. *Crit Care Med* 12:127-141 (1984).
7. JK Stoller. Respiratory effects of positive end-expiratory pressure. *Respir Care* 33:454-463 (1988).
8. G Hedenstierna, J Santesson, O Norlander. Airway closure and distribution of inspired gas in the extremely obese, breathing spontaneously and during anesthesia with IPPV. *Acta Anaesthesiol Scand* 20:334-340 (1976).
9. A Kumar, KJ Falke, B Geffin, CF Aldredge, et al. Continuous positive pressure ventilation in acute respiratory failure. *N Engl J Med* 283:1430-1436 (1970).
10. JA Katz, GM Ozanne, SE Zinn, HB Fairley. Time course and mechanisms of lung-volume increase with PEEP in acute pulmonary failure. *Anesthesiology* 54:9-16 (1981).
11. PM Suter, HB Fairley, MD Isenberg. Optimum end-expiratory airway pressure in patients with acute pulmonary failure. *N Engl J Med* 292:284-289 (1975).
12. J Malo, A Jameel, LDH Wood. How does positive end-expiratory pressure reduce intrapulmonary shunt in canine pulmonary edema? *J Appl Physiol: Respir Environ Exerc Physiol* 57:1002-1010 (1984).
13. R Dueck, PD Wagner, JB West. Effects of positive end-expiratory pressure on gas exchange in dogs with normal and edematous lungs. *Anesthesiology* 47:359-366 (1977).
14. RM Kacmarek. Positive end-expiratory pressure. In Fundamentals of respiratory care, Pierson DJ and Kacmarek RM (eds.), Churchill Livingstone Publishers, New York, pp 891-920 (1992).
15. DR Dantzker, JP Lynch, JG Weg. Depression of cardiac output is a mechanism of shunt reduction in the therapy of acute respiratory failure. *Chest* 77:636-642 (1980).
16. JL Manzano, MA Blazquez, J Villar, J Villalobos, JJ Manzano. Aplicacion de altos niveles de PEEP durante la ventilacion controlada. *Med Intensiva* 5:10-17 (1981).
17. DJ Pierson. Alveolar rupture during mechanical ventilation: Role of PEEP, peak airway pressure and distending volume. *Respir Care* 33:472-486 (1988).
18. HH Webb, DF Tierney. Experimental pulmonary edema due to intermittent positive pressure ventilation with high inflation pressures. Protection by positive end-expiratory pressure. *Am Rev Respir Dis* 110:556-565 (1974).
19. D Dreyfuss, P Soler, G Basset, G Saumon. High inflation pressure pulmonary edema. Respective effects of high airway pressure, high tidal volume, and positive end-expiratory pressure. *Am Rev Respir Dis* 137:1159-1164 (1988).
20. D Dreyfuss, G Saumon. Role of tidal volume, FRC, and end-inspiratory volume in the development of pulmonary edema following mechanical ventilation. *Am Rev Respir Dis* 148:1194-203 (1993).
21. CD Cook, J Mead, GL Schreiner, NR Frank, JM Craig. Pulmonary mechanics during induced pulmonary edema in anesthetized dogs. *J Appl Physiol* 14:177-186 (1959).
22. AS Slutsky, SM Scharf, R Brown, RH Ingram. The effect of oleic acid-induced pulmonary edema on pulmonary chest wall mechanics in dogs. *Am Rev Respir Dis* 121:91-96 (1980).
23. D Matamis, F Lemaire, A Harf, Brun-Buisson, JC Ansquer, G Atlan. Total respiratory pressure-volume curves in the adult respiratory distress syndrome. *Chest* 86:58-66 (1984).
24. L Gattinoni, A Pesenti. ARDS: the non-homogeneous lung; facts and hypothesis. *Intens & Crit Care Digest* 6:1-4 (1987).
25. TC Corbridge, LDH Wood, GP Crawford, MJ Chudoba, J Yanos, JI Sznajder. Adverse effects of large tidal volume and low PEEP in canine acid aspiration. *Am Rev Respir Dis* 142:311-315 (1990).
26. JG Muscedere, JBM Mullen, K Gan, AS Slutsky. Tidal volume at low airway pressures can augment lung injury. *Am Rev Respir Dis* 149:1327-1334 (1994).

27. J Mancebo. PEEP, ARDS, and alveolar recruitment. *Intens Care Med* 18:383-385 (1992).

28. AS Slutsky. Barotrauma and alveolar recruitment. *Intens Care Med* 19:369-371 (1993).

29. MK Sykes. Does mechanical ventilation damage the lung? *Acta Anaesthesiol Scand* 35 [suppl 95]:35-39 (1991).

30. M Lichtwarck-Aschoff, JB Nielsen, UH Sjöstrand, EL Edgren. An experimental randomized study of five different ventilatory modes in a piglet model of severe respiratory distress. *Intens Care Med* 18:339-347 (1992).

31. B Lachmann. Open up the lung and keep the lung open. *Intens Care Med* 18:319-321 (1992).

32. AS Slutsky. Consensus Conference on mechanical ventilation. Part I. *Intens Care Med* 20:64-79 1994).

33. RR Kirby, JB Downs, JM Civetta, JH Modell, FJ Dannmiller, EF Klein, M Hodges. High level positive end-expiratory pressure (PEEP) in acute respiratory inssuficiency. *Chest* 67:156-163 (1975).

34. RR Demers, RS Irwin, SS Bramen. Criteria for optimum PEEP. *Respir Care* 22:596-601 (1977).

35. M Walkinshaw, WC Shoemaker. Use of volume loading to obtain preferred levels of PEEP. *Crit Care Med* 8:81-86 (1980).

36. GC Carroll, KJ Tuman, B Braverman, et al. Minimal positive end-expiratory pressure (PEEP) may be "best PEEP". *Chest* 93:1020-1025 (1988).

37. RK Albert. Least PEEP: Primum non nocere. *Chest* 87:2-4 (1985).

38. S Benito, F Lemaire. Pulmonary pressure-volume relationship in acute respiratory distress syndrome in adults: Role of positive end-expiratory pressure. *J Crit Care* 5:27-34 (1990).

39. L Gattinoni, L D'Andrea, P Pelosi, G Vitale, A Pesenti, R Fumagalli. Regional effects and mechanism of positive end-expiratory pressure in early adult respiratory distress syndrome. *JAMA* 269:2122-2127 (1993).

40. PE Pepe, LD Hudson, CJ Carrico. Early application of positive end-expiratory pressure in patients at risk for the adult respiratory distress syndrome. *N Engl J Med* 311:281-286 (1984).

TRACHEAL GAS INSUFFLATION

Avi Nahum, M.D., Ph.D.

Assistant Professor of Medicine, University of Minnesota, Department of Pulmonary and Critical Care, St. Paul-Ramsey Medical Center

INTRODUCTION

Tracheal gas insufflation (TGI) is the continuous or phasic insufflation of fresh gas into the central airways for the purpose of improving the efficiency of alveolar ventilation and/or minimizing the ventilatory pressure requirement. TGI usually employs modest flow rates of 2 to 15 L/min. Two mechanisms are responsible for improving the efficacy of conventional tidal breaths during TGI (1-3). First, fresh gas introduced by the catheter during expiration can dilute the CO_2 stored in the series (anatomic) deadspace compartment proximal to its tip. Second, at high catheter flow rates, turbulence generated at the tip of the catheter can enhance gas mixing in regions distant to the catheter tip, thereby contributing to carbon dioxide removal. TGI is unlikely to be very effective when the alveolar as opposed to the series compartment dominates the total physiologic deadspace; yet at small tidal volumes (whenever series deadspace is especially high) or when alveolar ventilation is very low, TGI should be helpful (10). Many investigators have attempted to combine flow through an intratracheal catheter with conventional mechanical ventilation (CMV) techniques (5-13) as well as high frequency jet (14) and oscillatory (15) ventilation. This approach takes advantage of improved ventilatory efficiency due to anatomic deadspace washout as well as enhanced mixing due to catheter turbulence generated to permit reduction in tidal volume (V_T) (11). Combining small tidal volumes with constant flow ventilation, delivers fresh gas further down the airways from the catheter site above the carina, bypassing the high-resistance central zone to achieve eucapnia at low flow rates (16). Furthermore, fresh gas introduced by the catheter improves the "efficacy" of each tidal breath.

During TGI, delivery of fresh gas occurs either throughout the respiratory cycle (continuous flow) or only during a specific segment of it (phasic flow) (5-8). Continuous catheter flow has also been utilized in combination with a shutter or a mechanical ventilator, where closure of an expiratory valve forces catheter flow to delivery all or part of the inspired tidal volume (9-11, 17). Phasic TGI is delivered selectively during inspiration or expiration. During expiratory TGI, catheter flow is timed to occur during all or part of expiration, augmenting alveolar ventilation by flushing CO_2 from the central airways and apparatus deadspace. CO_2 elimination during TGI depends on catheter flow rate ($\dot{V}_c$)

Acute Respiratory Distress Syndrome: Cellular and Molecular Mechanisms and Clinical Management
Edited by Matalon and Sznajder, Springer Science+Business Media New York, 1998

275

because fresh gas flushes a greater portion of the proximal deadspace at higher flow rates. Moreover, at higher flow rates, turbulence generated at the catheter tip may enhance distal gas mixing. The volume of fresh gas introduced into the trachea during TGI depends on expiratory time (T_E) and $\dot{V}_c$. At a certain $T_E \times \dot{V}_c$, fresh gas completely sweeps the proximal anatomic dead space during expiration. At that point, increasing $\dot{V}_c$ most likely does not dilute the CO_2 residing in the series deadspace any further. This operational characteristic of TGI and the fact that the decrease in $PaCO_2$ caused by a reduction in total physiologic deadspace fraction (V_D/V_T) is much less at lower V_D/V_T, limits the decrement in $PaCO_2$ afforded by TGI at high $\dot{V}_c$ (14). Nevertheless, at high $\dot{V}_c$, $PaCO_2$ continues to decrease with increasing $\dot{V}_c$, but at a slower rate (4, 18, 19). Once the series deadspace is flushed completely by the fresh gas during expiration, the flow dependence of $PaCO_2$ is thought to be secondary to enhanced turbulent mixing in the airways distal to the catheter tip (1, 20-22).

OPERATIONAL CHARACTERISTICS

During TGI, more distal catheter placement augments ventilatory breaths as TGI can flush a greater portion of the deadspace proximal to the catheter tip during exhalation (1, 3). Advancing the catheter towards the carina also advanced the turbulence zone generated by the catheter closer to the periphery during both phases of the respiratory cycle, thereby improving the efficacy of TGI. In normal dogs that exact location of the catheter tip (within a few centimeters of the carina) did not prove to be crucial to the efficacy of the TGI in augmenting alveolar ventilation (1). If applicable to patients, this would simplify the clinical application of TGI as bronchoscopically guided catheter position in critically ill patients may not be necessary.

TGI increases end-expiratory lung volume in three ways (1). First, part of the momentum of the discharging jet stream is transferred to the alveoli (17). Second, placement of the catheter within the trachea decreases its cross-sectional area and increases expiratory resistance and delays emptying. Third, catheter flow through the endotracheal tube, expiratory circuit, and expiratory valve during expiration can build a back pressure that impedes expiratory flow from the lung.

Because the major mechanisms of TGI is flushing of the proximal anatomic deadspace free of CO_2, if lung deflation takes place during expiration secondary to dynamic hyperinflation, the CO_2 that is constantly exhaled by the lung can decrease the efficacy of TGI. We tested this possibility in normal dogs, and our data suggest that the volume of fresh gas insufflated by the catheter, rather than $\dot{V}_c$, determines the efficacy of TGI (8). This implies that if TGI is used to augment alveolar ventilation during inverse ratio ventilation, one needs to apply progressively higher flow rates as the expiratory time is shortened.

Currently there is not a standard method of introducing the insufflation catheter into the trachea. In most human studies a small caliber catheter was introduced through an angled side-arm adapter attached to the endotracheal tube (ETT) and positioned just above the main carina (23-25). Catheter placement was usually performed with bronchoscopic guidance or estimated from a recent chest roentgenogram. This type of a system is simple to construct and can be duplicated in most intensive care units but suffers from some drawbacks. Placement of a catheter through the endotracheal tube interferes with suctioning of patients. Moreover, the catheter is not fixed in space and may cause bronchial mucosal injury if it whips within the trachea at high flows. New ETT designs that incorporate channels within the endotracheal tube wall would solve these problems and simplify application of TGI. Isabey and colleagues embedded small capillaries in the walls of an ETT for tracheal injection of gas (27). The same group have also used this modified ETT to deliver high velocity jets of O_2 at the carinal orifice in order to prevent arterial O_2 desaturation during suctioning (28). The modified ETT has been used in animal models (29) and can be used to perform TGI in patients. Most likely future clinical applications of TGI will use a modified ETT that incorporates the catheter in its wall attached to a standardized circuit for gas delivery.

MONITORING DURING TGI

Airway Opening Pressure and Lung Mechanics

Under baseline conditions, mean airway opening pressure ($\bar{P}_{ao}$) and mean alveolar pressure ($\bar{P}_A$) are related by the following expression:

$$\bar{P}_{ao} = \bar{P}_A + \{ \dot{V}_E/60 \times (R_E - R_I)\}$$

where R_E and R_I are the expiratory and inspiratory resistances, respectively, and $\dot{V}_E$ is the minute ventilation (30). Consequently, during mechanical ventilation monitoring P_{ao} gives useful information regarding P_A. During TGI, however, the jet stream increases flow through the ventilator circuit during expiration and creates a region where bi-directional flows exist. Both effects change the resistance characteristics of the respiratory system and modifies the relationship between $\bar{P}_{ao}$ and $\bar{P}_A$ observed at baseline ($\dot{V}_c = 0$ L/min) (18). Since R_E increases during TGI, $\bar{P}_{ao}$ tends to underestimate $\bar{P}_A$ (and end-expiratory lung volume) when the system is switched from baseline to TGI conditions. In an experimental study, monitoring tracheal pressure 2 cm beyond the tip of the catheter seemed to gauge lung volume changes at end-expiration accurately, suggesting that tracheal pressures should be monitored beyond the jet stream during TGI (1). During pan-expiratory TGI, catheter flow ceases during inspiration and inspiratory P_{ao} provides useful information regarding P_A as during CMV. In contrast during continuous TGI, inspiratory P_{ao} (measured at the tip of the ETT) differs from the tracheal pressure (P_{trach}, measured distal to catheter orifice). The magnitude of $P_{ao} - P_{trach}$ during inspiration depends on the flow through the circuit spanning the two pressure measurement points and is usually less than 3 cmH$_2$O. As the ventilator set V_T is decreased with increasing $\dot{V}_c$ the difference between P_{ao} and P_{trach} approaches zero. However, during expiration under both continuous and expiratory TGI conditions catheter flow pressurizes the respiratory system and as a result P_{ao} underestimates P_{trach}. The magnitude of $P_{ao} - P_{trach}$ during expiration increases with $\dot{V}_c$ and most likely depends on the geometry of the system and the orientation of the catheter with respect to the trachea.

TGI may also interfere with the ability of the clinician to measure lung mechanics. Respiratory system compliance and auto-PEEP measurements require application of a pause at end-inspiration and at end-expiration, respectively. If catheter flow continues during these measurements, the pressure within the respiratory system builds up with time. Consequently, a plateau pressure cannot be obtained and if unnoticed alveolar pressure can increase to hazardous levels. During pan-expiratory TGI depending on the timing and nature of the signal that gates the solenoid to divert catheter flow to atmosphere these measurements can still be made safely. Nevertheless, it is advisable to test the TGI-ventilator system using a mechanical lung model under controlled conditions prior to measuring lung mechanics at the bedside.

Tidal Volume

Whenever catheter flow is delivered during inspiration it contributes to total inspired V_T. The contribution to total inspired V_T is eliminated if TGI is timed to occur only during expiration (37). Even then decompression of the TGI circuit into the ventilator circuit during inspiratory phase of solenoid closure contributes to total inspired V_T (31). In most TGI circuits, however, this volume is rather small ($\approx$ 10 - 20 ml at $\dot{V}_c$ of 10 L/min). These problems are obviated if an independent measure of V_T such as inductive plethysmography is used. The effect of TGI on total inspired V_T depends on mode of operation of the ventilator.

Flow-controlled volume-cycled ventilation. During continuous TGI total inspired V_T is composed of two components: that delivered by the ventilator (V_{Tv}) and the catheter (V_{Tc}). The total inspired V_T is then given by the sum of these two components ($V_T = V_{Tv} + V_{Tc}$). The contribution of continuous TGI to total inspired V_T can be estimated from the duration of inspiration (T_I) and $\dot{V}_c$ as: $V_{Tc} = T_I \times \dot{V}_c$. Consequently, during flow-controlled volume-cycled ventilation total inspired V_T can be maintained relatively constant during continuous TGI by decreasing the ventilator set V_T by an amount equal to V_{Tc} (23, 31).

Pressure-controlled ventilation. During pressure-controlled ventilation (PVC) application of TGI does not change the total inspired V_T provided TGI does not pressurize the respiratory system beyond the set-pressure. As $\dot{V}_c$ is increased the ventilator delivered V_T declines but the total inspired V_T remains the same (12, 19) The respiratory system behaves in this manner as long as V_{Tc} is less than the V_T generated by the set-pressure under PCV conditions without TGI. If V_{Tc} exceeds the V_T generated by PCV in the absence of TGI, then TGI will overpressurize the circuit and peak P_{ao} will be greater than that produced by the ventilator set-pressure. Almost all ventilators allow pressures higher than that generated by the set-pressure as long as P_{ao} remains below the high pressure limit of the ventilator. Consequently, excessive pressures can be produced within the respiratory system if the product $\dot{V}_c \cdot T_I$ is too large. When this happens P_{ao} time profile becomes a hybrid of PCV and constant-flow volume-cycled ventilation, resembling that generated during volume-assured pressure support ventilation. This problem can be circumvented by introducing a pressure-release valve into the ventilator circuit that dumps circuit pressure above a set threshold (32).

Monitoring Efficacy of TGI

The primary effect of TGI is on physiologic dead space due to a reduction in anatomic dead space. However, the clinically monitored parameter at the bedside is usually $PaCO_2$ or capnography. TGI modifies the profile of expired capnogram as fresh gas delivered by the catheter dilutes the CO_2 exhaled from the lungs. The capnogram by necessity measures exhaled CO_2 at the tip of the ETT and as a result monitors the exhaled CO_2 with a time (and volume) lag relative to the catheter tip. Nevertheless, exhaled capnogram can be used qualitatively to gauge the completeness of expiratory washout. Clearly, if end-tidal PCO_2 declines to very low values (< 3 mmHg) expiratory washout is most likely complete and further increments in $\dot{V}c$ may not impact $PaCO_2$ greatly. However, if more CO_2-laden gas is removed from the periphery of the lung by the distal effects of TGI as $\dot{V}_c$ is increased, end-tidal PCO_2 may remain elevated or actually rise. Consequently, the relationship between the efficacy of TGI and the capnographically measured CO_2 profile is quite complex. Nevertheless, certain observations can be made based on capnographically observed measurements.

The fraction of unperfused alveoli can be estimated in terms of $PaCO_2$ and average end-tidal PCO_2 ($P_{ET}CO_2$) as ($PaCO_2 - P_{ET}CO_2$)/$PaCO_2$ (4). This ratio can be readily measured at the bedside by capnography and arterial-blood gas analysis. Since increasing the fraction of unperfused alveoli decreases the efficacy of TGI (as measured by percentage change in V_D and $PaCO_2$ relative to baseline conditions), the ratio $(PaCO_2 - P_{ET}CO_2)/PaCO_2$ may provide useful clinical information. Indeed, in both animal (4) and human studies (25, 26) this ratio was closely correlated ($r \approx 0.70$) with observed percentage reduction in $PaCO_2$ from baseline during TGI. The exact role of capnography to monitor the efficacy of TGI remains to be established, but available data suggest that it may be used to optimize $\dot{V}_c$ and ventilator settings.

TGI and Ventilator Interactions

Since TGI introduces an external flow source independent of the ventilator it can adversely effect the ability of the ventilator to monitor pressures and volumes and may cause the ventilator to alarm incessantly. Presence of catheter flow during expiration disables the monitoring role of the expiratory pneumotachograph of the ventilator causing some ventilators to alarm when the difference between the measured inspired and exhaled volumes exceed a certain value. More importantly, presence of an external flow that can pressurize the ventilator circuit interferes with the ability of the ventilator to detect a leak.

CLINICAL APPLICATIONS

The effect of a given TGI-induced change in V_D and $PaCO_2$ depends on the $PaCO_2$ and V_D/V_T values prior to the initiation of TGI, and the effect of TGI on CO_2 production (4). For a given fractional change in V_D, the percentage change in $PaCO_2$ increases dramatically as V_D/V_T exceeds 0.70 (18). This implies that, the effect of TGI on $PaCO_2$ is amplified as the respiratory system is allowed to operate at higher V_D/V_T (i.e., permissive hypercapnia). Consequently, TGI becomes more effective in decreasing $PaCO_2$ in the setting of hypercapnia (18).

In patients with acute lung injury (ALI) part of the deadspace resides in the alveoli as alveolar deadspace. The alveolar gas originating from those ventilated but hypoperfused lung regions are CO_2-poor. Consequently, gas expired from alveolar dead space dilutes CO_2-laden gas residing in the proximal anatomic dead space. Consequently, the impact of washing proximal dead space free of CO_2 on alveolar ventilation diminishes. Adopting a permissive hypercapnia strategy increases the amount of CO_2 that can be removed from the proximal anatomic dead space and counterbalances the decreased CO_2 removal efficacy of TGI caused by increased alveolar dead space.

As an adjunct to mechanical ventilation, TGI may be an effective tool to limit the extent of hypercapnia and/or to control the rate of rise of $PaCO_2$ during the pressure-targeted mechanical ventilation strategy by increasing the efficiency of each tidal breath. In this role, TGI has been applied successfully to mechanically ventilated patients. Ravenscraft and colleagues administered TGI to 8 patients with acute respiratory failure while maintaining total inspired V_T and respiratory rate (f) constant (26). In this study, TGI decreased $PaCO_2$ by 15% (from 53.1 ± 3.4 mmHg to 45.0 ± 2.1 mmHg) at a flow rate of 6 L/min. Similarly, Nakos and colleagues (26) were able to decrease $PaCO_2$ by 25% (from 46.0 ± 4.8 mmHg to 34.5 ± 6.2 mmHg) by using TGI at 6 L/min while maintaining $\dot{V}_E$ constant. The same group found a similar flow-dependent reduction in $PaCO_2$ during TGI in mechanically ventilated COPD patients (24). In COPD patients with tracheostomies, however, TGI was ineffective in reducing $PaCO_2$ (from 56 ± 4 mmHg at baseline to 53 ± 5 mmHg at $\dot{V}_c$ of 6 L/min) due to the reduction in flushable anatomic deadspace (24). Recently, Kuo and colleagues (25) studied the effects of TGI in 20 ARDS patients who had baseline ($\dot{V}_c = 0$ L/min) $PaCO_2$ of

55 ± 8 mmHg. TGI decreased $PaCO_2$ by 13% and 17% (by 7.2 and 9.1 mmHg) at $\dot{V}_c$ of 4 and 6 L/min, respectively. Such modest reductions of $PaCO_2$ reflect the predominance of alveolar over series dead space in conditions of severe parenchymal lung disease. However, using smaller tidal volumes and permissive hypercapnia should increase baseline $PaCO_2$ and accentuate the relative efficacy of TGI. Belghith and colleagues studied the effects of TGI in six ARDS patients with significant hypercapnia at baseline (33). In their study TGI at 4 L/min on average decreased $PaCO_2$ by 24 mmHg from 108 ± 32 to 84 ± 26 mmHg. Similarly, Kalfon and colleagues (34), using pan-expiratory TGI __- 15 L/min decreased P_aCO_2 from 76 ± 4 mmHg to $53 \pm$ mmHg in seven ARDS patients. These studies indicate that TGI is an effective tool to reduce P_aCO_2 during permissive hypercapnia

As an adjunct to mechanical ventilation, TGI can also be used to limit ventilatory distending forces while maintaining $PaCO_2$ constant. Using TGI in a canine oleic acid-induced pulmonary edema model, adequate alveolar ventilation was maintained at much smaller V_T and pressures, where the valved TGI catheter (which functioned as the ventilator in this setting) achieved the same ventilatory task at 35% of the V_T and 70% of elastic end-inspiratory pressure. Using a similar strategy, Kolobow and colleagues demonstrated that they could adequately ventilate sheep that had undergone resection of 88% of their lung tissue, without resorting to excessive $\dot{V}_E$ or airway pressures (10). Nakos and colleagues (23) used TGI at 6 - 8 L/min to decrease V_T and peak P_{ao} by 25% and 20%, respectively, while maintaining $PaCO_2$ and f constant in 7 patients with ALI.

In normal dogs, TGI tended to reduce venous admixture ($\dot{Q}_V/\dot{Q}_T$) and increase PaO_2 (1, 12, 19), a tendency that was also observed in 8 critically ill patients (only one with ALI), however, the rise in PaO_2 was not significant (26). In the clinical setting, continuous-TGI (6 L/min) consistently increased end-expiratory lung volume (FRC) slightly (62 ± 16 ml at $\dot{V}c$ of 6 L/min in 4 patients) (26). The response in oxygenation, however, was variable: PaO_2/PAO_2 increased in 4 of 8 patients, remained the same in 2, and decreased in 2. In the study of Kuo and colleagues (25) TGI did not significantly affect PaO_2 of ARDS patients. Similarly, in 6 oleic-acid injured dogs, TGI did not improve PaO_2 and $\dot{Q}_V/\dot{Q}_T$ when V_T and FRC were kept constant (18). However, Nakos and colleagues (23) observed a significant increase in PaO_2 during TGI at 6 L/min (from 82 ± 6 mmHg to 89 82 ± 6 mmHg, $p < 0.05$). Similarly, Kalfon and colleagues observed a significant improvement in oxygenation with a concomitant decrease in cardiac output during TGI. In both studies, V_T and f was maintained constant but no attempt was made to control FRC during TGI. Current data suggest that TGI does not impact oxygenation provided total inspired V_T and FRC is not augmented during application of TGI.

Minute ventilation sparing effect of TGI may also be used to reduce work of breathing of some intubated patients. However, TGI may impair the ability of some patients to trigger the ventilator, because the patient's inspiratory effort must first outstrip catheter flow and overcome the dynamic hyperinflation caused by TGI in order to lower airway opening pressure below the trigger threshold (35). The net effect of TGI on work of breathing would depend on the interactions between TGI and the ventilator as well as the efficiency of TGI in decreasing deadspace and $\dot{V}_E$ requirements. Obviously, incorporating TGI into a flow-by system would preserve the combined benefits of flow-triggering and improved gas exchange associated with TGI. Patients with neuromuscular weakness that have relatively normal lung parenchyma and retain CO_2 because they can only generate small tidal volumes (i.e., proximal anatomic dead space contributes significantly to V_D/V_T) are excellent candidates for TGI.

POTENTIAL COMPLICATIONS

Although a promising adjunct to CMV, TGI is not without potential complications. When high flows are delivered into the airways, potentially any obstruction to outflow of gas could cause overinflation of the lungs within seconds and may cause pneumothorax, pulmonary venous air embolism, and/or hemodynamic compromise. Esophageal pressure and/or chest wall monitoring may be required to monitor changes in lung volume. A second concern is bronchial mucosal damage due to impact of the jet stream onto the bronchial mucosa as well as the possible physical impact of the catheter tips from oscillations secondary to high flows. The force created by the jet stream impacting on the surface can be quite high and account for the bronchial mucosal damage observed in experiments during constant flow ventilation (36). Proper humidification of the inspired gas is essential at such high flow rates. Long term use of TGI may result in inspissation or retention of secretions, especially if the insufflated gas is not adequately humidified.

REFERENCES

1.   Nahum A, Ravenscraft SA, Nakos G, Adams AB, Burke WC, Marini JJ. Effect of catheter flow direction on CO_2 removal during tracheal gas insufflation in dogs. J Appl Physiol 1993;75:1238-1246.
2.   Slutsky AS, Watson J, Leith DE, Brown R. Tracheal insufflation of 0_2 (TRIO) at low flow rates sustains life for several hours. Anesthesiology 1985;63:278-286.
3.   Nahum A, Ravenscraft SA, Adams AB, Marini JJ. Distal effects of tracheal gas insufflation: changes with catheter position and oleic acid lung injury. J Appl Physiol 1996;81:1121-1127.
4.   Nahum A, Shapiro RS, Ravenscraft SA, Adams AB, Marini JJ. Efficacy of expiratory tracheal gas insufflation in a canine model of lung injury. Am J Respir Crit Care Med 1995;152:489-495.
5.   Jonson B, Similowski T, Levy P, Viires N, Pariente R. Expiratory flushing of airways: A method to reduce deadspace ventilation. Eur Respir J 1990;3:1202-1206.
6.   Gilbert J, Larsson A, Smith B, Bunegin L. Intermittent-flow expiratory ventilation (IFEV): Delivery technique and principles of action - a preliminary communication. 1991;25:451-456.
7.   Burke WC, Nahum A, Ravenscraft SA, et al. Modes of tracheal gas insufflation. Comparison of continuous and phase-specific gas injection in normal dogs. Am Rev Respir Dis 1993;148:562-568.
8.   Ravenscraft SA, Shapiro R, Nahum A, et al. Tracheal gas insufflation: Catheter effectiveness is determined by expiratory flush volume. Am J Respir Crit Care Med 1996;153:1817-1824.
9.   Kolobow T, Powers T, Mandava S, et al. Intratracheal pulmonary ventilation (ITPV): control of positive end-expiratory pressure at the level of the carina through the use of a novel ITPV catheter design. Anesth Analg 1994;78:455-461.
10. Muller E, Kolobow T, Mandava S, et al. On how to ventilate lungs as small as 12% of normal. Intratracheal pulmonary ventilation (ITPV). A new mode of pulmonary ventilation (abstract). Am Rev Resp Dis 1991;143:A693.
11. Sznajder JI, Becker CJ, Crawford GP, Wood LDH. Combination of constant flow and continuous positive pressure ventilation in canine pulmonary edema. J. Appl. Physiol. 1989;67:817-823.

12. Nahum A, Ravenscraft SA, Nakos G, et al. Tracheal gas insufflation during pressure-control ventilation. Effect of catheter position, diameter, and flow rate. Am Rev Respir Dis 1992;146:1411-1418.

13. Lehnert BE, Oberdoerster EG, Slutsky AS. Constant-flow ventilation of apneic dogs. J Appl Physiol 1982;53:483-489.

14. Huafeng W, Shi-Ao J, Zhicheng M, Haosheng B, Xiuyun B. Experimental study of high-frequency two-way jet ventilation. Crit Care Med 1992;20:420-423.

15. Dolan S, Derdak S, Solomon D, Farmer C, Johanningman J, Gelineau J. Tracheal gas insufflation combined with high-frequency oscillatory ventilation. Crit Care Med 1996;24:458-465.

16. Venegas JG, Yamada Y, Hales CA. Contributions of diffusion jet flow and cardiac activity to regional ventilation in CFV. J Appl Physiol 1991;71:1540.

17. Nahum A, Sznajder JI, Solway J, Wood LDH, Schumacker PT. Pressure, flow, and density relationships in airway models during constant-flow ventilation. J Appl Physiol 1988;64:2066.

18. Nahum A, Chandra A, Niknam J, Ravenscraft SA, Adams AB, Marini JJ. Effect of tracheal gas insufflation on gas exchange in canine oleic acid-induced lung injury. Crit Care Med 1995;23:348-356.

19. Nahum A, Burke W, Ravenscraft SA, et al. Lung mechanics and gas exchange during pressure-controlled ventilation in dogs: augmentation of CO_2 elimination by an intratracheal catheter. Am Rev Respir Dis 1992;146:965-973.

20. Eckmann DM, Gavriely N. Intra-airway CO_2 distribution during airway insufflation in respiratory failure. J Appl Physiol 1995;78:546-554.

21. Gavriely N, Eckmann DM, Grotberg JB. Intra-airways gas transport during high-frequency chest vibration with tracheal insufflation in dogs. J Appl Physiol 1995;79:243-250.

22. Gavriely N, Eckmann DM, Grotberg JB. Gas exchange by intratracheal insufflation in a ventilatory failure dog model. J Clin Invest 1992;90:2376-2383.

23. Nakos G, Zakinthinos S, Kotanidou A, Roussos C. Tracheal gas insufflation reduces the tidal volume while $PaCO_2$ is maintained constant. Intensive Care Med 1994;20:407-413.

24. Nakos G, Lachana A, Prekates A, et al. Respiratory effects of tracheal gas insufflation in spontaneously breathing COPD patients. Intensive Care Med 1995;21904-912.

25. Kuo P-H, Wu H-D, Yu C-J, Yang S-H, Lai Y-L, Yang P-C. Efficacy of tracheal gas insufflation in acute respiratory distress syndrome with permissive hypercapnia. Am J Respir Crit Care Med 1996;154:612-616.

26. Ravenscraft SA, Burke WC, Nahum A, et al. Tracheal Gas Insufflation Augments CO_2 Clearance During Mechanical Ventilation. Am Rev Respir Dis 1993;148:345-351.

27. Isabey D, Boussignac G, Harf A. Effect of air entrainment on airway pressure during endotracheal gas injection. J Appl Physiol 1898;67:771-779.

28. Brochard L, Mion G, Isabey D, et al. Constant-flow insufflation prevents arterial oxygen desaturation during endotracheal suctioning. Am Rev Resp Dis 1991;144:395-400.

29. Pinquier D, Pavlovic D, Boussignac G, Aubier M, Beafils F. Benefits of low pressure multichannel endotracheal ventilation. Am J Respir Crit Care Med 1996;154:82-90.

30. Marini JJ, Ravenscraft SA. Mean airway pressure: physiologic determinants and clinical importance--Part 1: Physiologic determinants and measurements. Crit Care Med 1992;20:1461-72.

31. Nahum A, Ravenscraft SA, Adams AB, Marini JJ. Inspiratory tidal volume sparing effects of tracheal gas insufflation in dogs with oleic acid-induced lung injury. J Critical Care 1995;10:115-121.

32. Gowski D, Delgado E, Miro AM, et al. Tracheal gas insufflation during pressure-control ventilation: effect of using a pressure relief valve. Crit Care Med 1997;25:145-152.

33. Belghith M, Fierobe L, Brunet F, Monchi M, Mira J-P. Is tracheal gas insufflation an alternative to extrapulmonary gas exchangers in severe ARDS? Chest 1995;107:416-419.

34. Kalfon P, Umamaheswara R, Gallart L, et al. Permissive hypercapnia with and without expiratory washout in patients with severe acute respiratory distress syndrome. Anesthesiology 1997;87(1):6-17.

35. Hoyt JD, Marini JJ, Nahum A. Effect of tracheal gas insufflation on demand valve triggering and total work during continuous positive airway pressure ventilation. Chest 1996;110:775-783.

36. Sznajder Jl, Nahum A, Crawford G, Pollak ER, Schumacker PT, Wood LDH. Alveolar pressure inhomogeneity and gas exchange during constant-flow ventilation in dogs. J Appl Physiol 1989;67:1489-1492.

MECHANICAL VENTILATION IN ARDS: GOOD OR BAD NEWS?

PAOLO PELOSI, MD

Istituto di Anestesia e Rianimazione, Universita' di Milano, Ospedale Maggiore IRCCS, Milano, Italy

INTRODUCTION

Despite an enormous progress in the technology of ventilators, the improvement in the diagnostic tools and in pharmacological treatment the Adult Respiratory Distress Syndrome (ARDS) is characterized by high mortality rate [1,2,3]. Although necessary, mechanical ventilation is commonly considered to be one of the main causes of further lung injury to diseased lungs, reducing the possibility of recovery. For this reason the iatrogenic effect of mechanical ventilation has been named ventilator-associated lung injury (VALI). VALI is the consequence of a sustained increase in alveolar pressure ('barotrauma'), alveolar distension ('volotrauma') or alveolar collapse and decollapse with cycling during inspiration and expiration ('shear stress trauma'). Recent clinical studies suggest that the optimal ventilatory treatment should combine the use of a reduced tidal volume with consequent permissive hypercapnia to reduce volotrauma, low inspiratory pressures to reduce barotrauma and an adequate level of positive end-expiratory pressure to recruit as much collapsed parenchyma as possible to reduce shear stress trauma [4,5]. However, different etiologies leading to ARDS and time may produce different alterations in the lung structure with consequent different responses to the ventilatory treatment.

In this chapter, we will discuss: 1) the main causes of VALI; 2) the current recommendations regarding ventilatory strategies in ARDS; 3) recent insights in the physiopathology of ARDS and the effects of different ventilatory strategies on the lung structure; 4) the main clinical consequences of these new findings.

BAD NEWS: VENTILATOR-ASSOCIATED LUNG INJURY

Mechanical ventilation of the diseased lung is thought to be one of the most important factors inducing lung damage. VALI can be defined as the presence of extra-alveolar air, that is emphysema-like lesions, tension cysts, pneumothorax and so on, in locations where they are not normally found in patients receiving mechanical ventilation [6]. All forms of VALI develop after rupture of an over-distended alveolus and, in general, are the consequence of

Acute Respiratory Distress Syndrome: Cellular and Molecular Mechanisms and Clinical Management
Edited by Matalon and Sznajder, Springer Science+Business Media New York, 1998

285

a sustained increase in alveolar pressures (barotrauma), alveolar distension (volotrauma) or alveolar collapse and decollapse with cycling during inspiration and expiration (shear stress trauma).

Barotrauma

Barotrauma has been attributed to mean airway pressure, peak pressure or positive end-expiratory pressure (PEEP). A number of animal [7,8] and human studies [9,10,11] documented an association between the incidence of barotrauma and high inspiratory peak pressures. However, pressures alone may be an important risk only to the extent that they reflect the transalveolar pressure and thus alveolar distension. Transalveolar pressure depends on the difference between alveolar and pleural pressures. Pleural pressure, however, depends on the mechanical properties of the chest wall, that is, at the same alveolar pressure, high chest wall compliance implies high transalveolar pressure, whereas low chest wall compliance implies low transalveolar pressures.

Volotrauma

Recent experiments on animals showed that the large excursion in volume, and not the pressure itself, may be the main determinant of VALI. This process seems to be exacerbated by the presence of previous lung injury. Thus VALI barotrauma would be better named, by some authors, as volotrauma [12,13,14,15].

Shear Stress Trauma

At a transpulmonary pressure of 30 cmH$_2$O, the pressure tending to expand an atelectatic region surrounded by fully expanded lung would be approximately 140 cmH$_2$O [16]. Such forces may well be the major cause of structural damage and the cause of the release of mediators from disrupted lung during mechanical ventilation in ARDS [17]. During ventilation part of the lung may continuously collapse and decollapse, especially if inadequate levels of PEEP are applied. As a consequence, these collapsing and decollapsing regions should undergo very high stretch forces which, with time, may produce VALI. Table 1 summarizes the possible positive and negative effects on the lung structure caused by the application of different pressures or volumes and PEEP during mechanical ventilation in ARDS.

STATE OF THE ART

In 1993, an international panel of investigators proposed some principles for the ventilatory treatment on ARDS [18]. Their recommendations were based on the physiological principles exposed above and may be summarized as follows: 1) Clinicians should use only ventilatory modes that have been proven and with which they are familiar. 2) An acceptable arterial saturation should be targeted (usually higher or equal to 90%). 3) A plateau pressure higher than 35 cmH$_2$O should be avoided, whenever it is possible. When plateau pressures exceeds this value, reductions in tidal volume to as low as 5 ml/Kg should be considered. 4) Permissive hypercapnia is acceptable, unless contraindicated, if a "normal" PaCO$_2$ is not achievable within the limit for the plateau pressures. Rapid changes in PaCO$_2$ should be avoided. 5) PEEP may have a role in preventing lung damage, recruiting as much atelectatic lung parenchyma as possible. 6) FiO$_2$ should be minimized. 7) When oxygenation is inadequate, sedation and paralysis and position changes may be considered in the ventilatory management.

Table 1. Positive and negative effects of different tidal volumes (i.e., inspiratory pressures) and positive end-expiratory pressures (PEEP) on lung structure

	POSITIVE	NEGATIVE
HIGH VOLUME	alveolar recruitment at end-inspiration	volotrauma, barotrauma shear stress trauma
LOW VOLUME	low volotrauma, low barotrauma, low shear stress trauma	alveolar derecruitment at end-inspiration, alveolar derecruitment at end-expiration hypercapnia
HIGH PEEP	keep open the lung at end-expiration, low shear stress trauma	volotrauma, barotrauma
LOW PEEP	low volotrauma, low barotrauma	alveolar derecruitment at end-expiration high shear stress trauma
LOW PEEP and LOW VOLUME	low volotrauma, low barotrauma low shear stress trauma	alveolar derecruitment at end-expiration alveolar derecruitment at end-inspiration hypercapnia
LOW PEEP and HIGH VOLUME	alveolar recruitment at end-inspiration	high shear stress trauma alveolar derecruitment at end-expiration volotrauma, barotrauma
HIGH PEEP and LOW VOLUME	low volotrauma, low barotrauma low shear stress trauma keep open the lung at end-expiration alveolar recruitment at end-inspiration?	hypercapnia
HIGH PEEP + HIGH VOLUME	keep open the lung at end-expiration alveolar recruitment at end-expiration?	volotrauma, barotrauma
PEEP+ LOW and HIGH VOLUME	low volotrauma, low barotrauma low shear stress trauma alveolar recruitment at end-inspiration keep open the lung at end-expiration	volotrauma ? barotrauma? shear stress trauma?

RECENT INSIGHTS IN THE PHYSIOPATHOLOGY OF ARDS

Lung Structure in Early ARDS

ARDS is characterized by pulmonary infiltrates, which were previously considered to be rather homogeneously distributed throughout the lung, as evidenced by conventional chest radiograph. However using the computed tomography (CT) technology, it has been shown that in ARDS lung the localization of radiographic densities is primarily located in the dependent lung regions, i.e. the vertebral regions (lower) in supine position. In contrast, the non dependent regions, i.e. the sternal regions (upper) in supine position, seem, at least to visual inspection, quite normal [19]. Although not characteristic of all the patients these features may be considered representative of the majority of them [20].

This configuration of dishomogeneous lung is similar and only quantitatively different from what has been reported in normal subjects after anesthesia and paralysis. In fact during anesthesia, normal lungs appear rather dishomogeneous, since they are characterized by the presence of a variable amount of densities in the dependent part of the lung, occupying, on average, 8-10% of the entire lung field. The appearance of these densities has been attributed to the development of compression atelectasis, i.e. loss of gas in the most dependent part of the lung. This structural alteration is associated with increased right-to-left pulmonary shunting and dead space [21].

In ARDS patients the densities may account for up to 70-80% of the lung field, depending on the severity of the respiratory failure. The proportion of the lung which can be consequently ventilated may be reduced to almost 20-30% of a normal lung. Using a regional CT analysis of the lung, we found that the alveolar dimensions, expressed as gas-tissue ratio, were markedly reduced both in the ventral and in the dorsal regions, following a gravitational gradient. In contrast, all the lung parenchyma is homogeneously affected by the disease, with no part of the lung being healthy and edema did not exhibit a gravity dependent distribution along the vertical gradient in supine position. Consequently it did not accumulate preferentially in the dependent lung regions, as previously reported in animal models. As the total mass of the ARDS lung is more than twice that of the normal lung, the lung progressively collapses under its own weight, squeezing out the gas from the dependent regions with formation of compression atelectasis [22].

In summary, it is possible to model the ARDS lung, in the early phase of the disease, as composed of three compartments: one affected by the disease but continuously open to gases ('open diseased lung zone'), one fully diseased without any possibility of recruitment ('closed diseased lung zone'), and one composed of collapsed alveoli potentially recruitable with increasing pressure ('recruitable diseased lung zone').

Lung Structure in Late ARDS

Information on late ARDS is astonishingly scarce. It is well known from autopsy studies that the lung structure undergoes deep changes with time, progressing from a florid edema phase in which the basic structure of the lung is intact to a fibrotic phase characterized by reabsorption of edema and deep modifications in the lung structure, with formation of cysts and emphysema-like lesions as an hallmark. The presence of cysts and emphysema-like lesions correlates with the duration of mechanical ventilation and it is prevalent in the dependent regions [23]. However, other authors reported an alternative localization of these lesions in the upper lung [24].

In summary, with time the ARDS lung structure is modified with a reduction in the presence of dependent densities and an increase in emphysema-like lesions.

Effects of Peep and Tidal Volume on Alveolar Distension and Recruitment in Early Ards

Taking the CT scan at end-expiration and end-inspiration at different PEEP levels and tidal volumes, it is possible to measure the distribution of ventilation and the alveolar recruitment. In sedated and paralyzed ARDS patients in supine position the ventilation at a PEEP of 0 cmH$_2$O is preferentially distributed in the upper lung, the ratio between the amount of ventilation in the upper and lower lung being approximately 2.5:1. Increasing PEEP, the distribution of ventilation becomes progressively more homogeneous, the ratio being approximately 1:1 at 20 cmH$_2$O of PEEP. This implies that modifications in regional compliance occur with PEEP and in particular a decrease in compliance in the upper lung and an increase in the lower one. In the upper levels where the weight of the lung is low and atelectatic regions are few, the recruitment is negligible and over-distension is present, whatever the inspiratory pressure is applied. In contrast, in the lowest levels where the weight is higher and atelectatic regions are prevalent, application of PEEP may cause recruitment increasing regional compliance. During mechanical ventilation and tidal volume breath, a considerable part of the lung, thus, continuously collapses and decollapses, especially at low levels of PEEP, when a greater part of the lung is collapsed at end-expiration. This phenomenon occurs most often in the most dependent lung regions, where compression atelectasis are present. With higher PEEP levels, the reopening-collapsing tissue ratio is decreased because the amount of collapsed tissue at end-expiration is reduced [25].

However, these data were obtained maintaining a constant tidal volume, resulting in an increase in peak pressure when PEEP was changed. Thus it is difficult to determine the relative importance of the pressure level reached at end-inspiration (peak pressure) or at end-expiration (PEEP) to changes in ventilation and recruitment. More recently, the distribution of ventilation and recruitment has been evaluated during pressure controlled ventilation [26]. The lung inflation at end-expiration was not affected by application of PEEP, provided the peak inspiratory pressure was maintained constant. Moreover, as during volume controlled ventilation, PEEP redistributed ventilation more homogeneously. This means that PEEP similarly affects distribution of ventilation and recruitment both during volume controlled and pressure controlled mechanical ventilation. More importantly, it has been shown that the amount of recruitment obtained at a defined level of PEEP is a function of the inspiratory peak pressure. This means that more recruitment of lung tissue is obtained at the same PEEP after an inspiratory recruitment maneuver of 45 cmH_2O [27]. In summary, in early ARDS: 1) PEEP recruits part, but not all the atelectatic tissue prevalently located in the dependent lung regions. 2) Both in volume and pressure controlled mechanical ventilation, distribution of ventilation at low levels of PEEP is prevalent in the non dependent part of the lung, while is more homogeneously distributed at higher PEEP levels. 3) During tidal breath part of the lung tissue continuously collapses and decollapses, especially in the dependent lung regions and at lower PEEP levels. 4)An end-inspiratory pressure, up to 45 cmH_2O, is able to recruit new atelectatic regions and it is an important determinant of the amount of recruitment at PEEP.

The Role of Time and Individual Pathology in Determining Changes in the Lung Structure

All the considerations above refer to early phase of ARDS. In fact, the presence of atelectatic tissue, where it is possible to operate the favorable effect of pressures in recruitment, is prevalent in first 7-10 days from the onset of the disease. Since the lung structure deeply changes after some days of mechanical ventilation, it is obvious that the importance of PEEP in recruiting tissue diminish after this period. This does not mean that PEEP is useless in late ARDS to improve oxygenation, since other mechanisms, may be implicated in the improvement of oxygenation [28].

Moreover, ARDS is a syndrome coming from different etiologies, leading to similar symptoms. Recently, we found that patients with ARDS due to a 'direct' insult to the lung, i.e. pneumonia, etc., showed a different mechanical behavior compared to the patients in which ARDS was due to an 'indirect' insult, i.e. sepsis, trauma, peritonitis, etc. [29, 30]. The former group was characterized by lower lung compliance, normal chest wall compliance and low fast response in recruitment with PEEP. The second one was characterized by relatively normal lung compliance, low chest wall compliance, high intra-abdominal pressure, and marked fast response in recruitment with PEEP. This means that different etiologies of ARDS may lead, even in the early phases of the disease, to different pathophysiological changes in the lung structure and different responses to similar ventilatory managements.

GOOD NEWS: CLINICAL IMPLICATIONS

These findings may have an important role in determining the ventilatory strategies in ARDS patients and modifying some general accepted current recommendations. Here we try to summarize some general concepts: 1) ARDS is not a disease but a syndrome, i.e. a sum of symptoms due to different etiologies. Different etiologies may lead to different pathogenetic pathways leading to ARDS with different response in the mechanical ventilation. Thus, a general optimal ventilatory strategy applicable to patients probably does not exist. Indeed, it would be better to tailor it taking into account the pathophysiology of the individual patients and the primitive cause of the disease itself. 2) A plateau airway pressure of 35 cmH$_2$O may be not an absolute value. In fact, since in ARDS patients chest wall may present marked derangements, the airway pressure may not adequately reflect the transalveolar pressure. Lung distension is directly related to transpulmonary pressure. Normal lung tissue generally reach its maximum distension when transpulmonary pressure reaches 30-35 cmH$_2$O. A treatment strategy which uses the airway plateau pressure as an indicator of transpulmonary pressure may overestimate the true transpulmonary pressure. 3) During tidal breath, both during volume controlled and pressure controlled mechanical ventilation, part of the lung continuously collapse and decollapse, especially in the dependent lung regions. At higher PEEP levels the amount of collapsing and decollapsing tissue is reduced. However, application of PEEP is always a compromise, since the reopening of some lung regions is paralleled by over-distension of other regions. Moreover, PEEP alone is not enough to recruit all potentially recruitable alveolar units, at least up to 20 cmH$_2$O. On the other side application of too high levels of PEEP necessary to recruit all the atelectatic tissue, markedly limit the delivery of a tidal volume physiologically acceptable. During tidal breath, both during volume controlled and pressure controlled mechanical ventilation, application of PEEP reduces alveolar compliance in the non dependent regions while improves in the dependent ones. 4) The plateau inspiratory airway pressure, higher than 35 cmH$_2$O may an important determinant of new potential recruitment and may positively interact the recruitment due to PEEP (see table 1). In general, inspiratory pressure is the 'true' recruiting pressure, while PEEP should only maintain the recruitment obtained at end-inspiration. In this view periodic recruiting maneuvers, i.e. the sustained application of a transpulmonary pressure sufficient to approach the total lung capacity, may help to offset the tendency for progressive collapse that occurs when small tidal volumes are used. However, it is not clear the real role of periodic recruiting maneuvers on the lung structure and their effects with time.

In conclusion, many theoretical and practical questions have been raised from the more recent studies on the interactions between ventilator and the lung structure in ARDS. At the moment it is difficult to give a general good 'recipe' for the optimal ventilatory treatment. Everything (high volume, low volume, high PEEP, low PEEP, etc.) may be good or may be bad, the important thing is to think about it!

REFERENCES

1. Lewandowski K, Metz J, Deutschmann C, Preib H, Khufen R, Artigas A, Falke KJ (1995) Incidence, severity and mortality of acute respiratory failure in Berlin, Germany. Am J Respir Crit Care Med 107: 1121-1125
2. Vasilyev S, Schaap RN, Mortensen JD (1995) Hospital survival rates of patients with acute respiratory failure in modern respiratory intensive care units: an international, multicenter, prospective survey. Chest 107: 1083-1088
3. Bernard GR, Artigas A, Brigham KL, Carlet J, Falke K, Hudson L, Lamy M, Le Gall JR, Morris A, Spragg R (1994) The American-European Conference on ARDS. Definitions, mechanisms, relevant outcomes, and clinical trial coordination. Am J Respir Crit Care Med 149: 818-824
4. Hickling KG, Walsh J, Henderson S, Jackson R (1994) Low mortality rate in adult respiratory distress syndrome using low-volume, pressure-limited ventilation with permissive hypercapnia: a prospective study. Crit Care Med 22: 1568-1578

5. Amato MBP, Barbas CSV, Medeiros DM, Lorenzi FG, Kairalla RA, Deheinzelin D, Magaldi RB, De Carvalho CRR (1996) Improved survival in ARDS: beneficial effects of a lung protective strategy. Am J Respir Crit Care Med 153: A531 (abstract)

6. Manning HL (1994) Peak airway pressure: why the fuss? Chest 105: 242-247

7. Kolobow T, Moretti MP, Fumagalli R, Mascheroni D, Prato P, Chen V, Joris M (1987) Severe impairment in lung function induced by high peak airway pressure during mechanical ventilation. Am Rev Respir Dis 135: 312-315

8. Tsuno K, Prato P, Kolobow T (1990) Acute lung injury from mechanical ventilation at moderately high airway pressures. J Appl Physiol 89:956-961

9. de Latorre F, Tomasa A, Klamburg J (1977) Incidence of pneumothorax and pneumomediastinum in patients with aspiration requiring ventilatory support. Chest 72:141-144

10. Peterson G, Baier H (1983) Incidence of pulmonary barotrauma in a medical ICU. Crit Care Med 11:67-71

11. Shanapp LM, Chin DP, Szaflasrski N, Matthay MA (1995) Frequency and importance of barotrauma in 100 patients with acute lung injury. Crit Care Med 23:272-278

12 Dreyfuss D, Basset G, Soler P, Saumon G (1985) Intermittent positive-pressure hyperventilation with high inflation pressures produces pulmonary microvascular injury in rats. Am Rev Respir Dis 132: 880-884

13. Dreyfuss D, Soler P, Basset G, Saumon G (1988) High inflation pressure pulmonary edema. respective effects of high airway pressure, high tidal volume, and positive end-expiratory pressure. Am Rev Respir Dis 137:1159-1164

14. Dreyfuss D, Saumon G (1993) Role of tidal volume, FRC and end-inspiratory volume in the development of pulmonary edema following mechanical ventilation. Am Rev Respir Dis 148:1194-1203

15. Dreyfuss D, Soler P, Saumon G (1995) Mechanical ventilation-induced pulmonary edema. Interaction with previous lung alterations. Am J Respir Crit Care Med 151:1568-1675

16. Mead J, Takishima T, Leith D (1970) Stress distribution in lungs: a model of pulmonary elasticity. J Appl Physiol 28:569-608

17. Muscedere JC, Mullen JBM, Gan K, Slutsky AS (1994) Tidal volume at low pressures can augment lung injury. Am J Respir Crit Care Med 149:1327-1334

18. Slutsky AS (1993) Mechanical ventilation. Chest 104:1833-1859. Also, (1994) Intensive Care Med 20:64-79

19. Gattinoni L, Mascheroni D, Torresin A, Marcolin R, Fumagalli R, Vesconi S, Rossi GP, Baglioni S, Bassi F (1986) Morphological response to positive end-expiratory pressure in acute respiratory failure. Intensive Care Med 12:137-142

20. Tagliabue M, Casella TC, Zincone GE, Fumagalli R, Salvini E (1994) CT and chest radiography in the evaluation of adult respiratory distress syndrome. Acta Radiologica 35:230-234

21. Brismar B, Hedenstierna G, Lundquist (1985) Pulmonary densities during anesthesia with muscular relaxation: a proposal of atelectasis. Anesthesiology 62:422-428

22. Pelosi P, D'Andrea L, Vitale G, Pesenti A, Gattinoni L (1994) Vertical gradient of regional lung inflation in adult respiratory distress syndrome. Am J Respir Crit Care Med 149:8-13

23. Gattinoni L, Bombino M, Pelosi P, Lissoni A, Pesenti A, Fumagalli R, Tagliabue M (1994) Lung structure and function in different stages of severe adult respiratory distress syndrome. JAMA 271:1772-1779

24. Rouby JJ, Lhem E, Martin de Lassale E, Poete P, Bodin L, Finet JF, Callard P, Viars P (1993) Histologic aspects of pulmonary barotrauma in critically ill patients with acute respiratory failure. Intensive Care Med 19:383-389

25. Gattinoni L, Pelosi P, Crotti S, Valenza F (1995) Effects of positive end-expiratory pressure on regional distribution of tidal volume and recruitment in adult respiratory distress syndrome . Am J Respir Crit Care Med 151:1807-1814

26. Crotti S, Mascheroni D, Pelosi P, Valenza F, Lissoni A, Gattinoni L (1997) The inspiratory plateau pressure level affects end-expiratory lung inflation and densities during pressure controlled ventilation: a CT scan study in ARDS patients. Am J Respir Crit Care Med 155: A88 (abstract)

27. Crotti S, Mascheroni D, Pelosi P, Tubiolo D, Chiumello D, Gattinoni L (1997) Distribution of lung inflation and tidal volume during pressure controlled ventilation in ARDS: effects of PEEP. Am J Respir Crit Care Med 155: A87 (abstract).

28. Pelosi P, Crotti S, Brazzi L, Gattinoni L (1996) Computed tomography in adult respiratory distress syndrome: what has it taught us? Eur Respir J 9:1055-1062

29. Pelosi P, Cereda M, Foti G, Giacomini M, Pesenti A (1995) Alterations of lung and chest wall mechanics in patients with acute lung injury: effects of positive end-expiratory pressure. Am J Respir Crit Care Med 152:532-537

30. Pelosi P, Croci M, Chiumello D, Pedoto A, Gattinoni L (1996) Direct or indirect lung injury differently affects respiratory mechanics during acute respiratory failure. Intensive Care Med 22: 105 (abstract)

VENTILATOR-INDUCED LUNG INJURY
IN PATIENTS WITH ARDS

Lorraine N. Tremblay and Arthur S. Slutsky

Mount Sinai Hospital
University of Toronto
600 University Ave., Suite 656A
Toronto, Canada   M5G 1X5

INTRODUCTION

It is difficult to dissect out the contribution of ventilation versus other stimuli in the progression of lung injury in ARDS. Earlier this century, hyaline membranes and the histologic findings currently associated with ARDS were rarely found[1,2]. Following the widespread introduction of positive pressure ventilation into clinical practice, the term "respirator lung" was coined to describe findings of dense cellular infiltrates, pulmonary edema, and hyaline membranes upon post-mortem examination of patients subjected to respiratory support. Subsequent studies have characterized a number of histologic findings typical of lung injury in ventilated patients.

In animals, studies in a variety of species have clearly demonstrated that ventilation at either excessively high end inspiratory lung volumes, or excessively low end-expiratory lung volumes, can both initiate and/or exacerbate lung injury[3-6]. For example, young pigs ventilated with peak inspiratory pressures (PIP) of 40 cmH$_2$O, PEEP of 3-5 cmH$_2$O, RR 20 bpm, and FiO$_2$ of 0.4 for 22 hours developed alveolar hemorrhage, alveolar neutrophil infiltration, alveolar macrophage and type II pneumocyte proliferation, interstitial congestion and thickening, interstitial lymphocyte infiltration, emphysematous change, and hyaline membrane formation - very similar to the changes observed in early ARDS. With more prolonged ventilation (3-6 days) organized alveolar exudate was found, similar to the later stages of ARDS[5]. In contrast, those animals ventilated with lower PIP (18 cm H$_2$O) showed no notable changes in lung function or histology.

Other studies have shown underlying lung injury to have a synergistic effect, dramatically increasing the susceptibility of the lung to ventilator-induced lung injury (VILI). Although a number of inherent limitations (e.g. species differences in tolerance of ventilator induced lung injury, and differences in pathophysiology of the various animal lung injury models and that of ARDS) prevent direct extrapolation of these studies to the clinical

Acute Respiratory Distress Syndrome: Cellular and Molecular Mechanisms and Clinical Management
Edited by Matalon and Sznajder, Springer Science+Business Media New York, 1998

realm, there is no reason to believe that many of the basic principles (such as tissue disruption secondary to excessive force) should differ.

Thus, although the delineation between ventilator induced lung injury and injury secondary to other causes in ARDS is unclear, there is no doubt that mechanical ventilation *per se* can injure the lung, and for a number of reasons as will be discussed, the lung in ARDS is more susceptible to VILI. The task at hand, is to minimize such injury as our understanding of the mechanisms of VILI continues to evolve.

MANIFESTATIONS OF VILI IN ARDS

The spectrum of ventilator associated/induced injury seen clinically, extends from gross barotrauma resulting in air leaks to more subtle ultrastructural and biochemical abnormalities. With regards to the former, studies have reported an incidence of airleaks from 0.5 to 25% in various patients populations[7-10]. In patients with acute lung injury ($PaO_2/FiO_2 < 300$, bilateral CXR infiltrates), a recent prospective cohort study examining 100 consecutive patients found the incidence of barotrauma, as defined by subcutaneous emphysema, pneumothorax, or pneumomediastinum, to be 13%[7]. No significant difference in mortality was seen between those with and without barotrauma, although by logistic regression analysis barotrauma was associated with increased mortality (odds ratio 6.15).

Upon post-mortem examination of the lungs, Rouby et al. found evidence of airspace enlargement (alveolar overdistension in aerated lung areas or intraparenchymal pseudocysts in non-aerated lung areas) in 26/30 critically ill young patients[11]. A significantly higher incidence of pneumothorax occurred in patients with severe airspace enlargement, which in turn, correlated with exposure to higher PIP, larger tidal volumes, more prolonged use of high FiO_2. These patients were also those who sustained significantly greater weight losses over their ICU stay. Other features of barotrauma included pleural cysts, bronchiolar dilation, alveolar overdistension and intraparenchymal pseudocysts.

Using serial high resolution computer tomography, Finfer et al. found persistent abnormalities of lung structure principally in the anterior lung regions in 3 survivors of protracted ARDS (requiring ventilation for 86 to 97 days)[12]. In contrast, Gattinoni et al. noted cysts and emphysema-like lesions predominantly in the dependent lung zones[13]. The number and spatial distributions of lesions depended on the duration of ARDS.

FACTORS PREDISPOSING THE LUNG IN ARDS TO VILI

For structural disruption to occur secondary to mechanical ventilation, the magnitude of stress applied must exceed the strength or resilience of the underlying lung parenchyma. Invariably, patients with ARDS have a number of risk factors (e.g. underlying lung disease, malnutrition, oxygen toxicity, infection, age) that predispose them to such injury[3,8,11]. In a multivariate analysis of patients receiving mechanical ventilation for greater than 24 hours in an intensive care unit (n = 168 excluding patients with preexistent pneumothorax or chest tubes), only the presence of ARDS was found to independently correlate with the risk of developing pneumothorax[8]. This suggests that the association of air leaks with high airway pressures noted in prior studies, may have been a reflection of the severity of underlying lung injury necessitating use of high pressures for adequate ventilation, rather than the ventilatory pressures *per se*. Supportive of this, are studies in which the incidence of pneumothoraces was found to correlate with either the presence of increased numbers of bullae in dependent lung regions on CT (Figure 1) or the presence of bronchiolar, alveolar overdistension, pleural cysts, and intraparenchymal pseudocysts upon postmortem examination[11].

In addition to underlying tissue injury, ARDS invariably entails presence of a number of factors which may serve to magnify the forces placed on the lung by ventilation with pressures or volumes that otherwise could be borne by healthy lungs without consequence. As discussed elsewhere in this volume, surfactant dysfunction is commonly found in ARDS leading to increased surface tension at the gas-liquid interface within the alveoli and airways. This in turn results in an increased tendency for alveolar/airway collapse. Figure 1 illustrates the markedly reduced end expiratory lung volumes found at different PEEP levels in subjects with severe ARDS as compared to individuals with healthy lungs[14].

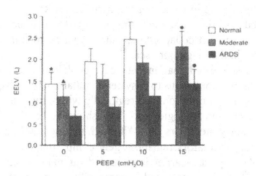

Figure 1: End-expiratory lung volume (EELV) at different PEEP levels in normal subjects (open bars) and in patients with moderate (gray bars) and severe (black bars) acute lung injury (ARDS). Data are expressed as mean ± SD. *p< 0.01, normal subjects compared with moderate and ARDS groups. ▲p<0.05, moderate group compared with ARDS group. *p<0.01, compared with corresponding value at PEEP 10 cmH₂O. Reproduced with permission [14].

In animal models, surfactant dysfunction has been shown to predispose the lung to further injury by mechanical ventilation[15-18]. Dreyfuss et al. demonstrated in rats that even very mild alterations, such as those produced by prolonged anesthesia (which deteriorates surfactant properties and promotes focal atelectasis) were sufficient to augment the deleterious effects of high volume ventilation in a synergistic manner[19]. The amount of edema produced by high volume mechanical ventilation inversely correlated with the quasi-static respiratory system compliance measured during the very first cycles of this ventilation (i.e. before ventilation increased lung injury significantly). Mechanical ventilation may also contribute to development of surfactant dysfunction[20-25].

In addition, in patients with early ARDS, Gattinoni et al. found significant collapse of dependent lung regions resulting in a smaller volume of aerated lung (e.g. as little as 20% of the volume of a normal lung)[26,27]. Alveolar edema, frequently present in ARDS, can further reduce the volume of aerated lung. Thus, delivery of even modest tidal volumes (e.g. 10-12 ml/kg) may result in overdistension of non-dependent alveoli equivalent to that that would be observed if healthy lungs were ventilated with tidal volumes of 40-48 ml/kg[28]. In animals, such volutrauma has been shown to result in ultrastructural breaks and increased permeability edema aggravated by increased filtration. Similar increases in epithelial permeability with lung volume has also been reported in humans[29,30].

Regional disparities in lung expansion can also significantly increase alveolar wall stress secondary to interdependence of the alveolar units. In a normal, uniformly expanded lung, the alveolar distending force can be simplified as the transpulmonary pressure (i.e. $P_{alv} - P_{pl}$). Once inhomogeneity is introduced, as is seen in ARDS[26], local distending forces are exerted by adjacent expanded alveoli act to restore lung expansion in atelectatic regions. These forces may be quite large. Mead et al. postulated that at a transpulmonary pressure of 30 cmH$_2$O, "the pressure tending to expand an atelectatic region surrounded by a fully expanded lung would be approximately 140 cmH$_2$O"[31].

Furthermore, secondary to a combination of factors including surfactant dysfunction, alveolar edema, and the weight of the overlying edematous lung, there is a tendency for dependent airways and alveoli to repetitively open/collapse with each tidal breath[12,32]. Gattinoni et al. demonstrated by computer tomography that P$_{inf}$ increases from ventral to dorsal dependent areas in supine patients with early ARDS[33]. Thus, although a given PEEP may be sufficient to splint open some non-dependent regions of the lung, other regions may either remain collapsed, or be subjected to tidal recruitment. This is thought to result in the observed pattern of cell disruption and hyaline membranes seen in distal airways due to shear injury generated by the repetitive ripping apart of apposed airway walls[34,35]. Infection or ischemic necrosis of collapsed dependent lung regions in ARDS, has also been postulated to play a role in the lung injury seen[13].

Slutsky et al. has suggested that mechanical ventilation may exacerbate both local injury, as well as potentially contribute to a systemic inflammatory response, via effects on the complex network of mediators produced regionally within the lung[36]. In animal models, injurious mechanical ventilation strategies have been shown to result in increased production of a number of inflammatory mediators[36,37]. Although in lungs with intact alveolar-capillary membranes, inflammatory mediators remain largely compartmentalized within the alveolar space[38,39], if the alveolar capillary barrier is disrupted (as may occur with either ventilator induced lung injury[40-44] or increased permeability as seen in ARDS), such compartmentalization may be lost[39]. The presence of inflammatory mediators in the circulation has been shown to play a critical role in the pathophysiology of multiple organ dysfunction and shock[45-48].

Clinically, a number of studies have shown a correlation between poor prognosis and elevated levels of lung lavage and plasma inflammatory mediators (e.g. TNFα, IL-1β, IL-8), or alternatively, a failure to mount adequate increases in anti-inflammatory mediators, e.g. IL10, IL1-ra[46,48-51]. Generalization of these studies is limited, however, by the diverse patient populations, small sample sizes, and differences in both timing and technique of cytokine measurements. Most studies suggest that the time course of cytokine concentrations, rather than the concentration at a single timepoint, provides the best predictor of patient outcome. Meduri et al found a persistent, excessive production of TNFα, IL1β, IL6, and IL-8 in the BAL followed by similar changes in the plasma in 10 nonsurvivors, while lower peak levels followed by a decline were seen in the 12 survivors[50]. Although far from conclusive, these findings do lend support to the postulate that spillover of an ongoing lung inflammatory response (as may occur with recurrent trauma from mechanical ventilation), might affect patient outcome, and could explain the observation that a minority of patients with ARDS die of respiratory failure[52]. Rather, lung injury appears to predispose to development of a systemic inflammatory response leading to multiple system organ failure and death[52-56].

STRATEGIES TO MINIMIZE VILI IN ARDS

The desire to minimize iatrogenic injury by avoiding presumed injurious factors formed the impetus behind the development of a number of novel ventilatory strategies (Table 1). Although a detailed overview of these techniques is covered elsewhere in this volume, and have been the subject of a number of recent reviews[57-59], several caveats pertinent to VILI bear closer examination.

Table 1. Examples of proposed "protective" ventilatory strategies

Avoidance of lung overdistension
- pressure and volume limited ventilation (± IVOX, ECCO$_2$R)
- proportional assist ventilation

Reduced tidal pressure-volume swings
- high frequency ventilation
- low tidal volume ventilation

Improved distribution of ventilation:
- "optimal" PEEP
- inverse-ratio ventilation
- surfactant supplementation
- partial liquid ventilation
- high frequency ventilation
- prone positioning

Avoidance of tidal opening and collapse of distal airways/alveoli
- appropriate PEEP
- partial liquid ventilation
- high frequency ventilation

To date, no particular ventilatory strategy has been demonstrated to be superior to another. In theory, the optimal strategy to prevent ventilator induced lung injury would maintain maximal recruitment of alveoli (preventing repetitive opening/collapse) without causing alveolar overdistension or adverse hemodynamic changes due to the degree of lung inflation (Figure 2)[60,61]. Early in the course of ARDS, many patients develop an abrupt increase in the slope of their respiratory system pressure-volume curves at approximately 10-15 cm H$_2$O. The pressure at which this occurs (P$_{inf}$) is thought to represent the opening pressure required to recruit collapsed alveolar units and small airways. Thus, use of PEEP ≥ P$_{inf}$ has been suggested to minimize lung injury. However, care must be taken to ensure that superimposed tidal ventilation does not occur on the upper less compliant portion of the PV curve, predisposing the patient to development of high volume/pressure injury[62].

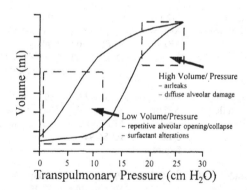

Figure 2: Figure illustrating the pressure volume curve of the respiratory system. In optimal ventilation strategies, volume excursions take place on the steepest part of the curve (i.e. region of maximal compliance). Ventilation occurring at either extreme of pressure or volume predisposes to lung injury.

As mentioned previously, a difficulty encountered in trying minimize VILI in ARDS is the presence of heterogeneous underlying lung injury. In supine patients with early ARDS, Gattinoni et al. demonstrated by computer tomography a progressive increase in P_{inf} as one goes from ventral non-dependent lung zones to dorsal dependent areas[33]. Thus, the "optimal" PEEP for one region, may be insufficient or cause overdistension in other lung regions. This is part of the rationale for partial liquid ventilation which may act to selectively recruit dependent lung regions due to the density of the liquid while enabling gas exchange secondary to the large O_2 and CO_2 carrying capacity of perfluorocarbons, in addition to having a low surface tension.

Patients with ARDS also represent a diverse population, whose clinical course (both from a respiratory as well as systemic standpoint) fluctuates with time. Thus, ongoing assessment of respiratory mechanics and function is required to optimize ventilation. The animal data suggest that certain aspects of ventilator induced injury occur rapidly[42]. Therefore, to prevent VILI, optimal therapy should ideally be implemented upon initiation of mechanical ventilation.

A number of the difficulties encountered in putting theory into practice are illustrated by the conflicting preliminary results of several recent clinical trials attempting to define "protective" conventional mechanical ventilation strategies in ARDS (Table 2).

Table 2. Prospective clinical studies examining the effect of pressure or volume limited conventional ventilatory strategies on mortality.

Group	Location	Design	Patients	Intervention	Controls	Outcome
Hickling et. al 1994[63]	1 I.C.U New Zealand	Prospective, observational	53 ARDS with LIS ≥ 2.5	PIP 30-40; Vt 4-7 ml/kg; PEEP prn to O_2 sat ≥ 90% SIMV	none	hospital mortality less than predicted by APACHE II (26.4% vs 53.3%)
Amato et al. 1996[64]	2 I.C.U.'s Brazil	Randomized, controlled	48 ARDS with LIS ≥ 2.5	PIP < 40; Vt < 6 ml/kg; PEEP ≥ Pflex	Vt=12 ml/kg PEEP prn to O_2 sat	decreased overall mortality (<40% vs >60%)
Stewart et al. 1997[65]	6 I.C.U.'s Canada	Randomized, controlled	112 pre-ARDS or ARDS	PIP ≤ 30; Vt ≤ 8 ml/kg; PEEP 5-20 to O_2 sat ≥ 90%	PIP ≤ 50; Vt 10-15 ml/kg PEEP 5-20 to O_2 sat ≥ 90%	no difference in mortality (48.3% vs. 46.3%)
Broccard et. al. 1997[66]	25 centers France	Randomized, controlled	108 ARDS single organ failure	Pplat = 25; Vt ≤ 8 ml/kg PEEP to O_2 sat 86-94%	PIP < 60; Vt ≥ 10 ml/kg $PaCO_2$ 38-42 PEEP ?	no difference in 60 day mortality (48% vs. 41%)
Brower et al. 1997[67]	2 centers U.S.A.	Randomized, controlled	52 ARDS	Pplat ≤ 30; Vt ≤ 8 ml/kg PEEP to O_2 sat 86-94%	Pplat ≤ 50; Vt 10-12 ml/kg PEEP to O_2 sat 86-94%	no difference in mortality (50% vs. 46%)

Considering the complexity of patients with ARDS, perhaps it is naive to expect to find dramatic differences in mortality in relatively small patient populations. The results of the study by Amato et al. (in which attention was paid to both the upper and lower limits of tidal ventilation) are encouraging, but require confirmation in light of the high mortality of the control group, and small patient numbers involved. To date, none of the other randomized trials in which PEEP was titrated by O_2 saturation have demonstrated a difference in mortality. Trials are also underway assessing the efficacy of a number of the other protective strategies mentioned in Table 1.

SUMMARY

Numerous animal studies have demonstrated that mechanical ventilation can not only exacerbate pre-existing lung injury, but can also initiate such injury *de novo*. Although the precise mechanisms are unclear, ventilation at either extreme of lung volume (i.e. overdistension or repetitive end-expiratory collapse) has been found to cause lung injury consisting of pulmonary edema, hyaline membranes, granulocyte infiltration, reduced lung compliance, hypoxemia, increased vascular permeability, increased production of inflammatory mediators, and pseudocyst formation.

In injured lungs, implementation of "protective" mechanical ventilation strategies becomes more difficult, secondary to regional disparities in lung injury and ventilation. In addition, underlying lung disease/dysfunction has been found to have a synergistic effect on increasing the susceptibility of the lung to injury.

At present, other than the single study by Amato and colleagues, no clear advantage has been shown for any particular ventilatory strategy in ARDS. Given the variety of mechanisms thought to play a role in VILI, it is improbable that a single ventilatory strategy will emerge as a panacea for all respiratory ailments. Prospective randomized clinical trials are needed clarify the optimal applications of current ventilatory strategies, and assess the efficacy of various cointerventions (e.g. novel anti-inflammatory therapies, surfactant supplementation). In the interim, care should be taken to tailor the particular ventilatory strategy to the patient so as to address specific physiologic concerns (e.g. increased surface tension, V/Q mismatch, inhomogeneous ventilation,...) while minimizing either regional lung over- or under- inflation.

REFERENCES

1. V.H. Moon. The pathology of secondary shock, *Am J Pathol.* 24:235 (1948).
2. C. Teplitz. The core pathobiology and integrated medical science of adult acute respiratory insufficiency, *Surg Clin North Am.* 56:1091 (1976).
3. D.J. Pierson. Alveolar rupture during mechanical ventilation: role of PEEP, peak airway pressure, and distending volume, *Resp Care.* 33:472 (1988).
4. T. Kolobow, M.P. Moretti, R. Fumagalli, D. Mascheroni, P. Prato, V. Chen, M. Joris, Severe impairment in lung function induced by high peak airway pressure during mechanical ventilation, *Am Rev Respir Dis.* 135:312 (1987).
5. K. Tsuno, K. Miura, M. Takeya, T. Kolobow, T. Morioka, Histopathologic pulmonary changes from mechanical ventilation at high peak airway pressures, *Am Rev Respir Dis.* 143:1115 (1991).
6. P.P. Hamilton, A. Onayemi, J.A. Smyth, J.E. Gillan, E. Cutz, A.B. Froese, A.C. Bryan, Comparison of conventional and high-frequency ventilation: oxygenation and lung pathology, *J Appl Physiol.* 55:131 (1983).
7. L.M. Schnapp, D.P. Chin, N. Szaflarski, M.A. Matthay, Frequency and importance of barotrauma in 100 patients with acute lung injury, *Crit Care Med.* 23:272 (1995).
8. R.B. Gammon, M.S. Shin, R.H. Groves, Jr., M.J. Hardin, C. Hsu, S.E. Buchalter, Clinical risk factors for pulmonary barotrauma: A multivariate analysis, *Am J Respir Crit Care Med.* 152:1235 (1995).

9. G.W. Petersen and H. Baier, Incidence of pulmonary barotrauma in a medical ICU. *Crit Care Med.* 11:67 (1983).

10. D.J. Cullen and D.L. Caldera, The incidence of ventilator-induced pulmonary barotrauma in critically ill patients, *Anesthesiology.* 50:185 (1979).

11. J.J. Rouby, T. Lherm, E.M. de Lassale, P. Poete, L. Bodin, J.F. Finet, P. Callard, P. Viars, Histologic aspects of pulmonary barotrauma in critically ill patients with acute respiratory failure, *Intensive Care Med.* 19:383 (1993).

12. S. Finfer and G. Rocker, Alveolar overdistension is an impoortant mechanism of persistent lung damage following severe protracted ARDS, *Anaesth Intens Care.* 24:569 (1996).

13. L. Gattinoni, M. Bombino, P. Pelosi, A. Lissoni, A. Pesenti, R. Fumagalli, M. Tagliabue, Lung structure and function in different stages of severe adult respiratory distress syndrome, *JAMA.* 271:1772 (1994).

14. P. Pelosi, M. Cereda, G. Foti, M. Giacomini, A. Pesenti, Alterations of lung and chest wall mechanics in patients with acute lung injury: effects of positive end-expiratory pressure, *Am J Respir Crit Care Med.* 152:531 (1995).

15. P.J. Coker, L.A. Hernandez, K.J. Peevy, K. Adkins, J.C. Parker, Increased sensitivity to mechanical ventilation after surfactant inactivation in young rabbit lungs, *Crit Care Med.* 20:635 (1992).

16. D.L. Bowton and D.L. Kong, High tidal volume ventilation produces increased lung water in oleic acid-injured rabbit lungs, *Crit Care Med.* 17:908 (1989).

17. L.A. Hernandez, P.J. Coker, S. May, A.L. Thompson, J.C. Parker, Mechanical ventilation increases microvascular permeability in oleic acid-injured lungs, *J Appl Physiol.* 69:2057 (1990).

18. Z. Bshouty, J. Ali, M. Younes, Effect of tidal volume and PEEP on rate of edema formation in in situ perfused canine lobes, *J Appl Physiol.* 64:1900 (1988).

19. D. Dreyfuss, P. Soler, G. Saumon, Mechanical ventilation-induced pulmonary edema. Interaction with previous lung alterations, *Am J Respir Crit Care Med.* 151:1568 (1995).

20. D. Massaro, L. Clerch, G.D. Massaro, Surfactant aggregation in rat lungs: influence of temperature and ventilation, *J Appl Physiol.* 51:646 (1981).

21. E.E. Faridy, S. Permutt, R.L. Riley, Effect of ventilation on surface forces in excised dogs' lungs. *J Appl Physiol.* 21:1453 (1966).

22. J. McClenahan and A. Urtnowski, Effect of ventilation on surfactant and its turnover rate, *J Appl Physiol.* 23:215 (1967).

23. I. Wyszogrodski, K. Kyei-Aboagye, W. Taeusch, M.E. Avery, Surfactant inactivation by hyperventilation: conservation by end-expiratory pressure, *J Appl Physiol.* 38:461 (1975).

24. M.J. Oyarzun and J.A. Clements, Control of lung surfactant by ventilation, adrenergic mediators, and prostaglandins in the rabbit, *Am Rev Respir Dis.* 117:879 (1978).

25. E.E. Faridy. Effect of ventilation on movement of surfactant in airways, *Respiration Physiology.* 27:323 (1976).

26. L. Gattinoni, A. Pesenti, A. Torresin, S. Baglioni, M. Rivolta, F. Rossi, F. Scarani, R. Marcolin, G. Cappelletti, Adult respiratory distress syndrome profiles by computed tomography, *J Thorac Imag.* 1:25 (1986).

27. L. Gattinoni, A. Pesanti, L. Avalli, F. Rossi, M. Bombino, Pressure-volume curve of total respiratory system in acute respiratory failure. Computed tomographic scan study. *Am Rev Respir Dis.* 136:730 (1987).

28. L. Gattinoni, A. Pesenti, S. Baglioni, M. Rivolta, P. Pelosi, Inflammatory pulmonary edema and positive end-expiratory pressure: correlation between imaging and physiologic studies, *J Thorac Imag.* 3:59 (1988).

29. J.D. Marks, J.M. Luce, N.M. Lazar, J.N. Wu, A. Lipavsky, J.F. Murray, Effect of increases in lung volume on clearance of aerosolized solute from human lungs, *J Appl Physiol.* 59:1242 (1985).

30. K.B. Nolop, D.L. Maxwell, D. Royston, J.M.B. Hughes, Effect of raised thoracic pressure and volume on clearance of aerosolized solute from human lungs, *J Appl Physiol.* 60:1493 (1986).

31. J. Mead, T. Takishima, D. Leith, Stress distribution in lungs: a model of pulmonary elasticity. *J Appl Physiol.* 28:596 (1970).

32. G. Enhorning. Photography of peripheral pulmonary airway expansion as affected by surfactant, *J Appl Physiol.* 42:976 (1977).

33. L. Gattinoni, L. D'Andrea, P. Pelosi, G. Vitale, A. Pesenti, R. Fumagalli, Regional effects and mechanism of positive end-expiratory pressure in early adult respiratory distress syndrome, *JAMA.* 269:2122 (1993).

34. B. Robertson. *Pulmonary Surfactant*, Elsevier, Amsterdam (1984).

35. T.J. Ryan. Stress failure of alveolar epithelial cells studied by scanning electron microscopy, *Journal American Academicx Dermatology.* 21:115 (1989).

36. L. Tremblay, F. Valenza, S.P. Ribeiro, J. Li, A.S. Slutsky, Injurious Ventilatory Strategies Increase Cytokines and *c-fos* M-RNA Expression in an Isolated Rat Lung Model, *Journal of Clinical Investigation*. 99:944 (1997).

37. Y. Imai, T. Kawano, K. Miyasaka, M. Takata, T. Imai, K. Okuyama, Inflammatory chemical mediators during conventional ventilation and during high frequency oscillatory ventilation, *Critical Care Med.* 150:1550 (1994).

38. S. Nelson, G.J. Bagby, B.G. Bainton, L.A. Wilson, J.J. Thompson, W.R. Summer, Compartmentalization of intraalveolar and systemic lipopolysaccharide-induced tumor necrosis factor and the pulmonary inflammatory response, *J of Infectious Diseases*. 159:189 (1989).

39. J.D. Tutor, C.M. Mason, E. Dobard, R.C. Beckerman, W.R. Summer, S. Nelson, Loss of compartmentalization of alveolar tumor necrosis factor after lung injury, *J respir Crit Care Med.* 149:1107 (1994).

40. E. John, M. McDevitt, W. Wilborn, G. Cassady, Ultrastructure of the lung after ventilation, *Br J Exp Pathol.* 63:401 (1982).

41. D. Dreyfuss, G. Basset, P. Soler, G. Saumon, Intermittent positive-pressure hyperventilation with high inflation pressures produces pulmonary microvascular injury in rats, *Am Rev Respir Dis.* 132:880 (1985).

42. D. Dreyfuss, P. Soler, G. Saumon, Spontaneous resolution of pulmonary edema caused by short periods of cyclic overinflation, *J Appl Physiol.* 72:2081 (1992).

43. D. Dreyfuss, P. Soler, G. Basset, G. Saumon, High inflation pressure pulmonary edema. Respective effects of high airway pressure, high tidal volume, and positive end-expiratory pressure, *Am Rev Respir Dis.* 137:1159 (1988).

44. Z. Fu, M.L. Costello, K. Tsukimoto, R. Prediletto, A.R. Elliot, O. Mathieu-Costello, J.B. West, High lung volume increases stress failure in pulmonary capillaries, *J Appl Physiol.* 73:123 (1992).

45. R.J. Debs, H.J. Fuchs, R. Philip, A.B. Montgomery, E.N. Brunette, D. Liggitt, J.S. Patton, J.E. Shellito, Lung-specific delivery of cytokines induces sustained pulmonary and systemic immunomodulation in rats, *J Immunol.* 140:3482 (1988).

46. R.C. Bone. Toward a theory regarding the pathogenesis of the systemic inflammatory response syndrome: What we do and do not know about cytokine regulation, *Crit Care Med.* 24:163 (1996).

47. R.M.H. Roumen, H. Redl, G. Schlag, G. Zilow, W. Sandtner, W. Koller, T. Hendriks, R.J. Goris, Inflammatory mediators in relattion to the development of multiple organ failure in patients after severe blunt trauma, *Crit Care Med.* 23:474 (1995).

48. S.C. Donnelly, R.M. Strieter, S.L. Kunkel, A. Walz, C.R. Robertson, D.C. Carter, I.S. Grant, A.J. Pollok, C. Haslett, Interleukin-8 and development of adult respiratory distress syndrome in at-risk patient groups, *The Lancet.* 341:643 (1993).

49. E. Borrelli, P. Roux-Lombard, G.E. Grau, E. Girardin, B. Ricou, J. Dayer, P.M. Suter, Plasma concentrations of cytokines, their soluble receptors, and antioxidant vitamins can predict the development of multiple organ failure in patients at risk, *Crit Care Med.* 24:392 (1996).

50. G. Meduri, Umberto, G. Kohler, S. Headley, E. Tolley, F. Stentz, A. Postlethwaite, Inflammatory cytokines in the BAL of patients with ARDS, *Chest.* 108:1303 (1995).

51. S.C. Donnelly, R.M. Strieter, P.T. Reid, S.L. Kunkel, M.D. Burdick, I. Armstrong, A. Mackenzie, C. Haslett, The association between mortality rates and decreased concentrations of interleukin-10 and interleukin-1 receptor antagonist in the lung fluids of patients with the adult respiratory distress syndrome, *Ann Intern Med.* 125:191 (1996).

52. L.D. Hudson. Survival data in patients with acute and chronic lung disease requiring mechanical ventilation, *Am Rev Respir Dis.* 140:S19 (1989).

53. A.B. Montgomery, M.A. Stager, C.J. Caricco, L.D. Hudson, Causes of mortality in patients with the adult respiratory distress syndrome, *Am Rev Respir Dis.* 132:485 (1985).

54. K.G. Hickling. Ventilatory management of ARDS: can it affect the outcome? *Intensive Care Med.* 16:219 (1990).

55. S.C. Donnelly and C. Haslett, Cellular mechanisms of acute lung injury:implications for future treatment of the adult respiratory distress syndrome, *Thorax.* 47:260 (1992).

56. A. Sauaia, F.A. Moore, E.E. Moore, J.B. Haenel, R.A. Read, Pneumonia: cause or symptom of postinjury multiple organ failure? *Am J of Surgery.* 166:606 (1993).

57. H. Burchardi. New strategies in mechanical ventilation for acute lung injury, *Eur Respir J.* 9:1063 (1996).

58. J.C. Ring and G.L. Stidham, Novel therapies for acute respiratory failure, *Pediatric Clinics of North America.* 41:1325 (1994).

59. A.S. Slutsky. Consensus conference on mechanical ventilation - January 28-30, 1993 at Northbrook, Illinois, USA, *Intensive Care Med.* 20:64 (1994).

60. T.B. Bezzant and J.D. Mortensen, Risks and hazards of mechanical ventilation: a collective review of published literature, *Disease-a-Month.* XL:581 (1994).

61. J. Biondi, D. Schulman, R. Matthay, Effects of mechanical ventilation on right and left ventricular function, *Clin Chest Med.* 9:55 (1988).

62. M.B. Amato, C.S. Barbas, D.M. Medeiros, G.d. Schettino, G.L. Filho, R.A. Kairalla, D. Deheinzelin, C. Morais, E.d. Fernandes, T.Y. Takagaki, et al. Beneficial effects of the "open lung approach" with low distending pressures in acute respiratory distress syndrome. A prospective randomized study on mechanical ventilation. *Am J Respir Crit Care Med.* 152:1835 (1995).

63. K.G. Hickling, J. Walsh, S. Henderson, R. Jackson, Low mortality rate in adult respiratory distress syndrome using low-volume, pressure-limited ventilation with permissive hypercapnia: A prospective study, *Critical Care Med.* 22:1568 (1994).

64. M.B. Amato, C.S. Barbas, G.L. Filho, R.A. Kairalla, D. Deheinzelin, R.B. Magaldi, C.R.R. Carvalho, Improved survival in ARDS: Beneficial effects of a lung protective strategy. *Am J Respir Crit Care Med.* 153:A531 (1996).

65. T.E. Stewart, M.O. Meade, J. Granton, S. Lapinsky, R. Hodder, R. McLean, D. Mazer, T. Rogevin, D. Schouten, T. Todd, et al. Pressure and volume limited ventilation strategy (PLVS) in patients at high risk for ARDS - results of a multicenter trial, *Am J Respir Crit Care Med.* 155:A505 (1997).

66. L. Brochard, F. Roudot-Thoraval, the collaborative group on VT reduction, Tidal volume reduction in acute respiratory distress syndrome (ARDS): A multicenter randomized study, *Am J Respir Crit Care Med.* 155:A505 (1997).

67. R. Brower, C. Shanholtz, D. Shade, H. Fessler, P. White, C. Wiener, J. Teeter, Y. Almog, J. Dodd-O, S. Piantadosi, Randomized controlled trial of small tidal volume ventilation (STV) in ARDS, *Am J Respir Crit Care Med.* 155:A93 (1997).

MECHANICAL VENTILATION FOR ARDS: SHOULD WE REDUCE TIDAL VOLUME?

François Lemaire, Eric Roupie

Service de Réanimation Médicale
Hôpital Henri Mondor
F - 94010 Créteil (France)

INTRODUCTION

For the last ten years, barotrauma has been proposed as a major cause for our failure in salvaging patiens having the Acute Respiratory Distress Syndrome (ARDS). After the first descriptions[1,2] of cysts, bullae and emphysematous-like lesions supposedly caused by mechanical ventilation, many authors related the incidence and severity of barotrauma -not only overdistension of airspaces, but also pneumothoraces- to the usual modalities of mechanical ventilation: large tidal volumes (VT), equal or higher than 12 mL/kg of body weight, and/or high levels of positive end expiratory pressure (PEEP)[3,4]. Dreyfuss and Saumon in a series of elegant studies, showed that the main determinant of alveolar damage -measured as capillary permeability increase- was in fact the end inspiratory lung volume[5]. In the clinical arena, end inspiratory volume can only be approximated as the airway pressure measured at the end of an end inspiratory plateau, during a zero flow period.

In this line of reasoning, Hickling et al.[6] proposed in 1990 to decrease the airway pressure by reducing VT size, and doing so, they demonstrated a lower mortality than expected -APACHE II predicted- in a series of 50 patients with severe ARDS, at the expense of what they called « permissive hypercapnia ». This concept was so well received that it diffused rapidly worldwide within the Critical Care community. A recent Consensus statement, for instance, claimed that maximal end inspiratory plateau pressure should not overpass 35 cmH$_2$O[7]. This number was derived from experimental studies, without clinical validation. It is only recently that this modality of mechanical ventilation has been tested in randomized trials, which we will review in this chapter.

CLINICAL BASIS

Several studies attempted to relate the incidence of barotrauma in ARDS patients to the modalities of mechanical ventilation. But none has succeeded so far in isolating one single factor: namely, PEEP level or VT size, mean, peak or plateau airway pressure. Rouby et al.[4], performing histological examination of the lungs of 30 young patients who died from (or with) ARDS found that pneumothoraces and enlarged airspaces were more commonly evidenced when peak airway pressure (56 ± 18 cmH$_2$O) and VT (> 12 mL/kg) were high. Similarly, in a series of 100 consecutive patients with acute lung injury (PaO2/FiO2 < 300),

Acute Respiratory Distress Syndrome: Cellular and Molecular Mechanisms and Clinical Management
Edited by Matalon and Sznajder, Springer Science+Business Media New York, 1998

barotrauma was present in 13% of them[8]. Mean peak inspiratory pressure was 51 cmH$_2$O in non trauma patients and 62 cmH$_2$O in patients who developed ultimately barotrauma. Gammon et al.[9] applied a multivariate analysis to identify the risk of developing ventilator induced pneumothorax in 168 patients. Twenty experienced pneumothorax, which was independently correlated only with the presence of ARDS. In the subgroup of patients having ARDS (41 patients), a correlation between barotrauma and airway pressure could not be evidenced.

Tracing of total respiratory system (lung + chest-wall) pressure-volume (P-V) curves can be used in order to define the mechanical characteristics of patients during mechanical ventilation [10]. P-V curves have also been used to titrate mechanical ventilation [11]. A typical curve is depicted in Figure 1[12].

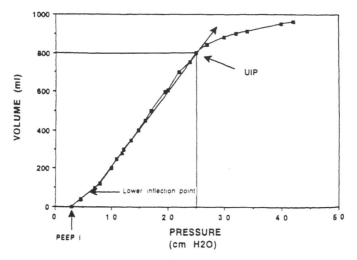

Figure 1.

This P-V curve has a typical curvi-linear pattern, with two inflections:

(a) *a lower « inflexion zone »*, just above end expiratory lung volume, suggesting recruitment (re-opening) of collapsed air-spaces when airway pressure is increased, either with PEEP of during tidal excursion. Figure 2 gives a typical example of gas exchange improvement when PEEP level is set up just above this inflection. Recent papers have drawn the attention on possible flaws of the tracing of total respiratory systems P-V curves, since inflection could represent abnormalities of the chest-wall, especially in surgical patients[13].

(b) *a second, upper and convex upward, inflection*, is evidenced in many ARDS P-V curves at higher lung volume. Roupie et al.[12] showed such an upper inflection point to be present in 25 patients with ARDS. In those patients, they compared the end inspiratory plateau pressure (P plat) obtained with a VT of 10 mL/kg to the level of upper inflection. With these mechanical ventilation settings, the end inspiratory plateau pressure was higher than upper inflection point in 80% of their patients, suggesting that the end inspiratory plateau volume and pressure were located in the «risky» zone of the P-V curve, as indicated by the flattening of the curve after upper inflection. Figure 3 shows that only when VT is equal to or lower than 7 mL/kg, the majority of patients (> 80%) have a tidal excursion limited to the linear segment of their P-V curve.

All these studies suggest that: *(a) barotrauma occurs commonly in ARDS, (b) despite it was not confirmed as an independent factor, occurrence of barotrauma aggravates the prognosis, (c) besides air leaks (pneumothorax, pneumo-mediastinum, etc.), «volutrauma», i.e. alveolar stretching and rupture, can induce capillaro-alveolar fluid leakage, perpetuating ARDS, and (d) accordingly, the reduction of airway pressure and volume should prevent the occurrence of barotrauma and improve outcome.*

304

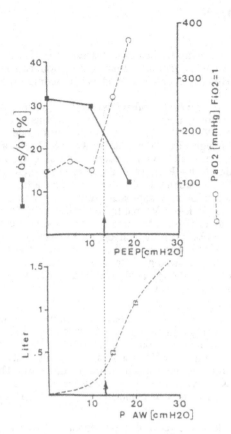

Figure 2. Relationship between P-V curve (bottom) and PaO2 and shunt (QS/QT) (upper), as a function of airway pressure (PAW) and PEEP in a typical patients (From Matamis et al. [10])

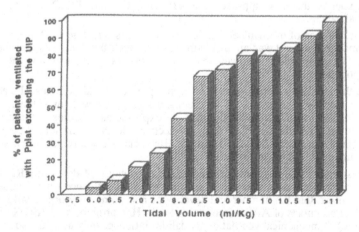

Figure 3. Graph summarizing the percentage of patients with ARDS in whom plateau pressure would exceed the upper inflection point as a function of tidal volume applied during ventilation with PEEP. The end-inspiratory plateau pressure reached for each tidal volume is compared to the upper inflection point defined on the P-V curve (see Figure 1).

For the last ten years, many studies have assessed the effects of VT reduction on gas exchange, respiratory mechanics, cardiac output and hemodynamics, oxygen consumption and oxygen transport. However, outcome studies are scarce, most of them still being unpublished as full articles.

Amato et al., published the first confirmative study[14,15] of this concept. They undertook a prospective randomized study, comparing in ARDS patients two strategies: conventional mechanical ventilation with VT of 12 mL/kg, an assist-control mode, VT high enough to keep $PaCO_2 < 40$ mmHg and enough PEEP to maintain $FiO_2 < 60\%$, and a new approach with VT ≤ 6 mL/kg, a PC-IRV mode, peak airway pressure < 40 cmH_2O, and PEEP titrated in any individual patient according to the Pflex of P-V curve. Forty-eight patients were enrolled in this trial (two ICUs) with 25 patients in the new treatment arm, and 23 in the other one. Survival was much higher in the new approach group, characterized also by a much higher PEEP level (mean: 18 cmH_2O), lower VT and moderate hypercapnia.

These results were not confirmed by three other randomized clinical trials[16,17,18], which failed to evidence differences between the conventional treatment and the low VT groups. In the largest series[16], 108 patients were enrolled in 25 centers. The two groups were similar after randomization, this study showed no difference in morbidity or mortality.

These four studies have only appeared as abstracts, which makes this analysis rather preliminary and hypothetical. However, we can henceforth propose a few tracks for discussion:

- In the three «negative» studies, the differences between groups (control and new treatment) appeared relatively thin, in terms of VT size and maximal airway pressure. For instance, the mean VT of the three conventional groups was around 10 mL/kg, which is not very far from 7 mL (in the low VT group), and certainly much lower than the classical 13-15 mL/kg which used to be applied in the recent past. The concept of low VT ventilation and permissive hypercapnia has already spread out so widely that it is difficult - if not impossible, ethically- to ask clinicians to apply high volume and pressures as controlled conditions.

- There are many possible side-effects of hypercapnia[19]; theoretically, beneficial effects of VT reduction could have been offset by negative effects of hypercarbia.

- If end-inspiratory volume (and pressure) is really the bottom line, it can be altered by manipulating VT, but also PEEP. In the three negative studies, PEEP remained the same at a relatively low level in the two groups. Conversely, in the Amato's study, PEEP was much higher in the new approach arm (18 cmH_2O), and PEEP was titrated for any individual patient.

- Individual titration of mechanical ventilation parameters (PEEP level, VT) can be preferred to systematic settings; it was another difference between the Amatos's and the three other studies. Our own study[13] showed that upper inflection point was not the same in our ARDS population.

- Besides the possible deleterious role of higher pressures and volumes, the accent has be put more recently on the risk of ventilating the lung at low lung volume: Corbridge[20], Muscedere[21] showed convincingly that in experimental studies, increasing the end expiratory lung volume by PEEP had a protective effects on the lung (assessed by P-V curves, shunt measurement and histology). How this effect combines with superimposed tidal excursion is still under investigation[22].

- As the respiratory failure is still debated as the main cause of death in ARDS, the role of mechanical ventilation is also under scrutiny as to its influence on outcome. All recent studies confirm that the number and severity of non pulmonary organ failures determine most of the prognosis of ARDS (see in this book: Has prognosis of ARDS improved? by F. Lemaire). If mechanical ventilation modalities influence only a minor part of the overall survival rate in ARDS, it will be difficult to demonstrate its specific role in outcome studies, at least as long as mortality remains the major end-point.

IN CONCLUSION

Barotrauma has been a major working hypothesis in the ARDS field for the last decade. The first confirmative studies are disappointing. Where to go now? Was the concept fallacious? Or have not we be so far unable to test it appropriately? It is difficult to answer at this time and we can only propose some tracks for future trials.

Primary versus secondary ARDS

As it is clearer and clearer that there are two forms of ARDS, one starting from and involving predominantly the lungs (usually called « primary ») and another one, where the lung lesions are secondary to (or associated with) a multiple organ system failure, it will be crucial in the future to define in which ARDS population the mechanical ventilation new modalities should be assessed. Schematically, broadening the inclusion criteria to secondary ARDS gives the opportunity to enroll a large number of patients, but expose to high mortality, which, in addition, is unlikely to be influenced by respiratory maneuvers alone. Conversely, restriction to patients with only one (lung) failure, slowers markedly the process of inclusion.

Low versus high volume lung injury

Two different types of iatrogenic lung lesion have been ascribed to mechanical ventilation: one caused by mechanical ventilation at low lung volume, which can be prevented by the use of some « optimal » level of PEEP (20, 21), and a second one, occurring at high lung volume, in relation to high tidal volume and/or PEEP. Both hypotheses are plausible, seductive, well-documented in the experimental literature. They are not necessarily exclusive, and could probably combine their deleterious effects. But how shall we be able to dissociate their effects in our future trials? This is crucial, as this could explain the different outcome between the Amato's and the three other trials with low VT.

Is there a « standard » management which could be applied to a control group?

So little has been validated in the treatment of ARDS so far (23) that definition of the standard management is always the matter of endless discussion. The randomized trial of ECCO2R by Morris et al. (24) has been severely questioned by the users of the technique because they claimed that mechanical ventilation with inverted I/E ratio and pressure limitation were not « standard » treatment. Similarly, we have seen tat a VT of 10 ml/kg used in the three low VT trials for the control arm was possibly itself too low.

Can one single manipulation be responsible for a significant outcome alteration and be assessed by a reasonably sized randomized trial when so many interventions can be used, such as nitric oxid inhalation, prone positioning, late corticoïds infusion, exogenous surfactant administration, partial liquid ventilation, etc. Kollef, in a recent editorial, raised the question as to whether we should try to test simultaneously several therapies. He mentioned that the NIH sponsored ARDS clinical Trial Treatment Group started such a new strategy, by testing the addition of low VT and ketokenazole administration (25).

Clearly, the Intensive Care Community should be able to answer these questions before launching the next large randomized trials, so expensive and, so far, so disappointing.

REFERENCES

1. F. Lemaire, J. Cerrina, F. Lange, A. Harf, J. Carlet, J. Bignon, PEEP induced airspace overdistension complicating a paraquat lung, *Chest* 81: 654 (1982).
2. A. Churg, J. Golden, S. Fligiel, J.C. Hogg, Bronchopulmonary dysplasia in adults, *Am. Rev. Respir. Dis.* 127: 117 (1983).
3. R.B. Gammon, M.S. Shin, S.E. Buchalter, Pulmonary barotrauma in mechanical ventilation: patterns and risk factors, *Chest* 102: 568 (1992).

4. J.J. Rouby, T. Lherm, E. Martin de Lassale, P. Poète, L. Bodin, J.F. Finet, P. Callard, Histologic aspects of pulmonary barotrauma in critically ill patients with acute respiratory failure, *Intens. Care Med.* 19: 383 (1993).

5. D. Dreyfuss and G. Saumon, Ventilator-induced injury, in *Principles and practice of mechanical ventilation*, Mc Graw-Hill Publisher, New York (1994).

6. K.G. Hickling, S.J. Henderson, R. Jackson, Low mortality associated with low volume pressure limited ventilation with permissive hypercapnia in severe ARDS, *Intens. Care Med.* 16: 372 (1990).

7. A.S. Slutsky, ACCP consensus conference: Mechanical ventilation, *Chest* 104: 1833 (1993).

8. L.M. Schnapp, D.P. Chin, N. Szaflarski, M. Matthay, Frequency and importance of barotrauma in 100 patients with ALI, *Crit. Care Med.* 23: 272 (1995).

9. B.R. Gammon, M.S. Shin, R.H. Groves, J.M. Hardin, H. Chanchieh, S.E. Buchalter, Clinical risk factors for pulmonary trauma: a multivariate analysis, *Am. J. Respir. Crit. Care Med.* 152: 1235 (1995).

10. D. Matamis, F. Lemaire, A. Harf, C. Brun-Buisson, J.C. Ansquer, G. Atlan, Total respiratory pressure volume curves in the ARDS, *Chest* 86: 58 (1984).

11. L. Brochard, Respiratory pressuve-volume curves (in press), in *Principles and practice of intensive care monitoring*, Mac Graw-Hill Publisher, New York (1997).

12. E. Roupie, M. Dambrosio, G. Servillo, H. Mentec, S. El Atrous, L. Beydon, C. Brun-Buisson, F. Lemaire, L. Brochard, Titration of tital volume and induced hypercapnia in ARDS, *Am. J. Respir. Crit. Care Med.* 152: 121 (1995).

13. S. Grasso, F. Puntillo, G. Perchiazzi, L. Mascia, N. Brienza, R. Giuliani, T. Fiore, A. rianza, A. Slutsky, M. Ranieri, Impairment of the elastic properties of the lung and chest wall in patients with ARDS: role of the abdomen, *Am. J. Respir. Crit. Care Med.* 155: A393 (1997).

14. M.B. Amato, C. Barbas, D. Medeiros, G. Schettino, G. Filho, R. Kairalla, D. Deheinzelin, C. Morais, E. Fernandes, T. Takagaki, C. De Carvalho, Beneficial effects of the « open lung approach » with low distending pressure in ARDS, *Am. J. Respir. Crit. Care Med.* 152: 1852 (1995).

15. M. Amato, C. Barbas, D. Medeiros, L. Filho, R.A. Kairalla, D. Deheinzelin, R. Magaldi, C. Carvalho, Improved survival in ARDS. Beneficial effects of a lung protective strategy. *Am. J. Respir. Crit. Care Med.* 153: A531 (1996).

16. L. Brochard, F. Roudot-Thoraval, et al., Tidal volume reduction in ARDS: a multicenter randomized study, *Am. J. Respir. Crit. Care Med.* 155: A505 (1997).

17. T.E. Stewart, M.O. Meade, J. Granton, S. Lapinsky, R. Hodder, R. Mc. Lean, D. Mazer, T. Rogevin, T. Todd, D.J. Cook, A.S. Slutsky, Pressure and volume limitated ventilation strategy in patients at risk for ARDS. Results of a multicenter trial, *Am. J. Respir. Crit. Care Med.* 155: A505 (1997).

18. R. Brower, C. Shanholtz, D. Shade, H. Fessler, D. White, C. Wiener, J. Teeter, Y. Almog, J.O. Dodd, S. Piandosi, Randomized controlled trial of small tidal volume ventilation in ARDS, *Am. J. Respir. Crit. Care Med.* 155: A93 (1997).

19. F. Feihl and C. Perret, Permissive hypercapnia: how permissive should we be? *Am. J. Respir. Crit. Care Med.* 150: 1722 (1994).

20. T.C. Corbridge, L.D.H. Wood, G.P. Crawford, M.J. Chudoba, J. Yanos, J.I. Sznadjer, Adverse effects of large tidal volume and low PEEP in canine acid aspiration, *Am. J. Respir. Crit. Care Med.* 142: 311 (1990).

21. J.G. Muscedere, J.B.M. Mullen, K. Gan, A.S. Slutsky, Tital ventilation at low airway pressure can augment lung injury, *Am. J. Respir. Crit. Care Med.* 149: 1327 (1994).

22. L. Gattinoni, P. Pelosi, S. Crotti, F. Valenza, Effects of PEEP on regional distribution of tidal volume and recruitment in ARDS, *Am. J. Respir. Crit. Care Med.* 151: 1807 (1995).

23. M.H. Kolleff, D.P. Schuster, The acute respiratory distress syndrome. *New Engl. J. Med.* 332: 27 (1995).

24. A.H. Morris, C.J. Wallace, R.L. Menlove, Randomized clinical trial of pressure-controlled inverse ratio ventilation and ECCO2 removal for ARDS, *Am. J. Respir. Crit. Care Med.* 149: 295 (1994).

25. M.H. Kollef, Rescue therapy for the ARDS, Chest 111: 845 (1994).

ENDOTRACHEAL TUBES RESISTANCE AND PASSIVE EXPIRATORY TIME DURING MECHANICAL VENTILATION IN ARDS

P.K Behrakis[1]   G.GH. Georgiadis[2] and M.P. Vassiliou [3]

1. Department of Experimental Physiology , School of Medicine, Athens University,Greece
2. Intensive Care Unit,Heppokrateion Hospital, Athens
3. Respiratory Department,St. Savas Hospital of Athens , Greece.

The appropriate regulation of the respirator serttings during mechanical ventilation in ARDS is influenced not only by the mechanical disturbances of the lung parenchyma but it also depends on the mechanical properties of the used Endotracheal Tube (ET). The intubation of the ARDS patient means practically the addition of an important resistor to the whole respiratory circuit. Indeed, several studies in vitro or/and in vivo during the last thirty years [1-12] have pointed out that the ET is the major site of the developped flow Resistance in the intubated patient.

In a recent work of ours [13] we have studied in detail the in vitro resistive behaviour of the adult ETs, with internal diameter ranging from 5.0 to 9.5 mm , with four different gases or gas mixtures: N_2O (100%), O_2 (100%), N_2 (80%)-O_2(20%) and He (80%)-O_2(20%) during both the inspiratory and the expiratory phases. Our study was carried out with a mechanical model (Fig 1) which mimicked the ET positioned in the trachea. Flow and pressure measurements were made under steady state conditions at different plateau points corresponding to gradually increasing and decreasing flow and pressure levels. The ET resistance (R_{tb}) expressed as the pressure drop between the points A and B of Fig.1 (ΔP) divided by the corresponding flow (V') was plotted versus V'. Linear regression analysis of R_{tb} versusV' enabled us to calculate the intercepts and slopes corrresponding to the linear (k_1) and the non linear (k_2) coefficients of resistance, respectively. The correlation coefficient (r) values were an efficacy measure of the selected model , according to Roehrer's formula : $R_{tb} = k_1 + k_2V'$ [14]. Our results, presented in Table 1, indicate the following:

1. There is an extremely high degree of fitness (r values ranging from 0.976 to 0.999, p<0.001) suggesting that the linear model is appropriate to express the flow resistive beahaviour of the adult ETs in vitro. Our data show that all studied ETs with all gases or gas mixtures behave as non linear resistors with a linear increase of R_{tb} with V'.

2. A gradual fall of k_1 & k_2 with the increase of ID of ETs is observed following different patterns in the inspiratory and the expiratory phase for each gas or gas mixture included in this

study (Table 2). Equations in this table appear with different intercepts and slopes in inspiratory and expiratory phases as well as among different gases or gas mixtures.

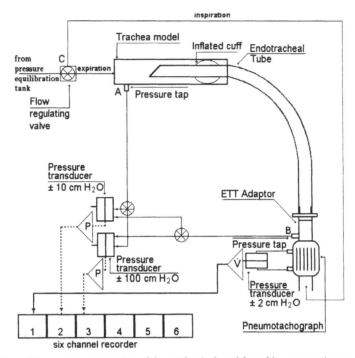

Fig. 1 Diagramatic presentation of the mechanical model used in our experiments.

3. The He(80%)-O_2(20%) shows the least values of k_2 for all studied ETs during both the inspiratory and the expiratory phase and this must be the reason that its R_{tb} values are the least ones, well appart from the other three gases or gas mixtures. The resistive pattern of O_2 (100%) is actually similar to the resistive pattern of N_2(80%) - O_2 (20%) for all ETs in both respiratory phases and the N_2O(100%) presents the higher values of R_{tb} during both respiratory phases.

4. Inspiratory k_1 are lower than expiratory k_1, while the opposite is true for k_2. Although, k_1 and k_2 differences between the inspiratory and expiratory phases are refered in the literature as non important (13), we believe that our experimental geometrical conditions, which mimic the in vivo intubation (ET connector at the begining of the inspiratory flow) must be the reason for the relative superiority of turbulance in the inspiratory than in the expiratory phase. This can also explain the higher inspiratory than expiratory R_{tb} observed, but the "counterbalancing" effect of the higher inspiratory k_1 versus the higher expiratory k_2 tends to restrict the difference between inspiratory and expiratory R_{tb}. It should be mentioned that, in spite of differences in k_1 and k_2 values among the four gases or gas mixtures studied for every ET, expiratory k_1 and k_2 are strongly correlated to their corresponding inspiratory coefficients. The obtained high r values (Table 3) confirm that the corresponding regression equations have an appreciable predictive value.

Table 1. Linear (k_1 : cm H_2O . s . l^{-1}) , non linear (k_2 :cm H_2O . s^2. l^{-2}) and regression coefficients (r) for all tested endotracheal tubes (ETs) with different Internal Diameters (ID = 6.0 - 9.5 mm) in inspiratory and expiratory phases , using : A. N_2(80%) -O_2(20%) , B. O_2 (100%) , C. N_2O(100%) and D. He(80%) - O_2(20%).

	ET Int. Diameter (mm)	Inspiratory Phase			Expiratory Phase		
		k_1	k_2	r	k_1	k_2	r
N_2(80%) -O_2(20%)	6.0	1.30	19.01	0.996	3.36	17.64	0.999
	6.5	1.10	14.32	0.996	2.37	11.03	0.999
	7.0	1.01	10.92	0.997	1.82	7.80	0.996
	7.5	0.75	8.19	0.996	1.31	5.22	0.997
	8.0	0.68	6.07	0.993	1.31	4.82	0.997
	8.5	0.47	5.01	0.998	1.19	3.16	0.995
	9.0	0.60	2.92	0.998	1.06	2.42	0.993
	9.5	0.43	2.56	0.999	0.71	2.08	0.991
O_2 (100%)	6.0	1.32	27.88	0.999	3.68	23.28	0.992
	6.5	1.10	18.66	0.998	3.46	12.90	0.990
	7.0	1.27	13.97	0.999	2.46	9.29	0.990
	7.5	0.95	8.97	0.998	2.07	6.01	0.990
	8.0	0.67	7.78	0.999	1.91	5.75	0.988
	8.5	0.64	6.03	0.999	1.28	3.92	0.991
	9.0	0.65	3.97	0.998	1.15	2.91	0.990
	9.5	0.50	3.08	0.995	1.08	2.28	0.994
N_2O(100%)	6.0	1.11	22.48	0.995	3.09	17.89	0.999
	6.5	0.91	15.54	0.991	2.21	11.18	0.999
	7.0	1.04	12.47	0.995	2.03	8.07	0.999
	7.5	1.08	7.23	0.976	1.59	5.05	0.998
	8.0	0.88	6.35	0.998	1.42	4.97	0.996
	8.5	0.59	5.28	0.997	1.12	3.28	0.993
	9.0	0.67	3.25	0.996	0.95	2.64	0.992
	9.5	0.53	2.67	0.997	0.87	1.98	0.992
He(80%) - O_2(20%)	6.0	2.25	7.14	0.995	2.55	6.10	0.990
	6.5	1.68	4.97	0.996	1.88	4.05	0.974
	7.0	1.21	4.00	0.997	1.53	3.14	0.989
	7.5	1.08	2.75	0.991	1.21	2.29	0.938
	8.0	0.78	2.71	0.999	1.24	2.05	0.977
	8.5	0.68	2.09	0.988	0.81	1.47	0.992
	9.0	0.46	1.68	0.998	0.65	1.18	0.984
	9.5	0.37	1.29	0.998	0.51	1.03	0.996

Table 2. Regression equations of k_1 & k_2 versus ET size (ID = 6.0 -9.5 mm) in inspiratory and expiratory phases for all tested gases or gas mixtures. Regression model: $\ln Y = \ln a + b \times \ln X$, where Y: the k_1 (cm H_2O . s . l^{-1}) or k_2 (cm H_2O . s^2. l^{-2}) coefficient . X: the ET Internal Diameter (mm)

Gas or gas mixture	Inspiratory k_1	Expiratory k_1
$N_2(80\%)$-$O_2(20\%)$	$\ln Y = 4.59 - 2.40 \times \ln X$, r =0.963	$\ln Y = 6.41 - 2.95 \times \ln X$, r =0.973
$O_2(100\%)$	$\ln Y = 2.90 - 1.52 \times \ln X$, r =0.853	$\ln Y = 5.98 - 2.73 \times \ln X$, r =0.994
$N_2O(100\%)$	$\ln Y = 4.19 - 2.14 \times \ln X$, r =0.943	$\ln Y = 6.64 - 2.93 \times \ln X$, r =0.986
$He(80\%)$-$O_2(20\%)$	$\ln Y = 7.72 - 3.84 \times \ln X$, r =0.993	$\ln Y = 6.93 - 3.34 \times \ln X$, r =0.985
Gas or gas mixture	**Inspiratory k_2**	**Expiratory k_2**
$N_2(80\%)$-$O_2(20\%)$	$\ln Y = 11.01 - 4.45 \times \ln X$, r= 0.995	$\ln Y = 11.09 - 4.62 \times \ln X$, r =0.995
$O_2(100\%)$	$\ln Y = 11.42 - 4.62 \times \ln X$, r =0.992	$\ln Y = 11.10 - 4.62 \times \ln X$, r =0.994
$N_2O(100\%)$	$\ln Y = 11.74 - 4.69 \times \ln X$, r =0.996	$\ln Y = 11.64 - 4.83 \times \ln X$, r =0.993
$He(80\%)$-$O_2(20\%)$	$\ln Y = 8.24 - 3.52 \times \ln X$, r =0.993	$\ln Y = 7.11 - 3.05 \times \ln X$, r =0.912

Differences between in vitro versus in vivo measurements of R_{tb} have been discussed in the relative literature [10,15-18]. R_{tb} values are higher in vivo than in vitro ones because of temperature variations , ET kinking or secretions deposition . These factors change the ET geometry and may induce important and varying deviations of the measured values from the real mechanical properties of ETs . Thus , comparative results and conclusions are confounding . Data obtained under our experimental conditions might be taken into account as a minimum R_{tb} value in any case of R_{tb} measurements in vivo.

Table 3. Linear regression equations between expiratory k_1 versus inspiratory k_1 and expiratory k_2 versus inspiratory k_2 for all studied gases or gas mixtures .

Gas or gas mixture	Expiratory k_1 versus inspiratory k_1	Expiratory k_2 versus inspiratory k_2
$N_2(80\%)$-$O_2(20\%)$	Exp k_1 = -0.878 + 3.618 x Insp. k_1 , r = 0.772	Exp. k_2 = -0.301 + 1.657 x Insp. k_2 , r = 0.997
N_2O (100%)	Exp. k_1 = -0.453 + 2.922 x Insp. k_1 , r = 0.954	Exp. k_2 = -2.063 + 0.936 x Insp. k_2 , r = -0.997
O_1 (100%)	Exp. k_1 = -1.526 +3.854 x Insp. k_1 , r = 0.902	Exp. k_2 = -0.908 + 0.834 x Insp. k_2 , r = 0.999
$He(80\%)$-$O_2(20\%)$	Exp. k_1 = 0.044 + 1.187 x Insp. k_1 , r =0.995	Exp. k_2 = -0.09 + 0.822 x Insp. k_2 , r = 0.999

Fig 2 refers to graphical presentation of passive expiratory Flow-Volume loops of a normal subject and an ARDS patient with inspired Tidal Volume of 1 L. Normal Ers and Rintr values where obtained from literature [6] whereas Rtb values were taken from our measurements during the expiratory phase. Figure 2 shows that:

1. All passive expiratory Flow-Volume loops are curvilinear due to the non linear resistive coefficient (k_2) of the ETs.

2. For all ETs and gas or gas mixtures studied, expiratory flow rates are higher in ARDS patients in comparison to normal subjects reflecting the higher Ers values of ARDS patients.

3. There is a clear effect of ET size on expiratory Flow-Volume loops in both normal subjects and in ARDS patients; for a given volume, small sized ETs are characterized by lower expiratory flow rates compared to the larger ones. This is attributed to the higher k_1 & k_2 coefficients and consequently higher pressure drop along the smaller ETs.

4. Expiratory Flow-Volume loops are also influenced by the inspired gas or gas mixture. For a given volume and ET size the He(80%)-O_2(20%) mixture shows the higher flow rates in both normal subjects and ARDS patients . This observation is related to the relatively lower k_1 & k_2 coefficients of the He(80%)-O_2(20%) mixture.

Our analysis of the effects of ETs and inspired gases in normal subjects and ARDS ppatients was extended to calculations of the T_E . The duration of the passive expiratiory phase in the mechanically ventilated ARDS patients is of particular importance with its specific biophysical relation to the mechanical properties of the respiratory system. Under these conditions, the T_E is strongly affected by the expiratory resistive properties of the ET. The ET serially added to the upper airways would increase the flow resistance for the whole: ET + Respiratory System, that should not be ignored or underestimated as a critical factor for proper mechanical ventilation of every ARDS patient. The duration of a complete expiration seems to be the only appropriate way to study the passive expiratory timing. The Time Constant (τ_{rs}) can not be used in studies of non linear phenomena like passive expiration through an ET - resistor which is characterized by a non ignorable k_2 turbulent resistive coefficient. Our expresimental data proove that the assumption of flow resistance linearity in intubated patients is not valid.

Passive expiration during mechanical ventilation is determined by the classical equation of motion of the respiratory system: ErsV = (Rintr+Rtb)V' (2) , where Ers: the Respiratory System elastance , Rintr: the intrinsic Respiratory System resistance , V: the inspiratory volume above FRC , V' : the Expiratory Flow.

Equation (2) allowed us to provide the effecs of ET size and inspired gasor gas mixture on the passive expiratory Flow-Volume loops in ARDS patients .

In order to study T_E , we transformed equation (2) as follow:

$$T_E = \frac{d + k_1 \ln|d - k_1|}{E_{rs}} \quad \text{where} \quad d = \sqrt{k_1^2 + 4k_2 E_{rs} V} \qquad (3)$$

Based on Equation (3) we studied the effects of ET size and inspired gas or gas mixture on T_E. Our results are presented in Table 4 refering to normal subjects and ARDS patients. Data of table 4 are in accordance with figure 2 depicting passive expiratory Flow-Volume loops. A relatively shorter T_E corresponds to a higher expiratory flow rate and vice versa.

In conclusion, the ARDS patient is aggrevated not only by the intrinsic elastic and resistive loading resulting by the disease itself but also by an additional flow resistance due to the endotracheal intubation. ETs are non linear resistors with a mechanical behaviour directly related to the their ID and to the inspired gas or gas mixture. Narrower ETs are characterized

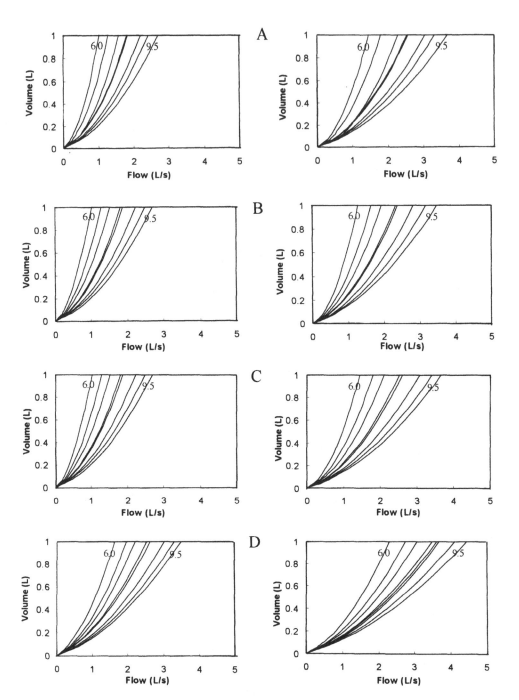

Fig 2. Expiratory Flow - Volume Curves through Endontracheal Tubes with Internal Diameter = 6.0 - 9.5 mm, using A: O_2 (100%) . B:N_2O (100%). C: N_2(80%)-O_2(20%). D:He(80%)-O_2(20%), in normal subjects (left) and in ARDS patients (right).

Table 4. Passive Expiratory time (in seconds) through ETs with I.D. of 6.0 - 9.5 mm , using four different gases or gas mixtures (Tidal Volume = 1 L) in normal subjects and in ARDS patients .

ETs ID (mm)	6.0	6.5	7.0	7.5	8.0	8.5	9.0	9.5
Gas or Gas mixture	Passive expiratory time (T_E) in normal subjects (s)							
N_2(80%)-O_2(20%)	2.54	1.99	1.67	1.37	1.33	1.12	1.00	0.91
O_2 (100%)	2.51	1.97	1.72	1.39	1.35	1.13	1.02	0.91
N_2O (100%)	2.87	2.27	1.87	1.54	1.50	1.22	1.08	0.98
He(80%)-O_2(20%)	1.61	1.31	1.15	1.00	0.96	0.81	0.74	0.69
	Passive expiratory time (T_E) in ARDS patients (s)							
N_2(80%)-O_2(20%)	1.82	1.46	1.24	1.04	1.01	0.86	0.78	0.72
O_2 (100%)	1.81	1.45	1.27	1.05	1.03	0.87	0.80	0.72
N_2O (100%)	2.05	1.63	1.38	1.15	1.12	0.94	0.84	0.76
He(80%)-O_2(20%)	1.19	0.99	0.88	0.77	0.75	0.65	0.59	0.56

by relatively higher flow resistance, whereas the He(80%)-O_2(20%) mixture offer substantially less resistance than room air, N_2O (100%) and O_2 (100%). Both, ET size and inspired gas or gas mixture have a strong effect on passive expiratory flow rate and T_E. ARDS patients show a shorter duration of passive expiration compared to normal subjects and this has to be taken into account for the proper setting of the ventilator in order to avoid a possible undesirable increase of end-expiratory pause. A further ventilatory improvement can be achieved by using larger ETs and He mixtures of Oxygen.

REFEENCES

1. Orkin L , Siegel M , Rovestine E ,1954 , Resistance to breathing by apparatus used in anesthesia . *Anesthesia Anagelsia Current Res* ; 33: 217 - 233.
2. Sahn SA , Laksminarayan S , Petty TL, 1976, Weaning from mechanical ventilation . *JAMA*; 235 : 2208 - 2212.
3. Sullivan M , Paliotta J , Saklad M, 1976, Endotracheal tube as a factor in measurement of respiratory mechanics. *J Appl Physiol* ; 40:590-592.
4. Demers RR , Sullivan MJ , Paliotta J,1977, Airflow Resistances of Endotracheal Tubes. *JAMA* ; 237 :1362 .
5. Loring SH , Elliot ES , Drazen JM,1979, Kinetic energy loss and convective acceleration in respiratory resistance measurements . *Lung* ; 156 :37 - 42 .
6. Behrakis PK, Higgs BD, Baydur A, Zin WA, Milic -Emili J,1983,Respiratory mechanics during halothane anesthesia and anesthesia-paralysis in humans.*J Appl Physiol*; 55: 1085-1092.

7. Gottfried SB , Rossi A , Higgs BD , Calverley MA , Zocchi K , Bosic C , Milic Emili J, 1985, Noninvasive Determination of Respiratory System Mechanics during Mechanical ventilation for acute respiratory failure. *Am Rev Respir Dis* ; 131: 414-429.

8. Bolder PM , Healy TEJ , Bolder AR , Beatty PCW , Kay B, 1986, The extra work of breathing through adult endotracheal tubes . *Anesth Analg* ; 65 : 853-859.

9. Baydur A , Sassoon SH , Stiles CM, 1987, Partitioning of Respiratory mechanics in Young Adults . Effects of duration of anesthesia . *Am Rev Respir Dis* ; 135: 165 - 172.

10. Wright PE, Marini JJ, Bernard GR,1989, In vitro versus in vivo comparison of endotracheal tube airflow resistance . *Am Rev Respir Dis* ; 140:10-18.

11. Dennison FH , Taft AA , Mishoe SC , Hooker LL , Eatherly AB , Beckham RW,1989, Analysis of resistance to gas flow in nine adult ventilator circuits.*Chest* ; 96: 1374 - 1379.

12. Prezant DJ, Aldrich TK, Karpel JP, Park SS, 1990, Inspiratory flow dynamics during mechanical vantilation in patients with respiratory failure. *Am Rev Respir Dis* ;142:1284- 1287.

13.Behrakis PK, Georgiadis G.CH, Vassiliou MP, Alivisatos GP and Milic Emili J, 1997, Inspiratory and expiratory flow resistance of adult endotracheal tubes. *Submitted.*

14. Rohrer R, 1915, Der Stroemungswiderstand in den menschlicen Atemwegen und der Einfluss der unregelmassigen Verzweigung des bronchial Systems auf den Atmungsverlauf in verschiedenen Lungenbezirken . *Pfluegers Arch Gesamte Physiol Menschen* Tiere . 162 : 225-299.

15.Peslin R, Felicio da Silva J, Chabot F, Duvivier C,1992, Respiratory mechanics studied by multiple linear regression in unsedated ventilated patients. *Eur Respir J* ; 5:871-878.

16.Conti G, De Blasi RA , Lappa A, Ferretti A, Antonelli M, Bufi M, Gasparetto A,1994, Evaluation of Respiratory System Resistance in Mechanically Ventilated Patients: The Role of the Endotracheal tube. *Intensive Care Medicine*; 20:421-424.

17.Chang HK, Mortola JP,1981, Fluid dynamic factors in tracheal pressure measurement. *J Appl Physiol* ; 51: 218 - 225 .

18. Lofaso F , Louis B , Brochard L , Harf A , Isabey D, 1992, Use of the Blasius resistance formula to estimate the effective diameter of endotracheal tubes.*Am Rev Respir Dis*;146 : 974-979.

USE OF LIQUID VENTILATION IN ANIMAL MODELS OF ARDS

Serge J. C. Verbrugge, Vera Šorm, Burkhard Lachmann

Dept. Anesthesiology
Erasmus University Rotterdam
Dr. Molewaterplein 50
3000 DR Rotterdam
The Netherlands

INTRODUCTION

Laplace, a French mathematician (1749-1827), was the first to draw attention to surface active forces in general, and described the relationship between force, surface tension, and radius of an air-liquid interface of a bubble:

$$P = 2\gamma/r$$

(P = pressure to stabilize a bubble; γ = surface tension at air-liquid interface; and r = radius of a bubble). A century later, von Neergaard[1] applied this law to pulmonary alveoli, by demonstrating that the pressures required to expand an air-filled lung were almost three times higher that required to distend a lung filled with fluid. In this way the surface tension effect at the air-liquid boundary was eliminated (Fig. 1).[1] From these findings he concluded that: (1) two-thirds of the retractive forces in the lung are caused by surface tension phenomena, which act at the air-liquid interface of the alveoli; and (2) the surface tension at the air-liquid interface must be reduced by the presence of a surface active material with a low surface tension to allow normal breathing. This surface active material has been recognized as pulmonary surfactant.[2]

Both Neonatal respiratory distress syndrome (RDS) and acute RDS are characterized by a diminished pulmonary surfactant system resulting in increased forces at the air-liquid interface with end-expiratory alveolar collapse, atelectasis, increases in right-to-left shunt and a decrease in PaO_2.

To overcome the elevated retractive forces of stiff lungs and to improve oxygenation Shaffer eliminated the elevated surface forces at the air-liquid interface in stiff lungs by filling them with perfluorocarbons (PFC).[3] This alternative technique was originally performed by totally filling the lungs with PFCs and giving tidal breaths of preoxygenated PFCs; this was

Acute Respiratory Distress Syndrome: Cellular and Molecular Mechanisms and Clinical Management
Edited by Matalon and Sznajder, Springer Science+Business Media New York, 1998

319

called total liquid ventilation. This technique is not encouraging as it requires sophisticated equipement in the form of a liquid ventilator and an extra-corporeal membrane oxygenator; thus the search began for a simpeler technique of liquid ventilation to support pulmonary gas exchange. Fuhrman was the first to demonstrate the feasibility of applying perfluorocarbons in healthy animals without the need for a specialized liquid breathing system[4]. In this technique the lung is filled with PFC to a volume equal to or below functional residual capacity (FRC), after which conventional mechanical ventilation is superimposed.

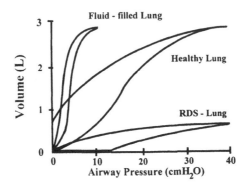

Figure 1. Pressure-volume diagram of a normal air-filled lung and a lung with respiratory distress (RDS). Von Neergaard [1] showed in 1929 that much larger pressures were required to expand an air-filled lung than a lung filled with fluid. In RDS even higher pressures are required to expand the lung because of the high surface tension at the air-liquid interface in the alveoli, caused by a diminished surfactant system.

PERFLUOROCARBONS

The most important physical properties of PFCs that make them suitable for ventilatory purposes are their unique ability to dissolve large amounts of oxygen and carbon dioxide, and their remarkable low surface tension (Table 1). PFC liquids are relatively simple organic compounds in which all hydrogen atoms have been replaced by fluorine, bromide, or iodine. These colorless, clear, and odorless liquids are insoluble in aqueous media. PFCs are very stable

Table 1. Physical properties of some perfluorocarbon liquids.

	Perflubron	FC-77	RM-101
Density (g/ml)	1.92	1.75	1.77
Vapor pressure (mmHg at 37 °C)	10.5	75	64
Surface tension (dyn/cm)	18	14	15
O_2 solubility (ml/100ml)	53	56	52
CO_2 solubility (ml/100ml)	210	198	160

320

biologically inert liquids and do not seem to be metabolized by biological systems. As an oxygen carrier, PFCs are used in various indications, either intravenously in emulsified form as a blood replacement, or intratracheally in neat (unemulsified) form. When administered into the lungs, systemic absorption and distribution of small amounts of PFC to other tissues has been demonstrated; however, the main elimination of PFC is through evaporation via the lungs. This elimination is dependent on the vapour pressure of the specific PFC used (Table 1).

PLV IN ANIMAL MODELS OF ARDS

Introduction

Experiments with PLV in animal models of ARDS have shown that:

1)   Higher doses of PFC lead to higher levels of oxygenation.[5] This is suggested to result from dose-dependent recruitment of collapsed atelectatic alveoli by PFC fluid. Thereby gas exchange also continues during the expiratory phase of the respiratory cycle.

2)   Oxygenation deteriorates over time if no additional doses of PFC are applied.[6] This is attributed to evaporation of PFC, which will cause affected alveoli to collapse.

3)   Lung mechanics and carbon dioxide elimination improve after an initial low dose of PFC and show no further improvements with subsequent higher doses of PFC.[5] This is attributed to the replacement of the alveolar air-liquid interface with a thin air-PFC interface. Evaporating PFC appears to cover the entire lung surface. As PFCs have a low constant surface tension, pulmonary compliance is increased after a low dose PFC and CO_2 elimination is higher. No further improvement is seen after additional PFC dosing.

4)   'Gas-PEEP' has to be applied during PLV to prevent bulk movement of PFC fluids from the alveoli into the airways and to prevent dangerously high airway pressures at the onset of inspiration.[5]

5)   PLV does not impair any cardiovascular parameter, even in animals with a large anterior-posterior thoracic diameter. Mean pulmonary artery pressure decreases when PFC is applied, due to reversal of hypoxic pulmonary vasoconstriction.[7]

6)   PLV can prevent the progress of histologically assessed lung injury.[8]

7)   PLV can be combined with other ventilatory support techniques in ALI such as nitric oxide administration and high frequency oscillation.[9, 10]

Working mechanism of partial liquid ventilation in RDS

Our group was the first to apply PLV to animals suffering on acute respiratory failure.[5] To answer the question how much PFC is necessary to improve gas exchange in acute respiratory failure, lung lavaged rabbits were treated with incremental doses of 3 ml PFC/kg bodyweight up to a total volume of 15 ml/kg bodyweight. The animals were ventilated in a volume-controlled mode and with 6 cm H_2O of PEEP, which remained unchanged during the 80-minute study period. PaO_2 in the PFC-treated animals improved significantly with each subsequent dose of PFC, eventually reaching prelavage values. $PaCO_2$, peak airway and plateau pressures decreased significantly after instillation of 3 ml PFC/kg and remained stable with the subsequent treated doses. These data clearly show that oxygenation improves in a dose-dependent way, in contrast to lung mechanics that improved already after low-dose PFC.

To see how long these effects last under similar experimental conditions, we conducted a study with PLV over a period of 3 hours at a dose of 18 ml PFC/kg. Oxygenation was almost restored to a healthy state immediately after instillation of PFC and remained stable throughout the study period. Alveolar ventilation and $PaCO_2$ could be kept at a constant level in the PFC group. Moreover, the PFC treated animals showed a persistent significant decrease in peak and mean airway pressure as compared with post-lavage data, and respiratory system compliance improved. More importantly, these improvements occurred without significant changes in mean arterial pressure, heart rate, or central venous pressure. Additionally, histologic examination of the lungs from PFC treated animals did not reveal significant morphological abnormalities, whereas the lungs of the conventionally ventilated group showed atelectasis, hyaline membranes, and overdistension and rupture of alveoli.[11]

Based on these results, we conducted another study in which the influence of different doses of PFC over a period of 6 hours on gas exchange was analysed.[7] Oxygenation showed a time and dose-dependency; deterioration of oxygenation was faster in those animals receiving smaller doses of PFC. The relationship between the dose of PFC administered and the time at which impairment of lung function was seen strongly suggests that during PLV evaporation of PFC over time would cause the affected alveoli to collapse and, therefore, limit the efficacy of pulmonary gas exchange. This impairment of gas exchange will occur sooner with smaller doses of PFC if the liquid is not replaced.

In the same study, we attempted to establish what amount of PFC can be applied intrapulmonary in PLV without having an adverse effect on peak inspiratory pressure. It was shown that a PFC dose exceeding 25 ml/kg (i.e., higher than FRC), resulted in higher peak airway pressures during PLV at volume-controlled ventilation than during conventional gas ventilation with the same tidal volume. This effect was attributed to tissue elasticity becoming the predominant component of the alveolar retractive forces.[7]

From the above-mentioned studies a hypothetical mechanism of action for PLV can be deducted (Fig 2). Panel A shows the atelectatic ARDS lung. After a small dose of PFC (3 ml/kg) a thin film with a low surface tension is formed at the air-liquid interface due to evaporation of the PFCs (Figure 2B) and covers the lung units of the whole lung. Due to this film the increased surface tension in the diseased lung is reduced to a low but constant value, which leads to a decrease of inflation pressure. Due to the constant surface tension of PFC, this pressure cannot further decrease with additional doses of PFC. Independent of this speculation, the dose-dependent improvement in oxygenation results from the filling and opening of the collapsed atelectic alveoli in the dependent part of the lung by the noncompressible PFC, thus preventing them from end-expiratory collapse. This leads to a continuation of gas exchange, even during the expiratory phase of the respiratory cycle. With increasing amounts of PFC in the lung, more collapsed atelectatic alveoli can be opened and prevented from end-expiratory collapse (Figure 2B vs. 2C) thus eliminating intrapulmonary shunt.

PLV does not impair hemodynamics

These above mentioned studies were conducted in rabbits with a small thoracic diameter. However, based on the high specific gravity of PFC (twice that of water), one could assume that PFC may lead to a hemodynamic interference by compression of the pulmonary capillaries in the dependent parts of the lung of an adult person with a large thoracic diameter. Therefore, we investigated in an experimental study the dose-related influence of PFC administration on hemodynamics and gas exchange in lung-lavaged pigs with a thoracic diameter of 24 cm and a bodyweight of 50 kg.[7] It was found that increasing doses of PFC had no deleterious effect on any of the hemodynamic parameters. After 5 ml/kg PFC, mean pulmonary artery pressure even decreased and was unaffected by further PFC administration. This can be explained by the relief of hypoxic pulmonary vasoconstriction. These results were confirmed by a study of Hernan et al. in adult sheep.[12] After inducing lung injury by instilling 2 ml/kg hydrochloric acid into the trachea, they examined whether PLV could be used to

Upper lung

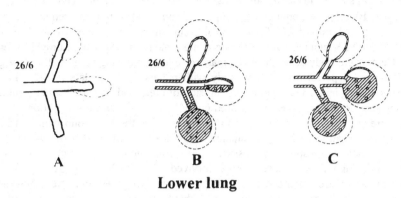

Lower lung

Figure 2. Shows the lining of the air-liquid barrier of the alveoli and the dose-dependent improvement in oxygenation. A; shows the atelectatic RDS lung. B; shows the partially perfluorocarbon-filled lung. C; shows the effect of instilling additional perfluorocarbon. See text for details (solid line = end-expiration; dashed line = end-inspiration)

enhance gas exchange in large animals, without adverse effects on hemodynamics. The sheep were randomly divided into a PLV group and a control group that received normal gas ventilation. During the 4 hour observation period, PaO_2 was higher and shunt was reduced in the PLV group, whereas the values remained unchanged in the control group. This study supports our findings that PLV can improve lung mechanics and gas exchange without hemodynamic impairment.

PLV can prevent the progress of histologically assessed lung injury

Other researchers were interested in investigating whether PFCs can influence the progress of morphologic damage after induction of respiratory insufficiency. Nesti et al. applied PLV in a piglet model of gastric aspiration.[13] Volume-controlled gas ventilation was compared to PLV. PFC was instilled 1 hour after acid instillation. In the PFC group oxygenation improved and became significantly better from 2.5 to 6 hours after injury, whereas $PaCO_2$ did not show any difference compared to the control group. Peak inspiratory pressure during fixed volume ventilation did not differ significantly between groups, but end-inspiratory pressure was lower and compliance consistently higher after 3 hours in the PLV group. Moreover, there was almost no histologic evidence for lung injury in the PLV group, in contrast to the gas-ventilated group.

Hirschl et al. compared lung histology after PLV and gas ventilation in a model of acute respiratory failure.[14] They applied PLV to adult sheep in which lung injury was induced by intravenous administration of oleic acid, followed by saline pulmonary lavage. After this, animals were put on bijugular venovenous extracorporeal life support (ECLS), when alveolar-arterial oxygen pressure difference was 600 mm Hg or more and PaO_2 was 50 mm Hg or less with fraction of inspired oxygen of 1.0. For the first 30 min on ECLS, all animals were ventilated with gas. The following 2.5 hours, ventilation with 15 ml/kg gas was continued with the addition of 35 ml/kg of PFC in the PLV group or without intervention in the control group. The extracorporeal blood flow was adjusted to maintain PaO_2 at 50 to 80 mm Hg. At 3 hours after initiation of ECLS, shunt was significantly reduced and pulmonary compliance was increased in the PLV group compared with the control group. The ECLS flow rate required to maintain the PaO_2 in the 50 to 80 mm Hg range was lower in the PLV group. After the 2.5 hours ECLS was stopped in each group and gas or liquid ventilation was continued for 1 hour or until death. Blood gases, pulmonary compliance and shunt were measured in both groups. After the discontinuation of ECLS, all control group animals died, most within 5 to 30 min. Four of the five PLV animals survived the predetermined 1 hour period. Lung biopsies were taken and light microscopy demonstrated a marked reduction in lung injury in the PLV group.

Papo et al. performed a PLV study with piglets.[15] The piglets (2.3-2.9 kg) were ventilated and lung injury was induced by infusion of oleic acid. Hemodynamics, pulmonary mechanics and blood gases were monitored every 15 minutes during the 3 hour study. The animals were randomised to a PLV group or a volume-controlled group with PEEP. They found that oxygenation and lung compliance significantly increased and peak inspiratory pressure and static end-inspiratory pressure decreased in the PLV group compared with the control group. Significantly more animals survived in the PLV group. After 3 hours, all surviving animals were sacrificed and their lungs were weighted and fixed. Normal lung histology was markedly preserved in the animals that had received PFC, whereas lungs of the gas ventilated animals appeared seriously damaged.

These data suggest that PLV with PFC might prevent acute respiratory distress syndrome (ARDS) after acute lung injury, induced by aspiration or oleic acid.

PLV IN CLINICAL TRIALS IN PATIENTS WITH ACUTE RESPIRATORY DISTRESS SYNDROME

Although preliminary evaluations of phase II and III clinical studies with PLV have shown that PFCs can be safely applied to patients with ALI,[16] these trials did not show the tremendous improvement of gas exchange that might have been anticipated from animal studies, nor could they demonstrate improvement in survival of patients treated with PLV compared to those treated with conventional mechanical ventilation only. There is a tendency towards an increased number of pneumothoraces in patients treated with PLV.

POSSIBLE SIDE-EFFECTS OF PLV

There are some unclear aspects of PLV that may have adverse consequences for the lung. First, although the alveolar air-liquid interface is eliminated when PFCs are instilled into the alveolus, a new interface with new surface tension is created at the air-PFC barrier. The law of Laplace ($P = 2\gamma/r$) describes the relationship between the pressure to stabilize an alveolus (P) and surface tension at the gas-liquid interface of an alveolus (γ) in relation to the radius of the alveolus (r). In the normal 'healthy' situation, surfactant present at the air-liquid interface in the lung has the unique property of lowering air-liquid surface tension in parallel with a decrease in alveolar radius (to almost 0 mN/m for low alveolar radii), thus keeping the ratio of γ/r of the alveolus constant and guaranteeing expiratory alveolar stability at low pressures (Fig.). PFC does not show this dynamic behaviour of surface tension. It maintains a constant air-PFC surface tension for any alveolar radius (Fig.). The consequences:

1)   At end-expiration, the alveolar radius is small, whereas the surface tension is high (12 mN/m). Alveoli will collapse if insufficient levels of 'gas-PEEP' are applied.[17] Therefore, 'liquid PEEP' always needs to be accompanied by a high enough level of 'gas PEEP' to prevent end-expiratory collapse of non-PFC filled alveoli.. Insufficient PEEP levels may subject the alveolar wall to shear forces that may promote alveolar disrupture and, successively, pneumothorax formations.

2)   As to the level of 'gas-PEEP', it depends on the disease state of the individual alveolus whether or not the presence of a thin layer of PFC is beneficial. In those alveoli that are unable to reduce surface tension to 12 mN/m or below at minimal alveolar radii, the level of gas-PEEP

to keep alveoli open at end-expiration will be lower during PLV than the level of 'gas-PEEP' during conventional mechanical ventilation. In those alveoli, however, that still have a surfactant system that is able to reduce surface tension to values below 12 mN/m, there will be an interaction between PFC and the normal alveolar lining fluid, leading to levels of 'gas PEEP' with PLV higher than those necessary to keep the alveoli open with 'gas-PEEP' only. Therefore, it may be important not to give PFC early in the disease process of ALI. For the same reason, the weaning period during the transition from PLV to normal gas breathing may be prolonged.[17]

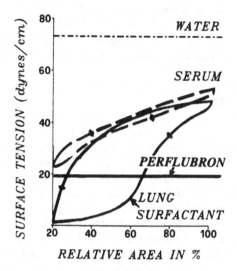

Figure 3. The surface tension behaviour of lung surfactant, serum perflubron and water. Lung surfactant shows dynamic surface tension behaviour, with low surface tension for lower surface areas and higher surface tension for higher surface areas. In ARDS, the lung surfactant at the air-liquid interface is replaced with serum, which displays much higher surface tensions for each surface area when compared to lung surfactant. Water does not display dynamic surface tension behaviour and neither does perflubron. They both show a constant surface tension for different surface areas.

Interactions with surfactant were confirmed in a recent study in which clearance of the radioactive tracer ^{99m}Tc-diethylene trianime pentaacetic acid in healthy rabbits afer three hours PLV. Clearance of this tracer was found to be increasedin PLV compared to rabbits treated with mechanical ventilation only at the same ventilatory parameters.[18] Increased clearance of this tracer was previously shown to be a sensitive indicator of changes in the lung surfactant system. It was found that the clearance of the tracer was increased after PLV.[18] We speculate that the increased surface tension at the alveolar air-liquid interface in the expiratory phase in the PFC treated animals might have contributed to the increased clearance rate of the radioactive tracer. These data indicate an interaction of PFCs with the pulmonary surfactant system.

To investigate whether this interaction with the pulmonary surfactant system is a long-lasting effect we exposed healthy rabbits to three hours of PLV with 12 mL/kg bodyweight PFC.[19] After seven days, the rabbits were anesthetized again and gas exchange and respiratory parameters were measured, including peak and mean airway pressures, and none of the measured parameters was changed. This indicates that the changes in the pulmonary surfactant system after PLV can be overcome in healthy rabbits.[19] The exact mechanism and effect of PFC on the pulmonary surfactant system remains to be elucidated.

Another uncertainty with PLV is where the gas of the tidal breaths superimposed on the PFC is going. Suggesting the air would go to the non-fluid filled parts of the lung, there could be a severe lung overinflation in those lung areas, due to the low surface tension of PFC, when conventional ventilation modes such as volume-controlled flow constant mechanical ventilation are used. This may result in a high incidence of pneumothoraces with PLV. To prevent such complications it is mandatory to combine fluid-PEEP with pressure-controlled modes of ventilation, in which the pressure level in any alveolus can never exceed the pressure level set on the ventilator. This may prevent dangerous alveolar overdistension.

Finally, there are important in vitro changes to the cells of the immune system in the lung after exposure to PFCs. Exposure of alveolar macrophages to PFC decreases production of reactive oxygen metabolites upon chemical stimulation;[20] neutrophils exposed to PFC have less H_2O_2 production and show less chemotactic activity.[21] Whether such changes are beneficial in terms of reducing lung injury due to toxic oxygen radicals, or negative in terms of reduced host defence against intrapulmonary pathogens, needs further investigation.

FUTURE CONSIDERATIONS

The question may be posed whether PLV will have any additional value to optimal mechanical ventilation with PEEP alone in improving ARDS patient survival. Moreover, given the non-dynamic surface tension behaviour of PFCs, restoring the physiological situation at the alveolar air-liquid interphase in ARDS by means of exogenous surfactant therapy may prove to be superior to the application of PFCs. Recent studies in surfactant-depleted lungs have shown that lung compliance can be improved to an even greater extent if intratracheal PFC administration is preceded by exogenous surfactant, which ensures a low interfacial surface tension at low alveolar volumes.[22] Future studies will have to demonstrate whether PLV or surfactant, or a combination, is the best therapy to overcome respiratory failure and improve patient survival in ARDS. Only the ongoing phase III patient studies will show whether or not PLV has an impact on patient survival

ACKNOWLEDGEMENT

We thank Mrs. L Visser-Isles for English language editing.

REFERENCES

1   von Neergaard K: Neue Auffassungen über einen Grundbegriff der Atemmechanik; Die Retraktionskraft der Lunge, ahängig von der Oberflächenspannung in den Alveolen, Z Ges Exp Med 1929; 66: 373-394.

2   Avery MA, Mead J. Surface properties in relation to atelectasis and hyaline membrane disease, Am J Dis Child 1959; 97, 517-23.

3   Shaffer TH, Rubenstein D, Moskowitz GD, Delivoria-Papadopoulos M. Gaseous exchange and acid-base balance in premature lambs during liquid ventilation since birth. Pediatr Res 1976; 10: 227-231.

4   Fuhrman BP, Paczan PR, DeFrancis M: Perfluorocarbon-associated gas exchange. Crit Care Med 1991; 19: 712-722.

5   Tütüncü AS, Faithfull S, Lachmann B. Intratracheal perfluorocarbon administration combined with mechanical ventilation in experimental respiratory distress syndrome: dose-dependent improvement of gas exchange. Crit Care Med 1993; 21: 962-969.

6   Tütüncü AS, Akpir K, Mulder P, Erdmann W, Lachmann B. Intratracheal perfluorocarbon administration as an aid in the ventilatory management of respiratory distress syndrome. Anesthesiology 1993; 79: 1083-1093.

7   Houmes RJM, Verbrugge SJC, Hendrik ER, Lachmann B. Hemodynamic effects of partial liquid ventilation with perfluorocarbon in acute lung injury. Intens Care Med 1995; 21: 966-972.

8   Hirschl RB, Parent RB, Tooley R, McCracken M, Jonson K, Shaffer TH, Wolfson MR, Bartlett RH. Liquid ventilation improves pulmonary function, gas exchange and lung injury in a model of respiratory failure. Ann Surg 1995; 21: 79-88.

9   Houmes RJM, Hartog A, Verbrugge SJC, Böhm SH, Lachmann B. Combining partial liquid ventilation with nitric oxide to improve gas exchange in acute lung injury. Intens Care Med 1997; 23: 162-168.

10   Baden HP, Mellama JD, Bratton SL, O'Rourke PP, Jackson JG. High-frequency oscillation with partial liquid ventilation in a model of acute respiratory failure. Crit Care Med 1997; 25: 299-302.

11   AS, Faithfull S, Lachmann B. Comparison of ventilatory support with intratrachealperfluorocarbon administration and conventional mechanical ventilation in animals with acute respiratory failure. Am Rev Resp Dis 1993; 148: 785-792.

12   Hernan LJ, Fuhrman BP, Kaiser ir RE, Penfil S, Foley C, Papo MC, Leach CL. Perfluorocarbon-associated gas exchange in normal and acid-injured large sheep. Crit Care Med 1996; 24: 475-481.

13   Nesti FD, Fuhrman BP, Steinhorn DM, Papo MC, Hernan LJ, Duffy LC, Fisher JE, Leach CL, Paczan PR, Burak BA. Perfluorocarbon-associated gas exchange in gastric aspiration. Crit Care Med 1994, 22: 1445-1452.

14   Hirschl RB, Tooley R, Parent AC, Johnson K, Bartlett RH. Improvement of gas exchange, pulmonary function, and lung injury with partial liquid ventilation: A study model in a setting of severe respiratory failure. Chest 1995; 108: 500-508.

15   Papo MC, Paczan PR, Fuhrman BP, Steinhorn DM, Hernan LJ, Leach CL, Holm BA, Fisher JE, Kahn BA. Perfluorocarbon-associated gas exchange improves oxygenation, lung mechanics, and survival in a model of adult respiratory distress syndrome. Crit Care Med 1996; 24: 466-474.

16   Hirschl RB, Pranikoff T, Gauger P, Schreiner RJ, Dechert R, Bartlett RH. Liquid ventilation in adults, children and full-term neonates. Lancet 1995; 346: 1201-1202.

17   Salman NH, Fuhrman BP, Steinhorn DM, Papo MC, Hernan LJ, Leach CL, Fisher JE. Prolonged studies of perfluorocarbon associated gas exchange and of the resumption of conventional mechanical ventilation. Crit Care Med 1995; 23: 919-924.

18   Tütüncü AS, Houmes RJ, Bos JAH, Wollmer P, Lachmann B. Evaluation of lung function after intratracheal perfluorocarbon administration in healthy animals: pulmonary clearance of ^{99m}Tc-DTPA. Crit Care Med 1996;24: 274-279.

19   Tütüncü AS, Lachmann B. Effects of partial liquid ventilation on gas exchange and lung mechanics in healthy animals. ACP 1995; 5: 195-199.

20   Smith TM, Steinhorn DM, Thusu K, Fuhrmann BP, Dandona P. A liquid perfluorochemical decreases the in vitro production of reactive oxygen species by alveolar macrophages. Crit Care Med 1995; 23: 1533-1539.

21   Rossman JE, Caty MG, Rich GA, Karamanoukian HL, Azizkhan RG. Neutrophil activation and chemotaxis after in vitro treatment with perfluorocarbon. J Pediatr Surg 1996; 31: 1147-1151.

22   Tarczy-Hornoch P, Hildebrandt J, Mates EA, Standaert TA, Lamm JE, Hodson WA, Jackson JG. Effects of exogenous surfactant on lung pressure-volume characteristics during liquid ventilation. J Appl Physiol 1996; 80: 1764-1771.

INHALED PROSTACYCLIN FOR THE TREATMENT OF ARDS

Nicola Brienza and Marco V. Ranieri

Istituto di Anestesiologia e Rianimazione, Università di Bari, Italy

Since 1967 the Acute Respiratory Distress Syndrome has represented a major cause of morbidity and mortality in the critical care setting. Many studies have been carried out over the past 30 years in the attempt to better define the pathophysiology of ARDS, but the overall result is that there is no single, clear-cut factor to which ARDS can be ascribed. At the present time, ARDS can be viewed as the pulmonary epiphenomenon of a systemic, multi-organ inflammatory response eliciting activation of a complex network of mediators. In the lung, such response causes non homogeneous permeability alterations and alveolar collapse with severe ventilation/perfusion mismatch. The mainstay of the inflammatory response is the pulmonary vascular endothelium activation and damage by lung injury mediators such as endotoxins, TNF-α, cytokines, arachidonic acid metabolites (1). The endothelial damage, with the loosening of the "tight junctions", causes an increase of capillary leakage and a protein-enriched fluid engorgement of interstitial and alveolar spaces.

Under physiological conditions, pulmonary endothelium modulates vascular tone through vasoactive substances release and uptake. When ventilation/perfusion (V/Q) mismatch occurs, a normal endothelium is able to modulate hypoxic pulmonary vasocostriction (HPV) by decreasing uptake of vasocostricting substances (e.g. thromboxane A_2 (TXA_2) and/or by increasing release of platelet-activating factor (PAF) and endothelins. This mechanism is able to divert flow from hypoventilated to normally ventilated alveoli and to correct hypoxemia (1). On the other side, the release of vasodilating substances such as Nitric Oxide (NO) and prostacyclin (PGI_2) may prevent an overwhelming vasocostriction in normoventilated areas. During ARDS, the endothelial damage causes imbalance between vasocostricting and vasodilating substances resulting in HPV abolition, inappropriate vasodilation where the inflammatory response occurs, and inappropriate vasocostriction where ventilation is normal. This imbalance causes ventilation/perfusion mismatch with coexistence of normal, high (normal ventilation, poor perfusion) and low (poor ventilation, excessive perfusion - shunt) V/Q areas. The clinical correlate is a severe hypoxemia associated with pulmonary hypertension.

The therapeutical approach of ARDS should be directed towards the treatment of the causative event. However, in most cases, the critical care physician faces ARDS after the primary insult has occurred. Indeed, even after the primary insult, there are often secondary

Acute Respiratory Distress Syndrome: Cellular and Molecular Mechanisms and Clinical Management
Edited by Matalon and Sznajder, Springer Science+Business Media New York, 1998

factors, both local and systemic, that perpetuate lung injury and whose neutralization could be necessary in order to improve lung injury. However, although many attempts have been made to reduce and stop the acute inflammatory response of ARDS, no real improvement of the proposed therapeutic stategies has been clearly demonstrated. The reported mortality rate of ARDS has arguably declined from 50-70% since its original description in 1967. The major reason can be due to the complexity of the phenomenon such as the balance of the mediators may vary between patients, and within the same patient, between different periods of ARDS.

Therefore, ARDS management is primarily supportive, and mechanical ventilation remains the mainstay. Application of positive airway pressure is commonly used to increase lung volume and recruit initially poorly-, or non-ventilated lung units, improving ventilation distribution to well perfused lung areas. However, besides the ventilatory approach, the better understanding of the lung injury pathophysiology has provided rational basis for improving V/Q mismatch and pulmonary hypertension by pharmacologic approach. While still supportive, there is an increasing number of reports suggesting a potential benefit of the pharmacologic therapy in improving hypoxemia and pulmonary hypertension.

When, in the eighties, pulmonary hypertension was recognized as a critical factor in the ARDS mortality, experimental and clinical researchers sought a magic vasodilating substance able to reduce pulmonary hypertension. Most of the researchers focused on vasodilating substances such nitroglycerin (2) and nitroprusside (3). More recently prostaglandin E_1 and PGI_2 have also been tested. The latter substances were chosen because they could provide both vasodilation and reduction of platelet aggregation, neutrophil chemotaxis, and macrophage activation. All these effects were thought to be beneficial in blunting the inflammatory response of ARDS. Both conventional and novel vasodilating substances were able to reduce pulmonary artery pressure and capillary leakage, and to improve right ventricular function (4, 5). However, the lack of selectivity during intravenous administration caused systemic and pulmonary vasodilation, and, through the loss of HPV, contributed to shunting of blood, V/Q mismatch and severe hypoxemia. As a matter of fact, as pointed out by Wetzel (6), pulmonary vasodilators should not cause systemic hypotension (selective for the pulmonary circulation) nor should they increase intrapulmonary shunt (selective for that part of the pulmonary circulation that serves ventilated lung units). There are no therapeutic agents that selectively dilate any part of the circulation. Even when short biologic half-life agents such as prostaglandins were delivered into the pulmonary artery with the hope that the effect was local and dissipated by the time the systemic circulation was reached, no selective effect could be obtained.

As no substance with the characteristics of the ideal pulmonary vasodilator exists, research has been oriented not towards the search for a new substance, but towards a different administration route. The administration through the transbronchial route of substances with short half-life, nebulized in few micron particles seems to be promising. The administration through the transbronchial route is able to restrict the vasodilating properties to the well-ventilated areas. This allows a selective reduction of vascular resistance in these lung units with flow-redistribution.

The use of nitric oxide has made it clear that the double selectivity is not impossible. NO is rapidly inactivated by the hemoglobin binding and does not reach systemic circulation, is selectively delivered to ventilated lung units resulting in increasing flow to these areas. Unfortunately, a complex and pretty expensive technology, the potential occurrence of toxic metabolites and methemoglobinemia make its use not entirely safe and effective.

Inhaled prostacyclin represents an alternative to NO. PGI_2 is a powerful vasodilator with short half-life, released by endothelial cells. Under physiological conditions it has a critical role in modulating pulmonary vascular tone. At physiological pH values, its half-life is 2-3 minutes because it is spontaneously hydrolized into an inactive metabolite, 6-keto-prostaglandin F1a. When prostacyclin is administered through the transbronchial route, it binds to pulmonary vessels smooth muscle cells receptors and increases intracellular level of cyclic adenosine monoposphate (cAMP) by activating adenylate cyclase. cAMP activates the protein kinase A and decreases free intracellular calcium which induces vasorelaxation of the vascular smooth muscle. Moreover, PGI_2 causes NO release by endothelial cells.

PGI_2 inhalation has been proven of benefit inducing selective pulmonary vasodilation in pediatric and adult RDS. Welte et al (7), in an experimental model of HPV-induced pulmonary hypertension, observed a 52% reduction of pulmonary vascular resistance, without any systemic vasodilation, during prostacyclin inhalation. Walmrath et al. (8) in severe ARDS patients reported both a reduction of pulmonary hypertension and an improvement of arterial oxygenation by shunt reduction. Such improvement was observed also in patients affected by severe pneumonia during mechanical ventilation (9). However, the effectiveness of prostacyclin inhalation was greatly reduced by the presence of lung fibrosis (9). Most of the studies have analyzed the effectiveness of prostacyclin inhalation on an on-off basis. However, this therapeutic approach may be appropriate also on a long term period. In 8 severe ARDS patients (Lung Injury Score = 2.9 ± 0.1), PGI_2 (mean dose of 21 ± 1 ng/kg/min) was continuously administered over a 24 hours period through the transbronchial route by a Servo 300 ventilator equipped with a Servo Ultra Nebulizer 345 (10). The prostacyclin administration resulted in a significant improvement of pulmonary hypertension and arterial oxygenation. The pulmonary arterial pressure decreased significantly after 1 hour of inhalation, and in the following 23 hours, its value did not further improve. On the other side, arterial oxygenation showed a progressive improvement peaking after 12 hours. The different timing of improvement might be related to different effects of prostacyclin. While the short term improvement might be due to smooth muscle vascular tone reduction resulting in selective vasodilation of the pulmonary vessel and reduction in both pulmonary hypertension and pulmonary shunt, the long term improvement (12-24 hours) might be related to the blunting of the inflammatory response. The latter effect could be associated to the anti-inflammatory properties of prostacyclin.

Prostacyclin represents a safe and valid alternative to NO. It seems to provide adequate improvement in pulmonary hypertension and hypoxemia, with no side effects. In opposite to NO, prostacyclin is spontaneously hydrolized to inactive metabolites (11), and continuous administration does not cause toxic cardiopulmonary effects neither platelet alterations (12). Moreover, it can play a significant role in locally preventing microcirculatory alterations. This potential effect has been emphasized by Eichelbronner et al (13). In their recent study, the authors suggest that, similar to intravenously administered PGI_2, aerosol administration of this drug allows for improvement of intramucosal pH in patients with septic shock. This effect may be due to the pharmacologic qualities of PGI_2 in association with its inapparent spillover into the systemic circulation.

In front of the beneficial effects of prostacyclin, certain practical aspects of its administration by inhalation route should be addressed. The first point is the uncertainty of the amount of drug that reaches alveoli. The amount is very small (5%) (14) and depends on the particle size. Moreover, in front of the very fast response to NO, the effects of prostacyclin may take time before becoming clinically apparent. Both these aspects make prostacyclin titration extremely cumbersome, and the inhalation dosage may be greater than the intravenous one. Dosages used for aerosol inhalation range from 5 to 200 ng/Kg/min (15). A second problem rises from the topical irritation of airway epithelial cells by prostacyclin and its alkaline buffer solution. Recently, long term inhalation has failed to produce any significant toxicity in healthy lambs (12). However, a sick lung might be unable to easily handle large amounts of prostacyclin.

The inhalational route of PGI_2 delivery is a new and promising application which can successfully replace intravenous vasodilator therapy for pulmonary hypertension. In view of its few side-effects and of its comparable efficacy, it can be considered as a valid alternative to NO.

REFERENCES

1.   Curzen NP, Jourdan KB, Mitchell JA. Endothelial modification of pulmonary vascular tone. Intensive Care Med, 22: 596-607, 1996.
2.   Radermacher P, Santak B, Becker H, et al. Prostaglandin E_1 and nitroglycerin reduce pulmonary capillary pressure but worsen ventilation/perfusion distributions in patient with ARDS. Anesthesiology, 70: 601-606, 1989.
3.   Sibbald WJ, Driedger AA, McCallum DM, et al. Nitroprusside infusion does not improve biventricular performance in patients with acute hypoxemic respiratory failure. J Crit Care, 1: 197-203, 1986.
4.   Mélot CP, Lejeune M, Leeman JJ, et al. Prostaglandin E_1 in the ARDS. Benefit for pulmonary hypertension and cost for pulmonary gas exchange. Am Rev Respir Dis, 139: 106-116, 1989.
5.   Radermacher P, Santak B, Wust J, et al. Prostacyclin and right ventricular function in patients with pulmonary hypertension associated with ARDS. Intensive Care Med, 16: 227-232, 1990.
6.   Wetzel R. Aerosolized Prostacyclin. In search of the ideal pulmonary vasodilator. Anesthesiology, 82: 1315-1317, 1995.
7.   Welte M, Zwissler B, Habazettl H, Messmer K. PGI_2 aerosol versus nitric oxide for selective pulmonary vasodilation in hypoxic pulmonary vasoconstriction. Eur Surg Res, 25: 329-340, 1993.
8.   Walmrath D, Schneider T, Pilch J, et al. Aerosolized prostacyclin in Adult Respiratory Distress Syndrome. Lancet, 342: 961-962, 1993.
9.   Walmrath D, Schneider T, Pilch J, et al. Effects of aerosolized prostacyclin in severe pneumonia. Impact of fibrosis. Am J Respir Crit Care Med, 151: 724-730, 1995.
10.   Brienza N, Grasso S, Bruno F, et al. Effects of continuous administration of aerosolized prostacyclin on hemodynamics and gas exchange in ARDS. Int Care Med, 22 (S3):261, 1996.
11.   Kerins DM, Murray R, Fitzgerald GA. Prostacyclin and prostaglandin E_1: molecular mechanisms and therapeutic utility. Prog Hemostasis Thrombosis, 10: 307-337, 1991.
12.   Habler O, Kleen M, Zwissler B, et al. Inhalation of prostacyclin (PGI_2) for 8 hours does not produce signs of acute pulmonary toxicity in healthy lambs. Intensive Care Med, 22: 426-433, 1996.
13.   Eichelbronner O, Reinelt H, Wiedeck H, et al. Aerosolized prostacyclin and inhaled nitric oxide in septic shock - different effects on splanchnic oxygenation? Intensive Care Med, 22: 880-887, 1996.
14.   Thomas SH, O'Doherty MJ, Fidler HM, et al. Pulmonary deposition of a nebulised aerosol during mechanical ventilation. Thorax, 48: 154-159, 1993.
15.   Scheeren T, Radermacher P. Prostacyclin (PGI_2): New aspects of an old substance in the treatment of critically ill patients. Intensive Care Med, 23:146-158, 1997.

HAS PROGNOSIS OF ARDS IMPROVED?

François Lemaire, Eric Roupie

Service de Réanimation Médicale
Hôpital Henri Mondor
F - 94010 Créteil (France)

INTRODUCTION

Recently, Milberg et al.[1] reported that, in their institution (five ICUs at Harborview Medical Center, Seattle), mortality of acute respiratory distress syndrome (ARDS) patients had steadily declined from 67% in 1983 to 40% in 1993, and even less in the following year. This information was well received, since many other studies supported such a trend, for a syndrome in which mortality is usually reported as high as 70 or 80%. Indeed, mortality was 54% in a paper by Sloane et al.[2], 53% by Suchyta et al.[3], or, even lower, 12% by Stocker[4] and 20% by Lewandowski[5]. However, when Kraft et al.[6] plotted mortality over time of published series of ARDS selected after a comprehensive search, between 1967 and 1994, they could not find a trend toward improvement (Figure 1). The reasons for such opposed views rely in fact on the heterogeneity of published series, in terms of case mix, ICU structure, definition of ARDS, scoring, cutoffs, end points, etc.

The prognosis of ARDS is clearly influenced by the *definitions* used for enrollment of patients into trials, and also by the *case mix* of all specific *ICUs* participating to these studies: medical vs surgical referral, co-morbidities, etiologies of ARDS, exclusion criteria, etc. After addressing these issues, we will consider in this chapter the value *of specific* and *global* indices as predictors of survival.

DEFINITION OF ARDS

For many years, there was no consensus as to what was called ARDS. This is probably the main reason why it is so difficult to use historical comparisons between series of ARDS to assess changes in mortality: the risk factors, entry criteria, PaO2/FiO2, cutoffs, exclusions, need for mechanical ventilation and pulmonary artery catheter insertion, all these factors and many others have actually varied over years.

The proposal of a new definition[7] of ARDS some years ago was well accepted by the Intensive Care community. It graded the respiratory failure in two categories: a milder one, the acute lung injury (ALI), with a P/F ratio ≤ 300 mmHg, and a more severe one (ARDS), with P/F ≤ 200 mmHg (Table 1).

This new definition needs to be compared to the classical criteria called «stricter» criteria by Moss et al.[8].

Acute Respiratory Distress Syndrome: Cellular and Molecular Mechanisms and Clinical Management
Edited by Matalon and Sznajder, Springer Science+Business Media New York, 1998

Table 1. Criteria for acute lung injury (ALI) and acute respiratory distress syndrome (ARDS).[7]

	Oxygenation	Chest radiograph	Pulmonary artery wedge pressure
ALI	$PaO_2/FiO_2 \leq 300$ mmHg (regardless of PEEP level)	Bilateral infiltrates seen on frontal chest radiograph	≤ 18 mmHg when measured or no evidence of left atrial hypertension
ARDS	$PaO_2/FiO_2 \leq 200$ mmHg (regardless of PEEP level)	Bilateral infiltrates seen on frontal chest radiograph	≤ 18 mmHg measured or no evidence of left atrial hypertension

Table 2. Strict criteria of ARDS [8].

All five criteria need to be met:
1. ARF from a known, at risk, diagnosis
2. A low static respiratory system compliance
3. A non increased PAOP
4. Diffuse bilateral infiltrates on chest radiograph
5. A decreased $P(A-Q)O_2$

Table 3. Lung injury score[16].

		Value
1. Chest roengenogram score		
No alveolar consolidation		0
Alveolar consolidation confined to 1 quadrant		1
Alveolar consolidation confined to 2 quadrants		2
Alveolar consolidation confined to 3 quadrants		3
Alveolar consolidation in all 4 quadrants		4
2. Hypoxemia score		
PaO_2/FiO_2	≥ 300	0
PaO_2/FiO_2	225-299	1
PaO_2/FiO_2	175-224	2
PaO_2/FiO_2	100-174	3
PaO_2/FiO_2	< 100	4
3. PEEP score (when ventilated)		
PEEP	≤ 5 cmH$_2$O	0
PEEP	6-8 cmH$_2$O	1
PEEP	9-11 cmH$_2$O	2
PEEP	12-14 cmH$_2$O	3
PEEP	≥ 15 cmH$_2$O	4
4. Respiratory system compliance score (when available)		
Compliance	≥ 80 mL/cmH$_2$O	0
Compliance	60-79 mL/cmH$_2$O	1
Compliance	40-59 mL/cmH$_2$O	2
Complicance	20-39 mL/cmH$_2$O	3
Compliance	≤ 19 mL/cmH$_2$O	4

The final value is obtained by dividing the aggregate sum by the number of components that were used

	Score
No lung injury	0
Mild-to-moderate lung injury	0.1-2.5
Severe lung injury (ARDS)	> 2.5

Table 4. Proposal of a modified LIS (Moss et al. [8])

- Oxygenation: $PaO_2/FiO_2 < 175$
- Chest radiograph: bilateral infiltration on frontal chest X-ray

When the American-European Consensus Conference met, numerous points have been discussed at length:

The need to *measure the actual pulmonary artery occluded pressure (PAOP)* is less mandatory than it used to be, possibly after Rinaldo[9] showed that the necessity of inserting PAC could introduce a strong bias in the selection of patients, sorting out only what she called a population of «moribunds». However, the new definition maintains the necessity to exclude the patients with left ventricular failure. Actually, in a multicenter incidence prospective study, Roupie et al.[10] found that mortality of hypoxemic patients (PaO$_2$ ≤ 300 mmHg) was 40%, but it was 31% in the subgroup of patients excluded from the ALI definition because PAOP was ≥ 18 mmHg.

Bilateral or unilateral chest X-ray opacities was also a matter of discussions. As shown in table 1, the bilateral criterion was finally left in the definition. Sloane et al.[2] found only 3% of their patients not meeting this criterion at enrollment, and the concept of ARDS with clear X-ray is still hypothetical (despite the point was raised in the Consensus Conference statement). Roupie et al., in their incidence study, have also shown that hypoxemic patients (≤ 200 mmHg) with unilateral lung disease had a much better prognosis than patients with diffuse bilateral opacities (ARDS): 40% vs 60%.

«Considerable discussion centered around the *cutoff between ALI and ARDS* with regards to the PaO$_2$/FiO$_2$ ratio» states the Consensus document[7], which finally recommended to choose a cutoff of 200 mmHg. Could the particularly high mortality of the NIH sponsored ECMO trial (# 90%) be attributed to more stringent entry criteria that were used: PaO$_2$ ≤ 50 mmHg under 100% oxygen breathing? Recent studies which addressed this issue confirmed that P/F was a relevant criterion[2,11], although not highly sensitive (see below). When Sloane compared his enrollment criteria (PaO$_2$ ≤ 250) to Bone's[12] (PaO$_2$ < 150 without PEEP, or < 200 with PEEP) and Montgomery's[13] (≤ 150 mmHg), he found that only few patients were excluded when entry criteria became more selective; most frequently, it merely delayed inclusion by a few days. The mean value of P/F in the Sloane series was actually 135 mmHg. In a prospective study (ARDS registry), Doyle et al.[14] evaluated the impact of shifting the cutoff from 300 to 150 on patients' enrollment. In this study, 54% of patients had at admission a ratio P/F between 150 and 300 mmHg, and 36% of those progressed further to develop a P/F ≤ 150 by 72 hours. The conclusion of the authors was also that applying more liberal oxygenation criteria resulted in an earlier inclusion of a significant group of patients. Conversely, in their incidence study, Roupie et al. have shown that those patients who fullfilled the criteria of ALI (P/F < 300 mmHg) and who never deteriorate were a very small minority, suggesting different ICU admission policies between Europe and the USA.

Role of PEEP on oxygenation. The Consensus panelists did not wish to modulate the oxygenation criterion by respect to the PEEP level, as done, for instance, by Bone et al.[12]. In a recent abstract, Tütüncü et al.[15] confirmed that application of PEEP («optimal») can modify the conditions of enrollment by altering PaO2 and shunt.

Five years before the Consensus Conference, Murray et al.[16] proposed an «expanded» definition of the ARDS. Its main objective was to provide a scale of severity of ARDS, ranging from 0 to 4 (see Table 3). It can also be used to identify ARDS patients, when the Lung Injury Score (LIS) is higher than 2.5. Moss et al.[8] assessed the accuracy of the American-European Consensus Conference definition, the LIS and a modified LIS (Table 4), against a more classical («stricter») definition of ARDS. They found that the modified LIS and the American-European Consensus Conference definition had an identical and good predictive value, slightly better than the original LIS. They showed also that these definitions have an optimal predictive value when patients are first selected on the presence of known risks (sepsis, pancreatitis, hypertransfusion, aspiration, trauma). When they are applied to patients without establishing these at-risk diagnoses, accuracy is much less.

CASE MIX

All predictive systems have an optimal value when used within the environment (ICUs, city, country) in which they have been validated. A major difficulty arises when they are applied in other institutions, structures, countries, with a different case mix.

Age is an important parameter to consider. Sloane[2], Knaus[11], Bartlett[17], Cohen[18], Ferring[19], Suchyta[20], have all shown that mortality of ARDS is much higher in older patients.

Accordingly, comparison between the Berlin patients[5], with a mean age of 32 years, and those from San Francisco[14] (mean age: 54 years) is certainly risky.

Etiologies play also a major role in determining the prognosis. The European collaborative study[21], for instance, showed that mortality varied markedly, in relation to the cause of ARDS, ranging from 38% for trauma to 86% for opportunistic infections (Figure 1).

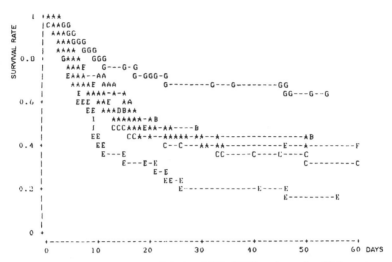

Figure 1. Percent survival of the various etiologie of 583 ARDS patients after 60 days. A, aspiration; B, extraabdominal infection; C, intraabdominal infection; D, miscellaneous; E, opportunistic pneumonia; F, pneumonia; G, trauma. *$p < 0.0001$.

Suchyta et al.[3] have noted that pneumonia was a causing factor for ARDS in 41% of their patients, but was nearly not found in the Seattle series[13]. This could explain why they found that respiratory failure was a cause for death in 40% of their patients, at odds with the 16% reported by Montgomery[13]. Patients with multiple trauma may represent up to 50% of surgical ICU series[1,5], and are not, or rarely, admitted to some medical ICUs[14]. The referrals and admission policies to ICUs may have also changed with time: pneumonia represented 65% of the patients enrolled in the ECMO trial[22]; no study these days has such a high percentage.

When *multivariate analysis* is applied to ARDS patients in order to define predictors for deaths, *sepsis* appears always as a major and independent factor[10,14,19]. In the subset of patients with sepsis (32/129) of the Brussels series[19], mortality was 69%, much higher than the average rate (52%). In the Doyle study[14], sepsis was the third most significant adverse prognostic factor, after non pulmonary organ system dysfunction and chronic liver disease. Mortality rate was also the highest in the sepsis subgroup of patients.

PRIMARY *vs* SECONDARY ARDS

Frequently associated with sepsis, ***non pulmonary organ dysfunction*** (OSD) have also an important role. Suchyta[3], Roupie[10], Doyle[14] Bartlett[17], Ferring[19], and Villar[23] have shown that the occurrence of several OSD was associated with increased mortality. Bartlett showed also that many patients progressed from isolated (lung) failure to multiple organ failure: one day before inclusion, 86% of 490 patients had isolated lung failure. On day 0 (inclusion), they were still 63%, 54% on day +2 and only 40% the last day of data collection. The link between sepsis, multiple organ failure and poor prognosis was first seen by Johanson[24] during the mid-eighties. In the Créteil series, mortality ranged from 40% (single lung failure) to 90% (multiple organ failure).

Such large discrepancy is in line with a proposal that ARDS should be divided into two main categories: a primary disease, when the lung is first and only damaged, and a secondary form of ARDS, when lung lesions participate to a broader multiple organ failure. As prognosis in these two forms of ARDS is markedly different, patients should not be included in the same trial, or at least stratified accordingly.

Immuno-suppression bears also a bad prognosis. Nearly absent in ICUs thirty or twenty years ago, immuno-suppressed patients (with cancer, AIDS, aplasia, transplants, etc.) are not nowadays uncommonly admitted to our ICUs. Twenty per cent of the ARDS patients reported by Doyle et al. had an organ transplantation. In our institution, an ARDS registry collected 129 ARDS patients from November 1992 to October 1995, with 35% of patients immuno-suppressed. The patients with single failure had a mortality of 42%, and those with more than one OSF at entry a mortality of 92%. Immuno-suppression was independently (multivariate analysis) related to mortality. Only 11% of patients with immuno-suppression survived. The very poor prognosis of patients having ARDS and MOSF in our series is predominantly caused by this large share of immuno-suppressed patients[25].

PREDICTORS OF SURVIVAL IN ARDS

Since the first description of prediction indices, 15 years ago, there has always been a competition between specific and global approaches.

Specific to the lung, for instance, is the PaO_2/FiO_2 ratio. We have already seen that, by and large, deeper the hypoxemia, worse the prognosis. Despite oxygenation indices (PaO_2/FiO_2, $P(A-A)O_2$), QS/QT) were not different at admission between survivors (S) and non survivors (NS) in the European study[21], at 24 hours, when the initial response to treatment and PEEP could be taken into account, these indices became discriminant. However, many recent studies have minimized this factor. Doyle et al.[14] found no difference of mortality in the ARDS patients wether they use a P/F ratio lower than 250 or lower than 150. In the APACHE III study[11], the predictive value of PaO_2/FiO_2 was much less significant (ROC: 0.68) than of the global indices (APACHE II and III scores, APACHE predicted risk) (ROC: 0.80), in a single and a multivariate analysis a well. As Artigas[22], Sloane et al.[2] reported similar gas exchange in S and NS at admission. Mortality was higher when P/F at admission was lower than 150, but there were no significant difference between < 100 and 100-149, and between 150-200 and 201-250. Then, survivors demonstrated daily improvement during the first week of treatment.

Knaus et al.[11], using the APACHE III database, were able to select 423 patients having a diagnosis of ARDS out of 17,440 admissions. When they applied a multivariate analysis using the admission APACHE III score, the ICU primary admission diagnosis and treatment location before ICU admission, prediction for survival was excellent. They showed also that, at any PaO_2/FiO_2 cutoff, the distribution of risks was very broad. In their conclusion, the authors recommanded that, in future trials on ARDS, patients should be stratified according to their predicted risk (by APACHE or SAPS-type systems).

CONCLUSION

Can we claim that we have evidence today that survival of ARDS patients have been improved during the last decade? Probably not, due to the numerous confounding factors that we have listed in this chapter. However, we can risk a touch of optimism.

The results from Seattle[2] are quite convincing, showing a large decline in mortality within the recent years, due to an homogeneous medical team and an unchanged admission policy during the last ten years. The rather perplexing issue is that there is no obvious clue to these results, besides a multifactorial and progressive improvement of care. The evaluation of $ECCO_2R$ made by Morris et al.[26] was disappointing, since they were unable to show benefit due to the new treatment. However, their pratients were remarkably well matched with the ECMO trial patients[22]. Despite both studies were negative, survival progressed from 10% during the seventies to 40% in the more recent study (1994).

It is thus conceivable that improvement of the respiratory failure itself (by virtue of VT reduction, optimal PEEP, NO inhalation, prone positioning, better fluid balance control,

337

better treatment of nosocomial pneumonias, etc.) improved survival of those patients having a unique or predominant lung failure. But we cannot measure this effect due to the large fraction, usually > 50%, of ARDS secondary to (or associated with) multiple organ failure, which has still a very bad prognosis. Another possible, and complementary reason, is also that we are now treating patients with malignancies, or immuno-suppression, who were not included in the earlier trials. Mortality may have remain the same, but patients did not...

REFERENCES

1. J.A. Milberg, D.R. Davis, K.P. Steinberg, L.D. Hudson, Improved survival of patients with ARDS: 1983-1993, *JAMA* 273: 306 (1995).
2. P.J. Sloane, M.H. Gee, J.E. Gottlieb, K.H. Alberting, S.P. Peters, P.J. Burns, G. Machiedo, J.E. Fish, A multicenter resgistry of patients with ARDS: physiology and outcome, *Am. J. Respir. Crit. Care Med.* 146: 419 (1992).
3. M.R. Suchyta, T.P. Clemmer, C.G. Elliott, J.E. Orme, L.K. Weaver, The ARDS: a report of survival and modifying factors, *Chest* 101: 1074 (1992).
4. R. Stocker, T. Neff, S. Stein, P. Ecknauer, O. Trentz, E. Russi, Prone positioning and low-volume pressure-limited ventilation improve survival in patients with severe ARDS, *Chest* 111: 1008 (1997).
5. K. Lewandowski, R. Rossaint, K. Slama, K.J. Falke, H. Weidemann, Reduced mortality in severe ARDS, *Chest* 103: 1309 (1993).
6. P. Kraft, P. Fridrich, R.D. Pernerstorfer, D. Koc, B. Schneider, A.F. Hammerle, H. Steitzer, The ARDS: definitions, severity and clinical outcome, an analysis of 101 clinical investigations, *Intens. Care Med.* 22: 519 (1996).
7. G.R. Bernard, A. Artigas, K.L. Brigham, J. Carlet, K. Falke, L. Hudson, M. Lamy, J.R. Le Gall, A. Morris, R. Spragg, The American-European consensus Conference on ARDS: definitions, mechanisms, relevant outcomes and clinical trial coordination, *Am. J. Respir. Crit. Care Med.* 149: 818 (1994).
8. M. Moss, P.L. Goodman, M. Heinig, S. Barkin, L. Ackerson, P.E. Parson, Establishing the relative accuracy of three new definitions of the ARDS, *Crit. Care Med.* 23: 1629 (1995).
9. J. Rinaldo, The prognosis of ARDS: an inappropriate pessimism? *Chest* 90: 470 (1986).
10. E. Roupie, L. Brochard, E. Lepage, Incidence des IRA et du SDRA: une étude multicentrique, *Réan. Med. Urg.* 5: A744 (1997).
11. W.A. Knaus, X. Sun, R.B. Hakim, D.P. Wagner, Evaluation of definitions for ARDS, *Am. J. Respir. Crit. Care Med.* 150: 311 (1994).
12. R.C. Bone, G. Slotman, R. Maunder, Randomized double-blind, multi-center study of prostaglandin E1 in patients with the ARDS, *Chest* 96: 114 (1989).
13. A.B. Mongomery, M.A. Stager, C.J. Carrico, L.D. Hudson, Causes of mortality in patients with the ARDS, *Am. Rev. Respir. Dis.* 132: 485 (1985).
14. R.L. Doyle, N. Szaflarski, G.W. Modin, J.P. Wiener-Kronish, M.A. Matthay, Identification of patients with ALI: predictors of mortality, *Am. J. Respir. Crit. Care Med.* 152: 1818 (1995).
15. A.S. Tütüncu, N. Cakar, F. Esen, L. Telsi, K. Akpir, The current definition of ARDS does not reflect the severity of lung injury, *Am. J. Respir. Crit. Care Med.* 155: A390 (1997).
16. J.F. Murray, M.A. Matthay, J.M. Luce, M.R. Flick, An expanded definition of the ARDS, *Am. Rev. Respir. Dis.* 138: 720 (1988).
17. R.H. Bartlett, A.H. Morris, H.B. Fairley, R. Hirsch, N. O'Connor, H. Pontoppidan, A prospective study of acute hypoxic respiratory failure, *Chest* 89: 684 (1986).
18. I.L. Cohen, J. Lambrinos, Investigating the impact of age on outcome of mechanical ventilation using a population of 41,848 patients from a Statewide database, *Chest* 107: 1673 (1995).
19. M. Ferring, J.L. Vincent, Is outcome from ARDS related to the severity of respiratory failure? *Eur. Respir. J.* 10: 1297 (1997).

20. M.R. Suchyta, D.P. Clemmer, C.G. Elliott, J.F. Orme, A.H. Morris, J. Jacobson, J. Menlove, Increased mortality of older patients with ARDS, *Chest* 111: 1334 (1997).

21. A. Artigas, J. Mancebo, Etiology and multiple organ failure as prognostic factors in ARDS, in: Update in intensive care and emergency medicine, *Jean-Louis Vincent Editor, Springer-Verlag Publisher*, Berlin (1987).

22. W.M. Zapol, M.T. Snider, J.D. Hill, R.J. Fallat, R.H. Barlett, L.H. Edmunds, A.H. Morris, E.C. Peirce, A.H. Thomas, H.J. Proctor, P.A. Drinker, P.C. Pratt, A. Bagniewski, R.G. Miller, Extracorporeal membrane oxygenation in severe ARF, *JAMA* 242: 2193 (1979).

23. J. Villar, J.J. Manzano, M.A. Blazquez, J. Quintana, S. Lubillo, Multiple system organ failure in ARF, *J. Crit. Care* 6: 75 (1991).

24. R.C. Bell, J. Coalson, J.D. Smith, W.G. Johanson, Multiple organ system failure and infection in ARDS, *Ann. Intern. Med.* 99: 293 (1983)

25. E. Roupie, H. Mentec, L. Brochard, C. Brun-Buisson, F. Lemaire, Relations between the Adult Respiratory Distress Syndrome (ARDS) and the multiple organ failure (MOF), Intens. Care Med. 18: A250 (1992).

26. A.H. Morris, C.J. Wallace, R.L. Menlove, Randomized clinical trial of pressure-controlled inverse ratio ventilation and ECCO2 removal for ARDS, *Am. J. Respir. Crit. Care Med.* 149: 295 (1994).

EFFECT OF VENTILATION STRATEGY ON CYTOKINE EXPRESSION IN AN *EX VIVO* LUNG MODEL

Lorraine Tremblay,[1] Debra Miatto,[2] Qutayba Hamid,[2] Franco Valenza,[1] and Arthur S. Slutsky[1]

[1]Departments of Surgery and Medicine
University of Toronto
Toronto, Canada M5G 1X5
[2]Meakins Christie Laboratories
McGill University
Montreal, Canada H2X 2P2

Although mechanical ventilation is an indispensable therapeutic intervention for the treatment of respiratory failure, it can also initiate and/or exacerbate underlying lung injury. The precise mechanisms remain unclear. We hypothesized that one mechanism by which injurious ventilation strategies lead to lung injury, is via increased production of inflammatory cytokines within the lung[1].

In order to assess the effect of mechanical ventilation strategy on the lung independent of hemodynamic effects or influx of systemic mediators, an isolated rat lung model was used. As 2 ventilation strategies known to cause lung injury are ventilation with high end inspiratory lung volumes (lung overdistension) or low end-expiratory lung volumes (such that cyclic alveolar collapse/recruitment occurs with each tidal breath), we randomized rat lungs to 2 hours of ventilation with either: a relatively non-injurious control strategy using a physiologic tidal volume and low positive end expiratory pressure (V_T = 7 cc/kg, PEEP = 3 cm H_2O); or one of the following 3 "injurious" ventilatory strategies : a high PEEP strategy with a high end inspiratory lung volume (V_T = 15 cc/kg, PEEP = 10); a strategy with an equivalent high end-inspiratory lung volume, but no PEEP (V_T = 40 cc/kg, PEEP = 0); or a strategy leading to intermediate degree of lung distension with no PEEP (V_T = 15 cc/kg, PEEP = 0)[1].

The endpoints assessed included: lung lavage concentrations of cytokines (TNFα,IL-1ß,IL-6,IL-10,MIP-2, and IFNγ) and tissue homogenate levels of mRNA for TNFα and *c-fos* after 2 hours of ventilation; as well as static pressure-volume loops both prior to, and following the ventilation period.

We found a significant effect of ventilatory strategy on production of cytokines by the lung[1]. Specifically, mechanical ventilation with high end inspiratory pressure/high PEEP increased lung lavage concentration of the inflammatory cytokine TNFα by a factor of 2, as compared to control ventilation. Zero PEEP ventilation at a lower end inspiratory volume

Acute Respiratory Distress Syndrome: Cellular and Molecular Mechanisms and Clinical Management
Edited by Matalon and Sznajder, Springer Science+Business Media New York, 1998

341

increased lung lavage concentration threefold as compared to control ventilation. However, when high end inspiratory lung volume ventilation in combination with zero PEEP was used, a synergistic 56 fold increase in lavage TNFα was observed relative to the control ventilation strategy. A similar trend was found for the other 5 cytokines, although with the exception of the high end inspiratory lung volume zero PEEP strategy, the increases in cytokines relative to the control group did not reach significance.

At the transcriptional level, both zero PEEP ventilatory strategies were found to significantly increase mRNA levels of the early response gene *c-fos*, and all 3 "injurious" strategies lead to a significant increase in TNFα mRNA as compared to the control ventilation strategy[1].

In addition, both zero PEEP ventilation strategies led to significant decreases in lung compliance over the ventilation period. However, no significant difference in lung compliance was found in either group ventilated with PEEP.

To determine the distribution and cell types involved in the ventilator induced changes in cytokines within the lung, we repeated the above protocol and carried out *in situ* hybridization for both TNFα and IL-6 mRNA on lung cryosections. For these studies, an additional control group consisting of lungs harvested immediately following sacrifice was included. Messenger RNA expression was assessed by a blinded observer, using a semi-quantitative scoring system consisting of the average number of positive cells/total cells/3 hpf's per lung section.

Significant changes in magnitude and distribution of IL-6 and TNFα mRNA were found depending on the ventilatory strategy used. For example, mRNA for TNFα and IL-6 was significantly increased in both the high end inspiratory lung volume/high PEEP group, and the intermediate lung volume/zero PEEP group after 2 hours of *ex vivo* ventilation. In contrast, no difference was found between lungs submitted to control ventilation or lungs freshly harvested. Double staining with keratin polyclonal antibody suggested that in the lungs undergoing injurious ventilation, the majority of TNFα positive cells were epithelial cells.

We conclude that one mechanism whereby injurious mechanical ventilation can both initiate, or exacerbate, local (and potentially systemic) injury, is via significant alterations in production of inflammatory mediators by the vast epithelial surface of the lung.

REFERENCES

1. L. Tremblay, F. Valenza, S.P. Ribeiro, J. Li, and A.S. Slutsky, Injurious ventilatory strategies increase cytokines and *c-fos* m-RNA expression in an isolated rat lung model, *J Clin Invest*. 99:944 (1997).

ABNORMAL TISSUE OXYGENATION AND CARDIOVASCULAR CHANGES IN ENDOTOXAEMIA: BENEFICIAL EFFECTS OF VOLUME RESUSCITATION.

Peter B. Anning, Mark Sair, *C. Peter Winlove, Timothy W.Evans.

Unit of Critical Care, and *Physiological Flow Studies Group,
Imperial College of Science, Technology and Medicine,
London, United Kingdom.

INTRODUCTION

Sepsis and its related syndromes develop frequently in hospitalized patients, with an associated mortality of 10-20%. In the presence of circulatory failure, this figure rises to over 60% and may account for up to 200,000 deaths per annum in the USA alone. Most patients succumb to a multiple organ dysfunction syndrome (MODS) rather than hypotension *per se*, but the reasons for this are not clear. Sepsis is known to disrupt microcirculatory flow and nutrient exchange. Intravascular leukaggregation, abnormal red blood cell deformability, increased microvascular permeability, interstitial protein loss and tissue oedema are frequently observed. These changes are promoted by proinflammatory mediators and are partly modulated by the endothelium. It is hypothesized that endothelial injury exacerbates maldistribution of regional blood flow and leads to cellular hypoxia and vital organ dysfunction.

In patients with sepsis, intravenous fluids and inotrope administration are widely used to support the systemic circulation. Although these interventions may reverse latent intravascular hypovolaemia and improve myocardial contractility, there are few data regarding their effects on tissue oxygen tension, and they have not been associated with any survival benefit in patients with septic shock. Clinically, a hyperdynamic circulatory response to fluid resuscitation is observed. We demonstrated recently that significant tissue hypoxia and abnormal microvascular regulation of tissue oxygenation is observed in endotoxaemic rats, despite apparently normal microcirculatory perfusion[1]. The aim of this study was to further investigate the effects of fluid resuscitation on tissue oxygenation and central cardiovascular indices in the endotoxaemic rat.

METHODS & RESULTS

Using a rat model of endotoxaemia, amperometric measurements were made of skeletal muscle (tissue) oxygen tension (PtO_2) and its response to changes in inspired oxygen concentration (FiO_2). Simultaneous measurements of systemic hemodynamic indices (mean arterial blood pressure (MAP) and cardiac output (CO)) and arterial blood gas tensions were observed.

Acute Respiratory Distress Syndrome: Cellular and Molecular Mechanisms and Clinical Management
Edited by Matalon and Sznajder, Springer Science+Business Media New York, 1998

343

In the presence of normal PaO_2, PtO_2 in endotoxaemic animals was significantly lower than sham animals (1.0 ± 0.2 vs. 3.7 ± 1.4kPa respectively, $p<0.05$). Muscle PtO_2 in endotoxemic rats was significantly lower than volume resuscitated groups, with a marked attenuation of the response to increasing FiO_2 (e.g. 1.13 ± 0.53 vs. 5.69 ± 1.47kPa at 0.95 FiO_2 respectively, $p<0.05$). These changes were associated with significant metabolic acidaemia compared to control and resuscitated endotoxaemic animals (e.g. 7.23 ± 0.02 for endotoxaemic vs. 7.34 ± 0.09 for controls at baseline, $p<0.05$). No significant differences in either MAP or CO were observed.

DISCUSSION

In summary, this study indicates that significant tissue hypoxia is a complication of experimental endotoxaemia. In this model, impaired tissue microvascular control and attenuation of response to changing FiO_2 was present. Patients with sepsis frequently require volume resuscitation, although few data exist regarding its effects on endotoxemia-induced tissue hypoxia. In rats with peritonitis, volume resuscitation reverses the observed fall in both PtO_2 and cardiac output, although experiments with dogs suggest the reversibility is organ specific. Our data demonstrate that volume resuscitation can reverse the effects of endotoxemia on PtO_2, without affecting haemodynamics.. Our model displayed a normodynamic circulation, suggesting that fluid deficit was limited to the peripheral microcirculation only. Coupled with the partial reversal of the endotoxaemia-induced acidaemia, these data emphasise the clinical importance of adequate fluid resuscitation in sepsis.

REFERENCES

1. Sair M, Etherington PJ, Curzen NP, Winlove CP Evans TW. Tissue oxygenation and perfusion in endotoxemia, *Am.J.Physiol.* 271: H1620-5 (1996).

MECHANISMS OF NITRIC OXIDE INDUCED INJURY TO THE ALVEOLAR EPITHELIUM

Sha Zhu[1], Machelle Manuel[2], Imad Haddad[3] and Sadis Matalon[1]

[1]Department of Anesthesiology, University of Alabama at Birmingham, Birmingham, AL 35233, USA
[2]Department of Pharmacology, Texas Tech University Health Sciences Center, Amarillo, TX 79106, USA
[3]University of Minnesota, Division of Pulmonary & Critical Care, Minneapolis, MN 55455, USA

INTRODUCTION

The major function of the lung is gas exchange. The movement of both oxygen and carbon dioxide across the blood-gas barrier is by simple diffusion. This process is optimized by the large alveolar surface area, the close proximity of the alveolar and pulmonary capillary membranes, and the lack of any significant amount of fluid in the alveolar space. The relative dryness of the alveolar space is thought to be the result of: (1) the low permeability of the alveolar epithelium to both electrolytes and plasma proteins; (2) the presence of pulmonary surfactant, which lowers the surface tension of the blood-gas interface; and (3) the ability of alveolar epithelial cells to actively transport sodium ions from the alveolar to the basolateral spaces[1].

The alveolar epithelium is continuously exposed to both endogenously and exogenously derived sources of reactive oxygen and nitrogen species. These reactive species are formed as intermediates in mitochondrial electron-transport systems and microsomal metabolism of endogenous compounds and xenobiotics, including drugs and environmental pollutants and various cytoplasmic sources[2]. In addition, neutrophils and other inflammatory cells generate and release reactive oxygen species via an NADPH oxidase-dependent mechanism, which is mediated by membrane receptor activation of protein kinase C and phospholipase C. Prolonged and continuous exposure to reactive oxygen species damages the pulmonary surfactant system and alveolar epithelium resulting in increased amounts of protein in the alveolar space, pulmonary atelectasis, arterial hypoxemia and eventually death from respiratory failure. Herein we will review the biochemistry of the reactive oxygen and nitrogen species, the basic mechanisms by which they interact with target molecules in the alveolar epithelium and the short and long-term sequelae of these interactions.

Acute Respiratory Distress Syndrome: Cellular and Molecular Mechanisms and Clinical Management
Edited by Matalon and Sznajder, Springer Science+Business Media New York, 1998

BIOCHEMISTRY OF REACTIVE OXYGEN SPECIES

Under normal oxygen tensions, approximately 98% of oxygen undergoes a four-electron catalytic reduction to form water by mitochondrial cytochrome c oxidase. The remaining 2% of oxygen, however, may undergo sequential incomplete reduction to form reactive oxygen species such as superoxide (O_2^{-}) and hydrogen peroxide (H_2O_2). Both O_2^{-} and H_2O_2 are relatively long-lived compounds in biologic systems. H_2O_2 can directly cross cell membranes by simple diffusion, while O_2^{-} crosses cell membrane via anion channels. However, the limited reactivity of O_2^{-} and H_2O_2 with many biological molecules and their very low intracellular concentrations (10 pM and 1-100 nM, respectively) have raised questions about their toxicity *per se*.

Several factors may exacerbate production of reactive oxygen species in acute and chronic lung diseases. First, increased oxygen concentration is commonly required to alleviate arterial and tissue hypoxemia in patients with pulmonary and cardial diseases. Exposure of lung cells, subcellular organelles, and tissue to hyperoxia (100% O_2) has been shown to increase mitochondrial hydrogen peroxide production 10-15 fold [3]. Second, in response to pro-inflammatory cytokines, activated neutrophils and macrophages migrate to the lungs and release reactive oxygen species by the membrane-bound enzyme-complex NADPH oxidase [4]. Third, under conditions of ischemia, decreased perfusion, low oxygen tension, or trauma, xanthine dehydrogenase, the innocuous form of the enzyme, is converted to xanthine oxidase, which utilizes xanthine and molecular oxygen to produce partially reduced oxygen species (PROS). The results of several studies suggest that xanthine oxidase may be released from the intestine or liver into the circulation and bind to pulmonary endothelium where it can serve as a locus for the intense production of reactive oxygen species[5].

A more potent reactive metabolite of O_2^{-} is the hydroxyl radical ($^{\cdot}OH$). Superoxide anions directly reduce H_2O_2 to give O_2, hydroxide ion (OH^{-}) and $^{\cdot}OH$. In the presence of trace metals (usually Fe^{3+}, sometimes Cu^{2+}), $^{\cdot}OH$ can be generated via the following pathways:

$$O_2^{-} + Fe^{3+} \quad \rightarrow Fe^{2+} + O_2$$
$$Fe^{2+} + H_2O_2 \quad \rightarrow Fe^{3+} + OH^{-} + {}^{\cdot}OH$$
$$\overline{O_2^{-} + H_2O_2 \quad \rightarrow O_2 + OH^{-} + {}^{\cdot}OH}$$

Hydroxyl radicals are much more potent oxidants than either O_2^{-} or H_2O_2 and are capable of producing extensive cellular damage. Their reactivity is so high and nonspecific that the site of target reaction is confined to within a few molecular radii of the site of its generation. There are no direct enzymatic scavenging systems present *in vivo* for this radical. *In vitro*, the presence of OH is revealed by the inhibition of its formation by the scavenging action of mannitol, ethanol or dimethyl sulfoxide (DMSO).

MOLECULAR BIOLOGY OF $^{\cdot}NO$

Nitric Oxide Synthase

Generation of $^{\cdot}OH$ by the Fenton reaction requires the interaction of three different species (O_2, H_2O_2 and Fe^{3+}). In the epithelial lining fluid the concentrations of reactive oxygen species are kept low due to the presence of the antioxidant enzymes superoxide dismutase and catalase, along with a number of non-enzymatic antioxidants, including vitamin E, reduced glutathione and ascorbate [6]. Furthermore, most iron is chelated in a non-catalytic form by transferrin and ceruloplasmin. Although the formation of ·OH via the Fenton reaction *in vivo* may still occur, especially in situations where the intracellular load of free iron has been increased [7,8], a second pathway for the generation of potential oxidants with the reactivity of ·OH without the need for metal catalysis, involving nitric oxide (·NO) has recently been described [9].

NO is synthesized from the five electron oxidation of either of the two equivalent guanidine nitrogens of L-arginine. The reaction is catalyzed by one of three isozymes of nitric oxide synthase (NOS), using reduced NADPH as the source of electrons and cofactors, including tetrahydrobiopterin (H_4B) and flavin nucleotides (FMN; FAD). Molecular oxygen is a co-substrate, N^G-hydroxy-L-arginine is formed as a short-lived intermediate and L-citrulline is a byproduct. All isoforms of NOS bind calmodulin and contain heme. Electrons are supplied by NADPH, transferred along the flavins and calmodulin, and presented to the catalytic heme[10].

Our current understanding of how NO performs an extraordinarily diverse array of physiologic and pathophysiologic functions remains quite rudimentary. Nonetheless, it appears that regulation of NO biosynthesis explains, in part, its diverse functions, and thus significant advances have been made studying unique characteristics of NOS isoforms. Collectively, NOS are homodimeric cytochrome P450-like hemoproteins that have oxygenase and reductase domains in their amino (NH_2) and carboxyl (COOH) termini respectively. The active dimeric form is hypothesized to be dependent upon H_4B. The domains are separated by a Ca^{2+}/calmodulin binding region. The reductase domain is homologous to NADPH-cytochrome P450 including binding sites for FMN, FAD and NADPH. The constitutive forms have similar phosphorylation sites and one of them (NOS-III; see below) has a unique NH_2 terminal myristolyation site.

Differences between the isozymes, however, underscore the function of NO in various biological systems. The nomenclature evolved from descriptions of cellular source of the enzyme (e.g., neuronal, macrophage or endothelial) or its expression (constitutive vs inducible) and by many conventions is now relegated to the chronology in which the enzyme was purified and cloned. NOS I and III are constitutively expressed and their activity is regulated by intracellular calcium; prototypic sources are neuronal and endothelial cell, respectively[10]. NOS II is induced by cytokines, its activity is largely independent of calcium and it is regulated at a transcriptional level. The prototypic source is the macrophage [11,12].

Although the human genes are present on discrete chromosomes, considerable homology exists between the three isoforms suggesting common ancestral origin with subsequent gene duplication and transposition. NOS-I and III derived NO is produced in small quantities for brief periods of time and underscore intra-and intercellular signaling events such as neurotransmission or vascular homeostasis. In contrast, NOS-II produces large amounts of ·NO for prolonged periods of time (assuming availability of substrate and cofactors) and contributes to more diffuse physiological roles associated with inflammation or infection.

PHYSIOLOGICAL EFFECTS OF ·NO

Biological Targets

The biologic actions of ·NO are dictated by the reactions it undergoes with different target molecules in cells, membranes, and extracellular milieu. For each target, depending on the amount and duration of induction, ·NO can exert beneficial or detrimental effects. Known targets for ·NO include:

Guanylate cyclase. NO binds the heme group of soluble guanylate cyclase leading to an increase in cyclic guanosine-3',5'-monophosphate (cGMP) levels. Many effects of cGMP are mediated via a group of enzymes, cGMP-associated protein kinases (PKG's). These PKGs act to reduce intracellular calcium causing smooth muscle relaxation. NO-mediated increased cGMP levels also prevents platelets aggregation, and decreases adhesion of neutrophils. However, excessive NO-mediated cGMP production has been implicated in sepsis-induced refractory hypotension and shock [13].

Hemoglobin. The major route for the destruction of NO *in vivo* is the fast and irreversible reaction with *oxy*-hemoglobin (Hb-Fe^{2+}O$_2$) or *oxy*-myoglobin to produce nitrate and methemoglobin, according to the following reactions:

$$Hb\text{-}Fe^{2+}O_2 + NO \rightarrow Hb\text{-}Fe^{2+}ONOO \rightarrow Hb\text{-}Fe^{3+} + NO_3^-$$

Because of its vasorelaxant properties and rapid inactivation in the blood by its reaction with hemoglobin, NO inhalation has been advocated as a means of selectively reducing pulmonary hypertension and improving systemic oxygenation in a variety of clinical situations including bronchopulmonary dysplasia, and the acute respiratory distress syndrome (ARDS) [14]. However, patients with diminished methemoglobin reductase activity, such as neonates, are unable to efficiently convert methemoglobin to ferrous hemoglobin, and could therefore be at greater risk for developing methemoglobinemia and decreased oxygen transport. Recent reports indicate that the reversible S-nitrosylation of a cysteine residue located in the β-chain (Cysβ93) plays an important role in optimizing oxygen delivery to the tissues [15].

Iron/sulfur (4Fe/4S) centers of enzymes. Production of NO by activated macrophages defends the host against infectious agents, including bacteria, parasites, viruses, and destroys tumor cells. The proposed mechanisms responsible for these effects involve the reaction of NO with the nonheme iron of iron-sulfur complexes, resulting in the inactivation of iron-sulfur containing enzymes including mitochondrial aconitase, cytochrome c oxidase [16], and the DNA synthesis rate limiting enzyme, ribonucleotide reductase. Inhibition of these critical enzymes leads to suppression of mitochondrial respiration, energy metabolism and cell replication. However, the NO effects are nonspecific and its overproduction may be cytotoxic not only for microbes, but also for the cells and tissues that produce it [17].

Other free radicals. NO has an unpaired electron, and thus can readily react with other free radicals, as specified below:

a. Reaction with oxygen. If a tank of NO is allowed to leak into air, a cloud of lethal and highly reactive orange-brown nitrogen dioxide (NO$_2$) is formed, according to the following reactions:

$$2\,NO + O_2 \rightarrow 2\,NO_2$$

At low NO concentrations (<1-2 μM), observed *in vivo* at most pathologic conditions, the low probability of any two NO molecules encountering each other makes the formation of NO$_2$ extremely slow. Instead, NO may react with a single molecule of oxygen is a second order reaction to form a nitrosyldioxyl radical (ONOO·).

b. Reaction with O$_2^-$. NO reacts with O$_2^-$ at a near diffusion-limited rate constant of about 7 x 10^9 M^{-1}s^{-1} to form ONOO$^-$. This reaction was mistakenly thought to be protective since it decreased the amount of O_2^- detected. ONOO- has a pK_a of 6.8 at 37°C and thus may remain stable for months in alkaline solutions. The protonated form of peroxynitrous acid (ONOOH) forms $\cdot NO_2$ and an intermediate with a reactivity equivalent to the hydroxyl radical, derived from the *trans* isomerization of ONOOH:

$$O_2^- + \cdot NO \rightarrow ONOO^- + H^+ \rightarrow ONOOH \rightarrow \begin{cases} "\cdot OH \cdots NO_2" \\ NO_3^- \end{cases}$$

Under physiological conditions, a minimum of 25% of ONOO⁻ decomposes to form the "$\cdot OH \cdots NO_2$", with the remainder recombining to form nitrate. Thus peroxynitrite may serve as a source for $\cdot OH$ type species without the requirement of metal catalysis[9].

While being highly reactive, its modest rate of decomposition under physiological conditions allows ONOO⁻ to diffuse for up to several cell diameters to critical cellular targets before, becoming protonated and decomposing. ONOO⁻ initiates iron-independent lipid peroxidation and oxidizes thiols, damages the mitochondria electron transport chain [18], and causes lipid peroxidation of human low density lipoproteins. In addition, metal ions, such as Fe^{3+}EDTA and copper in the active site of superoxide dismutase (SOD), catalyze the heterolytic cleavage of ONOO⁻ to form a nitronium ion-like species (NO_2^+) which nitrates phenolics including tyrosine in proteins [19].

Under normal conditions, intracellular O_2^- concentrations are kept at remarkably low levels (10 pM) because eukaryotic cells contain large amounts of SOD (4-10 μM). Under these conditions, ONOO⁻ formation is minimal. However, inflammatory cells produce large number of both $\cdot NO$ and O_2^- when stimulated by cytokines, interferon γ, LPS, and other inflammatory agents . When the concentration of $\cdot NO$ increases to the micromolar range, it can effectively compete with SOD for O_2^- to form ONOO⁻ (rate constant of the reaction of O_2^- with $\cdot NO$ is three times faster than that with SOD). Using luminol-dependent chemiluminescence, ONOO⁻ production has been demonstrated by human neutrophils [20], rat alveolar macrophages [12], and bovine aortic endothelial cells [21]. Furthermore, airway epithelial cells constitutively express iNOS[22].

One may speculate that low molecular weight antioxidants (such as reduced glutathione and ascorbate) present in the epithelial lining fluid may scavenge peroxynitrite thus preventing its interaction with biological targets. However, because of its high reactivity, ONOO⁻ will attack biological targets even in the presence of antioxidant substances [23]. Furthermore, physiological concentrations of carbon dioxide and bicarbonate enhance the reactivity of ONOO⁻ via the formation of the nitrosoperoxycarbonate anion ($O=N-OOCO_2^-$) and increase its nitration efficiency[24,25]. Equally important, bicarbonate reversed the inhibition of ONOO⁻-induced nitration by ascorbate and urate[25]. The detection of nitrotyrosine in the lungs of patients with adult respiratory distress syndrome (ARDS) [26] and lungs of rats exposed to endotoxin [27] or hyperoxia [26] indicates that nitration reactions occur *in vivo*.

Recently, a new pathway for the formation of nitrating species has been proposed. Eiserich et al. [28]showed that the reaction of nitrite (NO_2^-), the autoxidation product of $\cdot NO$, with hypochlorous acid (HOCl) forms reactive intermediate species with spectral characteristics similar to those of nitryl chloride ($Cl-NO_2$), that are also capable of nitrating phenolic substrates such as tyrosine and 4- hydroxyphenylacetic acid, with maximum yields obtained at physiological pH. This reaction may have considerable physiological significance since considerable amounts of HOCl are being produced by activated neutrophils via the action of myeloperoxidase on hydrogen peroxide. Recent observations indicate a six-fold elevation of 3-chlorotyrosine in atherosclerotic tissue obtained during vascular surgery as compared to normal aortic intima. The detection of 3-chlorotyrosine in human atherosclerotic lesions indicates that halogenation reactions catalyzed by the myeloperoxidase system of phagocytes constitute one pathway for protein oxidation in vivo[29].

c. Thiols. It has been suggested that various forms of NO (such as N_2O_3, NO^+, or ONOO⁻) may react with thiols to yield S-nitrosothiols (RS-NO) and that NO circulates in plasma mainly as an S-nitroso adduct of serum albumin. However, the exact biochemical pathways leading to S-nitrosothiol formation are not clear. NO will not react directly with a thiol (RSH) as the reaction is unbalanced and thermodynamically unfavorable. NO will react with a one electron acceptor, such as iron, nitrogen dioxide or as recently suggested, molecular oxygen[30], to form a nitrosonium ion (NO^+), which can then interact with thiols to form nitrosothiols. Micromolar concentrations of S-nitrosoglutathione have been detected in the airway fluid of normal subjects and significantly higher levels were observed in the lungs of patients with pneumonia or during inhalation of 80 ppm NO [31]. It has been suggested that formation of RS-NO adducts stabilizes NO, decreasing its cytotoxic potential, while maintaining its bioactive properties. NO can also be transported on cysteine residues of hemoglobin which may facilitate efficient delivery of oxygen to tissues [15].

ONOO⁻ Formation in ARDS

ARDS, triggered by a number of pathologic conditions, is a clinical syndrome that features severe lung inflammation with abnormal permeability of the alveolar epithelium. The edema is a result of injury to both endothelial and epithelial cells caused by reactive species and proteolytic enzymes released by activated neutrophils and alveolar macrophages. Despite the identification of many mediators which lead to neutrophil tissue infiltration and activation, overall mortality from ARDS remains at 50-70%. The lack of specific treatment for ARDS is due to the complex interplay between the different humoral mediators released by the initiating condition.

The following observations establish the potential involvement of ANO, ONOO⁻ and various reactive nitrogen species in the generation and propagation of pulmonary epithelial injury in a variety of ARDS-type pathological conditions: (1) Induction of immune complex alveolitis in rat lungs results in significant elevation of ANO decomposition products and albumin levels in the bronchoalveolar lavage (BAL), indicating the presence of increased alveolar permeability to solute. Alveolar instillation of L-NMMA mitigates ANO production and alveolar epithelial injury [32]. (2) Paraquat induced injury to the lung results in stimulation of ANO synthesis. All signs of injury, including increased airway resistance and alveolar permeability to solute, are mitigated by administration of selective and competitive inhibitors of nitric oxide synthase [33]. (3) Ischemia-reperfusion injury to isolated rat lungs is associated with an increase in protein nitrotyrosine in lung homogenates using amino acid analysis, increased nitrate and nitrite levels in perfusate fluid, and formation of tissue oxidized protein and lipid products. Administration of L-NAME (NOS inhibitor) 30 min prior to induction of ischemia abolishes the increases in both nitrotyrosine and nitrate and nitrite, and significantly reduces the formation of lung thiobarbituric acid reactive substances (TBARS) and protein carbonyl levels [34]. (4) Infecting hamster tracheal rings with *Bordetella pertussis in vitro* produces epithelial cytopathology. Destruction of ciliated cells and inhibition of DNA synthesis are associated with induction of NO synthesis by the tracheal epithelial cells. The cytopathology is dramatically attenuated by the NOS inhibitors L-NMMA and aminoguanidine. These results indicate that pertussis toxins elicits NO production in the same cells that suffer the subsequent deleterious effects [17]. (5) Pneumonia due to influenza virus involves increased production of both NO and O_2^- [35]. Increased expression and activity of NOS-II is observed in lungs infected with the influenza virus. L-NMMA, administered intraperitoneally daily to mice from day 3 after virus inoculation improves survival.

One way to demonstrate ONOO⁻ formation *in vivo* is to detect the presence of stable by-products of its reaction with various biological compounds. 3-Nitro-tyrosine, the product of the addition of a nitro group ($-NO_2$) to the *ortho* position of the hydroxyl group of tyrosine, is such a stable compound. Using a polyclonal antibody, which recognizes antigenic sites related to nitrotyrosine [26], we demonstrated increased immunostaining in the lung of pediatric patients who died with ARDS and in the lungs of rats exposed to sublethal hyperoxia (100% O_2 for 60 h). Immunostaining was specific since it was blocked by the addition of an excess amount of antigen, and was absent when the nitrotyrosine antibody was replaced with non-specific IgG. Nitrotyrosine formation was detected only in rat lung sections incubated *in vitro* with ONOO⁻, but not NO or reactive oxygen species. The most likely candidate capable of nitrating tyrosine residues is ONOO⁻. Thus, this data suggest that ONOO⁻ is formed in the lungs of patients and animals with acute lung injury.

However, ONOO⁻ may not be the only species capable of tyrosine nitration. NO_2 can also nitrate tyrosine, although it is much less efficient than ONOO⁻ because two molecules of NO_2 are required to nitrate one tyrosine. Another possible nitration pathway, as mentioned previously, is the reaction of NO-derived nitrite, under acidic conditions with oxidants such as H_2O_2 and hypochlorous acid to form the nitrating agents [36].

Physiological Consequences of Protein Nitration

Several reports indicate that protein nitration may lead to selective loss of protein function. Nitration of tyrosine residues of human IgG, abrogated their C_{1q}-binding activity [37]. The inactivation of *E. coli* dUTPase and the occurrence of a tyrosine residue in a strictly conserved sequence motif, suggest the critical importance of this residue for the function of the enzyme [38]. Nitration of tyrosine residues of α_1-proteinase inhibitor resulted in selective loss of elastase inhibitory activity, but not chymotrypsin or trypsin-inhibitory activity [39]. Exposure of surfactant protein A (SP-A) to peroxynitrite or specific nitrating agents led to nitration of a single tyrosine residue in its carbohydrate recognition domain and diminished the ability of SP-A to aggregate lipids and bind to mannose [40-42]. SP-A, isolated from the lungs of lambs breathing NO and O_2 also had decreased ability to aggregate lipids [43]. In contrast, exposure of SP-A to strong oxidizing agents, which by themselves do not nitrate tyrosines, did not alter SP-A function. Tyrosine nitration has also been shown to inhibit protein phosphorylation by tyrosine kinases, and thus interfere with intracellular signal transduction [44,45]

Injury to pulmonary surfactant and alveolar epithelium *in vivo*

Because of its vasorelaxant properties and its rapid inactivation in the blood by its reaction with hemoglobin, NO inhalation has been advocated as a means of selectively reducing pulmonary hypertension and improving systemic oxygenation in a variety of clinical situations including bronchopulmonary dysplasia and ARDS. However there is concern that inhalation of NO in the presence of acute inflammation may lead to the formation of reactive oxygen-nitrogen species that may damage the alveolar epithelium and pulmonary surfactant system.

A number of recent reports seem to indicate that NO inhalation may indeed damage the lungs. Exposure of newborn piglets to 100 ppm NO and 95% O_2 for 48 h resulted in significant injury to pulmonary surfactant, manifested by inhibition of surface activity and worsened pulmonary inflammation [46]. Pulmonary surfactant samples isolated from newborn lambs exposed to NO gas (200 ppm) for 6 h exhibited abnormal surface properties. SP-A, isolated from the lungs of lambs that breathed 200 ppm NO, exhibited a small, but significant decrease in the ability to aggregate lipids *in vitro*[43]. Hallman et al. [47] reported that exposure of rats to 100 ppm NO in 95% O_2 for 24 h developed surfactant dysfunction caused in part by alterations of proteins in the epithelial lining fluid.

It may be argued that the concentrations of inhaled NO in these studies were outside the range used clinically. However, due to the short exposure period utilized in these experiments, the value of the product of *concentration x time* of inhaled NO is comparable to the corresponding value in a patient who breathes 20 ppm NO for 3 d. Exposure of rats to 0.5 ppm NO for 9 weeks resulted in significantly higher injury to lung interstitial cells and matrix than an equivalent exposure to NO_2, implicating NO as an agent more toxic than NO_2 [48]. None of the animals showed overt evidence of pulmonary injury such as arterial hypoxemia, increased albumin content in the BAL or respiratory failure. Accordingly, prolonged inhalation of NO in ARDS may lead to subacute lung injury that may compound the existing pathology.

SUMMARY AND CONCLUSIONS

Oxidant stress affects virtually all aspects of biologic existence by reaction with, and modification of, structural, metabolic, and genetic material. Protective mechanisms have evolved to defend cell components, but disease states, and other environmental stresses can overwhelm defense mechanisms and cause cytotoxicity. The discovery of the L-arginine-NO pathway has modified our understanding of the nature of the injurious species and the role NO plays in oxidant stress in the lung. There is no argument that inhalation of NO has been proven efficacious in improving oxygenation in infants with idiopathic pulmonary hypertension and in a variety of other disorders. Furthermore, in some cases, NO may actually reduce oxidant lung injury by preventing the propagation of lipid peroxidation. However, the reaction of NO with superoxide has been shown to produce peroxynitrite, a very reactive species capable of damaging the pulmonary surfactant and the alveolar epithelium. Thus the biological action of NO may depend on what type of molecule it is interacting with. Recent studies indicating that inhaled NO injury to the alveolar epithelium may be reduced by intratracheal instillation of antioxidant enzymes [49] offer promise in developing novel approaches to minimize its toxicity and thus enhance its therapeutic potential.

Acknowledgments

This work was supported by grants from the National Institutes of Health (HL31197 and HL51173), and a grant from the Office of Naval Research (N00014-97-1-0309).

References

1. Matalon, S., Benos, D.J. & Jackson, R.M. *Am.J.Physiol.* **271**, L1-22 (1996).
2. Freeman, B.A. & Crapo, J.D. *Lab.Invest.* **47**, 412-426 (1982).
3. Turrens, J.F., Freeman, B.A. & Crapo, J.D. *Arch.Biochem.Biophys.* **217**, 411-421 (1982).
4. Babior, B.M. *Environ.Health Perspect.* **102 Suppl 10**, 53-56 (1994).
5. Weinbroum, A., Nielsen, V.G., Tan, S., et al. *Am.J.Physiol.* **268**, G988-96 (1995).
6. Cantin, A.M., North, S.L., Hubbard, R.C. & Crystal, R.G. *J.Appl.Physiol.* **63**, 152-157 (1987).

7. Chao, C.C., Park, S.H. & Aust, A.E. *Arch.Biochem.Biophys.* **326**, 152-157 (1996).

8. Hardy, J.A. & Aust, A.E. *Carcinogenesis* **16**, 319-325 (1995).

9. Beckman, J.S., Beckman, T.W., Chen, J., Marshall, P.A. & Freeman, B.A. *Proc.Natl.Acad.Sci.USA* **87**, 1620-1624 (1990).

10. Ignarro, L.J. *Annu.Rev.Pharmacol.Toxicol.* **30**, 535-560 (1990).

11. Kwon, N.S., Stuehr, D.J. & Nathan, C.F. *J.Exp.Med.* **174**, 761-767 (1991).

12. Ischiropoulos, H., Zhu, L. & Beckman, J.S. *Arch.Biochem.Biophys.* **298**, 446-451 (1992).

13. Wei, X.Q., Charles, I.G., Smith, A., et al. *Nature* **375**, 408-411 (1995).

14. Rossaint, R., Falke, K.J., Lopez, F., Slama, K., Pison, U. & Zapol, W.M. *N.Engl.J.Med.* **328**, 399-405 (1993).

15. Jia, L., Bonaventura, J. & Stamler, J.S. *Nature* **380**, 221-226 (1996).

16. Cleeter, M.W., Cooper, J.M., Darley-Usmar, V.M., Moncada, S. & Schapira, A.H. *FEBS Lett.* **345**, 50-54 (1994).

17. Heiss, L.N., Lancaster, J.R., Jr., Corbett, J.A. & Goldman, W.E. *Proc.Natl.Acad.Sci.USA* **91**, 267-270 (1994).

18. Radi, R., Rodriguez, M., Castro, L. & Telleri, R. *Arch.Biochem.Biophys.* **308**, 89-95 (1994).

19. Beckman, J.S., Ischiropoulos, H., Zhu, L., et al. *Arch.Biochem.Biophys.* **298**, 438-445 (1992).

20. Carreras, M.C., Pargament, G.A., Catz, S.D., Poderoso, J.J. & Boveris, A. *FEBS Lett.* **341**, 65-68 (1994).

21. Kooy, N.W. & Royall, J.A. *Arch.Biochem.Biophys.* **310**, 352-359 (1994).

22. Kobzik, L., Bredt, D.S., Lowenstein, C.J., et al. *Am.J.Respir.Cell Mol.Biol.* **9**, 371-377 (1993).

23. van der Vliet, A., Smith, D., O'Neill, C.A., et al. *Biochem.J.* **303**, 295-301 (1994).

24. Denicola, A., Freeman, B.A., Trujillo, M. & Radi, R. *Arch.Biochem.Biophys.* **333**, 49-58 (1996).

25. Gow, A., Duran, D., Thom, S.R. & Ischiropoulos, H. *Arch.Biochem.Biophys.* **333**, 42-48 (1996).

26. Haddad, I.Y., Pataki, G., Hu, P., Galliani, C., Beckman, J.S. & Matalon, S. *J.Clin.Invest.* **94**, 2407-2413 (1994).

27. Wizemann, T.M., Gardner, C.R., Laskin, J.D., et al. *J.Leukoc.Biol.* **56**, 759-768 (1994).

28. Eiserich, J.P., Cross, C.E., Jones, A.D., Halliwell, B. & van der Vliet, A. *J.Biol.Chem.* **271**, 19199-19208 (1996).

29. Hazen, S.L. & Heinecke, J.W. *J.Clin.Invest.* **99**, 2075-2081 (1997).

30. Gow, A.J., Buerk, D.G. & Ischiropoulos, H. *J.Biol.Chem.* **272**, 2841-2845 (1997).

31. Gaston, B., Reilly, J., Drazen, J.M., et al. *Proc.Natl.Acad.Sci.USA* **90**, 10957-10961 (1993).

32. Mulligan, M.S., Hevel, J.M., Marletta, M.A. & Ward, P.A. *Proc.Natl.Acad.Sci.USA* **88** , 6338-6342 (1991).

33. Berisha, H.I., Pakbaz, H., Absood, A. & Said, S.I. *Proc.Natl.Acad.Sci.USA* **91**, 7445-7449 (1994).

34. Ischiropoulos, H., al-Mehdi, A.B. & Fisher, A.B. *Am.J.Physiol.* **269**, L158-64 (1995).

35. Akaike, T., Noguchi, Y., Ijiri, S., et al. *Proc.Natl.Acad.Sci.USA* **93**, 2448-2453 (1996).

36. van der Vliet, A., Eiserich, J.P., O'Neill, C.A., Halliwell, B. & Cross, C.E. *Arch.Biochem.Biophys.* **319**, 341-349 (1995).

37. McCall, M.N. & Easterbrook-Smith, S.B. *Biochem.J.* **257**, 845-851 (1989).

38. Vertessy, B.G., Zalud, P., Nyman, O.P. & Zeppezauer, M. *Biochim.Biophys.Acta* **1205**, 146-150 (1994).

39. Feste, A. & Gan, J.C. *J.Biol.Chem.* **256**, 6374-6380 (1981).

40. Zhu, S., Haddad, I.Y. & Matalon, S. *Arch.Biochem.Biophys.* **333**, 282-290 (1996).

41. Haddad, I.Y., Crow, J.P., Hu, P., Ye, Y., Beckman, J. & Matalon, S. *Am.J.Physiol.* **267**, L242-9 (1994).

42. Haddad, I.Y., Zhu, S., Ischiropoulos, H. & Matalon, S. *Am.J.Physiol.* **270**, L281-8 (1996).

43. Matalon, S., DeMarco, V., Haddad, I.Y., et al. *Am.J.Physiol.* **270**, L273-80 (1996).

44. Kong, S.K., Yim, M.B., Stadtman, E.R. & Chock, P.B. *Proc.Natl.Acad.Sci.U.S.A.* **93**, 3377-3382 (1996).

45. Gow, A.J., Duran, D., Malcolm, S. & Ischiropoulos, H. *FEBS Lett.* **385**, 63-66 (1996).

46. Robbins, C.G., Davis, J.M., Merritt, T.A., et al. *Am.J.Physiol.* **269**, L545-50 (1995).

47. Hallman, M., Waffarn, F., Bry, K., et al. *J.Appl.Physiol.* **80**, 2026-2034 (1996).

48. Mercer, R.R., Costa, D.L. & Crapo, J.D. *Lab.Invest.* **73**, 20-28 (1995).

49. Robbins, C.G., Horowitz, S., Merritt, T.A., et al. *Am.J.Physiol.* **272**, L903-L907(1997).

MACROPHAGE KILLING OF MYCOPLASMAS:
INVOLVEMENT OF SURFACTANT PROTEIN A AND NITRIC OXIDE

Judy M. Hickman-Davis[1], Sadis Matalon[2,3,4] and J. Russell Lindsey[1]

Departments of Comparative Medicine[1] and Anesthesiology[2],
Physiology and Biophysics[3] and Pediatrics[4],
Schools of Medicine and Dentistry,
University of Alabama at Birmingham,
Birmingham, AL 35294

Mycoplasma pneumoniae is one of the leading causes of pneumonia worldwide. Infections due to this agent occur in smoldering, year round endemics and in cyclic 4-7 year epidemics. In the U.S., *M. pneumoniae* accounts for 20-30% of all pneumonias in the general population, for a total of 8-15 million cases a year[1]. *M. pneumoniae* frequently exacerbates asthma[2], and chronic obstructive pulmonary disease (COPD)[3]. Furthermore, it is becoming increasingly apparent that because of the great diversity of clinical manifestations and the special testing required to distinguish active infection, the role of *M. pneumoniae* as a cause of severe disease in the lungs and other organs is much under diagnosed[4].

Efforts to understand protective immunity against mycoplasma infections of the respiratory tract through studies of specific immune mechanisms have proved disappointing for both human[1] and animal[5] mycoplasma infections. It remains unknown whether specific immunity has a role in protection against respiratory mycoplasmas, with one exception. Specific antibody clearly is important in the late stages of *Mycoplasma pneumoniae* infection as patients with hypogammaglobulinemia develop chronic mycoplasma lung disease and systemic infections such as arthritis[6]. Similarly, other mycoplasmas, including *Mycoplasma hominis*, *Mycoplasma salivarium* and *Ureaplasma urealiticum*, also cause chronic infections with arthritis in hypogammaglobulinemic patients (reviewed in Cassell[7]). Thus innate immunity is important for control of acute infection while specific immunity appears to be important for the control of dissemination of organisms to other organs of the body.

Mycoplasma pulmonis infection in mice provides an excellent animal model that reproduces the essential features of human respiratory mycoplasmosis. Mouse strains differ markedly to resistance to *M. pulmonis*, with C57BL/6 and C3H/He mice representing the extremes in response to this infection. C57BL/6 mice have a 100-fold higher 50% lethal dose, 50% pneumonia dose, and 50% microscopic lesion dose than C3H/He mice. During the first 72 hours postinfection, the numbers of mycoplasmas decrease by more than 83% in the lungs of C57BL/6 mice but increase by 18000% in the lungs of C3H/He mice. In

Acute Respiratory Distress Syndrome: Cellular and Molecular Mechanisms and Clinical Management
Edited by Matalon and Sznajder, Springer Science+Business Media New York, 1998

C57BL/6 mice, maximal mycoplasmacidal activity occurs within 8 hours postinfection although mechanical clearance does not differ between the two strains of mice during this time[8, 9]. Demonstration of specific antibody in serum, as well as an increase in the number of macrophages, neutrophils, or lymphocytes in the lungs does not occur until > 72 hours postinfection[10, 11]. Thus, nonspecific intrapulmonary killing of *M. pulmonis* occurs and is most likely mediated by rapidly activated resident alveolar macrophages (AMs).

There is strong indirect evidence that innate immunity involving AMs is of major importance in antimycoplasmal defense of the lungs. In C57BL/6 mice infected with *M. pulmonis* and exposed to nitrogen dioxide, intrapulmonary killing decreased as AM viability decreased and subsequently increased as AM viability was restored[12]. Macrophage depletion has been used to investigate the protective roles of AMs in the lungs[13] and the resident macrophages in the liver and spleen [14]. To further delineate the role of the AM in early clearance of mycoplasmas from the lungs, intratracheal insufflation of liposome encapsulated dichloromethylene bisphosphonate (L-Cl$_2$MBP) has been used to selectively deplete AMs in mice[15].

Cl$_2$MBP is a compound used clinically for the treatment of osteolytic bone diseases. The drug itself is not toxic, does not easily cross cell membranes and has an extremely short half life in circulation[16]. When encapsulated into multi lamellar liposomes, Cl$_2$MBP is highly specific for phagocytic cells. Phagocytes ingest the L-Cl$_2$MBP which are degraded in the lysosomes releasing free Cl$_2$MBP into the cytoplasm and causing cell death. However, neutrophils appear to be functionally and morphologically unaffected by L-Cl$_2$MBP both *in vivo* and *in vitro* presumably because of their low liposome ingestion. The precise mechanism of Cl$_2$MBP cytotoxicity for macrophages is unknown, but it may be due to depletion of iron or other metal complexes in the cell, or a direct effect on ATP metabolism (reviewed by Van Rooijen[17]). L-Cl$_2$MBP has been shown to have little effect on alveolar epithelium or interstitial macrophages[18], although free Cl$_2$MBP has been shown by electron microscopy to cause edema of the alveolar epithelium[13]. Liposomes of different cholesterol compositions have been shown to alter macrophage function by blocking or activating phagocytosis. To show that the observed effects are not merely due to macrophage depletion, PBS containing liposomes (L-PBS) can be incorporated into experiments for control purposes.

We have shown that AM depletion prior to infection with *M. pulmonis* reduced mycoplasma killing in resistant C57BL/6 mice to a level comparable to that in susceptible C3H/He mice without AM depletion. In contrast, AM depletion did not alter killing of mycoplasmas in the lungs of infected C3H/He mice[15]. These results directly identify the AM as the main effector cell in early mycoplasmal resistance of C57BL/6 mice. The fact that mycoplasma numbers in AM-depleted C3H/He mice remained unchanged suggests that C3H/He mice have (i) a defective macrophage activation pathway, (ii) a functional defect in one of their AM subset populations, or (iii) a defect in nonspecific opsonization in their lungs[15].

In vitro, studies on host defense against mycoplasmas have concentrated on the role of phagocytes. In the absence of antibody it has been demonstrated that mycoplasmas can attach to the surface of phagocytes but are not ingested[19]. The mechanisms by which most mycoplasmas resist ingestion are unknown, although resistance of *M. pulmonis* has been shown to be trypsin sensitive, suggesting the presence of an antiphagocytic surface protein. In the presence of specific antibody, however, mycoplasmas are rapidly ingested and 90 to 99% of cell associated mycoplasmas are killed within 4 hours[20]. Previous studies have also shown that the concentrated noncellular portion of lavages from *M. pulmonis*-infected C57BL/6 mice, although unable to kill mycoplasmas alone, could initiate killing of mycoplasmas when introduced into AM cultures[21]. The requirement of opsonins for *in vitro* mycoplasmacidal activity coupled with the *in vivo* data showing significant clearance of organisms by resistant animals within hours of infection suggests the presence of some nonspecific opsonin.

As Pison et al.[22] have suggested, surfactant proteins may provide "in the alveolar lining fluid a first line of defense against infection that would act quickly before enough time had elapsed to acquire specific immunity." Surfactant proteins SP-A and SP-D are hybrid molecules termed 'collectins' that belong to the Ca^{2+}-dependent animal lectin superfamily. Collectins are carbohydrate-binding proteins other than antibodies and enzymes that share the common structural characteristics of an amino-terminal collagen-like domain connected to a Ca^{2+}-dependent carbohydrate-recognition domain (CRD). Mannose-binding protein (MBP), conglutinin and CL-43 complete the collectin subfamily of plasma lectins[23]. Collectins exist as multimers of trimers with SP-A and MBP able to assemble into hexamers of trimers. SP-A and MBP resemble the first complement protein C1q in the arrangement of eighteen CRDs with kinked collagen stalks. SP-D and conglutinin arrange their globular heads around collagen spokes as cruciform structures, and CL-43 has the simplest arrangement of a single unit consisting of three polypepetides. The collectins are ideally suited to the role of first line defense in that they are widely distributed, capable of antigen recognition and can discern self versus non-self[24]. The collectins recognize bacteria, fungi and viruses by binding mannose and N-acetylglucosamine residues on microbial cell walls[23]. The collectins bind a wide range of pathogens *in vitro* (Table 1), however, it is likely that regulatory and microbial specificity exists for collectin binding *in vivo*.

SP-A and SP-D are thought to participate in two major physiologic processes: (i) the regulation of surfactant homeostasis associated with tubular myelin formation and (ii) nonspecific innate immune responses of the lung[25, 26]. With the creation of SP-A deficient mice, there has been a greater emphasis placed on the role of SP-A following infectious or toxic insult. These SP-A knockout mice have an increased susceptibility to infection with Streptococcus although there are no apparent breeding or survival abnormalities when housed under pathogen free conditions[27, 28]. The presence of SP-D in SP-A knockout mice may account for the lack of gross abnormalities in respiratory function. Although SP-A and SP-D probably have discrete functions in the normal animal, similar anatomic distribution and structure may allow SP-A and SP-D to be functionally interchangeable under certain conditions. SP-A interacts with AMs in a highly specific manner through a cell surface SP-A receptor[29]. SP-A has been shown to: (i) effect release of reactive oxygen species from AMs[30]; (ii) stimulate chemotaxis of AMs[31]; (iii) enhance phagocytosis and killing of bacterial, viral and fungal pathogens by AMs; and (iv) enhance FcγR- and C1qR-mediated phagocytosis[32] *in vitro*.

We investigated the role of SP-A in early antimycoplasmal defense characterizing the *in vitro* interactions of SP-A with resistant C57BL/6 AMs and the *Mycoplasma pulmonis* strain UAB CT. Human SP-A was used in these studies as it has been shown to be capable of effecting the functions of AMs from many species[33-35]. We found that human SP-A bound C57BL/6 mouse AMs in a concentration-, time- and temperature-dependent manner. Consistent with receptor mediated binding, there was saturation of AM-associated SP-A at the 15 µg/ml concentration after 2 hours at 4°C. It has been demonstrated that SP-A is capable of stimulating phagocytosis of bacteria to which it binds[36]. We found that SP-A was capable of binding to mycoplasmas in a concentration- and partially Ca^{2+}-dependent manner. While significant binding to mycoplasmas was seen in the absence of Ca^{2+}, this binding was increased by as much as 70% in the presence of Ca^{2+}. SP-A did not bind to mycoplasmas at low concentrations (0 to 5 µg/ml) of SP-A, however, this is not surprising considering that mycoplasmas lack cell walls and are the smallest of the self-replicating organisms[7].

To determine the capability of SP-A to stimulate phagocytosis and killing of mycoplasmas, isolated AMs were activated with IFNγ, incubated with SP-A and infected

Table 1. Collectin Target Pathogens

Collectin	Target	Reference
Conglutinin	Escherishia coli	40
	HIV 1	41
	Influenza A virus	42, 43
	Salmonella typhimurium	40
MBP	Cryptococcus neoformans	44
	Klebsiella pneumoniae	45
	HIV 1	46
	HIV 2	47
	Influenza A virus	48, 49
	Mycobacterium tuberculosis	50
	Pneumocystis carinii	51, 52
	Saccharomyces cervisiae	53
	Salmonella montevideo	54
	Trypanosoma cruz	55
CL-43	Cryptococcus neoformans	24
SP-A	Aspergillus fumigatus	56
	Candida tropicalis	57
	Escherishia coli	58, 59
	Haemophilus influenzae	36
	Herpes simplex virus	60, 61
	Influenza A virus	62, 63
	Klebsiella pneumoniae	33
	Mycobacterium tuberculosis	64, 65
	Pneumocystis carinii	66-68
	Pseudomonas aeruginosa	58
	Staphylococcus aureus	34, 58, 69
	Streptococcus pneumoniae	36
SP-D	Aspergillus fumigatus	56
	Cryptococcus neoformans	70
	Escherishia coli	71
	Influenza A virus	72
	Klebsiella pneumoniae	73
	Pneumocystis carinii	74

Table 2. Nitric oxide/ Reactive Nitrogen Species Target Pathogens

Pathogen	Reference
Candida albicans	75-78
Chlamydia trachomatis	79
Cryptococcus neoformans	80, 81
Escherishia coli	37, 38
Klebsiella pneumoniae	39
Legionella pneumophilia	82
Leishmania major	83-87
Mycobacterium avium subsp. paratuberculosis	88
Mycobacterium bovis	89
Mycobacterium tuberculosis	90-92
Pseudomonas aeruginosa	93
Toxoplasma gondii	94
Trypanosoma cruzi	95, 96
Schistosoma mansoni	97
Vesicular stomatitis virus	98

with *M. pulmonis*. AMs were ruptured by sonication at various time points (0 to 8 hours) and quantitative mycoplasma cultures were performed. SP-A significantly enhanced the killing of mycoplasmas with a maximal decrease of 83% in total recoverable organisms by 6 hours postinfection. Our data showed that the killing of mycoplasmas was SP-A- and time-dependent, and that the SP-A-mediated mycoplasmacidal effect was lost by 8 hours postinfection presumably when SP-A became depleted in the media. The role of SP-A in mycoplasma killing by AMs probably involves modulation of AM function by SP-A rather than as a nonspecific opsonin for several reasons: (i) *M. pulmonis* "opsonized" with SP-A prior to the addition to IFNγ-activated AM cultures did not result in significant killing; (ii) SP-A was adhered to AMs and the excess SP-A removed prior to infection with *M. pulmonis* and (iii) mycoplasmas did not preferentially bind to SP-A-treated AMs.

SP-A also is known to effect the release of reactive oxygen species[30], another mechanism that could account for the SP-A mediated mycoplasmacidal activity of AMs. In the present study, the addition of the inducible nitric oxide synthase inhibitor, N^G-monomethyl-L-arginine (*N*-MeArg), to AM cultures abrogated the SP-A mediated mycoplasmacidal activity, indicating that nitric oxide (•NO) may be involved in mycoplasma killing. •NO production following IFNγ stimulation of macrophages has been shown to play an important role in the control of intracellular and extracellular pathogens (Table 2). Peroxynitrite, a strong oxidant formed by AMs as a reaction product of superoxide and •NO, also has been shown to be highly bactericidal[37, 38], and could have an important role in mycoplasma killing. Nitrate and nitrite, the decomposition products of •NO, were significantly increased in cultures containing SP-A and decreased in cultures containing *N*-MeArg, further implicating •NO as a factor involved in SP-A mediated mycoplasma killing.

While •NO is a well recognized molecule of microbicidal macrophages, the mechanism(s) by which •NO aids in host defense remain undefined. •NO may have a direct microbicidal effect through (i) the reaction with iron or thiol groups on proteins forming iron-nitrosyl complexes that inactivate enzymes important in DNA replication or mitochondrial respiration[39], or (ii) the formation of such reactive oxidant species as peroxynitrite discussed above.

In summary, we have demonstrated *in vitro* that SP-A binds specifically to C57BL/6 mouse AMs in culture, mediates killing of mycoplasmas by activated AMs, and that •NO and/or its toxic metabolites are involved in mycoplasma killing. SP-A mediated killing of mycoplasmas by AMs may be the primary mechanism of innate host defense against early mycoplasmal infections in the lungs. In general, the understanding of pulmonary host defense mechanisms against mycoplasmas lags behind that of other systems because of the poor accessibility and complexity of the lung environment. If the innate mechanisms of the lung are of primary importance in defense against mycoplasmal and other bacterial infections, it may be possible to develop practical therapies or to increase the lungs protective capacity through a better understanding of the early immune response.

References

1. Krause, D.C., and D. Taylor-Robinson, Mycoplasmas which infect humans, *Mycoplasmas: molecular biology and pathogensis*, ed. J. Maniloff et al., Washington DC: American Society for Microbiology, (1992).

2. Gil, J.C.,et al., Isolation of *Mycoplasma pneumoniae* from asthmatic patients, *Ann. Allergy* 70:23-25. (1993).

3. Melbye, H., J. Kongerud, and L. Vorland, Reversible airflow limitation in adults with respiratory infection, *Eur. Respir. J.* 7:1239-1245. (1994).

4. Cassell, G.H., Severe mycoplasma disease--rare or underdiagnosed?, *Western J. Med.* 162: 172-175. (1995).

5. Simecka, J.W.,et al., Mycoplasma diseases of animals, *Mycoplasmas: Molecular Biology and Pathogenisis*, ed. J. Maniloff et al., Washington, DC: American Society for Microbiology, Chapter 24:391-415. (1992).

6. Foy, H.M., Infections caused by *Mycoplasma pneumoniae* and possible carrier state in different populations of patients, *Clin. Infect. Dis.* 17(S1):S37-46. (1993).

7. Cassell, G.H.,et al., Mycoplasma infections, *Harrison's Principles of Internal Medicine*, ed. Fauci, and E. Pack. 14 ed, New York, NY: McGraw-Hill, Chpt 180:1-29 (1997). In press.

8. Davis, J.K.,et al., Strain differences in susceptibility to murine respiratory mycoplasmosis in C57BL/6N and C3He/HeN mice, *Infect. Immun.* 50:647-654 (1985).

9. Parker, R.F.,et al., Pulmonary clearance of *Mycoplasma pulmonis* in C57BL/6N and C3H/HeN mice, *Infect. Immun.* 55:2631-2635 (1987).

10. Cartner, S.C.,et al., Chronic Respiratory mycoplasmosis in C3H/HeN and C57BL/6N mice: lesion severity and antibody response, *Infect. Immun.* 63:4138-4142 (1995).

11. Parker, R.F.,et al., Short term exposure to nitrogen dioxide enhances susceptibility to murine respiratory mycoplasmosis and decreases intrapulmonary killing of *Mycoplasma. pulmonis*, *Am. Rev. Respir. Dis.* 140:502-512 (1989).

12. Davis, J.K.,et al., Decreased intrapulmonary killing of *Mycoplasma pulmonis* after short-term exposure to NO_2 is associated with damaged alveolar macrophages., *Am. Rev. Respir. Dis.* 145:406-411 (1992).

13. Berg, J.T.,et al., Depletion of alveolar macrophages by liposome-encapsulated dichloromethylene diphosphate, *J. Appl. Physiol.* 74:2812-2819 (1993).

14. Bette, M.,et al., Distribution and kinetics of super-antigen induced cytokine gene expression in mouse spleen, *J. Exp. Med.* 178:1531-1540 (1993).

15. Hickman-Davis, J.M.,et al., Depletion of alveolar macropahges exacerbates respiratory mycoplasmosis in mycoplasma-resistant C57BL mice but not mycoplasma-susceptible C3H mice, *Infect. Immun.* 65:2278-2282 (1997).

16. Fleisch, H., Biphosphonates: a new class of drugs in diseases of bone and calcium metabolism," *Handbook Exp. Pharmacol.* 83:441-446 (1988).

17. Van Rooijen, N., and A. Sanders, Liposome mediated depletion of macrophages: mechanism of action, preparation of liposomes and applications, *J. Immunol. Methods* 174:83-93 (1994).

18. Thepen, T.,et al., Regulation of immune response to inhaled antigen by alveolar macrophages: differential effects of *in vivo* alveolar macrophage elimination on the induction of tolerance vs. immunity, *Eur. J. Immunol.* 21:2845-2850 (1991).

19. Cassell, G.H., W.A. Clyde Jr, and J.K. Davis, Mycoplasma respiratory infections, *The Mycoplasmas*, ed. et al. S. Razine, NY: Acad. Press, Chpt.IV:65-106 (1985).

20. Howard, C.J., and G. Taylor, Interaction of mycoplasmas and phagocytes, *Yale J. Biol. Med.* 56:643-648 (1983).

21. Davis, J.K., M. Davidson, and T.R. Schoeb, Murine respiratory mycoplasmosis: a model to study effects of oxidants, 29 p. Research Report No. 47, Health Effects Institute, Cambridge, MA. (1991).

22. Pison, U.,et al., Host defence capacities of pulmonary surfactant: evidence for "non-surfactant" functions of the surfactant system, *Eur. J. Clin. Invest.* 24:586-599 (1994).

23. Holmskov, U.,et al., Collectins: collagenous C-type lectins of the innate immune defense system, *Immunol. Today* 15:67-73 (1994).

24. Epstein, J.,et al., The collectins in innate immunity, *Curr. Opinion Immunol.* 8:29-35 (1996).

25. Hawgood, S., and F.R. Poulain, Functions of the surfactant proteins: a perspective, *Ped. Pulmonol.* 19:99-104 (1995).

26. Johansson, J., T. Curstedt, and B. Robertson, The proteins of the surfactant system, *Eur. Respir. J.* 7:372-391 (1994).

27. LeVine, A.M.,et al., Surfactant protein A-deficient mice are susceptible to group B streptococcal infection, *J. Immunol.* 158:4336-4340 (1997).

28. Korfhagen, T.R.,et al., Altered surfactant function and structure in SP-A gene targeted mice, *Proc. Natl. Acad. Sci.* 93:9594-9599 (1996).

29. Chroneos, Z.C.,et al., Purification of a cell-surface receptor for surfactant protein A, *J. Biol. Chem.* 271:16375-16383 (1996).

30. Weissbach, S.,et al., Surfactant protein A modulates the release of reactive oxygen species from alveolar macrophages, *Am. J. Physiol. (Lung Cell. Mol. Physiol. 11)* 267 (1994):L660-L666.

31. Wright, J.R., and D.C. Youmans, Pulmonary surfactant protein A stimulates chemotaxis of alveolar macrophages, *Am. J. Physiol. (Lung Cell Mol. Physiol. 8)* 264:L338-L344 (1993).

32. Tenner, A.,et al., Human pulmonary surfactant protein A (SP-A) a protein structurally homologous to C1q can enhance FcR- and CR1-mediated phagocytosis, *J. Biol. Chem.* 264:13923-13928 (1989).

33. Kabha, K.,et al., SP-A enhances phagocytosis of *Klebsiella* by interaction with capsular polysaccharides and alveolar macrophages, *Am. J. Physiol. (Lung Cell. Mol. Physiol. 16)* 272:L344-L352 (1997).

34. McNeely, T.B. and J.D. Coonrod, Comparison of the opsonic activity of human surfactant protein A for *Staphylococcus aureus* and *Streptococcus pneumoniae* with rabbit and human macrophages, *J. Infect. Dis.* 167:91-97 (1993).

35. Pison, U., J.R. Wright, and S. Hawgood, Specific binding of surfactant appoprotein SP-A to rat alveolar macrophages, *Am. J. Physiol. (Lung Cell. Mol. Physiol. 6)* 262:L412-L417 (1992).

36. Tino, M. J. and J.R. Wright, Surfactant protein A stimulates phagocytosis of specific pulmonary pathogens by alveolar macrophages, *Am. J. Physiol. (Lung Cell. Mol. Physiol. 14)* 270:L677-L688 (1996).

37. Zhu, L., C. Gunn, and J.S. Beckman, Bacteriocidal activity of peroxynitrite," *Arch. Biochem. Biophys.* 298:452-457 (1992).

38. Brunelli, L., J.P. Crow, and J.S. Beckman, The comparative toxicity of nitric oxide and peroxynitrite to *Escherichia coli*, *Arch. Biochem. Biophys.* 316:327-334 (1995).

39. Tsai, W.C.,et al., Nitric oxide is required for effective innate immunity against *Klebsiella pneumoniae*, *Infect. Immun.* 65:1870-1875 (1997).

40. Friis-Christiansen, P.,et al., In vivo and in vitro bacteriocidal activity of conglutinin, a mammalian plasma lectin, *Scand. J. Immunol.* 31:453-460 (1990).

41. Andersen, O.,et al., Conglutinin binds HIV 1 envelope glycoprotein gp160 and inhibits its interaction with cell membrane CD4, *Scand. J. Immunol.* 33:81-88 (1991).

42. Anders, E.M., C.A. Hartley, and D.C. Jackson, Bovine and mouse serum beta inhibitors of influenza A viruses are mannose-binding lectins, *Proc. Natl. Acad. Sci* 87:4485-4489 (1990).

43. Hartley, C.A., D.C. Jackson, and E.M. Anders, Two distinct serum mannose-binding lectins function as beta inhibitors of influenza virus: identification of bovine serum beta inhibitor as conglutinin, *J. Virol.* 66:4358-4363 (1992).

44. Levitz, S.M., A. Tabuni, and C. Tresseler, Effect of mannose-binding protein on binding of *Cryptococcus neoformans* to human pagocytes, *Infect. Immun.* 61:4891-4893 (1993).

45. Jiang, G.,et al., Binding of mannose-binding protein to *Klebsiella O3* lippopolysaccharide possessing the mannose homopolysaccharide as the *O*-specific polysaccharide and its relation to complement activation, *Infect. Immun.* 63:2537-2540 (1995).

46. Ezekowitz, R.A.,et al., A human serum mannose-binding protein inhibits in vitro infection by the human immunodeficiency virus, *J. Exp. Med.* 169:185-196 (1989).

47. Haurum, J.S.,et al., Complement activation upon binding of mannan-binding protein to HIV envelope glycoproteins, *AIDS* 7:1307-1313 (1993).

48. Hartshorn, K.L.,et al., Human mannose-binding protein functions as an opsonin for influenza A viruses, *J. Clin. Invest.* 91:1414-1420 (1993).

49. Malhotra, R.,et al., Binding of human collectins (SP-A and MBP) to influenza virus, *Biochem. J.* 305:455-461 (1994).

50. Garred, P.,et al., Dual role of mannan-binding protein in infections: another case of heterosis, *Eur. J. Immunogenet.* 21:125-131 (1994).

51. Ezekowitz, R.A.B.,et al., Uptake of *Pneumocystis carinii* mediated by the mannose macrophage receptor, *Nature* 351:155-158 (1991).

52. O'Riordan, D.M., J.E. Standing, and A.H. Limper, *Pneumocystis carinii* glycoprotein A binds macrophage mannose receptors, *Infect. Immun.* 63:779-784 (1995).

53. Super, M.,et al., Association of low levels of mannan-binding protein with a common defect of opsonization, *Lancet* 2:1236-1239 (1989).

54. Kuhlman, M., K. Joiner, and R.A. Ezekowitz, The human mannose-binding protein functions as an opsonin, *J. Exp. Med.* 169:1733-1745 (1989).

55. Kahn, S.,et al., *Trypanosoma cruzi* amastigote adhesion to macrophages is facilitated by the mannose receptor, *J. Exp. Med.* 182:1243-1258 (1995).

56. Madan, T.,et al., Binding of pulmonary surfactant proteins A and D to *Aspergillus fumigatus* conidia enhances phagocytosis and killing by human neutrophils and alveolar macrophages, *Infect. Immun.* 65:3171-3179 (1997).

57. Weissbach, S.,et al., Surfactant protein A (SP-A) stimulates phagocytosis of *Candida tropicalis* by alveolar macropahges, *FASEB J.* 6:A1270 (1992).

58. Manz-Keinke, H., H. Plattner, and J. Schlepper-Schafer, Lung surfactant protein A (SP-A) enhances serum-independent phagocytosis of bacteria by alveolar macrophages, *Eur. J. Cell Biol.* 57:95-100 (1992).

59. Pikaar, J.C., et al., Opsonic activities of surfactant proteins A and D in phagocytosis of gram-negative bacteria by alveolar macrophages, *J. Infect. Dis.* 172:481-89 (1995).

60. Van Iwaarden, J.F., et al., Binding of surfactant protein A (SP-A) to herpes simplex virus type 1-infected cells is mediated by the charbohydrate moeity of SP-A, *J. Biol. Chem.* 267:25039-43 (1992).

61. Van Iwaarden, J.F.,et al., Surfactant protein A as opsonin in phagocytosis of herpes simplex virus type 1 by rat alveolar macrophages, *Am. J. Physiol. (Lung Cell. Mol. Physiol. 5)* 261:L204-209 (1991).

62. Benne, C.A.,et al., Surfactant protein A, but not surfactant protein D, is an opsonin for influenza A virus phagocytosis by rat alveolar macrophages, *Eur. J. Immunol.* 27:886-890 (1997).

63. Benne, C.A.,et al., Interactions of surfactant protein A with influenza A viruses: binding and neutralization, *J. Infect. Dis.* 171:335-41 (1995).

64. Gaynor, C.D.,et al., Pulmonary surfactant protein A mediates enhanced phagocytosis of *Mycobacterium tuberculosis* by a direct interaction with human macrophages, *J. Immunol.* 155:5343-5351 (1995).

65. Downing, J.F.,et al., Surfactant protein A promotes attachment of *Mycobacterium tuberculosis* to alveolar macrophages during infection with human immunodeficiency virus, *Proc. Natl. Acad. Sci.* 92:4848-4852 (1995).

66. McCormack, F.X.,et al., The charbohydrate recognition domain of surfactant protein A mediates binding to the major surface glycoprotein of *Pneumocystis carinii, Biochem.* 36:8092-8099 (1997).

67. Williams, M.D.,et al., Human surfactant protein A enhances attachment of *Pneumocystis carinii* to rat alveolar macrophages, *Am. J. Respir. Cell Mol. Biol.* 14:232-238 (1996).

68. Zimmerman, P.E.,et al., 120-kD surface glycoprotein of *Pneumocystis carinii* is a ligand for surfactant protein A, *J. Clin. Invest.* 89:143-149 (1992).

69. Van Iwaarden, F.,et al., Pulmonary surfactant protein A enhances the host-defense mechanism of rat alveolar macrophages, *Am. J. Respir. Cell. Mol. Biol.* 2:91-98 (1990).

70. Schelenz, S.,et al., Binding of host collectins to the pathogenic yeast *Cryptococcus neoformans*: human surfactant protein D acts as an agglutinin for acapsular yeast cells, *Infect. Immun.* 63:3360-3366 (1995).

71. Kuan, S.F., K. Rust, and E. Crouch, Interactions of surfactant protein D with bacterial lipopolysaccharides. SP-D is an *E. coli*-binding protein in bronchoalveolar lavage, *J. Clin. Invest.* 90:97-106 (1992).

72. Hartshorn, K.L.,et al., Evidence for a protective role of pulmonay surfactant protein D (SP-D) against influenza A viruses, *J. Clin. Invest.* 94:311-319 (1994).

73. Lim, B.L.,et al., Expression of the charbohydrate recognition domain of lung surfactant protein D and demonstration of its binding to lipopolysaccharides of gram-negantive bacteria, *Biochem. Biophys. Res. Comm.* 202:1674-1680 (1994).

74. O'Riordan, D.M.,et al., Surfactant protein D interacts with *Pneumocystis carinii* and mediates organism adherence to alveolar macrophages, *J. Clin. Invest.* 95:2699-2710 (1995).

75. Cenci, E.,et al., Interleukin-4 and interleukin-10 inhibit nitric oxide-dependent macrophage killing of *Candida albicans, Eur. J. Immunol.* 23:1034-1038 (1993).

76. Jones-Carson, J.,et al., γδ T cell-induced nitric oxide production enhances resistance to mucosal candidiasis, *Nature Med.* 1:552-557 (1995).

77. Vazquez-Torres, A.,et al., Nitric oxide enhances resistance of SCID mice to mucosal candidiasis, *J. Infect. Dis.* 172:192-198 (1995).

78. Vazquez-Torres, A., J. Jones-Carson, and E. Balish, Peroxynitrite contributes to the candidicidal activity of nitric oxide -producing macrophages, *Infect. Immun.* 64:3127-3133 (1996).

79. Mayer, J.,et al., Gamma-interferon-induced nitric oxide reduces *Chlamydia trachomatitis* infectivity in McCoy cells, *Infect. Immun.* 61:491-497 (1993).

80. Alspaugh, J.A., and D.L. Granger, Inhibition of *Cryptococcus neoformans* replication by nitrogen oxide supports the role of these molecules as effectors of macrophage-mediated cytostasis, *Infect. Immun.* 59:2291-2296 (1991).

81. Granger, D.,et al., Metabolic fate of L-arginine in relation to microbiostatic capability of murine macrophages, *J. Clin. Invest.* 85:264-273 (1990).

82. Brieland, J.K.,et al., In vivo regulation of replicative *Legionella pneumophila* lung infection by endogenous tumor necrosis factor alpha and nitric oxide, *Infect. Immun.* 63:3253-3258 (1995).

83. Green, S.J.,et al., *Leishmania major* amastigotes initiate an L-arginine-dependent killing mechanism in IFNγ stimulated macrophages by induction of tumer necrosis factor, *J. Immunol.* 145:4290-4297 (1990).

84. Green, S.J.,et al., Activated macrophages destroy intracellular *Leishmania major* amastigotes by an L-arginine dependent killing mechanism, *J. Immunol.* 144:278-283 (1990).

85. Liew, F.Y.,et al., Macrophage killing of *Leishmania* parasite *in vivo* is mediated by nitirc oxide from L-arginine, *J. Immunol.* 144:4794-4797 (1990).

86. Liew, F.Y., Interactions between cytokines and nitric oxide, *Adv. Neuroimmunol.* 5:201-209 (1995).

87. Mauel, J., A. Ransijn, and Y. Buchmuller-Rouiller, Killing of Leishmania parasites in activated murine macrophages is based on an L-arginine-dependent process that produces nitrogen derivatives, *J. Leukocyte. Biol.* 49:73-82 (1991).

88. Zhao, B., M.T. Collins, and C.J. Czuprynski, Effects of gamma interferon and nitric oxide on the interaction of *Mycobacterium avium* subsp. *paratuberculosis* with bovine monocytes, *Infect. Immun.* 65:1761-1766 (1997).

89. Hanano, R., and S.H.E. Kaufmann, Nitric oxide production and mycobacterial growth inhibition by murine alveolar macrophages: the sequence of rIFN-gamma stimulation and *Mycobacterium bovis* BCG infection determines macrophage activation, *Immunol. Lett.* 45:23-27 (1995).

90. Chan, J.,et al., Killing of virulent *Mycobacterium tuberculosis* by reactive nitrogen intermediates produced by activated murine macrophages, *J. Exp. Med.* 175:1111-1122 (1992).

91. Chan, J.,et al., Effects of nitric oxide synthase inhibitors on murine infection with *Mycobacterium tuberculosis*, *Infect. Immun.* 63:736-740 (1995).

92. Denis, M., Interferon-gamma-treated murine macrophages inhibit growth of tubercle bacilli via the generation of reactive nitrogen species, *Cell. Immunol.* 132:150-157 (1991).

93. Gosselin, D.,et al., Role of tumor necrosis factor alpha in innate resistance to mouse pulmonary infection with *Psuedomonas aeruginosa*, *Infect. Immun.* 63:3272-3278 (1995).

94. Adams, L.B.,et al., Microbiostatic effect of murine macrophages for *Toxoplasma gondii*: role of synthesis of inorganic nitrogen oxides from L-arginine, *J. Immunol.* 144:2725-2729 (1990).

95. Petray, P.,et al., Role of nitric oxide in resistance and histopathology during experimental infection with *Trypanosoma cruzi*, *Immunol. Lett.* 47:121-126 (1995).

96. Denicola, A.,et al., Peroxynitrite-mediated cytotoxicity to *Trypanosoma cruzi*, *Arch. Biochem. Biophys.* 304:279-286 (1993).

97. James, S.L., and J. Glaven, Macrophage cytotoxicity against schistosomula of *Schistosoma mansoni* involved arginine-dependent production of reactive nitrogen intermediates, *J. Immunol.* 143:4208-4212 (1989).

98. Bi, Z., and C.S. Reiss, Inhibition of vesicular stomatitis virus infection by nitric oxide, *J. Virol.* 69:2208-2213 (1995).

EXHALED NITRIC OXIDE IN PATIENTS UNDERGOING CARDIOTHORACIC SURGERY: A NEW DIAGNOSTIC TOOL?

Nándor Marczin[1,2], Bernhard Riedel[2], David Royston[2] and Magdi Yacoub[1]

[1]Department of Cardiothoracic Surgery, National Heart and Lung Institute, Imperial College of Science Technology and Medicine
[2]Department of Anaesthetics, Harefield Hospital , Harefield, United Kingdom

INTRODUCTION

NO is produced by many cells within the lung and appears to play a critical role in the physiological control of the pulmonary vascular bed and airways and in the pathophysiology of lung diseases [1-3]. Pulmonary vascular endothelial cells and airway epithelial cells continuously generate low amounts of NO from the amino acid L-arginine via a Ca^{++}-calmodulin-dependent and constitutively active NO synthase (cNOS) [4,5]. NO serves as an intercellular signalling molecule stimulating the production of a second messenger (cGMP) via activation of soluble guanylate cyclase (sGC) in the neighbouring vascular and airway smooth muscle cells to elicit relaxation and to regulate blood flow through the lungs [5,6]. Constitutive NO release appears to counteract vasoconstriction, airway hyper-reactivity and is believed to inhibit smooth muscle proliferation [7,8]. Accordingly, reduced biological activity of NO may contribute to increased pulmonary vascular resistance, airway hyper-reactivity and to cellular proliferation of the smooth muscle cells and fibroblasts. In addition to endothelial and epithelial NO synthesis, neurones appear to express neuronal NO synthase (NANC system) contributing to constitutive production of NO in the lungs [9,10].

In disease states, cells of the respiratory tract including epithelial, smooth muscle cells and immune cells may express a distinct inducible NO synthase (iNOS) in response to inflammatory mediators such as bacterial lipopolysaccharide, interleukin-1 and tumour necrosis factor alpha [11-13]. Following the induction of iNOS large quantities of NO are produced that may cause tissue injury through increased interaction with other simultaneously available free radicals [14,15].

Generation of endogenous NO in the lungs can be detected in the exhaled air of animals and man [16,17]. Monitoring exhaled NO levels in patients has become a valuable diagnostic tool to assess a variety of lung pathologies [18]. Recent studies indicate that exhaled NO becomes elevated under inflammatory conditions such as sepsis, adult respiratory distress syndrome (ARDS), asthma and rejection of transplanted lungs [19-24]. These observations have important clinical implications such as using exhaled NO as an early marker of inflammation and a monitor of the effects of therapeutic modalities [2,18]. However, exhaled NO may be subject to variability due to changes in ventilation and blood flow heterogeneity, treatment with NO donors and medical intervention. To date, the influence of common clinical and physiologic variables on exhaled NO in humans have not been examined.

This study was aimed at 1) exploring the influence of common intraoperative variables on concentrations of NO in the exhaled air of patients undergoing cardiothoracic surgery, 2) evaluating exhaled NO in the setting of open heart surgery utilising cardiopulmonary bypass,

Acute Respiratory Distress Syndrome: Cellular and Molecular Mechanisms and Clinical Management
Edited by Matalon and Sznajder, Springer Science+Business Media New York, 1998

365

a condition which could lead to systemic inflammatory response and 3) in the setting of human lung transplantation where NO appears to regulate early graft function and survival.

METHODS

NO concentrations were measured intraoperatively using the chemiluminescence principle, with detection of the photochemical reaction between NO and ozone generated in the analyser. Measurements were made using a real-time, computer-controlled and integrated system (Logan Research Ltd 2000 series) [25]. A thin Teflon tube was inserted through a side arm of the endotracheal or the endobronchial tube and positioned at the distal end of the tube. Inspired and expired samples for analysis of NO and CO_2 were withdrawn at a flow rate of 150 ml/min and continuously monitored for 80 seconds. Five representative cycles were analysed for peak or mean exhaled NO levels.

EFFECTS OF VENTILATION PARAMETERS

Since concentration of NO in the exhaled air depends on both the production rate and minute ventilation [26,27], ventilation was initially standardised for inspired gas (100% O_2), tidal volume (5 ml/kg) and respiratory rate (10 breath/min). In all patients investigated, NO was detectable in the exhaled air with a characteristic oscillating signal which appeared to increase with expiration (as judged by the CO_2 signal) reaching peak NO levels of 3-13 ppb. In contrast, no NO signal was detected when the patient was replaced by a reservoir bag suggesting that the NO signal was derived from the patient lungs.

Composition of inspired gases

To investigate the potential influence of inspired NO on exhaled levels of NO, we investigated the effects of ventilating the patient with 50% compressed medical air made available through central pipelines in the hospital. As shown in Figure 1A, ventilation with 50 % medical air caused dramatic changes in the waveform of NO signal. Peak levels of NO increased immediately from 7 to 22 ppb when switching from 100% O_2 to medical air. Restoring fractional inspired concentrations of O_2 to 100% restored NO waveforms to baseline pattern within 30 sec. Analysis of the waveforms during medical air ventilation reveals that end expiratory concentrations of NO is only minimally changed and most of the detected increase of NO takes place during inspiration. This suggests that exhaled NO can be differentiated from inhaled NO and that inspired NO contributes only minimally to concentrations of NO in the expired air. On the basis of measurements in spontaneously breathing subjects, a similar conclusion was reached recently by other investigators [28]. Serial measurements of NO concentrations in the medical air supply in our hospital revealed variable levels (2-50 ppb), therefore, in the following studies patients were ventilated with 100% O_2 for the duration of the intraoperative NO measurements.

Minute volume

Increasing minute volume by increasing tidal volumes from 5 to 7.5 and 10 ml/kg at constant respiratory rate of 10 breath/min, reduced peak exhaled NO concentration from 8 ppb to 6.5 and 5 ppb respectively (Figure 1B). Similarly, increasing respiratory rate from 10 to 15 and 20 breath/min at constant tidal volume caused a reduction in peak exhaled NO from 6 ppb to 3.5 and 2 ppb, respectively (Figure 1C). These observations might suggest that increasing minute volume decreases exhaled NO concentration mainly by a dilution effect, however, other factors such as changes in NO production rate by increasing air flow rate could also contribute [26]. Nevertheless, changes in minute ventilation over a physiologic range significantly affect the measurement of NO in expired gas and therefore the measurement of expired NO should be discussed in conjunction with data on ventilation.

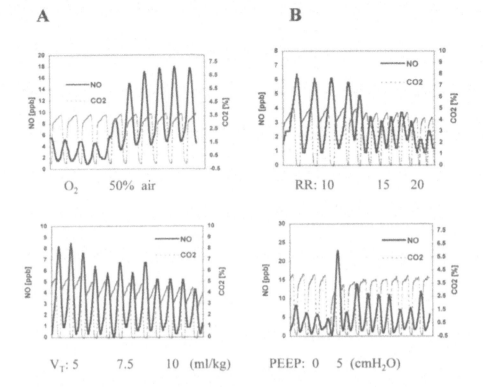

Figure1. Influence of ventilation parameters on exhaled NO and CO_2 levels. Panel A shows the effects of inspired NO in medical air, panel B depicts influence of increased minute volume by increasing tidal volume(V_T) at constant respiratory rate of 10 breath/min. Panel C exhibits influence of increased minute volume by increasing respiratory rate at constatnt tidal volume(5 ml/kg), and panel C shows the effects of 5 cmH$_2$O PEEP.

Positive End Expiratory Pressure (PEEP)

Results from a few animal studies suggest that ventilation with PEEP influences concentrations of NO in the exhaled air [29-31]. We investigated the effects of 5 cmH$_2$0 PEEP on exhaled NO in humans. As shown in figure 1D, application of PEEP resulted in an immediate increase in exhaled NO. Although the increase after delivering the first breath during PEEP could be an artefact due to reduced tidal volume, increased NO levels were detected after stabilisation of tidal volume. After cessation of PEEP, exhaled NO returned to baseline levels. Although some animal studies suggest a vagally mediated and stretch-induced change in airways leading to increased NO production, our data are most consistent with an explanation which considers increased alveolar surface area as the underlying mechanism of the PEEP effect [31]. The changes in NO concentrations appear to be immediate and last for the duration of PEEP application and observed at PEEP levels between 1 and 5 cm H$_2$0, when most alveolar recruitment occurs [30]. However, a reduction in pulmonary capillary blood volume by PEEP on exhaled NO levels could also be a plausible explanation (see below).

INFLUENCE OF INTRAVENOUS NITROVASODILATORS

Nitrovasodilators such as nitroglycerine (GTN), and sodium nitroprusside (SNP) have been used effectively for decades to treat ischemic and congestive heart disease and arterial hypertension. Their biological effect is thought to be mediated by NO and subsequent activation of vascular

367

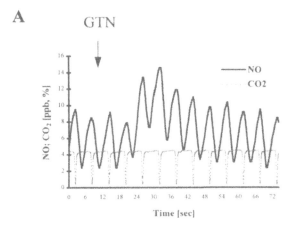

A

GTN

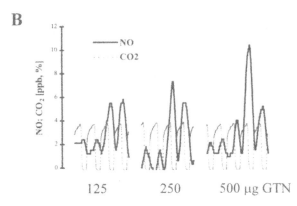

B

125   250   500 µg GTN

Figure 2. Influence of nitroglycerine (GTN) on exhaled NO. **Panel A** is a representative trace obtained after injecting 250 µg GTN into the central vein. **Panel B** depicts dose-dependent effects of GTN.

smooth muscle soluble guanylate cyclase. The release of NO from these drugs is postulated to occur through different mechanisms involving nonenzymic release in the case of SNP and thiol-dependent intracellular enzymatic conversion with GTN [32].

Recent studies demonstrated that NO was formed from GTN and SNP in vivo by detecting increased levels of exhaled NO in rabbits and lambs treated with intravenous nitrovasodilators [32-35]. In an attempt to reveal the significance of these findings in humans, we investigated the release of NO into exhaled air after intravenous administration of GTN and SNP.

Figure 2A depicts a representative signal in a cardiac patient before and after bolus injection of 250 µg of GTN. This patient exhibited baseline peak exhaled NO levels of 8 ppb which increased to 14 ppb 12 seconds after injecting GTN into a central vein. Figure 2B depicts dose dependent effects of GTN boluses on exhaled NO levels. Increases of 3.1, 6.0 and 8.1 ppb from baseline in peak exhaled NO levels were observed after injecting 125, 250 and 500 µg GTN, respectively. The relationship between GTN induced increase in exhaled NO and the effect of GTN on arterial blood pressure was investigated in five different patients. Injection of 250 µg GTN produced an increase in peak exhaled NO levels of 12.6+/-2.6 ppb from 4.3+/-1.7 ppb. The mean time from bolus injection to elicit maximal NO levels was 12.6 sec. Coinciding with NO release, systolic arterial blook pressure fell from 149+/-11 mmHg to 118+/-12 mmHg and diastolic arterial pressure from 80+/-9 mmHg to 67+/-7 mmHg.

To investigate the capability of SNP to release NO into exhaled air, we tested clinically equipotent doses of SNP and GTN. Injection of 150 μg SNP into a central vein caused only a small increase in exhaled NO levels, whereas injection of 125 μg GTN produced a considerable increase in exhaled NO in the same patient.

In summary, we observed transient, proportionate and dose-dependent formation and release of NO into exhaled air after administration of GTN into the central circulation of humans. This event coincided with a reduction of arterial blood pressure, supporting the hypothesis that NO release from GTN is responsible for the hypotensive effect. We hypothesise that intravenously administered GTN is transported to vascular endothelial cells where it is metabolised to NO. A portion of NO is released into the blood where it becomes inactivated by haemoglobin. In the systemic circulation abluminally released NO produces vasorelaxation, whereas in the lung NO diffuses into exhaled air. These mechanisms are less pronounced with SNP due to spontaneous release of NO in the blood. These findings may provide a non-invasive method to monitor a novel metabolic function of the pulmonary endothelium, as well as investigating nitrate pharmacology and tolerance, in vivo.

IMMEDIATE AND DELAYED EFFECTS OF CPB

During institution of CPB, venous return to the right atrium is drained and directed away from the pulmonary vasculature leading to a substantial reduction of blood flow in the pulmonary arteries and blood volume in the capillaries. Since a significant portion of NO can be bound and inactivated by haemoglobin in the lungs, pulmonary blood flow and blood volume might influence NO concentrations in exhaled air [27,30,36,37]. Furthermore, pulmonary blood flow might increase endothelial NO production via a shear-stress mediated mechanism and could contribute to exhaled NO levels. We thus speculated that institution of CPB and rapid decline in pulmonary artery pressure, flow and blood volume in the capillaries could have an effect on exhaled NO levels. To clarify this issue, ventilation was maintained during institution of CPB, cross clamping of the aorta and fibrillation of the heart.

In one representative patient, institution of CPB resulted in a decrease of mean pulmonary arterial blood pressure from 15 to 3 mmHg and in a gradual decrease of exhaled CO_2 (5.4; 2.9;1.6 and 1.3%) within 2 minutes. Associated with these events, a gradual increase in exhaled NO (10.8;10.8;15.9;and 18.6 ppb) was observed (Figure3A). In 5 patients studied, onset of CPB caused a transient increase in exhaled NO from 7.4 ±1.2 ppb to 13.7 ± 1.5 ppb. The increase in exhaled NO at onset of CPB appeared to be transient and exhaled NO levels returned to pre CPB levels within 5 min of CPB onset. These data suggest that exhaled NO is preserved and transiently increased during acute reduction in pulmonary blood flow. This could be explained by reduced binding to haemoglobin and therefore, increased delivery of NO to the alveoli.

CPB has been implicated in systemic inflammatory response after open heart surgery [38]. This condition is characterised by systemic release of inflammatory cytokines and endotoxin, known inducers of inducible NO synthase enzyme both in vitro and in vivo [39,40]. Since iNOS induction has been shown in various models of lung injury [13], we hypothesised that CPB might cause a delayed increase in exhaled NO levels and that increased NO in the airways could contribute to post-CPB complications. Peak end tidal NO concentration was 6.7 ± 0.9 ppb before CPB in 15 patients. After discontinuation of mechanical ventilation, airway NO concentration rose rapidly during CPB achieving 200-300 ppb. After restoration of ventilation towards the end of CPB, NO levels returned to pre CPB levels. Within the first hour after termination of CPB, exhaled NO was not different from baseline levels (Figure 3B). However, increased exhaled NO levels were detected in the early postoperative period (11 ± 1.6 ppb and 9.4 ± 1.2 ppb 3 and 6 hours after CPB, respectively). These data show that the early post CPB period is characterised by increased NO in the exhaled air in patients undergoing routine open heart surgery utilising CPB. This may represent transient augmentation of constitutive NO production or induction of NO synthase in the lungs.

EFFECTS OF PROLONGED ISCHEMIA AND REPERFUSION

Lung transplantation is an effective treatment for patients with end-stage lung disease.

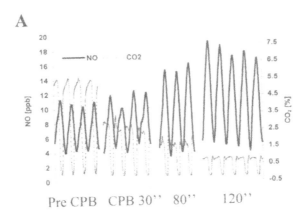

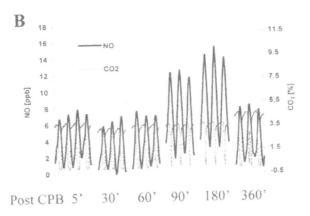

Figure 3. Immediate and delayed effects of cardiopulmonary bypass (CPB) on exhaled NO and CO_2. **Panel A** shows sequential representative traces obtained at the indicated times after instrumentation of CPB. **Panel B** exhibits traces obtained following CPB.

Despite recent advances in methods of lung preservation, the lungs remain extremely vulnerable to injury during harvesting, preservation and reperfusion [3]. The cellular targets and mechanisms of this injury may involve a variety of components in the lung including the capacity to generate NO. The crucial importance of NO in lung transplantation is supported by animal studies where administration of NO donors or inhalation of NO during the perioperative period significantly improves graft function and survival [41-43]. Inhalation of NO in human lung transplant recipients is also associated with favourable pulmonary haemodynamics, gas exchange, reduced duration of mechanical ventilation and mortality [44]. Whereas the effects of exogenous NO in transplantation are well documented, little is known regarding endogenous NO production. Although alterations in endogenous NO generation following transplantation have been documented in a few animal models, kinetics and significance of these changes in human transplantation remain totally unknown. The purpose of the study was to clarify this issue by monitoring exhaled NO levels intraoperatively and during the immediate postoperative period following lung transplantation.

Fifteen patients undergoing lung transplantation suffering from cystic fibrosis, Eisenmenger's syndrome, primary pulmonary hypertension, sarcoidosis, histocytosis X or emphysema were studied.

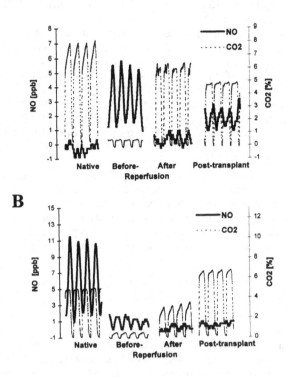

A

B

Figure 4. Kinetics of exhaled NO in human lung transplant recipients. Figure depicts representative signals in the native lungs of patients with cystic fibrosis (**Panel A**, Native) and patients with Eisenmenger's syndrome (**Panel B**, Native), and Type II and Type III responses, respectively. Traces fro donor lungs were measured prior to, or after reperfusion and 24 hrs post-transplantation. Please note efficient NO production in the absence of pulmonary blood flow (**Panel A**, Before Reperfusion)

They received either heart-lungs (8), sequential bilateral lungs (5) or single lung allografts (2) from braindead cadavers. The donors (mean age: 32.5 years) suffered either from subarachnoid haemorrhage, primary brain tumour or traumatic head injury. To investigate the effects of harvesting, preservation and reperfusion on exhaled NO levels, measurements were made after completion of the airway anastomoses both before and after reperfusion of the pulmonary vascular bed.

Baseline intraoperative exhaled NO levels in the native lungs of the transplant recipients were variable. All five of our patients with cystic fibrosis and one patient with histiocytosis exhibited very low NO levels. In contrast, patients with Eisenmenger's syndrome or primary pulmonary hypertension exhibited significant release of NO into exhaled air (6.7 ± 1.6 ppb).

Analysis of exhaled NO levels before and after reperfusion of the pulmonary vascular bed revealed three major types of responses. Type I response was characterised by a detectable NO signal prior to reperfusion (n=2/15; 2.2 ± 0.4 ppb) which remained unchanged during reperfusion and early post-operative period (2.9 ± 0.3 ppb). In patients with Type II responses (n=4/15) NO was detected prior to reperfusion (4.8+/-1.2 ppb), however reperfusion diminished NO release into exhaled air (1.1+/-0.4, p<0.01, Figure 4A). This attenuation was transient, as NO levels tended to increase towards pre-reperfusion levels within 24 hours (2.5+/-0.3 ppb, Figure 4A, post-transplantation). Type

III patients (n=9/15) exhibited extremely low levels of exhaled NO both prior to and after reperfusion (0.7 ± 0.1 ppb, Figure 4B). In these patients, a gradual recovery of exhaled NO levels was observed in the postoperative period (3-14 days).

We compared gas exchange parameters, mortality, duration of mechanical ventilation, Intensive Care Unit stay and requirement of inhaled NO therapy among the patients with different endogenous NO responses. Alveolar-arterial oxygen concentration gradient was 69 (41-97); 326 (135-537) and 352 (100-642) mmHg for patients with Type I, II and III responses. No patient died in category I, whereas mortality was 1/4 (25%) and 2/8 (25%) in Type II and III. Among the surviving patients, the average requirement for mechanical ventilation was 19.5 (10-29); 37 (4-79) and 152 (1-500) hours for Type I; II and III category, respectively. Similarly, average ICU stay was 43 (34-52); 63 (21-91) and 262 (32-1032) in these patients. Two patients required NO inhalation for severe allograft dysfunction in Type III category while this was not judged to be necessary in any of the patients with type I and II responses.

This preliminary study suggests that during the perioperative period, endogenous production of NO by cadaver lung allografts is markedly diminished and follows one of three patterns which appears to correlate to clinical behaviour. Perioperative monitoring of exhaled NO might provide a useful novel method of assessing donor preservation and evaluating strategies to modulate reperfusion injury. Furthermore, the loss of NO in the perioperative period may provide a rationale for preservation methods aimed at restoring endogenous NO production or replenishment by intravenous NO donors or inhaled exogenous NO.

SUMMARY

Recent technological developments made it possible to monitor breath to breath changes of NO levels in the exhaled air of anaesthetised and artificially ventilated human beings. The anaesthetist is in a privileged position to perform these investigations because of the capabilities to manipulate the sampling probe in the airways, alter ventilatory settings and administer vasoactive agents including NO donors. Investigations of exhaled NO in patients undergoing cardiothoracic surgery might provide exciting new insights into the mechanisms of NO production in the lungs and determine the pathophysiological variables which could influence NO concentrations in the exhaled air. Cardiothoracic surgical procedures include severe alterations of pulmonary vascular pressures, flows and blood volume, conditions which have been reported to alter NO concentrations in the exhaled air of various animals. Finally, cardiothoracic surgery is associated with a variety of biological processes such as systemic inflammatory response, ischemia and reperfusion where changes in NO production, release and reactivity could be a major pathogenic mechanism and target of therapeutic modalities with important clinical implications. However, caution should be taken in interpretation of findings for a variety of reasons. First, NO and related molecules react with a variety of tissue components such as haem iron, protein sulfhydrils and tyrosine residues. Furthermore, it is also rapidly inactivated by haemoglobin in the blood. Therefore, only a portion of NO is expected to be released into the air from the airway epithelial cells and only a small percentage from vascular endothelial cells. Second, this window of NO release may not represent tissue NO production equally during the various pathological conditions. Finally, the measured concentration of NO in the exhaled air is greatly affected by ventilation parameters. However, standardised perioperative monitoring of exhaled NO could be a fascinating new addition to the diagnostic tools to understand the pathogenesis of clinical problems associated with cardiothoracic surgery and to help in the anaesthetic and surgical management.

ACKNOWLEDGEMENTS

Dr. Marczin has been partially supported by the British Heart Foundation during these studies (BHF Project Grant PG/96031). Magdi Yacoub is a British Heart Foundation Professor of Cardiothoracic Surgery. The contribution of the Julia Polak Transplant Fund is greatly appreciated. We thank the consultants, junior doctors and theatre staff of Harefield Hospital for support and helpful discussions.

REFERENCES

1. Zapol WM, Rimar S, Gillis N, Marletta M, Bosken CH. Nitric oxide and the lung. **Am J Respir Crit Care Med** 1994; **149**: 1375-1380.
2. Barnes PJ. Nitric oxide and airway disease. **Ann Med** 1995; **27**: 389-393.
3. Pinsky DJ. The vascular biology of heart and lung preservation for transplantation. **Thromb Haemost** 1995; **74**: 58-65.
4. Moncada S, Higgs EA. Endogenous nitric oxide: Physiology, pathology and clinical relevance. **Eur J Clin Invest** 1991; **21**: 361-374.
5. Moncada S, Palmer RMJ, Higgs EA. Nitric oxide: Physiology, pathophysiology, and pharmacology. **Pharmacol Rev** 1991; **43**: 109-142.
6. Ignarro LJ, Ross G, Tillisch J. Pharmacology of endothelium-derived nitric oxide and nitrovasodilators. **Western Journal of Medicine** 1991; **154**: 51-62.
7. Curran AD. The role of nitric oxide in the development of asthma. **Int Arch Allergy Immunol** 1996; **111**: 1-4.
8. Garg UC, Hassid A. Inhibition of rat mesangial cell mitogenesis by nitric oxide-generating vasodilators. **Am J Physiol** 1989; **257**: 26/1) (F60-F66.
9. Ward JK, Barnes PJ, Tadjkarimi S, Yacoub MH, Belvisi MG. Evidence for the involvement of cGMP in neural bronchodilator responses in humal trachea. **J Physiol Lond** 1995; **483**: 525-536.
10. Ward JK, Barnes PJ, Springall DR, et al. Distribution of human i-NANC bronchodilator and nitric oxide-immunoreactive nerves. **Am J Respir Cell Mol Biol** 1995; **13**: 175-184.
11. Robbins RA, Barnes PJ, Springall DR, et al. Expression of inducible nitric oxide in human lung epithelial cells. **Biochem Biophys Res Commun** 1994; **203**: 209-218.
12. Robbins RA, Springall DR, Warren JB, et al. Inducible nitric oxide synthase is increased in murine lung epithelial cells by cytokine stimulation. **Biochem Biophys Res Commun** 1994; **198**: 835-843.
13. Belvisi M, Barnes PJ, Larkin S, et al. Nitric oxide synthase activity is elevated in inflammatory lung disease in humans. **Eur J Pharmacol** 1995; **283**: 255-258.
14. Radi R, Beckman JS, Bush KM, Freeman BA. Peroxynitrite-induced membrane lipid peroxidation: the cytotoxic potential of superoxide and nitric oxide. **Arch Biochem Biophys** 1991; **288**: 481-487.
15. Radi R, Beckman JS, Bush KM, Freeman BA. Peroxynitrite oxidation of sulfhydryls. The cytotoxic potential of superoxide and nitric oxide. **J Biol Chem** 1991; **266**: 4244-4250.
16. Gustafsson LE, Leone AM, Persson MG, Wiklund NP, Moncada S. Endogenous nitric oxide is present in the exhaled air of rabbits, guinea pigs and humans. **Biochem Biophys Res Commun** 1991; **181**: 852-857.
17. Leone AM, Gustafsson LE Francis PL, Persson MG, Wiklund NP, Moncada S. Nitric oxide is present in exhaled breath in humans. Direct GC-MS confirmation. **Biochem Biophys Res Commun** 1994; **201**: 883-887.
18. Barnes PJ, Kharitonov SA. Exhaled nitric oxide: a new lung function test [editorial]. **Thorax** 1996; **51**: 233-237.
19. Hussain SN, Abdul Hussain MN, el Dwairi Q. Exhaled nitric oxide as a marker for serum nitric oxide concentration in acute endotoxemia. **J Crit Care** 1996; **11**: 167-175.
20. Kharitonov SA, Barnes PJ. Nitric oxide in exhaled air is a new marker of airway inflammation. **Monaldi Arch Chest Dis** 1996; **51**: 533-537.
21. Mizuta T, Fujii Y, Minami M, et al. Increased nitric oxide levels in exhaled air of rat lung allografts. **J Thorac Cardiovasc Surg** 1997; **113**: 830-835.
22. Stewart TE, Valenza F, Ribeiro SP, et al. Increased nitric oxide in exhaled gas as an early marker of lung inflammation in a model of sepsis. **Am J Resp Crit Care Med** 1995; **151**: 713-718.
23. Alving K, Weitzberg E, Lundberg JM. Increased amount of nitric oxide in exhaled air of asthmatics. **Eur Respir J** 1993; **6**: 1368-1370.
24. Kharitonov SA, Yates D, Robbins RA, LoganSinclair R, Shinebourne EA, Barnes PJ. Increased nitric oxide in exhaled air of asthmatic patients. **Lancet** 1994; **343**: 133-135.
25. Marczin N, Riedel B, Royston D, Yacoub M. Intravenous nitrate vasodilators and exhaled nitric oxide]. **Lancet** 1997; **349**: 1742
26. Silkoff PE, McClean PA, Slutsky AS, et al. Marked flow-dependence of exhaled nitric oxide using a new technique to exclude nasal nitric oxide. **Am J Resp Crit Care Med** 1997; **155**: 260-267.
27. Hyde RW, Geigel EJ, Olszowka AJ, et al. Determination of production of nitric oxide by lower airways of humans--theory. **J Appl Physiol** 1997; **82**: 1290-1296.
28. Taylor JH, Williams TJ, Johns DP, Thien F, Ward C, Walters EH. NO concentration changes within the respiratory tract during spontaneous breathing in healthy adults. **Am J Resp Crit Care Med** 1997; **155**: A579 (Abstract).
29. Persson MG, Lonnqvist PA, Gustafsson LE. Positive end-expiratory pressure ventilation elicits increases in endogenously formed nitric oxide as detected in air exhaled by rabbits. **Anesthesiology** 1995; **82**: 969-974.
30. Carlin RE, Ferrario L, Boyd JT, Camporesi EM, McGraw DJ, Hakim TS. Determinants of nitric oxide in exhaled gas in the isolated rabbit lung. **Am J Resp Crit Care Med** 1997; **155**: 922-927.
31. Stromberg S, Lonnqvist PA, Persson MG, Gustafsson LE. Lung distension and carbon dioxide affect pulmonary nitric oxide formation in the anaesthetized rabbit. **Acta Physiol Scand** 1997; **159**: 59-67.
32. Husain M, Adrie C, Ichinose F, Kavosi M, Zapol WM. Exhaled nitric oxide as a marker for organic nitrate tolerance. **Circulation** 1994; **89**: 2498-2502.
33. Persson MG, Agvald P, Gustafsson LE. Rapid tolerance to formation of authentic NO from nitroglycerin in vivo. **Agents and Actions** 1995; **45**: 213-217.
34. Persson MG, Agvald P, Gustafsson LE. Detection of nitric oxide in exhaled air during administration of nitroglycerin in vivo. **Br J Pharmacol** 1994; **111**: 825-828.
35. Cederqvist B, Persson MG, Gustafsson LE. Direct demonstration of no formation in vivo from organic nitrites and nitrates, and correlation to effects on blood pressure and to in vitro effects. **Biochem Pharmacol** 1994; **47**:

1047-1053.

36. Rimar S, Gillis CN. Selective pulmonary vasodilation by inhaled nitric oxide is due to hemoglobin inactivation. **Circulation** 1993; **88**: 2884-2887.

37. Cremona G, Higenbottam T, Takao M, Hall L, Bower EA. Exhaled nitric oxide in isolated pig lungs. **J Appl Physiol** 1995; **78**: 59-63.

38. Hill GE, Whitten CW, Landers DF. The influence of cardiopulmonary bypass on cytokines and cell-cell communication. **J Cardiothorac Vasc Anesth** 1997; **11**: 367-375.

39. Engelman RM, Rousou JA, Flack JE, Deaton DW, Kalfin R, Das DK. Influence of steroids on complement and cytokine generation after cardiopulmonary bypass. **Ann Thorac Surg** 1995; **60**: 801-804.

40. Kennedy DJ, Butterworth JF, 4th. Clinical review 57: Endocrine function during and after cardiopulmonary bypass: recent observations. **J Clin Endocrinol Metab** 1994; **78**: 997-1002.

41. Dotsch J, Demirakca S, Terbrack HG, Huls G, Rascher W, Kuhl PG. Airway nitric oxide in asthmatic children and patients with cystic fibrosis. **Eur Respir J** 1996; **9**: 2537-2540.

42. Naka Y, Chowdhury NC, Liao H, et al. Enhanced preservation of orthotopically transplanted rat lungs by nitroglycerin but not hydralazine. Requirement for graft vascular homeostasis beyond harvest vasodilation. **Circ Res** 1995; **76**: 900-906.

43. Naka Y, Chowdhury NC, Oz MC, et al. Nitroglycerin maintains graft vascular homeostasis and enhances preservation in an orthotopic rat lung transplant model. **J Thorac Cardiovasc Surg** 1995; **109**: 206-210.

44. Date H, Triantafillou AN, Trulock EP, Pohl MS, Cooper JD, Patterson GA. Inhaled nitric oxide reduces human lung allograft dysfunction. **J Thorac Cardiovasc Surg** 1996; **111**: 913-919.

ARDS: NITRIC OXIDE AS CAUSE AND THERAPY IN MULTIPLE ORGAN FAILURE (MOF)

Suveer Singh BSc.MRCP, Andrew Jones MRCP, Timothy W. Evans BSc.MD.FRCP.PhD.EDICM

Unit of Critical Care, Imperial College School of Medicine,
Royal Brompton Hospital, London, U.K.

INTRODUCTION

The acute respiratory distress syndrome (ARDS) is characterized by refractory hypoxemia in the presence of bilateral pulmonary infiltrates on chest radiography[1]. It was recognised that ARDS probably represents only the extreme manifestation of a range of pulmonary insults. Hence, acute lung injury (ALI) was similarly defined but with less severe indices of hypoxemia (Table 1). A wide variety of serious medical and surgical conditions can precipitate the syndrome, only some of which involve the lung (Table 2). Although the mortality associated with ARDS has fallen in recent years (from approximately 70% to 50%) in a number of large centres, it is undoubtedly influenced by the underlying pathology. Furthermore, epidemiological studies have suggested for many years that the majority of patients succumb not to refractory hypoxemia, but rather to sepsis and associated multiple organ failure (MOF). MOF is defined by a constellation of symptoms and clinical signs of which multiple organs (e,g, kidney, heart, GIT, etc) are in a state of dysfunction or failure. Indeed, given that respiratory failure accounts for only 16% of ARDS deaths, improvements in outcome of ARDS may depend more on treatment of sepsis and further understanding/prevention of MOF than on improvements in oxygenation and pulmonary haemodynamics [2,3].

PATHOPHYSIOLOGY OF ARDS

ARDS is characterised by high permeability alveolar oedema, pulmonary capillary microthrombosis, a loss of hypoxic pulmonary vasoconstriction (HPV) and mismatching of alveolar perfusion (Q) with ventilation (V). These processes account in the main for the characteristic clinical features of pulmonary oedema, refractory hypoxaemia and pulmonary hypertension. Despite the wide range of precipitating factors, it is thought that the underlying pathophysiological mechanisms are the same, i.e that ARDS represents the common endpoint of a wide variety of inflammatory pathways.

Acute Respiratory Distress Syndrome: Cellular and Molecular Mechanisms and Clinical Management
Edited by Matalon and Sznajder, Springer Science+Business Media New York, 1998

Table 1. Definitions: ALI and ARDS. (Adapted from [1])

	Timing	Oxygenation	Chest radiograph	Pulmonary artery occlusion pressure
ALI	Acute onset	PaO_2/FiO_2 <300 mmHg (39.5kPa) (regardless of PEEP level)	Bilateral infiltrates seen on frontal CXR	<18 mmHg when measured, or no clinical evidence of left atrial hypertension
ARDS	Acute onset	PaO_2/FiO_2 <200 mmHg (26.3 kPa) (regardless of PEEP level)	Bilateral infiltrates seen on frontal CXR	<18 mmHg when measured, or no clinical evidence of left atrial hypertension

Table 2. Clinical conditions associated with the development of ARDS

Respiratory	Non-respiratory
Pneumonia (Bacterial/Viral/Fungal)	Sepsis syndrome
Aspiration of gastric contents	Major Trauma/Shock
Pulmonary contusion	Massive burns
Post-pneumonectomy	DIC
Inhalation of smoke or toxins	Transfusion reactions
Near-drowning	Fat embolism
Thoracic irradiation	Pregnancy-associated
Oxygen toxicity	(eg.amniotic fluid embolism)
Ischaemia-reperfusion	Pancreatitis
Vasculitis (eg. Goodpasture's)	Drug/Toxin reactions
	(eg.paraquat, heroin)
	Post-CP bypass
	Head injury/ raised ICP
	Tumour Lysis syndrome

DIC:disseminated intravascular coagulopathy; CP: cardiopulmonary; ICP: intracranial pressure.

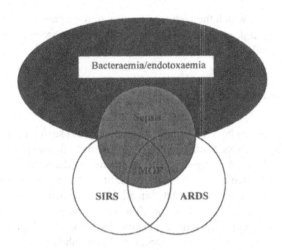

Figure 1. The overlapping clinical syndromes associated with Sepsis, MOF and ARDS.
SIRS systemic inflammatory response, MOF multiple organ failure, ARDS acute respiratory distress syndrome

Since the majority of patients with ARDS die a non-respiratory death, principally from MOF, ARDS probably represents only the pulmonary manifestation of a pan-endothelial insult[4] (Figure 1) which may result in interstitial oedema formation in most organ systems, leading to impaired tissue oxygenation, partly through a direct effect on diffusion, but also through adverse effects on microvascular control within the regional microcirculation.

MOF and ARDS

In a large number of cases, the condition that precipitates ARDS is associated itself with organ failure (e.g. trauma leading to blood loss, sepsis-induced renal failure), and it is frequently difficult to ascertain whether or not lung injury represents a primary initiating event in the development of MOF; or whether ARDS merely represents pulmonary insufficiency in the context of a wide range of organ dysfunction precipitated by a single initiating event. Either way, the association of ARDS with MOF remains high[3,4] (Table 3). As many as 40% of patients with sepsis develop ARDS. Thus, the progression from a state of relative health to one of organ dysfunction, organ failure and death in critically ill patients admitted to the ICU was initially thought to be a result of secondary infection following an insult to the host. However, large numbers of cases of MOF develop without demonstrable evidence of an infectious link, causing researchers to re-think the *infectious model*. Furthermore, the concept of a systemic inflammatory

Table 3. Incidence of Organ Dysfunction associated with ARDS (adapted from[5] with permission)

System	Failure (%)	System	Failure (%)
Cardiac	10-23	Haematologic	0-26
Renal	40-55	Central Nervous System	7-30
Gastrointestinal	7-30	Hepatic	12-95

response syndrome (SIRS) in which the features of sepsis may be present without identifiable infection has reinforced the *inflammatory model* that has been proposed to describe the evolution of ARDS to MOF in critically ill patients (reviewed in detail elsewhere[5]).

The Role of the Micro-circulation in MOF and ARDS

A number of hypotheses have been put forward to explain such inflammatory models in the evolution of MOF from ARDS. Some are based on macrophage activation, others by wide ranging neutrophil and inflammatory cascade activity, including the release of free radicals; and others around the concept of ischemic-reoxygenation injury occuring at the time of the initial insult. Diffuse microvascular injury is postulated in all, often occurring in advance of biochemical or clinical evidence of organ dysfunction. The consequences of pan-endothelial injury in MOF are multiple. Within such microcirculations, the endothelium releases a wide variety of vasoactive mediators, such as nitric oxide (NO), endothelins (ETs), and the products o cyclooxygenase metabolism (COX); through their actions a loss of microvascular control develops in addition to a diminished number of perfused capillaries and reduced oxygen uptake. Inflammatory cell activation and mediator release may also influence cellular metabolism at a mitochondrial level so diminishing oxygen utilisation; thus microvascular and indeed cellular derangement at the microvascular bed is likely to account for the progression from, and link between ARDS and MOF.

THE PULMONARY CIRCULATION IN ARDS

Increased pulmonary microvascular permeability is a reflection of damage to the pulmonary endothelium. The resulting alveolar oedema may directly impair oxygenation, but it appears that V/ Q mismatch accounts for the majority, if not all of the refractory hypoxemia that characterises ARDS. This in turn is almost certainly attributable to a loss of HPV, the physiological reflex that ensures V/Q matching occurs even in the presence of localized pulmonary damage. Moreover, there is evidence in both animal models and clinical investigations that patients with ALI/ARDS develop increased pulmonary vascular resistance (PVR), the extent of which may have prognostic significance[6,7]. The ideal therapeutic intervention in ALI/ARDS would therefore diminish PVR, improve V/Q matching by reducing intrapulmonary shunt and lead to a diminution in alveolar-capillary membrane permeability.

PULMONARY VASCULAR MANIPULATION IN ARDS

It has been long recognised that the increased PVR associated with ARDS may lead to diminished right ventricular ejection fraction (RVEF) leading in turn to intraventricular shift and impaired left ventricular performance. The use of intravenous pulmonary vasodilator agents has, however, been fraught with difficulty. Firstly, the actions of the majority of these agents is not confined to the pulmonary circulation, resulting in a reduction in left ventricular after load and (frequently) catastrophic systemic hypotension. Secondly, the adverse effects of these agents on residual HPV may lead to worsening shunt and increasing hypoxemia. By contrast, the application of an agent delivered by inhalation to ventilated alveolar units would divert blood to those areas of lung receiving effective ventilatory support, thereby improving gas exchange and diminishing shunt. Should this agent be rapidly metabolised, its effects might be limited to the pulmonary circulation, thereby leading to a reduction in PVR. A number of agents have been used in this role in recent years but the most significant of which is NO[8].

NITRIC OXIDE

Nitric oxide, a potent endogenous vasodilator, is synthesized from L-arginine by nitric oxide synthases (NOS) in a number of cell types. There are at least three distinct isoforms; Endothelial (e) and neuronal (n) NOS are constitutive but a third isoform is induced (iNOS) by endotoxin and inflammatory cytokines. All are inhibited by a variety of L-arginine analogs including N-monomethyl-L-arginine (L-NMMA), N^G-nitro-L-arginine (L-NNA) and N^G-nitro-L-arginine methylester (L-NAME). Such inhibitors have been shown to cause a rapid increase in blood pressure in animal and man, implicating NO as an important determinant of systemic vascular tone in health. Appropriate agonists, or mechanical shear stress of the endothelium can activate cNOS: NO is released within seconds or minutes. After an intravascular stimulus such as cytokine production or endotoxin application, iNOS is induced in the relevant cell type by gene transcription. Inducible NOS generates up to 1,000 times more NO than cNOS and cellular production continues for a number of hours following a single inflammatory stimulus. Molecular studies in animal models of sepsis suggest that up-regulation of iNOS in vascular tissue is accompanied simultaneously by down-regulation of eNOS which may have important implications for the disruption of physiological vascular control mechanisms. iNOS is expressed in a wide variety of vascular beds in rats treated with endotoxin; in the lung this is localised not only to vascular tissue but also airway epithelium.

The biological actions of NO are attributable to the activation of guanylyl cyclase which catalyzes the conversion of GTP to cGMP. The rise in cGMP, an intracellular messenger, causes a reduction in intracellular calcium concentration and smooth muscle relaxation. NO has a wide variety of other vascular and non-vascular effects including inhibition of platelet adherence and aggregation, inhibition of leucocyte chemotaxis and signal transduction in the nervous system. NO is extremely rapidly bound to heme and has an in vivo half life measured in seconds[9].

Role of Endogenous NO in Modulating Pulmonary Vascular Tone

In health, NO is an important mediator of pulmonary blood flow and of V/Q matching within the lungs, and eNOS probably contributes to the maintenance of a low resting pulmonary vascular tone and may modulate HPV. Moreover, diminution of pulmonary eNOS mRNA expression has been shown to correlate inversely with total PVR and the morphological severity of pulmonary hypertension, implicating diminished basal NO production as a mechanism of the elevation in PVR[10].

Inhaled NO in ARDS

The concept of inhaled NO as a selective pulmonary vasodilator is attractive, and has been explored extensively since 1990. These properties were first demonstrated in spontaneously breathing lambs: 40-80 ppmNO reversed acute HPV without altering systemic haemodynamics[11]. Subsequent improvements in oxygenation and V/Q following inhaled NO were demonstrated in a variety of animal models although varying dose response relationships have been identified in different species and between different models[12-15]. In healthy human volunteers with hypoxia-induced pulmonary hypertension, the addition of NO to inhaled gas reduces Ppa to base line levels without influencing mean systemic arterial pressure[16].

Clinical Studies

Initial results in animal studies using inhaled NO give raise to therapeutic trials in ALI/ARDS. The effects of NO inhalation (18 and 36 ppm) compared with intravenously-infused prostacyclin were first tested in nine patients with ARDS. Both agents reduced PVR by up to 20%, but NO alone selectively reduced mean Ppa and improved PaO_2/FiO_2 ratio without affecting systemic pressures[17]. The beneficial effects of NO on oxygenation were subsequently shown to be related to a reduction in intrapulmonary shunt. A variety of studies have since been published, the vast majority of which have been carried out in a small number of patients (less than 40), with ALI/ARDS of varying severity and precipitated by a range of conditions. Not all studies have been designed to assess the effects of NO on pulmonary haemodynamics on oxygenation. A number were designed to assess interaction with a variety of different types of mechanical ventilation, all of which are likely to influence pulmonary vascular tone (e.g. presence or absence of PEEP, presence or absence of permissive hypercapnia). A number of common themes emerge. Firstly, the efficacy of NO in terms of its effects on oxygenation are highly variable and a successful response has been variably defined. It seems likely that 40-60% of patients with established ARDS respond in a clinically significant fashion to inhaled NO (> 20% improvement in PaO_2/FiO_2 ratio). Secondly, any beneficial effects of NO on PVR are less easy to ascertain. A 20% improvement in PVR is rarely obtained, and frequently effects on oxygenation are seen with minimal alteration in pulmonary hemodynamics. Thirdly, prediction of a beneficial effect in individual patients is impossible. There seems to be no dose-response relationship in terms of improved gas exhange or PVR to inhaled NO. Overall, it seems that patients likely to benefit from NO in terms of an improvement in either parameter occur at doses of 20 ppm NO or less, although occasionally patients respond at doses up to 40 ppm. Finally, the time course of NO-induced changes is variable. Thus, enhanced oxygenation may be achieved within one to two minutes of starting the administration of NO and baseline conditions re-achieved within 5-8 minutes of stopping[18]. Moreover, there is some evidence that responsiveness to NO may vary from day-to-day. The influence of histopathological changes in the evolving injury, and the effects of mechanical ventilatory support are difficult to determine. It seems likely that NO may be more effective in patients seen during the early stages of the condition, in which circulating vasomotor substances may account for adverse changes in PVR[19], rather than later in the course of the syndrome by which time structural changes in the pulmonary circulation have occurred.

Despite the utility of NO demonstrated under different conditions in patients with lung injury, no overall positive survival effect has been shown from studies of patients with ARDS. Moreover, no placebo-controlled, randomized, prospective study has been published. Nevertheless, a trend towards reduced mortality has been identified in a subgroup of fluid-resuscitated vasopressor dependent patients with ARDS secondary to sepsis who responded to inhaled NO. A positive response was associated with increased RV function, cardiac index and improved oxygen delivery, whilst a lack of response and higher mortality were characteristic of patients with depressed cardiac reserves[20].

Adverse Effects of NO

The poisonous nature of the nitrogen oxide gases was highlighted in the mid 1960s when at least three patients died on induction of anaesthesia due to the contamination of anaesthetic gases by nitrous oxide and nitrogen dioxide $(NO_2)^{21}$. Despite its potential toxicity, NO exists in the atmosphere and is present normally in the expired air of humans, perhaps representing host defence activity for the respiratory tract, whilst autoinhalation of NO may be important in V/Q matching. The toxicity of inhaled NO is attributable to the formation of NO_2, peroxynitrite radicals (in the presence of superoxide anions) and methaemoglobinaemia (Met-Hb) (reviewed in detail[22]). The balance of deleterious proxidant, versus cytoprotective antioxidant effects may be critically dependent on relative concentrations of oxygen, and individual reactive species in the presence of inhaled NO. Fatality due to acute NO_2 toxicity is due to pulmonary oedema in doses above 150ppm, although such concentrations are extremely rare at clinically-relevant levels of NO. Nevertheless, administration of NO needs to be conducted with care, using appropriate apparatus and monitoring systems- specifically electrochemical detectors which will constantly monitor the levels of inhaled NO and NO_2. Regular measurements of Met-Hb may be necessary in the early stages of administration and when doses are changed. Adverse effects of inhaled NO on coagulation are unlikely to be significant in clinical circumstances.

NOS INHIBITION IN ARDS / MOF

A large number of studies in animal models of sepsis have suggested that the application of NOS inhibitors can attenuate hypotension in LPS or tumour necrosis factor-induced shock. Subsequently, low doses of NOS inhibitors (L-NMMA, L-NAME) were used in 2 patients with inotrope-refractory septic shock. In both cases, systemic vascular tone was restored[23]. A further series in 12 patients with septic shock showed that L-NMMA increased BP, SVR, PVR and filling pressures but diminished cardiac output[24]. These effects were abolished by L-arginine which itself caused transient vasodilation and an increase in cardiac output. Methylene Blue, a guanylyl cyclase antagonist, has been used with similar haemodynamic benefit in patients. The results of these promising early trials have resulted in the initiation of a large multi-centre trial of L-NMMA in septic shock, preliminary communication regarding the primary (resolution of shock) and secondary (removal of vaso-pressor agents) end-points at 72 hours have proved encouraging and adverse events remarkably few. However, with evidence that the balance between constitutive and inducible NOS isoforms is an important determinant of local vascular tone in the septic circulation, any sweeping disruption of this balance by partially selective NOS inhibitors may be disadvantageous[25,26]. In particular, the pulmonary arterial hypertension that is characteristically seen in ALI/ARDS may be exacerbated. A theoretically desirable goal might be the specific inhibition of iNOS-induced NO while preserving the presumed beneficial, physiological effects of cNOS. To date, highly selective iNOS inhibitors such as aminoguanidine have not been applicable in the clinical arena due to toxicity, although they may prove useful in clarifying conflicting laboratory data regarding the effects of NO induction and blockade in various animal models of sepsis[27]. In sepsis associated ARDS, the application of selective iNOS inhibition in combination with exogenous NO or NO donors, to maintain the basal requirement of NO, is an area of ongoing study in the experimental setting.

SUMMARY & CONCLUSIONS

The use of inhaled NO should be considered in patients with severe ARDS and marked hypoxemia (e.g. PaO2<12kPa on FO_2 1.0), in whom deterioration in oxygenation continues despite appropriate management of the underlying cause and optimisation of ventilatory support. Inhaled NO is currently an unlicenced therapy with potential toxicity and should be used only on compassionate grounds until evidence of definite benefit emerges from randomised controlled trials. A dose-response study, with full pulmonary and systemic hemodynamic monitoring should be performed prior to its implementation, probably in the range up to 20 ppm although doses up to 40 ppm may be used in exceptional circumstances. A positive response to NO may be usefully defined as a greater than 10% improvement in PaO_2/FiO_2 ratio and/or a greater than 20% reduction in PVR. A back up means of administration should be available, particularly with regard to monitoring equipment, since acute removal of NO may lead to rebound pulmonary hypertension with adverse changes on right ventricular function and gaseous exchange.

An increasing body of evidence of the epidemiological link between sepsis/MOF and ARDS, and of the poor prognostic association, is emerging. Hence, the pursuit of improved outcomes in this group of ARDS patients may lie in further understanding the mechanisms underlying septic shock/ MOF: In particular, investigation into the initiating events of microvascular dysfunction (i.e., the role of the endothelium, vascular smooth muscle and their vasoactive substances in determining regional micro-perfusion) and primary mitochondrial dysfunction (not reviewed here). Molecular, pharmacological, and (latterly) clinical studies suggest that a range of agents including nitric oxide synthase inhibitors (either non-selective or of the inducible enzyme) may prove useful adjuncts in the clinical support of this difficult ARDS subset.

REFERENCES

1. Bernard GR, Artigas A, Brigham KL et al. The American-European consensus conference on ARDS: Definitions, mechanisms, relevant outcomes and clinical trial co-ordination. Am J Respir Crit Care Med 1994;149:818-824
2. Montgomery B, Stager MA, Carrico CJ, Hudson LD. Causes of mortality in patients with the adult respiratory distress syndrome. Am Rev Respir Dis 1985;132:485-9
3. Ferring M, Vincent JL. Is outcome from ARDS related to the severity of respiratory failure? Eur Respir J 1997;10:1297-1300
4. Bone RC, Balk R, Slotman G et al. Adult respiratory distress syndrome. Sequence and importance of development of multiple organ failure. The Prostaglandin E1 Study Group. Chest 1992;101:320-326
5. Gill RS, Sibbald WJ. Systemic manifestations. In ARDS Acute Respiratory Distress Syndrome in Adults,(ed. TW Evans and C Haslett), Chapman and Hall, London 1996
6. Bernard GR, Rinaldo J, Harris T et al). Early predictors of ARDS reversal in patients with established ARDS. Am Rev Respir Dis 1985;131:A143(Abstract)
7. Pallares LCM, Evans TW. Oxygen transport in the critically ill. Respir Med 1992;86:289-295
8. Brett SJ, Evans TW. Nitric oxide:physiological roles and therapeutic implications in the lung. Br J Hosp Med 1995;S5(2):1-4
9. Singh S and Evans TW. Nitric oxide, the biological mediator of the decade:fact or fiction? Eur Respir J 1997;10, 699-707
10. Giaid A and Saleh D. Reduced expression of endothelial nitric oxide synthase in the lungs of patients with pulmonary hypertension. N Eng J Med 1995;333:214-21
11. Frostell CG, Fratacci MD, Wain JC et al. Inhaled nitric oxide:a selective pulmonary vasodilator reversing hypoxic pulmonary vasoconstriction. Circulation 1991;83:2038-47
12. Frattaci MD, Frostell CG, Chen TY et al. Inhaled nitric oxide: a selective pulmonary vasodilator of heparin-protamine vasoconstriction in sheep. Anaesthesiology 1991;75:990-9
13. Pison U, Lopez FA, Heidelmeyer CF et al. Inhaled nitric oxide reverses hypoxic pulmonary vasoconstriction without impairing gas exchange. J Appl Physiol 1993;74:1287-92

14. Ogura H, Cioffi WG, Offner PJ et al. Effect of inhaled nitric oxide on pulmonary function after sepsis in a swine model. Surgery 1994;116:313-21.

15. Weitzberg E, Rudehill A and Lundberg JM. Nitric oxide inhalation attenuates pulmonary hypertension and improves gas exchange in endotoxin shock. Eur J Pharmacol 1993; 233:85-94

16. Frostell CG, Blomquist H, Hedenstierna G et al. Inhaled nitric oxide selectively reverses human hypoxic pulmonary vasoconstriction without causing systemic vasodilation. Anaesthesiology 1993;78:427-35

17. Rossaint R, Falke KJ, Lopez F et al. Inhaled nitric oxide for the adult respiratory distress syndrome. N Eng J Med 1993;328:399-405

18. Gerlach H, Rossaint R, Pappert D et al. Time-course and dose-response of nitric oxide inhalation for systemic oxygenation and pulmonary hypertension in patients with adult respiratory distress syndrome. Eur J Clin Invest 1993;23:499-502

19. Furchgott RF and Vanhoutte PM. Endothelium-derived relaxing and contracting factors. FASEB J 1989;3:2007-18

20. Krafft P, Freidrich P, Fitzgerald RD et al. Effectiveness of Nitric Oxide Inhalation in Septic ARDS Chest 1996;109:486-93.

21. Clutton-Brock J.Two cases of poisoning by contamination of nitrous oxide with higher oxides of nitrogen during anaesthesia. Br J Anaesth 1967;39:388-92.

22. Singh S, Evans TW. Uses and Abuses of Nitric Oxide in ARDS. In Recent Advances in Anaesthesia and Analgesia (20), (ed.AP Adams and JN Cashman), Churchill Livingstone, Edinburgh (in press)

23. Petros A, Bennett D and Vallance P. Effect of nitric oxide inhibitors on hypotension in patients with septic shock. Lancet 1991;338,1557-8.

24. Petros,A, Lamb,G, Leone,A et al. Effects of a nitric oxide synthase inhibitor in humans with septic shock . Cardiovasc.Res 1994; 28(1): 34-39.

25. Liu SF, Adcock IM, Old RW et al. Differential regulation of the constitutive and inducible nitric oxide synthase mRNA by lipopolysaccharide treatment in vivo in the rat. Crit Care Med 1996;24:1219-1225

26. Mitchell JA, Kohlhaas KL, Sorrentino R et al. Induction by endotoxin of nitric oxide synthase in the rat mesentery: lack of effect on action of vasoconstrictors Br J Pharmacol 1993;109:265-270

27. Griffiths,M.J, Messent,M, MacAllister,R.J, Evans,T.W. Aminoguanidine selectively inhibits inducible nitric oxide synthase. Br.J.Pharmacol 1993;110(3): 963-968

ROLE OF REACTIVE NITROGEN SPECIES IN ASBESTOS-INDUCED PLEURO-PULMONARY INJURY

Shogo Tanaka,[1] Nonghoon Choe,[1] David R. Hemenway,[2] Sadis Matalon,[3] and Elliott Kagan[1]

[1]Department of Pathology, Uniformed Services University of the Health Sciences, Bethesda, Maryland 20814
[2]Department of Civil and Environmental Engineering, University of Vermont, Burlington, Vermont 05405
[3]Department of Anesthesiology, University of Alabama at Birmingham, Birmingham, Alabama 35233

INTRODUCTION

Asbestos is a generic term for a group of naturally-occurring, hydrated fibrous silicates. There are two main mineralogic varieties of asbestos minerals: amphiboles (which have a stiff and brittle texture) and serpentine (which has a softer, more flexible texture and a wavy configuration). Three varieties of asbestos have been used commercially in North America and Europe for most of this century: chrysotile (a serpentine asbestos), crocidolite and amosite (both amphiboles). Asbestos fibers have long been a focus of scientific interest because of their capacity to induce pleural and pulmonary diseases characterized by fibrosis (parietal pleural plaques, visceral pleural fibrosis and asbestosis) and/or neoplasia (malignant pleural mesothelioma and bronchogenic carcinoma).[1] Although all commercial varieties of asbestos are capable of inducing these diseases, there has been considerable scientific debate regarding the relative potential of the different fiber types (chrysotile vs. amphiboles) to induce pleuro-pulmonary injury.[2,3] The pathogenesis of asbestos-related disorders is complex, but there is evidence implicating cytokines and reactive oxygen species (ROS) including superoxide anion (O_2^{*-}), H_2O_2 and hydroxyl radical (OH*) in the induction of asbestos-induced cellular injury.[1-5] The putative role of ROS in mediating asbestos-related cytotoxicity is predicated on the notion that asbestos fibers induce cellular injury via O_2^{*-}-driven, iron-catalyzed Haber-Weiss (Fenton) reactions that generate the OH* radical.[5] Studies demonstrating that the injurious effects of asbestos can be abrogated by scavengers of ROS such as superoxide dismutase, mannitol and catalase are supportive of this hypothesis.[6-9] However, while Fenton reactions may explain some of the actions of crocidolite (which is an iron-containing silicate), they cannot account for those of chrysotile (which is a magnesium-containing silicate whose chemical structure is devoid of iron).

Considerable attention has focused at this NATO Advanced Study Workshop on the putative role of reactive nitrogen species (RNS) including nitric oxide (*NO) and peroxynitrite anion (ONOO⁻), the reaction product of O_2^{*-} with *NO, in mediating acute pulmonary injury and the adult respiratory distress syndrome. It is conceivable that RNS also may be implicated in the pathogenesis

Acute Respiratory Distress Syndrome: Cellular and Molecular Mechanisms and Clinical Management
Edited by Matalon and Sznajder, Springer Science+Business Media New York, 1998

of asbestos-related injury. An earlier study demonstrated that both crocidolite and chrysotile asbestos fibers up-regulated the production of *NO by cultured rat alveolar macrophages in the presence of interferon-γ (IFN-γ).[10] As an extension of those observations, the present study has demonstrated that asbestos exposure generates RNS in cultured rat pleural mesothelial cells (RPMC) which are considered the primary targets of asbestos-induced pleural injury.[2] Furthermore, we have shown that inhaled asbestos fibers stimulate *NO production by rat alveolar and pleural macrophages and induce nitrotyrosine formation within both the pleura and the lung parenchyma of exposed rats.

MATERIALS AND METHODS

Mineral Dust Samples and Reagents

NIEHS crocidolite and NIEHS chrysotile asbestos fibers were obtained from the National Institute of Environmental Health Sciences (Research Triangle Park, NC). Carbonyl iron spheres were purchased from Sigma Chemical Co. (St. Louis, MO). All of the mineral samples comprised a significant respirable fraction and the fiber geometry of the asbestos samples have been published previously.[11] Sodium nitrite, 3-nitro-L-tyrosine, 3-amino-L-tyrosine, O-phospho-L-tyrosine, Harris's hematoxylin, and bacterial lipopolysaccharide (LPS) were purchased from Sigma, human recombinant interleukin-1β (IL-1β) was purchased from Genzyme (Cambridge, MA), and recombinant rat IFN-γ was obtained from GIBCO (Gaithersburg, MD). Cell culture media were purchased from Biofluids, Inc. (Rockville, MD)and the mouse macrophage cell line, RAW 264.7, was obtained from American Type Culture Collection (ATCC, Rockville, MD).

Rat Pleural Mesothelial Cell Cultures

Rat parietal pleural mesothelial cell cultures derived from Fischer-344 rats (Charles River Laboratories, Wilmington, MA) were established and maintained as described previously.[11] The mesothelial cell phenotype and purity of the cultures were confirmed by ultrastructural characteristics and by positive immunoreactivity for cytokeratin and vimentin.[11] The cultures employed for these experiments were from passages 10 to 12.

In Vitro Exposure Experiments

For nitrite (NO_2^-) measurements, 5×10^5 RPMC were added to 6-well culture plates (Costar) and, when confluent, were incubated with or without added particulates (4.2 μg/cm^2) and in the presence or absence of recombinant human IL-1β (50 ng/ml) for 12 to 48 h in 2.5 ml of DMEM + 10 fetal bovine serum (FBS) in 5% CO_2 at 37° C. For dose-response experiments, particulates were added at final concentrations of 1.05 to 8.4 mg/cm^2 in the presence or absence of IL-1β for 48 h. In other experiments, rat pleural mesothelial cells were seeded in 25 cm^2 flasks (Costar, 2×10^6 cells/flask) and, when confluence was attained, the cultures were used for measurement of iNOS gene expression. For these studies, the cells were incubated with IL-1β, in the presence (4 μg/cm^2) or absence of crocidolite or chrysotile asbestos fibers, for 2-12 h at 37° C in DMEM + 5% CO_2. RAW 264.7 cells stimulated with a combination of LPS (10 ng/ml) + IFN-γ (10 U/ml) for 24 h served as a positive control for iNOS gene induction. For immunocytochemical studies, confluent cells were incubated in 4-well slide chambers (LabTek, Napierville, IL), in the presence (0.5, 2 or 4 μg/cm^2) or absence of particulates and in the presence (50 ng/ml) or absence of IL-1β, for 36 h at 37° C in DMEM in 5% CO_2.

Inhalation Exposure Experiments

For in vivo experiments, adult male Fischer-344 rats were placed in inhalation chambers and exposed to either crocidolite (time-weighted average concentration: 6.88 mg/m^3), chrysotile (time-weighted average concentration: 8.25 mg/m^3), or filtered room air (sham-exposed group) as reported previously.[12] Ten rats from each group were exposed for 6 hr/day on 5 days/week over 2 weeks and were sacrificed at 1 week after the cessation of exposure, as described previously.[12]

Rat Alveolar Macrophage and Pleural Macrophage Cultures

Bronchoalveolar and pleural lavages were performed on exposed rats at the time of sacrifice. Bronchoalveolar lavage was performed and alveolar macrophages obtained, as described previously.[10] For pleural lavage analyses, an 18-gauge Teflon cannula was inserted into the pleural cavity via a small diaphragmatic aperture. Thereafter, each pleural cavity was lavaged twice *in situ* with 5 ml of Ca^{++}- and Mg^{++}-free HBSS. Washed bronchoalveolar or pleural cells from 2-3 rats per group were pooled, suspended, and added to 24-well tissue culture plates (2 x 10^6 cells/well). After allowing the cells to attach for 1 h at 37° C, non-adherent cells were removed, 1 ml of DMEM was added, and the adherent cells were incubated for an additional 24 h at 37° C. Conditioned medium samples were collected and analyzed for NO$_2^-$ content by the Griess reaction.[13]

Measurement of iNOS Gene Expression and Nitrite Production

The inducible form of *NO synthase (iNOS) gene expression was measured by the reverse transcription-polymerase chain reaction (RT-PCR), whereas *NO formation was assayed measuring its oxidation product, NO$_2^-$, in conditioned medium samples from mesothelial cell cultures by the Griess reaction. For RT-PCR analyses of iNOS gene expression, extracted total cellular RNA from each sample was reverse transcribed, and the cDNA products were determined by PCR amplification by using the 1st Strand cDNA Synthesis kit for RT-PCR (Boehringer Mannheim, Indianapolis, IN). Glyceraldehyde-3-phosphate dehydrogenase (GAPDH) was evaluated as an internal control gene. PCR primers were designed from the published sequences for iNOS,[14] and determined by using a primer analysis software program (OLIGOTM, National Bioscience, Inc., Plymouth, MN). Primers for GAPDH used in this experiment were published ones.[15] The PCR product length of iNOS was 631 bp and that of GAPDH was 441 bp.

Table 1. Time course of NO$_2^-$ production in RPMC cultured in the presence or absence of IL-1β (50 ng/ml) with or without added asbestos fibers (4.2 µg/cm^2).

Exposure Category	NO$_2^-$ Formation (µM)		
	12-Hour Cultures[†]	24-Hour Cultures[†]	48-Hour Cultures[†]
Unstimulated RPMC	0.06 ± 0.01	0.22 ± 0.04	0.18 ± 0.10
IL-1β alone	1.04 ± 0.34	6.46 ± 0.27	20.27 ± 1.78
IL-1β + crocidolite	1.01 ± 0.47	7.96 ± 1.25	32.87 ± 1.07*
IL-1β + chrysotile	1.20 ± 0.35	10.5 ± 2.56	53.02 ± 5.01*

[†]N= 3 experiments/category; *P < 0.005 vs. RPMC cultured in the presence of IL-1β alone.

Immunocytochemistry for iNOS Protein and Nitrotyrosine

For immunocytochemical studies, the sample cells on 4-well chamber slides were fixed with acetone, and were immunostained by using a monoclonal anti-iNOS protein antibody (Transduction Laboratories, Lexington, KY), or a polyclonal anti-nitrotyrosine antibody (Upstate Biotechnology, Lake Placid, NY) in 0.05 M Tris-HCl/0.15 M sodium chloride buffer, pH 7.4. The cells were labeled using a DAKO LSBA2 Kit and were counterstained with Harris's hematoxylin. For assessment of iNOS protein expression, the proportions of iNOS positive cells were determined by counting the number of iNOS positive cells in 10 microscopic fields at a magnification of x 400.

The lungs from asbestos-exposed and sham-exposed rats were fixed with 10% buffered formalin, embedded in paraffin, and then sectioned. Following paraffin removal and hydration, the sections were immunostained by the anti-nitrotyrosine antibody.

Statistical Analyses

Statistical significance of the data was determined by the Student's t test.

RESULTS

Requirement of IL-1β for iNOS Gene Expression in RPMC

The level of iNOS gene expression in RPMC was evaluated by RT-PCR. Neither crocidolite nor chrysotile asbestos fibers, *per se*, were capable of inducing iNOS gene expression in RPMC. However, cultures stimulated with IL-1β (50 ng/ml) in the presence or absence of added asbestos fibers induced iNOS mRNA in RPMC. When RPMC were incubated for 2-12 h with IL-1β, increasing levels of iNOS mRNA were detected with time. These findings indicate that the presence of IL-1β was required for the activation of iNOS in RPMC. The level of GAPDH mRNA expression was unchanged in all samples.

Asbestos Exposure Up-Regulates *NO Formation by RPMC and Rat Macrophages

Table 1 shows the time course for the induction of *NO formation by RPMC (measured as NO_2^- in conditioned medium samples). Although only negligible amounts of NO_2^- were detected in 12-hour cultures, NO_2^- formation increased progressively with time in all IL-1β-stimulated

Table 2. Effect of particulate dose on NO_2^- production in RPMC cultured for 48 h in the presence or absence of IL-1β (50 ng/ml).

Exposure Category	NO_2^- Formation (μM)		
	1.05 μg/cm² Dose[†]	2.1 μg/cm² Dose[†]	8.4 μg/cm² Dose[†]
Crocidolite alone	Not done	Not done	1.47 ± 0.40
Chrysotile alone	Not done	Not done	1.80 ± 0.17
IL-1β + crocidolite	24.14 ± 3.75	29.45 ± 4.00	56.35 ± 2.97**
IL-1β + chrysotile	39.14 ± 4.86*	50.12 ± 4.14**	59.90 ± 3.62**
IL-1β + carbonyl iron	27.64 ± 2.40	26.40 ± 1.91	31.36 ± 1.97

[†]N = 3 experiments/category; *P < 0.005 vs. RPMC cultured in the presence of IL-1β alone; **P < 0.0001 vs. RPMC cultured in the presence of IL-1β alone.

388

cultures. Thus, 48 h after RPMC were exposed to IL-1β and either chrysotile or crocidolite fibers (4.2 μg/cm²), significantly greater amounts of NO_2^- were generated than in cultures stimulated with IL-1β in the absence of asbestos. Notably, the effects of chrysotile were more potent than those of crocidolite.

As shown in Table 2, a significant dose-response relationship was demonstrated between asbestos exposure and the induction of *NO formation by RPMC. This was only evident in IL-1β-containing cultures since only negligible amounts of NO_2^- were produced by maximal doses of asbestos (8.4 μg/cm²) when the fibers were added to RPMC in the absence of IL-1β.

Table 3. Effect of asbestos inhalation on NO_2^- production by rat alveolar and pleural macrophages obtained 1 week after the cessation of inhalation exposure.

Type of Exposure	NO_2^- Content (μM)	
	Alveolar Macrophages[†]	Pleural Macrophages[†]
Sham exposure	4.03 ± 0.20	1.44 ± 0.36
Crocidolite exposure	9.73 ± 0.73*	6.12 ± 0.18**
Chrysotile exposure	13.79 ± 0.82*	10.72 ± 0.64**

[†]N = 3 experiments/category; *P < 0.0001 vs. sham exposure; **P < 0.0005 vs. sham exposure.

The addition of carbonyl iron particles to IL-1β-containing cultures did not stimulate *NO synthesis by RPMC. Furthermore, no dose-response relationship was demonstrated when increasing doses of carbonyl iron (a non-fibrogenic and non-carcinogenic particulate) were added to cultures co-stimulated with IL-1β. It is of interest that chrysotile was observed to be more potent than crocidolite in stimulating *NO production by RPMC at all dosages tested.

To determine whether asbestos exposure also induced *NO formation *in vivo*, the amount of NO_2^- production was measured in cultured alveolar and pleural macrophages obtained from either sham-exposed rats or from rats sacrificed one week after the cessation of asbestos inhalation. As shown in Table 3, significantly greater NO_2^- formation was detected in conditioned medium from both alveolar and pleural macrophages after asbestos inhalation than after sham exposure to filtered air. In this regard the effects of chrysotile inhalation were greater than those observed after crocidolite inhalation.

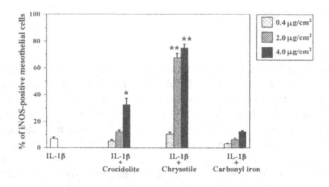

Figure 1. Dose response effects of particulates on iNOS protein expression by RPMC. *P < 0.05 vs. RPMC cultured with IL-1β alone; **P < 0.01 vs. RPMC cultured with IL-1β alone.

Immunohistochemical Studies for iNOS Protein and Nitrotyrosine

In Vitro **Particulate Exposures**. In order to determine the effects of asbestos fibers and IL-1β on iNOS protein expression, immunohistochemical studies were performed on RPMC after stimulation for 36 h with asbestos fibers or carbonyl iron particles (0.4, 2.0 and 4.0 μg/cm^2) in the presence (50 ng/ml) or absence of IL-1β.

None of the particulates induced iNOS protein expression in cultured RPMC in the absence of IL-1β. However, strong cytoplasmic staining for iNOS was evident in RPMC stimulated with IL-1β. Nevertheless, as illustrated in Figure 1, there were distinct differences in the proportions of iNOS-positive cells observed after exposure to IL-1β + asbestos fibers when compared with the effects of exposure to IL-1β alone or IL-1β + carbonyl iron. Furthermore, a dose-response relationship was noted between particulate exposure and the induction of iNOS protein expression in RPMC (Figure 1).

Formation of ONOO⁻ in RPMC was assessed by immunostaining the cells with a specific anti-nitrotyrosine antibody. For these studies, RPMC were stimulated for 36 h with particulates (4 μg/cm^2) in the presence or absence of IL-1β. Immunoreactivity for nitrotyrosine was not detected within RPMC stimulated with IL-1β alone or with particulates in the absence of IL-1β. However, strong, diffuse cytoplasmic immunoreactivity for nitrotyrosine was observed in the majority of cells challenged with IL-1β + chrysotile. In contrast, weaker immunoreactivity for nitrotyrosine was evident in cells exposed to IL-1β + crocidolite.

Inhalation Exposures. The lungs of rats that had inhaled either crocidolite or chrysotile asbestos fibers demonstrated strong, diffuse cytoplasmic immunoreactivity for nitrotyrosine throughout the epithelium of the large and small airways. Similar immunoreactivity was evident at alveolar duct bifurcations in these animals. Strong cytoplasmic immunoreactivity for nitrotyrosine also was a frequent finding within alveolar macrophages from crocidolite- and chrysotile-exposed rats. Another notable finding was the presence of strong immunostaining for nitrotyrosine in the visceral pleural mesothelial cells. In contrast, weak, focal immunoreactivity for nitrotyrosine was observed in occasional airway epithelial cells in the lungs of sham-exposed rats. Furthermore, only scanty alveolar macrophages demonstrated immunoreactivity for nitrotyrosine in the lungs of sham-exposed animals, and no such immunoreactivity was detected in their visceral pleural mesothelium. To confirm the immunospecificity of the anti-nitrotyrosine antibody, appropriate absorption experiments were performed with 10 mM concentrations of 3-nitro-L-tyrosine, 3-amino-L-tyrosine or O-phospho-L-tyrosine. These studies showed that positive immunostaining in lung sections could be abolished by 3-nitro-L-tyrosine but not by 3-amino-L-tyrosine or O-phospho-L-tyrosine. As a positive control for nitrotyrosine staining, lung sections from unexposed rats treated with 1 mM ONOO⁻ for 15 min demonstrated diffuse, strong immunoreactivity of the lung parenchyma.

DISCUSSION

One of the objectives of the present study was to determine whether asbestos exposure stimulates the formation of RNS in cultured RPMC. We found that asbestos fibers, *per se*, did not activate iNOS gene expression nor induce the formation of *NO in RPMC. Since certain cytokine combinations previously have been shown to induce *NO production in RPMC,[16] the effects of asbestos exposure then were evaluated in the context of co-stimulation of RPMC with IL-1β. The presence of IL-1β induced iNOS mRNA expression in RPMC after 2 h, and the amount of iNOS message increased progressively over 12 h. Although there was no obvious synergism between IL-1β and asbestos, both crocidolite and chrysotile fibers significantly up-regulated *NO synthesis (measured as NO$_2^-$) in RPMC co-stimulated with IL-1β in a dose-dependent and time-dependent fashion. Maximal effects were seen in 48-hour cultures at fiber concentrations of 8.4 μg/cm^2). In contrast, the effects of IL-1β + carbonyl iron particles (which are neither fibrogenic nor carcinogenic) were not significantly different from those of IL-1β alone. This indicates that the ability to stimulate *NO production in IL-1β-containing cultures was unique for asbestos fibers and was not simply a non-specific effect produced by any particulate phagocytic stimulus in RPMC.

Immunohistochemical studies for the detection of iNOS protein demonstrated that asbestos fibers augmented iNOS expression in RPMC co-stimulated with IL-1β in a dose-dependent fashion. In this regard, the effect of chrysotile was considerably greater than that of crocidolite. In contrast, the addition of carbonyl iron particles did not alter iNOS protein expression appreciably in the cultured cells. It is of interest that chrysotile fibers showed a greater capacity to up-regulate iNOS protein expression and *NO synthesis than did crocidolite fibers, since some investigators have suggested that amphibole asbestos (especially crocidolite) has a greater potential to induce pleural carcinogenesis than does chrysotile.[3,17]

These observations, coupled with those of an earlier study on rat alveolar macrophages,[10] clearly indicate that chrysotile and crocidolite fibers can stimulate the formation of RNS in cultured RPMC and rat alveolar macrophages. As a logical extension of these findings, we utilized a defined rat inhalation model of asbestosis to determine whether asbestos fibers might induce similar effects in vivo.[6] It is noteworthy that asbestos inhalation significantly up-regulated *NO production in both alveolar and pleural macrophages. As was noted in the experiments involving RPMC, the biologic effects of chrysotile inhalation were more striking than those observed after crocidolite inhalation. Immunohistochemical studies on rat lung sections from asbestos-exposed rats demonstrated strong immunoreactivity for nitrotyrosine at alveolar duct bifurcations and within alveolar macrophages. Immunostaining also was noted within bronchial epithelium and visceral pleural mesothelium. Similar findings were not observed in the lungs of sham-exposed rats. Immunohistochemical detection of nitrotyrosine in tissues has been used as a surrogate marker of nitration of tyrosine residues by ONOO- and other RNS in a variety of pulmonary disorders including adult respiratory distress syndrome.[18-20] Thus, the immunolocalization of nitrotyrosine in lung sections of asbestos-exposed rats in the present study provides evidence that asbestos inhalation can generate the formation of ONOO- in both the pleura and pulmonary parenchyma.

The production of RNS may have important significance for the development of asbestos-induced pleural and pulmonary injury. RNS may aggravate asbestos-related bronchoalveolar and pleural space inflammation by stimulating the release of inflammatory mediators in an autocrine or paracrine fashion from macrophages, mesothelial cells and epithelial cells. Notably, *NO has been shown to augment the release of IL-1 and tumor necrosis factor-α from mouse peritoneal macrophages.[21] Moreover, *NO may induce DNA strand breakages or DNA base modification by mobilizing iron from asbestos bodies or from crocidolite fibers.[22,23] Also, by generating ONOO- within cells, asbestos fibers may initiate pleuro-pulmonary injury via lipid peroxidation, oxidation of protein sulfhydryl moieties, reduction of cellular oxygen consumption and sodium uptake through amiloride-sensitive channels, and induction of cellular energy depletion by activation of the enzyme, poly (ADP-ribose) polymerase.[24-28] Apoptotic mechanisms also may be implicated in asbestos-related injury, since ONOO- can mediate apoptosis[29] and asbestos fibers have been shown to induce apoptosis in pleural mesothelial cells.[9]

In summary, this study has demonstrated that asbestos exposure can stimulate the production of RNS both in vitro and in vivo. We have shown in several cell types that chrysotile (serpentine asbestos) is more potent in this regard than is crocidolite (an amphibole asbestos), a finding that has important clinical significance. Furthermore, since the chemical structure of chrysotile is devoid of iron, the role of O_2*--driven Fenton reactions in mediating chrysotile asbestos-related injury needs to be re-evaluated.

ACKNOWLEDGMENT

This study was supported by grant HL-54196 from the National Institutes of Health.

REFERENCES

1. E. Kagan, Current perspectives in asbestosis, *Ann. Allergy* 54:464, 1985.
2. M. Kuwahara and E. Kagan, The mesothelial cell and its role in asbestos-induced pleural injury, *Int. J. Exp. Pathol.* 76:163, 1995.
3. B.T. Mossman, D.W. Kamp, and S.A. Weitzman, Mechanisms of carcinogenesis and clinical features of asbestos-associated cancers, *Cancer Invest.* 14:466, 1996.
4. A.R. Brody, Asbestos-induced lung disease, *Environ. Health Perspect.* 100:21, 1993.
5. M.B. Grisham. *Reactive Metabolites of Oxygen and Nitrogen in Biology and Medicine*, R.G. Landes Co., Austin, TX (1992).
6. B.T. Mossman, J.P. Marsh, A. Sesko, *et al.*, Inhibition of lung injury, inflammation, and interstitial pulmonary fibrosis by polyethylene glycol-conjugated catalase in a rapid inhalation model of asbestosis, *Am. Rev. Respir. Dis.* 141:1266, 1990.
7. M.A. Shatos, J.M. Doherty, J.P. Marsh, *et al.*, Prevention of asbestos-induced cell death in rat lung fibroblasts and alveolar macrophages by scavengers of active oxygen species, *Environ. Res.* 44:103, 1987.
8. V.A. Vallyathan, J.F. Mega, X. Shi, *et al.*, Enhanced generation of free radicals from phagocytes induced by mineral dusts, *Am. J. Respir. Cell Mol. Biol.*, 6:404, 1992.
9. V.C. Broaddus, L. Yang, L.M. Scavo, *et al.*, Asbestos induces apoptosis of human and rabbit pleural mesothelial cells via reactive oxygen species, *J. Clin. Invest.* 98:2050, 1996.
10. G. Thomas, T. Ando, K. Verma, *et al.*, Asbestos fibers and interferon-γ up-regulate nitric oxide production in rat alveolar macrophages, *Am. J. Respir. Cell Mol. Biol.* 11:707, 1994.
11. M. Kuwahara, M. Kuwahara, K.E. Bijwaard, *et al.*, Mesothelial cells produce a chemoattractant for fibroblasts: role of fibronectin, *Am. J. Respir. Cell Mol. Biol.* 5:256, 1991.
12. N. Choe, S. Tanaka, D.R. Hemenway, *et al.*, Pleural macrophage recruitment and activation in asbestos-induced pleural injury, *Environ. Health Perspect.* 105:(in press), 1997.
13. G. Thomas and P.W. Ramwell, Vasodilatory properties of mono-L-arginine-containing compounds, *Biochem. Biophys. Res. Commun.*, 154:332, 1988.
14. C.J. Lowenstein, C.S. Glatt, D.S. Bredt, *et al.*, Cloned and expressed macrophage nitric oxide synthase contrasts with the brain enzyme, *Proc. Natl. Acad. Sci. USA*, 89:6711, 1992.
15. K.E. Driscoll, J.M. Carter, B.W. Howard, *et al.*, Pulmonary inflammatory, chemokine, and mutagenic responses in rats after subchronic inhalation of carbon black, *Toxicol. Appl. Pharmacol.* 136:372, 1996.
16. M.W. Owens and M.B. Grisham, Nitric oxide synthesis by rat pleural mesothelial cells: induction by cytokines and lipopolysaccharide, *Am. J. Physiol.* 265:L110, 1993.
17. A.R. Gibbs, Role of asbestos and other fibres in the development of diffuse malignant mesothelioma, *Thorax* 45:649, 1990.
18. I.Y. Haddad, G. Pataki, P. Hu, *et al.*, Quantitation of nitrotyrosine levels in lung sections of patients and animals with acute lung injury, *J. Clin. Invest.* 94:2407, 1994.
19. K.M. Setoguchi, M. Takeya, T. Akaike, *et al.*, Expression of inducible nitric oxide synthase and its involvement in pulmonary granulomatous inflammation in rats, *Am. J. Pathol.* 149:2005, 1996.
20. I.Y. Haddad, S. Zhu, J. Crow, *et al.*, Inhibition of alveolar type II cell ATP synthesis and surfactant synthesis by nitric oxide, *Am. J. Physiol.* 270:L898, 1996.
21. J. Marcinkiewicz, A. Grabowska, and B. Chain, Nitric oxide up-regulates the release of inflammatory mediators by mouse macrophages, *Eur. J. Immunol.* 25:947, 1995.
22. L.G. Lund and A.E. Aust, Iron mobilization from crocidolite asbestos greatly enhances crocidolite-dependent formation of DNA single-strand breaks in φX174 RFI DNA, *Carcinogenesis*, 13:637, 1992.

23. C.-C. Chao, S.-H. Park, and A.E. Aust, Participation of nitric oxide and iron in te oxidation of DNA in asbestos-treated human lung epithelial cells, *Arch. Biochem. Biophys.* 326:152, 1996.

24. R. Radi, J.S. Beckman, R.J. Bridges, *et al.*, Peroxynitrite oxidation of sulfhydryls, *J. Biol. Chem.* 266:4244, 1991.

25. M.L. Bauer, J.S. Beckman, R.J. Bridges, *et al.*, Peroxynitrite inhibits sodium uptake in rat colonic membrane vesicles, *Biochem. Biophys. Acta* 1104:87, 1992.

26. P. Hu, H. Ischiropoulos, J.S. Beckman, *et al.*, Peroxynitrite inhibition of oxygen consumption and sodium transport in alveolar type II cells, *Am. J. Physiol.* 266:L628, 1994.

27. C. Szabó, C. Saunders, M. O'Connor, *et al.*, Peroxynitrite causes energy depletion and increases permeability via activation of poly (ADP-ribose) synthetase in pulmonary epithelial cells, *Am. J. Respir. Cell Mol. Biol.* 16:105, 1997.

28. H.Y. Dong, A. Buard, F. Lévy, *et al.*, Synthesis of poly (ADP-ribose) in asbestos treated rat pleural mesothelial cells in culture, *Mut. Res.* 331:197, 1995.

29. H. Ford, S. Watkins, K. Reblock, *et al.*, The role of inflammatory cytokines and nitric oxide in the pathogenesis of necrotizing enterocolitis, *J. Pediatr. Surg.* 32:275, 1996.

BIOMARKERS OF OXIDATIVE STRESS IN ADULT RESPIRATORY DISTRESS SYNDROME.

Stuart Malcolm, Raymond Foust III and Harry Ischiropoulos

Institute for Environmental Medicine and Department of Biochemistry and Biophysics, University of Pennsylvania, School of Medicine, Philadelphia, Pennsylvania 19104.

INTRODUCTION

Oxidative stress is considered as a significant component in the pathogenesis of the adult respiratory distress syndrome (ARDS) (1-3). Oxidative stress can be defined as the pathogenic outcome created by the oxidation of critical tissue targets by reactive species which are generated at rates that exceed tissue antioxidant capacity. Evidence for the presence of oxidative stress in ARDS patients is scarce. A major limitation for measuring reactive species in biological systems is their short half life. Since reactive species modify biological molecules such as proteins, lipids and DNA measurement of the modified targets provide the experimental tools for their detection and quantification. Previously we identified plasma proteins as a suitable target for quantification of oxidative stress in humans. Therefore the purpose of this study was to measure modified plasma proteins in ARDS patients as well as patients with sepsis. Two protein modifications were measured; protein carbonyl adducts and 3-nitrotyrosine. Protein carbonyls are derived by the direct oxidation of amino acid residues or conjugation of aldehydes that are formed by the oxidation of unsaturated lipids or sugars. Overall, plasma protein carbonyls indicate the formation of oxidants. Nitration of protein tyrosine residues by nitrating species results in the formation of 3-nitrotyrosine. The data collected indicated that elevation in protein carbonyls is associated with ARDS but not sepsis whereas plasma protein 3-nitrotyrosine was found to be elevated in both ARDS and septic patients.

Evidence that Oxidative Stress is Part of the Pathogenic Mechanism in Adult Respiratory Distress Syndrome.

Acute lung injury results in diffuse structural damage to the gas exchange alveocapillary unit. The effect on the endothelial barrier leads to an increase in permeability, the formation of non-cardiogenic pulmonary edema and the development of ARDS. The mortality rate for ARDS approaches 50% with the majority of deaths occurring due to multi-organ failure, sepsis or an underlying predisposing illness. Cell specific markers of acute lung injury such as von Willebrand factor antigen and various adhesion molecules have been found to correlate with the outcome in both human and animal models (4).

Acute Respiratory Distress Syndrome: Cellular and Molecular Mechanisms and Clinical Management
Edited by Matalon and Sznajder, Springer Science+Business Media New York, 1998

Relatively few studies have measured products of oxidative metabolism in patients with ARDS or at risk in developing ARDS. Hydrogen peroxide (H_2O_2) levels were found to be increased in the urine of patients with acute lung injury (5). Patients with sepsis-induced ARDS that expired had increased levels of urinary H_2O_2 as compared to survivors (5). Plasma levels of catalase an enzyme that converts H_2O_2 to water was found to be increased in sepsis-induced ARDS patients as compared to sepsis patients only (6). Increased levels of lipofuscin or 4-hydroxy-2-noneal products of lipid peroxidation were shown to predict the development of the syndrome in patients at risk of developing ARDS (7,8).

Biochemical and Molecular Aspects of Oxidative Stress.

Oxygen metabolism although essential for life imposes a potential threat to cells and tissue because of the formation of partially reduced oxygen species (1). One electron reduction of oxygen produces superoxide (O_2^-), a radical, whereas two electron reduction produces hydrogen peroxide H_2O_2, a reactive species. Therefore, electron flow through oxygen utilizing processes such as the mitochondrial electron transport chain, flavoproteins, cytochrome P450 and oxidases is tightly coupled to avoid partial reduction of oxygen. Normal cellular homeostasis is a delicate balance between the rate and magnitude of oxidant formation and the rate of oxidant elimination. Oxidative stress is the imbalance created by either overproduction of oxidants or diminished elimination of oxidants. In some pathological conditions such as hyperoxia, both overproduction and diminished detoxification can occur and account for tissue injury (1-3). It is important to make a clear distinction between oxidants and radicals. The term free radicals has been equated with reactive species or oxidants. By definition a radical is a molecule possessing an unpaired electron. Superoxide, nitric oxide, hydroxyl, alkoxyl and alkylperoxyl lipid radicals are radicals. However, with the exception of hydroxyl radical ($^.OH$) none of these radicals is a strong oxidant. Thus, not all radicals are strong oxidants and not all oxidants are radicals. The term reactive species may be more appropriate to describe all species because reactions between radicals and reactions between reactive species and redox active metals will generate strong oxidants.

The most reasonable biochemical explanation for the reactive species-mediated injury is the modification of critical cellular targets. Published data implicated superoxide as the key mediator of oxidative stress in cells and tissues (1). Iron-sulfur enzymes are direct targets for superoxide and toxicity can be derived from the inactivation of these enzymes (1). Hydrogen peroxide at low µM levels does not react with many biological targets at an appreciable rate. However, the reaction of hydrogen peroxide with reduced divalent redox active metals such as iron can lead to the formation of strong oxidants. This reactivity of H_2O_2 may be important in biological oxidations of proteins and lipids that take place at the sites of metal binding (9). Divalent redox active metals can also catalyze the formation of the highly reactive $^.OH$ radical by the metal-catalyzed Haber-Weiss reaction. However, $^.OH$ radical reacts with almost all biological targets at rates exceeding 10^9 M^{-1} s^{-1} and therefore its diffusion distance inside a cell is minimal. Thus, in order for hydroxyl radical to cause toxicity it must be formed within a few Angstroms from a biological target.

In addition to the metal catalyzed Haber Weiss reaction superoxide-derived oxidants can be generated by alternative pathways. One such pathway is the reaction of superoxide with nitric oxide to form peroxynitrite. Nitric oxide is synthesized by nitric oxide synthases (NOS). Nitric oxide, like O_2^-, is a radical that can diffuse among cells, where it principally reacts with guanylate cyclase and/or other iron-containing heme proteins. The second order rate constant of the reaction between nitric oxide and superoxide is 6.7×10^9 M^{-1} sec^{-1} (10). This is an extremely fast rate that is approximately thirty times faster than the reaction of $^.NO$ with oxyhemoglobin or guanylate cyclase and three times faster than the reaction of superoxide with superoxide dismutase. This implies that the formation of peroxynitrite can outcompete the major scavenging pathways for $^.NO$ and O_2^-. Moreover the rate of $ONOO^-$ formation is at least 10^5 times faster than the iron catalyzed Haber-Weiss reaction and increases by 100 times with only a ten fold increase in the concentration of $^.NO$ and O_2^-. Evidence for the contribution of peroxynitrite in cellular and tissue injury has been reported in a number of pathological disorders; pulmonary injury from immune-complexes, smoke inhalation, sepsis, ischemia-reperfusion, myocardial ischemia- reperfusion, kidney ischemic injury, neuronal injury by NMDA-receptor activation, 1-methyl-4-phenyl-1,2,3,6-tetrahydropyridine (a model of Parkinson's disease), 3-nitropropionic acid and malonate. (11-19). Recent data also provided evidence that peroxynitrite is the proximal mediator of apoptotic death following downregulation of Cu,Zn superoxide dismutase (20).

Detection of Reactive Species: Utility of Biomarkers.

Quantification and detection of reactive species in biological milieu is relatively difficult due to their short half life. This has led to the emergence of indirect measurements of modified targets. Modification of proteins, lipids and DNA represent suitable indices for detection and quantification of oxidants. The ability to detect and quantify reactive species is a function of the amount of modified molecules present at a given time (steady state) and the sensitivity of the assay. The steady state concentration of a modified target is a function of the rate of formation minus the rate of repair or removal and clearance. In turn, the rate of formation of the biological marker is proportional to the steady state concentration of reactive species. A biological marker can be of value because it may; 1) predict the development of the disease, 2) predict the outcome and 3) provide basic biochemical and molecular information regarding the mechanism of the disease. Biological markers for reactive species are the end-products of oxidation that can be derived by the reaction with more than one reactive species. As a result the ability of several biological markers to distinguish between the different reactive species is limited. Therefore development of selective, reactive species-specific biological marker(s) such as 3-nitrotyrosine may be useful.

The Biochemical Origin of Modified Plasma Proteins.

Serum Protein Reactive Carbonyls: Oxidation of proteins results in the introduction of carbonyl groups to the side chains of proteins which can be measured as a relatively specific marker of oxidative modification. The biochemical pathway for protein carbonyl formation can take three forms; Schiff based formation and Amidori rearrangement of an oxidized sugar, oxidation of polyunsaturated fats leading to the conjugation to a protein by a Michael type addition of an aldehyde and amino acid residues such as arginine, histidine and proline which are oxidized directly (9).

Serum Protein Nitrotyrosine: ·NO derived enzymatically although not a nitrating agent itself has been strongly implicated in the endogenous nitration of proteins (21). The nitration of tyrosine residues in vitro with ·NO as an essential component include; oxidation of nitrite by H_2O_2 and myeloperoxidase, oxidation of nitrite by hypochlorous acid, nitrogen oxides and nitryl halides (22). Utilizing plasma as a substrate and the solid phase immunoradiochemical assay to measure 3-nitrotyrosine the aforementioned methods of nitration were investigated. At high pathophysiological concentrations of the nitrating agents there was no detectable level of 3-nitrotyrosine. However, nitration of the plasma proteins via ONOO⁻ or the simultaneous generation of ·NO and O_2^- was achieved in a concentration dependent manner (23). Moreover, the intermediate which is formed as a result of the CO_2 catalyzed reaction with ONOO⁻ appears to be a plausible nitrating agent even at reasonable pathophysiological concentrations. The CO_2 seems to have a dual effect on ONOO⁻ reactivity; inhibiting other reactions, for example the oxidation of cysteine and tryptophan whilst increases nitration of tyrosine by more than two-fold (23,24,25). These findings would seem to indicate that the levels of 3-nitrotyrosine would reflect the generation of ONOO⁻.

Serum Protein Carbonyls and 3-Nitrotyrosine in ARDS Patients.

Oxidatively modified proteins such as protein carbonyls are identified through the conjugation of 2,4 dinitrophenylhydrazine (9). The plasma protein carbonyl levels in ARDS patients were (3.07±1.77, n=12) significantly higher (p<0.05, students t-test) than the levels measured in septic patients (0.66±0.57, n=5). Blood samples were taken from ARDS patients within 3 days of diagnosis and generally during the final week prior to the patient's death. There were no ARDS patients who had sepsis when the samples were collected A patient who survived ARDS had a protein carbonyl level of 1.3 nmol/mg protein. Observing the levels of 3-nitrotyrosine it was clear that in both ARDS (2.77±0.48, n=4) and sepsis (2.83±0.48, n=3) patients the levels were elevated. The increased level of

NO production associated with sepsis could be indicative of nitration due to the fact that a study on lipopolysaccharide (LPS) treatment of rat aorta and lung led to increased levels of nitration. The measurement of plasma protein 3-nitrotyrosine utilized the solid phase immunoradiochemical assay as described previously (26). As a control fresh and frozen (-80 °C for one month) plasma protein from healthy adult subjects was studied with no detectable amounts of either protein carbonyls or 3-nitrotyrosine.

The Molecular Basis for the Detection of Plasma Modified Proteins.

A steady state level of modified proteins is one in which there is an increase in production or a decrease in repair and removal or both. Protein carbonyls and 3-nitrotyrosine are assays in which increased levels of modified protein would indicate an individual who was experiencing oxidative stress. A study involving nitrated bovine serum albumin (BSA) indicated an increase in degradation via proteolysis as compared to non-nitrated BSA (27). The relationship in the removal of modified proteins indicates that in healthy adult rats nitrated albumin half life is approximately 4 days whereas in LPS treated rats the removal from circulation takes 7-8 days. This indicates that for the treated rats the removal pathway of modified proteins is slower increasing the length of the half life. The location of protein modification is unclear, however, if the disease is primarily lung based and blood passes through the pulmonary circulation then it maybe indicative of oxidative manipulations in lung.

Physiological Significance of Modified Protein Accumulation.

Although this study contains a relatively small number of patients, we attempted to identify specific plasma proteins that have been modified. Utilizing an affinity purified anti-nitrotyrosine and anti-DNPH antibodies, western blot analysis revealed that 3 major protein bands are nitrated (molecular mass approximately 120, 60 and 30 kd) and 5 protein bands showed reactivity with the anti-DNPH antibodies (molecular mass approximately 145, 120, 90, 60 and 30 kd). It has been previously shown that not all plasma proteins are nitrated or contain carbonyls (25,28). Based on data in human, rat plasma, tissue homogenates and cell lysates it is apparent that nitration and carbonyl formation are not random events. Consistently the same proteins were either nitrated or contain carbonyls. Although it is not known what predisposes a protein to be modified, identification of the modified proteins in these patients may provide useful information in terms of reactive species targeting of proteins. To date only a few protein targets for nitration have been identified in human disease such as mitochondrial Mn superoxide dismutase in rejected kidney allografts (29).

Overall, in a relatively small number of ARDS patients the data indicated the production of both oxidants and nitrating species in ARDS patients. The relatively lower levels of oxidatively modified proteins but the presence of equal 3-nitrotyrosine levels in sepsis suggests that the generation is mostly of nitrating species. The predictive value of these biomarkers for the development of the disease and outcome requires further evaluation in a large number of patients.

References

1. Freeman, B.A. and Crapo, J.D. Biology of the disease. Free radicals and tissue injury. Lab. Invest. 47: 412-426, 1982.
2. Frank, L., Bucher, J. R. and Roberts, R. J. Oxygen toxicity in neonatal and adult animals of various species. J. Appl. Physiol. 45:699-704, 1978.
3. Repine, J. E. Scientific perspectives on adult respiratory distress syndrome. Lancet 339:466-469, 1992
4. Pittet, JF, Mackersie, RC, Martin, TR and Matthay, MA. Biological markers of acute lung injury: Prognosis and pathogenic significance. Am. J. Resp. Crit. Care Med. 155: 1187-1205, 1997.
5. Marthru, M., Rooney, M.W., Dries, D.J. et al. Urine hydrogen peroxide during ARDS in patients with and without sepsis. Chest 105:232-236, 1994.
6. Leff, J.A., Parsons, P.E., Day, C.E. et al. Increased serum catalase activity in septic patients with ARDS. Am. Rev. Respir. Dis. 146:985-989, 1991.

7. Roumen, R.M., Hendriks, T., Man, BMD, and Goris, RJ. Serum lipofuscin as a prognostic indicator of ARDS and multiple organ failure. Br. J.Surg. 81:1300-1305, 1994.

8. Quinlan, GJ, Lamb, NJ, Evans, TW, Gutteridge, JMC,. Plasma fatty acid changes and increased lipid peroxidation in patients with ARDS. Crit. Care Med. 24:241-246, 1996

9. Stadtman, E.R. Protein oxidation and Aging. Science 257:1220-1224, 1992.

10. Huie, R.E. and S. Padjama. The reaction of NO with superoxide. Free Rad. Res. Comm. 18: 195-199, 1993.

11. Mulligan, M.S., Hevel, J.M., Marletta, M.A., and Ward, P.A. Tissue injury caused by deposition of immune complexes is L-arginine dependent. Proc. Natl. Acad. Sci. USA 88:6338-6342, 1991.

12. Ischiropoulos, H., Mendiguren, I., Fisher, D., Fisher, A.B. and Thom, S.R. Role of neutrophils and nitric oxide in lung alveolar injury from smoke inhalation. Am. J. Resp. Crit. Care Med. 150: 337-341, 1994

13. Wizemann, T.M., Gardner, C.R., Laskin, J.D., Quinones, S., Durham, K.D., Golle, N.L., Ohnishi, S.T. and Laskin, D.L. Production of nitric oxide and peroxynitrite in the lung during acute endotoxemia. J. Leuk. Biol. 56: 759-768, 1994.

14. Ischiropoulos, H., Al-Mehdi, A. B. and Fisher, A.B. Reactive species in rat lung injury: contribution of peroxynitrite. Am. J. Physiol. 269:L158-L164, 1995.

15. Wang P, Zweir JL (1996) Measurement of nitric oxide and peroxynitrite generation in the postischemic heart. J. Biol. Chem. 271:29223-29230

16. Szabo, C. Salzman, A. L. and Ischiropoulos, H. Endotoxin triggers the expression of an inducible isoform of nitric oxide synthase and the formation of peroxynitrite in the rat aorta in vivo. FEBS Lett. 363:235-238, 1995.

17. Noiri E, Peresleni T, Miller F, Goligorsky MS (1996) In vivo targeting of inducible NO synthase with oligodeoxynucleotides protects rat kidney against ischemia. J. Clin. Invest. 97:2377-2383

18. Schulz, J.B., Matthews, Jenkins, B.G., Ferrante, R.J., Siwek, D., Henshaw, D.R., Cipolloni, P.B., Mecocci, P., Kowall, N.W. Rosen, B.R. and Beal, M.F. Blockade of neuronal nitric oxide synthase protects against excitotoxicity in vivo. J. Neuroscience 15: 8419-8429, 1995.

19. Xia Y. Dawson VL. Dawson TM. Snyder SH. Zweier JL. Nitric oxide synthase generates superoxide and nitric oxide in arginine-depleted cells leading to peroxynitrite-mediated cellular injury. Proc. Nat. Acad. Sci. USA 93:6770-4, 1996.

20. Troy CM, Derossi D, Prochiantz A, Greene LA and Shelanski M. Down-regulation of copper/zinc superoxide dismutase leads to cell death via the nitric oxide-peroxynitrite pathway. (1996) J. Neuroscience 16: 253-261

21. Ischiropoulos, H., Zhu, L., Chen, J., Tsai, J-H.M., Martin, J.C., Smith, C.D. and Beckman, J.S. Peroxynitrite-mediated tyrosine nitration catalyzed by superoxide dismutase. Arch. Biochem. Biophys. 298: 431-437, 1992.

22. Van der Vliet, A., Eiserich, J.P., Halliwell, B. and Cross, C.E. Formation of reactive nitrogen species during peroxidase-catalyzed oxidation of nitrite. J. Biol. Chem. 272: 7617-7625, 1997.

23. Gow, A., Duran, D., Thom, S.R., and Ischiropoulos, H. (1996) Carbon dioxide enhancement of peroxynitrite-mediated protein tyrosine nitration. Arch. Biochem. Biophys. 333, 42-48.

24. Uppu, R.M., Squadrito, G.L. and Pryor, W.A. Acceleration of peroxynitrite oxidations by carbon dioxide. Archives Biochem. Biophys. 327: 335-343, 1996.

25. Denicola, A. Trujillo, M., Freeman, B.A. and Radi, R. Peroxynitrite reaction with carbon dioxide/bicarbonate: Kinetics and influence on peroxynitrite-mediated oxidation reactions.Arch. Biochem. Biophys. 333: 48-54, 1996.

26. Ischiropoulos, H., M.F. Beers, S.T. Ohnishi, D. Fisher, S.E. Garner and S.R. Thom. Nitric oxide production and perivascular tyrosine nitration in brain following carbon monoxide poisoning in the rat. J. Clin. Invest. 97:2260-2267, 1996.

27. Gow, A., Duran, D., S. Malcolm, and H. Ischiropoulos. Effects of peroxynitrite induced modifications to signal transduction and protein degradation. FEBS Lett. 385: 63-66, 1996.

28. Shacter, E, Williams, JA, Lim, M and Levine RL. Differential susceptibility of plasma proteins to oxidative modification: Examination by western blot immunoassay. Free Rad. Biol. Med. 17: 429-437, 1994.

29. MacMillan-Crow LA, Crow JP, Kerby JD, Beckman JS, Thompson JA (1996) Nitration and inactivation of manganese superoxide dismutase in chronic rejection of human renal allografts. Proc. Natl. Acad. Sci. 93:11853-11858.

REACTIVE OXYGEN AND NITROGEN SPECIES AND ADULT RESPIRATORY DISTRESS SYNDROME (ARDS): NEW MECHANISMS TO BE CONSIDERED

Albert van der Vliet[*], Carroll E. Cross[*] and Jason P. Eiserich[+]

[*]Division of Pulmonary/Critical Care Medicine, Department of Internal Medicine, University of California, Davis, California 95616, USA.
[+]Department of Anesthesiology, and the Center for Free Radical Biology, University of Alabama at Birmingham, Birmingham, Alabama 35233, USA.

1. INTRODUCTION

The adult respiratory distress syndrome (ARDS) represents a common response of the lung to a variety of different and often unrelated insults, most frequently sepsis, trauma, aspiration and shock. Important consequences are the priming and activation of many components of the inflammatory-immune system accompanied by alterations in the permeability of endothelial and epithelial cell membrane barriers manifest by accumulations of inflammatory pulmonary edema fluid accompanied by phagocyte infiltrations in the interstitial and alveolar compartments of the lung. The net physiologic result is severe hypoxemic acute respiratory failure due to lung ventilation-perfusion mismatching and extensive intrapulmonary shunts. As with inflammatory-immune activation states involving other systems (e.g., rheumatoid arthritis in joints), reactive oxygen species (ROS) and reactive nitrogen species (RNS) are likely to play a significant role in the pathobiology of the lung injury seen in ARDS (Louie et al., 1997; Forni et al., 1997).

There are numerous potential sources for the generation of ROS and RNS in the pathobiology of acute lung injury, not the least of which are the high concentrations of oxygen and of nitric oxide (NO•) which are often administered to subsets of patients with ARDS (Mankelow et al., 1997; Troncy et al., 1997). Some of these potential pathways are listed in Table 1. In the present paper we will discuss possible oxidative mechanisms involving phagocytes, focusing on pathways related to myeloperoxidase, ROS and RNS.

Table 1. Possible Pathways for ROS/RNS Generation in ARDS

● Xanthine oxidase	● P-450 systems
● Mitochondrial respiration	● Catecholamine oxidation
● Phagocyte/NADPH oxidase/peroxidase	● NOS pathways
● Non-phagocytic NADPH-oxidases/peroxidases	● Proinflammatory cytokines
● Lipoxygenases	● Redox-cycling metals
● NO• administration	● O_2 toxicity

Acute Respiratory Distress Syndrome: Cellular and Molecular Mechanisms and Clinical Management
Edited by Matalon and Sznajder, Springer Science+Business Media New York, 1998

2. PHAGOCYTES AND ARDS

Although phagocytes are critical for our survival, it has been long recognized that migration and activation of polymorphonuclear neutrophils (PMNs) in the lungs is almost surely a major contributing factor to the acute lung injury that occurs in patients with ARDS (Boxer et al., 1990; Repine and Beehler, 1991). This argument is buttressed by observations that PMN depletion prevents animal models of ARDS (Heflin and Brigham, 1981), that PMNs and their oxidative products contribute to endothelial cell injury in vitro and lung injury in vivo (Shasby et al., 1983); and that PMNs and their products accumulate in the lungs of patients with ARDS (Weiland et al., 1986).

However, PMN migration alone may not injure the lungs of normal humans. For example, activation and priming of their proteolytic and oxidant injury mechanisms may be required (Martin et al., 1989, 1991; Martin, 1997). The multitude of cytokines found to be present in ARDS (Chollet-Martin et al., 1992) would be expected to further potentiate PMN activation, thereby increasing their production of reactive oxidant species . . . even the circulating PMNs have been found to be activated in patients with ARDS (Zimmerman et al., 1983; Martin et al., 1991; Chollet-Martin et al., 1992).

Myeloperoxidase represents an important phagocytic enzyme, more plentiful in PMNs than monocytes and macrophages (Odeberg et al., 1974; Bos et al., 1978; Kettle & Winterbourn, 1997), and has been heavily used as a biomonitor of PMN traffic in numerous organs undergoing various stages of cell injury and related inflammatory-immune system activations including ARDS (Fantone & Ward, 1985; Weiland et al., 1986; Denis et al., 1994; Shayeritz et al., 1995; Sinclair et al., 1995; Okabayashi et al., 1996; Koh et al., 1996; Kushimoto et al., 1996), as depicted in Table 2.

Table 2: Lung Myeloperoxidase in ARDS

- A monitor of PMN traffic
- Elevated in following ARDS models:
    - Endotoxin, sepsis, IP zymosan
    - Hemorrhagic shock, gut ischemia, pancreatitis
    - IL-1, TNF, Complement (C') and PMN-activation
    - Transplant and cardiopulmonary bypass

---

3. OXIDES OF NITROGEN AND THE LUNG

Nitric oxide (NO•) is now well-recognized for its participation in diverse biological processes in nearly all aspects of life (Moncada and Higgs, 1991), including in inflammatory-immune processes in the lung (Gaston et al, 1994). Multiple lung cells synthesize NO• under both physiological and pathophysiological settings (Barnes and Belvisi, 1993), utilizing a family of enzymes termed NO• synthases, which use arginine as their substrate. NO• can travel significant distances to reach target cells neighboring the NO• generating cells (Lancaster, 1994; Malinski and Taha, 1992). Along this migration, NO• can interact with other oxidative molecules, including molecular O_2, to form higher nitrogen oxides (e.g. NO_2•, N_2O_3) which can either react with other biomolecules (e.g., thiols, amines) or simply hydrolyze to form NO_2^- and NO_3^-. Furthermore, and importantly for the context of this paper, NO• and its metabolites (e.g. NO_2^- specifically), themselves generated in increased amounts at inflammatory-immune reaction sites, can react with several phagocyte-derived oxidants, to form more reactive RNS (e.g., Koppenol, 1994). The extent of either of these reactions depends on the microenvironmental conditions under which NO• is released, and the concentration of the other bioreactants present.

Concerning the interactive role of NO• with other active pro-oxidant species, most recent research has focused emphasis on NO• reaction with O_2•⁻, to yield the powerful oxidant species peroxynitrite (ONOO⁻) (Beckman, 1995, 1996). Based primarily on the nearly diffusion limited reaction kinetics of this reaction (Huie and Padmaja, 1993), it is predicted that ONOO⁻ is generated whenever O_2•⁻ and NO• are produced simultaneously. The potential *in vivo* importance of such a reaction is supported by findings that superoxide dismutase (SOD) prolongs the biological half-life of NO• and increases its biological actions, putatively by lowering O_2•⁻ levels and minimizing degradation of NO• via reaction with O•⁻ (e.g., Gryglewski et al, 1986).

A second argument for *in vivo* generation of ONOO⁻ (especially during inflammatory conditions) is detection of 3-nitrotyrosine in proteins from a large number of diseased or inflamed tissues, often in conjunction with induction of iNOS and increased production of NO• (Beckman, 1995, 1996; Halliwell, 1997). As NO• itself is unable to nitrate tyrosine residues (e.g., Eiserich et al., 1995), more reactive NO•-derived nitrogen oxides (NO_2•, ONOO⁻) are thought responsible for tyrosine nitration, and based on kinetic considerations, it is commonly assumed that 3-nitrotyrosine *in vivo* is caused by ONOO⁻. Using largely immuno-histochemical techniques, 3-nitrotyrosine has also been detected at sites of inflammatory-immune processes in the lung (Saleh et al, 1997), including ARDS (Haddad et al, 1994; Kooy et al, 1995). However, the pathophysiological importance of 3-nitrotyrosine formation and the precise oxidative and nitrosative mechanisms involved in its formation are relatively unknown. These issues will be the focus of the remainder of this paper.

4.   TYROSINE NITRATION VIA RADICAL MECHANISMS

Chemical studies of tyrosine nitration by ONOO⁻ have indicated that nitration can be promoted in the presence of superoxide dismutase or Fe(III)EDTA, as well as heme peroxidases (Beckman *et al.*, 1992; Sampson *et al.*, 1996), presumably by formation of an NO_2^+- like intermediate, which is known to be capable of nitrating aromatic rings by electrophilic aromatic substitution (Olah *et al.*, 1989). However, another major product of tyrosine oxidation by ONOO⁻ is 3,3'-dityrosine, indicative of formation of intermediate tyrosyl radicals (van der Vliet et al., 1995). Hence, nitration of tyrosine by ONOO⁻ appears to occur via a one-electron mechanism, involving initial formation of tyrosyl radical and NO_2•, and dityrosine and 3-nitrotyrosine are subsequently formed by radical combination reactions (e.g., Prutz *et al.*, 1985; Lymar *et al.*, 1996), as illustrated in reactions 1-4.

$$
\begin{array}{llll}
ONOOH + Tyr\text{-}OH & \rightarrow & Tyr\text{-}O^• + NO_2^• + H_2O & (1) \\
NO_2^• + Tyr\text{-}OH & \rightarrow & Tyr\text{-}O^• + NO_2^- + H^+ & (2) \\
NO_2^• + Tyr\text{-}O^• & \rightarrow & 3\text{-nitrotyrosine} & (3) \\
2\ Tyr\text{-}O^• & \rightarrow & 3,3'\text{-dityrosine} & (4)
\end{array}
$$

Reaction (2) becomes more favorable over reaction (3) when tyrosine concentrations are relatively high, resulting in relatively more dityrosine formation and less nitration, as was indeed observed in studies with tyrosine in solution (van der Vliet *et al*, 1995), as well as in studies in freshly obtained human plasma. In the latter case, ONOO⁻-induced nitration and dimerization of free or protein-associated tyrosine was markedly affected when plasma was supplemented with free tyrosine. The overall formation of 3-nitrotyrosine was decreased, whereas more dityrosine was formed (unpublished results), consistent with a radical reaction mechanism (reactions 1-4). Formation of tyrosyl radicals in plasma by ONOO⁻ has recently been detected by ESR (Pietraforte and Minetti, 1997), which gives further support to this nitration mechanism.

It has become clear that tyrosine nitration by NO_2^+ in aqueous systems also occurs by a one-electron mechanism via intermediate tyrosyl radicals, rather than via direct electrophilic substitution. However, free NO_2^+ is extremely unstable in aqueous solution and hydrolizes rapidly to nitrate (NO_3^-), and is thus unlikely to be involved in nitration reactions in vivo (van der Vliet et al, 1996; 1997). Both tyrosine nitration and dimerization by $ONOO^-$ are enhanced in the presence of bicarbonate (van der Vliet et al, 1994; Lymar et al., 1996; Gow et al., 1996; Lemercier et al., 1997), which was discovered to be due to reaction of $ONOO^-$ with CO_2 to form $ONOOCO_2^-$, which appears more efficient as a nitrating species (Lymar et al., 1996). Collectively, irrespective of the nature of the nitrating species (ONOOH, $NO_2^{\bullet}$ or NO_2^+), the mechanism of tyrosine nitration appears to involve intermediate formation of tyrosyl radicals, and 3,3'-dityrosine is formed as an additional product.

5. ALL THAT NITRATES IS NOT PEROXYNITRITE

As tyrosine nitration is radical-mediated, involving formation of tyrosyl radicals and $NO_2^{\bullet}$, formation of these intermediates from other sources would be expected to contribute to tyrosine nitration in biological systems. Indeed, several such mechanisms can be envisioned to occur in vivo, especially during inflammatory-immune processes. $NO_2^{\bullet}$ can be generated via autoxidation of NO•. This reaction is slow, especially at physiological NO• levels, hence formation of $NO_2^{\bullet}$ by NO• autoxidation is expected to be minimal (e.g., Beckman and Koppenol, 1996).

Additionally, $NO_2^{\bullet}$ can be generated via one-electron oxidation of NO_2^-, and various biological oxidants can be expected to promote such reactions. For instance, it has long been recognized that heme peroxidases or pseudoperoxidases, such as methemoglobin or metmyoglobin, are able to oxidize NO_2^- in the presence of H_2O_2, and this has been postulated to occur via one-electron mechanisms (van der Vliet et al., 1997 and refs therein).

Although $NO_2^{\bullet}$ is capable of directly nitrating tyrosine residues in proteins (Prütz et al., 1985), this process is considered relatively inefficient, as two $NO_2^{\bullet}$ molecules are necessary to nitrate one tyrosine residue. However, tyrosyl radicals can also be generated by various mechanisms, including peroxidases (Kettle and Winterbourn, 1997), and radical combination between tyrosyl radicals and $NO_2^{\bullet}$ is very rapid, yielding 3-nitrotyrosine.

Our recent studies have shown that NO_2^- can be oxidized by various heme peroxidases, including horseradish peroxidase, myeloperoxidase (MPO), and lactoperoxidase (LPO), in the presence of hydrogen peroxide (H_2O_2), to most likely form NO_2•, and was found capable of promoting tyrosine nitration, which may be relevant during inflammatory processes (van der Vliet et al., 1997). The physiological importance of such mechanisms depends on whether NO_2^- is a competitive substrate for MPO or other peroxidases in vivo. Phenolic nitration by MPO-catalyzed NO_2^- oxidation was found to be only partially inhibited by chloride (Cl^-), the presumed major physiological substrate for MPO, and low concentrations of NO_2^- (2-10 μM) were in fact demonstrated to **catalyze** MPO-mediated oxidation of Cl^-, indicated by increased chlorination of aromatic substrates, and simultaneously cause aromatic nitration. The observed enhanced MPO-mediated Cl^--oxidation by NO_2^- was similar to that observed by other reductants, such as ascorbate or 5-amino salicylate (Bolscher et al., 1984; Zuurbier et al., 1990), and can be attributed to reduction of MPO compound II, which is inactive with respect to Cl^--oxidation, thereby recycling MPO (Fig. 1).

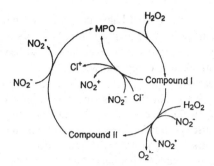

Fig. 1

Peroxidase-catalyzed oxidation of NO_2^-, as indicated by phenolic nitration, could also be detected in the presence of thiocyanate (SCN^-), an alternative physiological substrate for mammalian peroxidases, suggesting that NO_2^- may act as a pathophysiological substrate for the MPO (and perhaps other peroxidases). Thus, formation of $NO_2^{\cdot}$ via peroxidase-catalyzed oxidation of NO_2^- may provide an additional pathway contributing to **aromatic nitration *in vivo*.**

Interestingly, 3-nitrotyrosine and large amounts of active MPO have both been detected in atherosclerotic lesions (Beckman et al, 1994; Buttery et al, 1996; Daugherty et al, 1994), in joints of patients with rheumatoid arthritis (Farrell et al, 1992; Kaur and Halliwell, 1994), as well as in the lungs of patients with acute pulmonary inflammation (Haddad et al, 1994; Kooy et al, 1995), a condition characterized by activation and lung infiltration of PMNs as well as increased production of NO• and NO_2^- (e.g., Hunt et al, 1995; Kharitonov et al, 1996). Plasma nitrotyrosine levels have been noted to be elevated in septic shock (Fukuyama *et al.*, 1997), a condition known to be often associated with ARDS. However, it should be also noted that studies performed on inducible NO synthase-deficient mice suggest a significant role for augmented NO• levels in sepsis as being important as a homeostatic regulator in PMN activation and recruitment to endothelial surfaces (Hickey *et al.*, 1997).

Formation of Novel Nitrating and Chlorinating Intermediates During Reaction of NO_2^- with HOCl/OCl$^-$

One of the most potent and plentiful oxidants produced by phagocytes is hypochlorous acid (HOCl/OCl$^-$), which is formed via MPO-catalyzed oxidation of Cl$^-$ (Weiss et al, 1983). As NO• is also present in increased quantities at sites of inflammatory-immune reactions, interactions between NO•-derived RNS with the inflammatory oxidant HOCl/OCl$^-$ can be expected to occur under inflammatory conditions. Results from our laboratory have indicated that NO_2^-, the major metabolite of NO• in extravascular fluids, reacts with HOCl/OCl$^-$ to form nitrate (NO_3^-) via intermediate formation of nitryl chloride ($ClNO_2$) and/or chlorine nitrite (ClONO). These intermediates are powerful nitrating and chlorinating species, hence, formation of $ClNO_2$/ClONO by this reaction may represent a previously unrecognized mechanism of inflammation-mediated biological damage, and offer an additional or alternative mechanism of tyrosine nitration independent of ONOO$^-$ formation (Eiserich et al., 1996). This argument is strengthened by the recent finding that hypochlorous acid and nitrate cause oxidative modification and nitration of human lipoproteins (Panasenko et al, 1997).

Normally, NO_2^- is present at levels of 0.5-3.6 μM in plasma (Leone et al, 1994; Ueda et al, 1995), ~15 μM in respiratory tract lining fluids (Gaston et al, 1993), 30-210 μM in saliva, and 0.4-60 μM in gastric juice, but extracellular NO_2^- levels are markedly increased during inflammatory processes, reflecting increased NO• production. For instance, increased NO_2^- levels have been detected in synovial fluids of patients with rheumatoid arthritis (Farrell et al, 1992), and serum NO_2^- levels of 36 μM have been reported in human immunodeficiency virus-infected patients with interstitial pneumonia (Torre et al, 1996), dramatically higher than normal serum NO_2^- levels. Increased NO$^-$ levels have also been detected in condensed exhalates from patients with asthma compared with those of healthy subjects (Hunt et al, 1995), consistent with increases in expired NO• by asthmatics compared with healthy control subjects (Kharitonov et al, 1996). Nitrite levels are reportedly very high in respiratory tract surfaces (Goviadaraju et al, 1997), far exceeding plasma levels (e.g., possibly 1000x higher in the rat), presumably because epithelial cells generate NO• and the airway surface does not have hemoglobin degradation systems oxidizing NO• (or nitrite) to nitrate (NO_3^-). [However, it has been shown that most inhaled NO• results in $*NO_3^-$, as detected in blood and in urine (Westfelt et al, 1995).]

Do Human PMNs Utilize these Pathways?

We have obtained recent evidence indicating that activated human PMNs can utilize these above-described MPO-dependent pathways to form both nitrating and chlorinating intermediates (Eiserich et al., 1997). Addition of NO_2^- (1-50 μM) to PMA-stimulated PMN was found to cause nitration of phenolic substrates and enhance PMN-mediated chlorination reactions. The enhanced chlorination can be explained by: i) NO_2^--mediated recycling of inactive MPO Compound II to the native ferric enzyme; and ii) the ability of NO_2^- to compete with taurine released from PMN for reaction with HOCl/OCl$^-$ to form a more potent electrophilic chlorinating intermediate ($ClONO/ClNO_2$). Furthermore, we have obtained evidence that exposure of activated PMN to pathophysiologic fluxes of NO• resulted in nitration and chlorination reactions that in some conditions were **dependent** on active MPO, rather than formation of $ONOO^-$. Under identical conditions, addition of $^{15}NO_2^-$ led to ^{15}N enrichment of nitrated phenolic substrates, unequivocally confirming contribution of NO_2^- in PMN-mediated reaction pathways.

In summary, formation of NO•-derived reactive nitrogen species that are capable of inducing aromatic nitration appears to occur by multiple mechanisms, especially during inflammation and PMN activation, as schematically depicted below:

PHAGOCYTE

Fig. 2

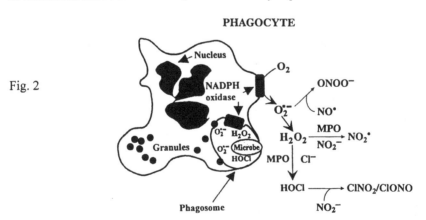

There is likewise an accumulating documentation of increased formation of 3-chlorotyrosine in tissues subjected to chronic inflammatory-immune reactions (Halliwell, 1997), and indeed in some of the same tissues where increased nitrotyrosine has been found (e.g., Beckman et al, 1994; Hazell et al., 1996; Hazen et al, 1997a). With increased sensitivity of techniques designed to quantitate chlorotyrosine (Hazen et al, 1997b), it can be expected that chlorotyrosine will be found in many tissues where elevated nitrotyrosine has been found.

6.   POSSIBLE CONSEQUENCES OF THESE IRREVERSIBLE TYROSINE MODIFICATIONS

Although there is ample evidence for the formation of 3-nitrotyrosine (and 3-chlorotyrosine or 3,3'-dityrosine) in a number of inflammatory diseases, the potential contribution of these modifications to the development of tissue injury has received less documentation. Nitration of tyrosine residues using the NO_2^+-donor tetranitromethane has been used extensively to investigate the location and essentiality of tyrosine residues in a large number of proteins (e.g., Riordan and Vallee, 1972; Mierzwa and Chan, 1987; Haddad et al., 1996), and has indicated that nitration of tyrosine residues is often associated with a loss of either enzyme or protein function. Increasingly, investigators are using peroxynitrite to selectively inactivate proteins via tyrosine nitration mechanisms (Zou et al., 1997). Furthermore, studies with isolated tyrosine kinase systems have indicated that nitration of critical tyrosine residues in tyrosine kinase substrates causes inhibition of tyrosine phosphorylation (Martin et al., 1990; Kong et al., 1996; Gow et al., 1996). Hence, formation of reactive nitrogen species may importantly affect signaling pathways involving (receptor) tyrosine kinases, however, this possibility has not yet been convincingly documented in intact cellular systems.

Cytoskeletal proteins such as actin or neurofilaments may represent important targets for tyrosine nitration reactions, as they are abundant proteins and contain several tyrosine residues which appear to be involved in structural assembly of these proteins. Chemical nitration of actin or neurofilaments has been demonstrated to disrupt assembly of these proteins, and modification of only a few subunits appears necessary to cause disruption of a structure involving thousands of subunits (Beckman, 1996 and refs. therein). Although extensive tyrosine nitration has been shown to occur in the myocardium during inflammatory forms of myocarditis (Kooy et al., 1997), it is still speculative whether or not the nitration plays a pathologic role in mediating the myocardial dysfunction.

More relevant to ARDS, several studies have indicated that $ONOO^-$ or related reactive nitrogen species are capable of nitrating tyrosine residues in surfactant proteins, and nitration of tyrosine residues in surfactant protein A has been causatively linked to decreases in its ability to aggregate lipids or decreases in binding to mannose receptors. Hence, such modification may disturb functions of SPA by diminishing its function to lowering alveolar surface tension (Zhu et al., 1996; Haddad et al., 1996), or by compromising its function to facilitate phagocytic uptake and killing of bacteria.

Similar to nitration, tyrosine chlorination or dimerization may affect cellular pathways involving critical tyrosine residues. As techniques to measure and characterize these compounds become increasingly available (Shigenenaga et al., 1997; Leeuwenburgh et al., 1997a, 1997b; Yi et al., 1997), their role in tissue pathobiology will become apparant. It has been recently described that 3-nitro-tyrosine attenuates hemodynamic responses to adrenoreceptor agonists and to angiotensin II (Kooy and Lewis, 1996).

7. REMAINING IMPORTANT QUESTIONS

One of the implications of our recent findings is that NO_2^- may not be a stiochiometric marker of $NO\bullet$ production by phagocytes or at sites of inflammation, as it is potentially removed by reaction with inflammatory oxidants. Determination of $NO\bullet$ production in tissues and fluids of patients with acute and chronic inflammation, or from isolated phagocytes as measured by NO_2^- may likely be an underestimate, and should include analysis of both NO_2^- and NO_3^-. Moreover, these NO_2^--oxidation mechanisms may also modulate PMN function or affect PMN-dependent tissue injury. For instance, NO_2^- has been shown to inhibit the bactericidal activity of HOCl/OCl⁻, proposedly by direct reaction of these two species (Kono, 1995; Klebanoff, 1993). The reaction product $ClNO_2/ClONO$, although a strongly oxidizing species and potentially an antimicrobial agent in its own right, is a short-lived intermediate and appears less efficient in bacterial killing compared to HOCl in bacterial suspensions.

However, it is difficult to extrapolate such findings to the situation present in the phagolysosome, where oxidant reactions with bacterial constituents are less likely to be limited by chemical stability or diffusion. Moreover, NO_2^- may catalytically enhance $MPO/H_2O_2/Cl^-$ -dependent bacterial killing, via the mechanism depicted in Fig. 1. Furthermore, peroxidases have been found capable of converting NO_2^- itself into a bactericidal agent, presumably $NO_2\bullet$ (Klebanoff, 1995; Kono, 1995). It remains unclear, therefore, to what extent NO• or NO_2^- affect PMN function with respect to bactericidal activity. Interestingly, cytokine-stimulated human PMNs were found to contain increased levels of iNOS, which was co-localized with MPO in primary granules and tyrosine nitration could be detected around ingested bacteria (Evans et al, 1996), which is likely to involve MPO-mediated pathways.

Taken together, augmented NO• generation at inflammatory-immune activation sites may generate an expanding number of potential nitrosating and/or nitrating species and potentially play a role in augmenting the production of chlorinating species, which collectively could result in (ir)reversible modifications in proteins, lipids or nucleic acids. Although irreversible tyrosine modifications may affect various cellular processes, the relevance of such modifications to the pathobiology of ARDS or other localized or systemic inflammatory-immune processes still needs to be established.

REFERENCES

Barnes PJ and Belvisi MG. Nitric oxide and lung disease. *Thorax* 1993; 48: 1034-1043.

Beckman JS. Oxidative damage and tyrosine nitration from peroxynitrite. *Chem Res Rox* 1996; 9: 836-844.

Beckman JS. Reactions between nitric oxide, superoxide, and peroxynitrite: footprints of peroxynitrite in vivo. *Adv Pharmacol* 1995; 34: 17-43.

Beckman JS, Chen J, Ischiropoulos H and Crow JP. Oxidative chemistry of peroxynitrite. *Meth Enz* 1994; 233: 229-240.

Beckman JS and Koppenol WH. Nitric oxide, superoxide, and peroxynitrite: the good, the bad, and the ugly. *Am J Physiol* 1996; 271: C1424-C1437.

Beckman JS, Ye YZ, Anderson P, Chen J, Accavetti MA, Tarpey MM and White CR. Extensive nitration of protein tyrosines in human atherosclerosis detected by immunohistochemistry. *Biol Chem Hoppe-Seyler* 1994; 375: 81-88.

Bolscher BG, Zoutberg GR, Cuperus RA and Wever R. Vitamin C stimulates the chlorinating activity of human myeloperoxidase. *Biochim Biophys Acta*, 1984; 784: 189-191.

Bos A, Wever R and Roos D. Characterization and quantification of the peroxidase in human monocytes. *Biochim et Biophys Acta* 1978; 525: 37-44.

Boxer LA, Axtell R. Suchard S. The role of the neutrophil in inflammatory diseases of the lung. *Blood Cells* 1990; 16: 25-42.

Buttery LD, Springall DR, Chester AH, Evans TJ, Standfield EN, Parums DV, Yacoub MH and Polak JM. Inducible nitric oxide synthase is present within human atherosclerotic lesions and promotes the formation and activity of peroxynitrite. *Lab Invest* 1996; 75: 77-85.

Chollet-Martin S, Montravers P, Gibert C, *et al.* Sub-population of hyperresponsive polymorphonuclear neutrophils in patients with adult respiratory distress syndrome. Role of cytokine production. *Am Rev Respir Dis* 1992; 146: 990-996.

Daugherty A, Dunn JL, Rateri DL and Heinecke JW. Myeloperoxidase, a catalyst for lipoprotein oxidation, is expressed in human atherosclerotic lesions. J Clin Invest 94 (1994): 437-444.

Denis M, Guojian L, Widmer M. A mouse model of lung injury induced by microbial products; implication of tumor necrosis factor. *Am J Resp Cell Molecul Biol* 1994; 10: 658-664.

Eiserich JP, Butler J, van der Vliet A, Cross CE and Halliwell B. Nitric oxide rapidly scavenges tyrosine and tryptophan radicals. *Biochem J* 1995; 310: 745-749.

Eiserich JP, Cross CE, Jones AD. Formation of nitrating and chlorinating species by reaction of nitrite with hypochlorous acid. *J Biol Chem* 1996; 271(32): 19199-19208.

Eiserich JP, Hristova M, Cross CE, Jones AD, Halliwell B, Freeman BA and van der Vliet A. Formation of nitric oxide derivatives catalysed by myeloperoxidase in neutrophils. *Nature* 1997 (in press).

Evans TJ, Buttery LD, Carpenter A, Springall DR, Polak JM and Cohen J. Cytokine-treated human neutrophils contain inducible nitric oxide synthase that produces nitration of ingested bacteria. *Proc Natl Acad Sci USA* 1996; 93: 9553-9558.

Fantone JC and Ward PA. Polymorphonuclear leukocyte-mediated cell and tissue injury: oxygen metabolites and their relations to human disease. *Human Path* 1985; 16: 973-8.

Farrell AJ, Blake DR, Palmer RMJ and Moncada S. Nitric oxide. *Ann Rheum Dis* 1992; 51: 1219-1222.

Forni LG, Kelly FJ and Leach RM. Radical approach to the acute respiratory distress syndrome. *Redox Report* 1997; 3: 85-97.

Fukuyama N, Takebayashi Y, Hida M, Ishida H, Ichimori K and Nakazawa H. Clinical evidence of peroxynitrite formation in chronic renal failure patients with septic shock. *Free Rad Bio Med* 1997; 22: 771-774.

Gaston B, Drazen JM, Loscalzo J and Stamler JS. The biology of nitrogen oxides in the airways. *Am J Respir Crit Care Med* 1994; 149: 538-551.

Gaston B, Reilly J, Drazen JM, Fackler J, Ramdey P, Arnelle D, Mullina ME, Sugarbaker DJ, Choe C, Singel DJ, Loscalzo J and Stamler JS. Endogenous nitrogen oxides and bronchodilator 5-nitrosothiols in human airways. *Proc Natl Acad Sci USA* 1993; 90: 10957-10961.

Govindaraju K, Cowley EA, Eidelman DH and Lloyd DK. Microanalysis of lung airway surface fluid by capillary electrophoresis with conductivity detection. *Anal Chem* 1997; 69: 2793-2797.

Gow AJ, Duran D, Malcolm S and Ischiropoulos H. Effects of peroxynitrite-induced protein modifications on tyrosine phosphorylation and degradation. *FEBS Lett* 1996; 385: 63-66.

Gow A, Duran D, Thom SR and Ischiropoulos H. Carbon dioxide enhancement of peroxynitrite-mediated protein tyrosine nitration. *Arch Biochem Biophys* 1996; 333: 42-48.

Gryglewski RJ, Palmer RM and Moncada S. Superoxide anion is involved in the breakdown of endothelium-derived vascular relaxing factor. *Nature* 1986; 320: 454-456.

Haddad IY, Pataki G, Hu P and Galliani C. Quantitation of nitrotyrosine levels in lung sections of patients and animals with acute lung injury. *J Clin Invest* 1994; 94: 2407-2413.

Haddad IY, Zhou S, Ischiropoulos H and Matalon S. Nitration of surfactant protein A results in decreased ability to aggregate lipids. *Am J Physiol* 1996; 270: L281-L288.

Halliwell B. What nitrates tyrosine? Is nitrotyrosine specific as a biomarker of peroxynitrite formation *in vivo*? *FEBS Lett* 1997, 157-160.

Hazell LJ, Arnold L, Flowers D, Waeg G, Malle E and Stocker R. Presence of hypochlorite-modified proteins in human atherosclerotic lesions. *J Clin Invest* 1996; 97: 1535-1544.

Hazen SL, Crowley JR, Mueller DM and Heinecke JW. Mass spectrometric quantification of 3-chlorotyrosine in human tissues with attomole sensitivity: a sensitive and specific marker for myeloperoxidase-catalyzed chlorination at sites of inflammation. *Free Rad Biol Med* 1997b; 23: 001-008.

Hazen SL, Heinecke JW. 3-chlorotyrosine, a specific marker of myeloperoxidase-catalyzed oxidation, is markedly elevated in low density lipoprotein isolated from human atherosclerotic intima. *J Clin Invest* 1997a; 99: 2075-2081.

Heflin AC Jr, Brigham KL. Prevention by granulocyte depletion of increased vascular permeability of sheep lung following endotoxemia. *J Clin Invest* 1981; 68: 1253-1260.

Hickey MJ, Sharkey KA, Sihota EG, Reinhardt PH, Macmicking JD, Nathan C and Kubes P. Inducible nitric oxide synthase-deficient mice have enhanced leukocyte-endothelium interactions in endotoxemia. *FASEB J* 1997; 955-964.

Huie RE and Padmaja S. The reaction of NO with superoxide. *Free Rad Res Comm* 1993; 18: 195-199.

Hunt J, Byrns RE, Ignarro LJ and Gaston B. Condensed expirate nitrite as a home marker for acute asthma. *Lancet* 1995; 346: 1235-1236.

Kaur H and Halliwell B. Evidence for nitric oxide-mediated oxidative damage in chronic inflammation. Nitrotyrosine in serum and synovial fluid from rheumatoid patients. *FEBS Lett* 1994; 350: 9-12.

Kettle AJ, Winterbourn CC. Myeloperoxidase: a key regulator of neutrophil oxidant production. *Redox Report* 1997; 3: 3-15.

Kharitonov SA, Chung KF, Evans D, O'Connor BJ and Barnes PJ. Increased exhaled nitric acid in asthma is mainly derived from the lower respiratory tract. *Am J Respir Crit Care Med* 1996; 153: 1773-1780.

Klebanoff SJ. Reactive nitrogen intermediates and antimocrobial activity: Role of nitrite. *Free Rad Biol Med* 1993; 14: 351-360.

Koh Y, Hybertson BM, Jepson EK and Repine JE. Tumor necrosis factor induced acute lung leak in rats: less than with interleukin-1. *Inflamm* 1996; 20: 461-469.

Kong S-K, Yim MB, Stadman ER and Chock PB. Peroxynitrite disables the tyrosine phosphorylation regulatory mechanism: Lymphocyte-specific tyrosine kinase fails to phosphorylate nitrated cdc2(6-20)NH$_2$ peptide. *Proc Natl Acad Sci USA* 1996; 93: 3377-3382.

Kono Y. The production of nitrating species by the reaction between nitrite and hypochlorous acid. *Biochem Molecul Biol Internat* 1995; 36: 275-283.

Kooy NW and Lewis SJ. The peroxynitrite product 3-nitro-L-tyrosine attenuates the hemodynamic responses to angiotensin II in vivo. *Eur J Pharmacol* 1996; 315: 165-170.

Kooy NW and Lewis SJ. Nitrotyrosine attenuates the hemodynamic effects of adrenoceptor agonists in vivo: Relevance to the pathophysiology of peroxynitrite. *Eur J Pharmacol* 1996; 310: 155-161.

Kooy NW, Lewis SJ, Royall JA, Ye YZ, Kelly DR and Beckman JS. Extensive tyrosine nitration in human myocardial inflammation: evidence for the presence of peroxynitrite. *Crit Care Med* 1997; 25: 812-819.

Kooy NW, Royall JA, Ye YZ, Kelly DR and Beckman JS. Evidence for in vivo peroxynitrite production in human acute lung injury. *Amer J Resp Crit Care Med* 1995; 151: 1250-1254.

Koppenol WH. Thermodynamic considerations on the formation of reactive species from hypochlorite, superoxide and nitrogen monoxide. Could nitrosyl chloride be produced by neutrophils and macrophages? FEBS Lett 1994; 347: 5-8.

Kushimoto S, Okajima K, Uchiba M, Murakami K, Okabe H and Takatsuki K. Pulmonary vascular injury induced by hemorrhagic shock is mediated by P-selectin in rats. *Thromb Res* 1996; 82: 97-106.

Lancaster, Jr. JR. Simulation of the diffusion and reaction of endogenously produced nitric oxide. *Proc Natl Acad Sci USA* 1994; 91: 8137-8141.

LeeuwenburghC, Hardy MM, Hazen SL, Wagner P, Oh-ishi S, Steinbrecher UP and Heinecke JW. Reactive nitrogen intermediates promote low density lipoprotein oxidation in human atherosclerotic intima. J Biol Chem 1997a; 272: 1433-1436.

Leeuwenburgh C, Rasmussen JE, Hsu FF, Mueller DM, Pennathur S and Heinecke JW. Mass spectrometric quantification of markers for protein oxidation by tyrosyl radical, copper and hydroxyl radical in low density lipoprotein isolated from human atherosclerotic plaques. J Biol Chem 1997b; 272: 3520-3526.

Lemercier J-N, Padmaja S, Cueto R, Squadrito GL, Uppu RM and Pryor WA. Carbon dioxide modulation of hydroxylation and nitration of phenol by peroxynitrite. *Arch Biochem Biophys* 1997; 345: 160-170.

Leone AM, Francis PL, Rhodes P and Moncada R. A rapid and simple method for the measurement of nitrite and nitrate in plasma by high performance capiallary electrophoresis. *Biochem Biophys Res Commun* 1994; 200: 951-957.

Louie S, Halliwell B, Cross CE. Adult respiratory distress syndrome: a radical perspective. *Advances in Pharmacol* 1997; 38: 457-491.

Lymar SV, Jiang Q and Hurst JK. Mechanism of carbon dioxide-catalyzed oxidation of tyrosine by peroxynitrite. *Biochem* 1996; 35: 7855-7861.

Malinski T and Taha Z. Nitric oxide release from a single cell measured *in situ* by a porphyrinic-based microsensor. *Nature* 1992; 358: 676-678.

Manktelow C, Bigatello LM, Hess D and Hurford WE. Physiologic determinants of the response to inhaled nitric oxide in patients with acute respiratory distress syndrome. *Anesth* 1997; 87: 297-307.

Martin BL, Wu D, Jakes S and Graves DJ. Chemical influences on the specificity of tyrosine phosphorylation. *J Biol Chem* 1990; 265: 7108-7111.

Martin TR. Leukocyte migration and activation in the lungs. *Eur Respir J* 1997; 10: 770-771.

Martin TR, Pistorese BP, Chi EY, Goodman RB and Matthay MA. Effects of B_4 in the human lung. Recruitment of neutrophils into the alveolar spaces without a change in protein permeability. *J Clin Invest* 1989; 84: 1609-1619.

Martin TR, Pistorese BP, Hudson LD and Maunder RJ. Function of lung and blood neutrophils in patients with the adult respiratory distress syndrome. Implications for the pathogenesis of lung infections. *Am Rev Respir Dis* 1991; 144: 254-262.

Mierzwa S and Chan SK. Chemical modification of human alpha 1-proteinase inhibitor by tetranitromethane. Structure-function relationship. *Biochem J*, 1987; 246: 37-42.

Moncada S and Higgs EA. Endogenous nitric oxide: Physiology, pathology, and clinical relevance. *Eur J Clin Invest* 1991; 21: 361-374.

Odeberg H, Olofsson T, Olsson I. Myeloperoxidase-mediated extracellular iodination during phagocytosis in granulocytes. *Scand J Haemat* 1974; 12: 155-160.

Okabayashi K, Triantafillou AN, Yamashita M, Aoe M, DeMeester SR, Cooper JD and Patterson GA. Inhaled nitric oxide improves lung allograft function after prolonged storage. *J Thor Cardiovasc Surg* 1996; 112: 293-299.

Olah GA, Malhorta R and Narang SC. Nitration: methods and mechanisms. VCH Publishers, New York, NY, 1989.

Panasenko OM, Briviba K, Klotz L-O and Sies H. Oxidative modification and nitration of human low-density lipoproteins by the reaction of hypochlorous acid with nitrite. Arch Biochem & Biophys 343 (1997): 254-259.

Pietraforte D and Minetti M. One-electron oxidation pathway of peroxynitrite decomposition in human blood plasma: evidence for the formation of protein tryptophan-centred radicals. *Biochem J* 1997; 321: 743-750.

Prutz WA, Monig H, Butler J and Land EJ. Reactions of nitrogen dioxide in aqueous model systems: Oxidation of tyrosine units in peptides and proteins. *Arch Biochem Biophys* 1985; 243: 125-134.

Riordan JF and Vallee BL. The functional roles of metals in metalloenzymes. *Adv Exper Med & Biol* 1974; 48: 33-57.

Repine JE, Beehler CJ. Neutrophils and the adult respiratory distress syndrome: two interlocking perspectives. *Am Rev Respir Dis* 1991; 144: 251-252.

Saleh D, Barnes PJ, Giaid A. Increased production of the potent oxidant peroxynitrite in the lungs of patients with idiopathic pulmonary fibrosis. *Am J Respir Crit Care Med* 1997; 155: 1763-1769.

Sampson JB, Rosen H and Beckman JS. Peroxynitrite-dependent tyrosine nitration catalyzed by superoxide dismutase, myeloperoxidase, and horseradish peroxidase. *Meth Enz* 1996; 269: 210-218.

Shasby DM, Shasby SS, Peach MJ. Granulocytes and phorbol myristate acetate increase permeability to albumin of cultures endothelial monolayers and isolates perfused lungs. *Am Rev Respir Dis* 1983; 127: 72-76.

Shayevitz JR, Rodriguez JL, Gilligan L, Johnson KJ and Tait AR. Volatile anesthetic modulation of lung injury and outcome in a murine model of multiple organ dysfunction syndrome. *Shock* 1995; 4: 61-67.

Shigenaga MK, Lee HH, Blount BC, Christen S, Shigeno ET, Yip H and Ames BN. Inflammation and No_x-induced nitration: Assay for 3-nitrotyrosine by HPLC with electrochemical detection. *Proc. Natl. Acad. Sci. USA* 1997; 94: 3211-3216.

Sinclair DG, Haslam PL, Quinlan GJ, Pepper JR and Evans TW. The effect of cardiopulmonary bypass on intestinal and pulmonary endothelial permeability. *Chest* 1995; 108: 718-724.

Torre D, Ferrario G, Speranza F, Orani A, Fiori GP and Zeroli C. Serum concentrations of nitrite in patients with HIV-1 infection. *J Clin Pathol (Lond.)* 1996; 49: 574-576.

413

Troncy E, Collet J-P, Shapiro S, Guimond J-G, Blair L, Charbonneau M and Blaise G. Should we treat acute respiratory distress syndrome with inhaled nitric oxide? *Lancet* 1997; 350: 111-112.

Ueda T, Maekawa T, Sadamitsu D, Oshita R, Ogino K and Nakamura K. The determination of nitrite and nitrate in human blood plasma by capillary zone electrophoresis. *Electrophoresis* 1995; 16: 1002-1004.

van der Vliet A, Eiserich JP, Halliwell B and Cross CE. Formation of reactive nitrogen species during peroxidase-catalyzed oxidation of nitrite. *J Biol Chem* 1997; 272(12): 7617-7625.

van der Vliet A, Eiserich JP, Kaur H, Cross CE and Halliwell B. Chapter 16, Nitrotyrosine as a marker for reactive nitrogen species. p. 175-184. In L. Packer (Ed.), Oxygen Radicals in Biological Systems, Part E: Nitric Oxide, Academic Press, Inc., San Diego, 1996.

van der Vliet A, Eiserich JP, O'Neill CA, Halliwell B and Cross CE. Tyrosine modification by reactive nitrogen species: a closer look. *Arch Biochem Biophys* 1995; 319: 341-349.

van der Vliet A, O'Neill CA, Halliwell B, Cross CE and Kaur H. Aromatic hydroxylation and nitration of phenylalanine and tyrosine by peroxynitrite. Evidence for hydroxyl radical production from peroxynitrite. *FEBS Lett* 1994; 339: 89-92.

Weiland JE, Davis WB, Holter JF, Mohammed JR, Dorinsky PM and Gadek JE. Lung neutrophils in the adult respiratory distress syndrome. Clinical and pathophysiologic significance. *Am Rev Respir Dis* 1986; 133: 218-225.

Weiss SJ, Lampert MB and Test ST. Long-lived oxidants generated by human neutrophils: characterization and bioactivity. *Science* 1983; 22: 625-628.

Westfelt UN, Benthin G, Lundin S, Stenqvist O and Wennmalm A. Conversion of inhaled nitric oxide to nitrate in man. *Brit J Pharmacol* 1995; 114: 1621-1624.

Yi D, Smythe GA, Blount BC and Duncan MW. Peroxynitrite-mediated nitration of peptides: characterization of the products by electrospray and combined gas chromatography-mass spectrometry. Arch Biochem & Biophys 1997; 344: 253-259.

Zhu S, Haddad IY and Matalon S. Nitration of surfactant protein A (SP-A) tyrosine residues results in decreased mannose binding ability. *Arch Biochem Biophys* 1996; 333: 282-290.

Zimmerman GA, Renzetti AD, Hill HR. Functional and metabolic activity of granulocytes from patients with adult respiratory distress syndrome. *Am Rev Respir Dis* 1983; 127: 290-300.

Zou M, Martin C and Ullrich V. Tyrosine nitration as a mechanism of selective inactivation of prostacyclin synthase by peroxynitrite. *Biol Chem* 1997; 378: 707-713.

Zuurbier KW, Bakkenist AR, Wever R and Muijsers AO. The chlorinating activity of human myeloperoxidase: high initial activity at neutral pH value and activation by electron donors. *Biochim Biophys Acta* 1990; 1037: 140-146.

HEALTH AND SCIENCE LEGISLATION IN THE UNITED STATES

William J. Martin II, M.D.[1,2]

[1]Floyd and Reba Smith Professor of Medicine
Division of Pulmonary, Critical Care and Occupational Medicine
Department of Medicine
Indiana University School of Medicine
Indianapolis, IN 46202-2879

[2]Health Policy Fellow, United States Senate (1995-1996)

INTRODUCTION

In the United States, the U.S. government through the legislative process and departmental structures greatly influences the future of health care and scientific investigation. Most clinicians and scientists are bewildered by the myriad complexities that effect health and science policy at the federal level; and as a result, we often remain amazingly passive as new policies and regulations evolve that greatly affect our professional lives. I say "amazingly passive" because in almost every other facet of our careers, clinicians and scientists are notoriously aggressive in pursuit of new knowledge and new approaches to problem solving. Although the reasons for this passivity are multifactorial, some relate to the relative lack of understanding of how the Federal government functions and how we as clinicians or scientists might positively influence the process.

With this in mind, I would like to provide an overview of how health and science policy evolves into specific laws and regulations at the federal level and how interested clinicians or scientists might become involved to better understand and to influence federal legislation and rule making.

For those of us from the U.S., much of what we need to know about the federal government, we learned in grade school. Unfortunately, together with state capitals, algebra and our designated "foreign language", this valuable information has long since faded from our memories. I will attempt to provide a succinct overview of the federal government and how it relates to health and science policy. The federal government consists of three branches: 1) the Executive branch with the President and various federal departments and agencies, 2) the Legislative branch including the Senate and the House of Representatives and 3) the Judicial branch consisting of the Supreme Court of the United States and the federal court system (Figure 1).

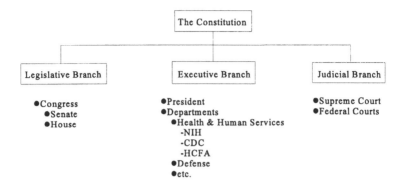

Figure 1. Overview of the U.S. federal government and examples for each branch of government.

In Article I, Section 1 of the United States Constitution, it states that "all legislative Powers herein granted shall be vested in a Congress of the United States, which shall consist of a Senate and House of Representatives". Each Congress lasts two years and is divided into two sessions numbered consecutively. The 105th Congress that began in January 1997 consists of 100 Senators and 435 members of the House.

CONGRESS AND ITS COMMITTEES

Budget Committee
- Sets fiscal policy for coming year
- Sets "functional priorities"
- Budget resolution sets out spending targets

Authorization Committee
- Creates programs and authorizes appropriations
- Writes laws creating entitlement
- Writes tax laws

Appropriations Committee
- Appropriates dollars for discretionary programs
  for fiscal year in 13 different appropriations bills

Figure 2. Examples of different types of congressional committees.

The development of health and science legislation occurs within the committee structure of the U.S. Senate and U.S. House of Representatives. Three broad categories of committees include: 1) budget committees, 2) authorization committees and 3) appropriation committees (Figure 2). Budget committees in both the Senate and House are typically busiest in the spring of the year setting fiscal policies and priorities for the coming year. Appropriation committees receive these functional priorities from the budget committees in April and May and have much of their task completed by July 1st. Actual passage of appropriation bills, however, typically takes place in September just prior to the next fiscal year beginning October 1st. In contrast, authorizing committees are those that create the new programs and authorize appropriations. For example, in the US Senate the Labor and Human

Resources Committee is responsible for the authorization of the National Institutes of Health (NIH) and the Center for Disease Control (CDC). The analogous committee in the U.S. House of Representatives is the Commerce Committee (Figure 3). The authorizing committees that deal with all tax-related health and science issues include the Finance Committee (Senate) and the Ways and Means Committee (House). However, the actual appropriations of funds to support these laws occurs within the Appropriations Committee prior to voting by the full House and Senate.

HOW HEALTH AND SCIENCE LEGISLATION EVOLVES

Senate
- Labor & Human Resources
- Finance
- Appropriations

House
- Commerce
- Ways and Means
- Appropriations

Figure 3. Major committees in both houses of Congress that have analogous legislative functions.

Following introduction of a bill or a resolution by a member of Congress, the bill is numbered with a prefix of HR (reference to House Resolution) or with an S (reference to the Senate). Usually, the bill is referred to the relevant committee. Failure of the committee or its sub-committee to act on a bill is equivalent to "killing it". In order to gain further information regarding the subject of a bill, the committee will frequently have hearings and solicit input from the public and experts. Following the various hearings and exhaustive review by appropriate committee staffs, the committee will initiate a "mark up" where the members of the committee will literally mark line by line all recommended changes or amendments. If the bill passes the committee, it is "reported out" then sent to the "floor" of the Senate or House. Further debate and amendments may occur at that time. The sponsors of a bill often dread amendments from the floor to an otherwise "clean bill" as this may further jeopardize the bills passage by either the Senate or the House. Most frequently, a successful bill in one chamber will have a similar "companion bill" that has been introduced in the other body. If both bills successfully pass, the minor differences are settled in "conference". The final bill must be approved by both the Senate and the House. The bill is then sent to the President for his signature. In the 104th Congress, a law was passed that permits the President to have a "line item veto" and thus the President under certain circumstances can select parts of the bill for passage and veto other parts.

HOW CONGRESS LEARNS ABOUT ISSUES

Congress has multiple ways to gain insight and information regarding specific legislative issues. As is apparent from above, Congressional committees and subcommittees can initiate any numbers of studies or hearings to learn more about legislative issues. Additionally, there are three congressional agencies which provide unbiased and nonpartisan information for congressional representatives and their staffs: 1) the Congressional Research

Services, 2) the General Accounting Office and 3) the Congressional Budget Office. The Congressional Research Services is part of the Library of Congress and represents a wonderful resource for a congressional office to seek information on virtually any topic. The General Accounting Office is the investigative arm of Congress and conducts various monitoring activities within federal agencies and initiates special studies at the request of individual members or committees of Congress. The Congressional Budget Office provides for Congess a budgetary analysis related to alternative fiscal and budgetary approaches to support programmatic needs. In recent years, this is frequently in the context of seeking alternative budgetary means to balance the federal budget.

Congressional offices frequently seek expertise and advice from federal agencies. Concerns about areas of inadequate scientific investigation might be referred to the NIH or CDC; whereas, issues related to drug approval process would be appropriately referred to the Food and Drug Administration (FDA). These agencies, as part of the Executive branch, have a carefully orchestrated interaction with members of Congress and with their respective congressional staffs. If a congressional office contacts a federal agency such as the FDA, NIH or CDC, one is mandated to go through the "Office of Legislative Affairs" at the agency so that each contact is documented and the type of information transferred to the congressional office has been approved by the specific agency. Finally, Congress and congressional staff frequently learn about issues through the highly organized process known as lobbying. Organized lobbying regarding health care or scientific issues can represent everything from specific business interests e.g. insurance, pharmaceuticals, or biotechnology to scientific organizations such as FASEB (Federation of American Societies for Experimental Biology) to patient advocacy groups such as the American Cancer Society or the American Lung Association.

Although lobbyists are often viewed in a disparaging way outside of the beltway, inside the beltway, lobbyists provide a highly important service to both members of Congress and congressional staffs. Effective lobbyists represent a wealth of important information and in most cases and steadfastly attempt to provide accurate and true information to Congress. Regardless of the specific agenda of a lobbyist or a lobbying organization, the accurate transfer of information to a member of Congress or to a congressional staff represents the best and often times the only means of establishing credibility. Providing inaccurate information that results in a loss of credibility is the worst possible outcome for a lobbyist. Thus, lobbyists often gain access to influence congressional actions based on a long and successful track record of providing good solid information, upon which legislation can be based.

SO YOU WANT TO TALK TO YOUR CONGRESSMAN

Whereas organizations and businesses often have access to Congress through their lobbyists, individuals may not. It is often intimidating to approach Congress regarding issues and one can easily feel as if a single voice will not have an impact. Do not believe it. Congressmen and their staffs will listen, especially to a member of their constituency. If you have the opportunity to visit with your own congressman, do your homework. It is important to understand what committee and sub-committee assignments your congressman has. Suppose you are interested in a health issue related to the Veterans Administration. Your Senator has a strong health background and serves on the Senate Finance Committee. If you do not have special access to this Senator, despite the Senator's interests and background in health issues, you would be wise to consider an alternative congressional office whose member serves on an appropriate sub-committee over seeing Veteran's activities. On the other hand, health care and scientific issues are often of a global nature and you may have an opportunity to discuss these with more than one congressman from your state. The better informed you are about the background of your congressman, the more likely you will be an effective communicator and your concerns will be effectively addressed in the congressional office.

Working with a congressional staff member can be intimidating and a bit mysterious. The role of the staff member is to inform his or her "boss", which is the term used on the Hill, and to protect the interest of the congressman. Even if you know the congressman personally, you are likely to deal with the staff member on an ongoing basis. It is frequently said in Washington that you "never bypass the staff", as any short term gain doing an "end run" around the staff, is likely to be a Pyrrhic victory. Virtually every successful lobbyist works hard to communicate to the staff and to work through the established chain of command in the particular office.

It is critical to know how to work with the congressional staff and to hopefully influence a specific congressman's thinking regarding a piece of health or science legislation. However, one difficulty clinicians and scientists have in dealing with federal legislation is that many of us have suboptimal communication skills. Perhaps it is a reflection of our impassioned views regarding the issues or perhaps we are used to telling people what to do. Nonetheless, health care providers and scientists can be perceived as arrogant and self righteous. It perhaps goes without saying, that you should always be courteous to everyone you meet or interact with in a congressional office. This includes everyone from the intern opening the mail to the Senior Staff Director. Learning to communicate effectively with a congressional office is paramount if you want to have a positive impact on legislation.

One of my most lasting impressions of working on the "Hill" is the high quality of the congressional staffs. Do not underestimate the ability of staff to quickly understand the key features of an argument and the impact this will have on the legislative process. Be appreciative that you are likely dealing with a knowledgeable audience. If you have the opportunity to visit with a "staffer" bring a single page of "talking points" that succinctly describes the issues and your perspectives regarding these issues. This is always appreciated. In addition, it provides a focus and allows you the opportunity to frame the discussion in terms of your perspectives.

Do not expect the staff to provide you with a definitive answer one way or the other. It is difficult for even seasoned staffers to express the opinion of their bosses accurately, and thus, they are very careful to couch their opinions with multiple qualifiers. Sometimes this is because the member has not yet made a public statement regarding a specific piece of legislation and remains in the information-gathering stage. It is important that you recognize that it is unlikely you will receive a yes or no answer to your questions. A good congressional staff, however, will provide you with meaningful feedback and guidance as to how an issue may evolve over the next several months.

Remember that influencing health and science legislation is an ongoing process. You should follow up your visit with a note or letter reiterating your perspectives and thanking the staff member for access to their valuable time. It is also imperative that you leave the staff member with the impression that you are available for any follow up discussions about this issue. Keep in mind that the congressional staff member is constantly in the process of solving problems and gathering information on an urgent basis. If you make yourself available (as all high quality lobbying groups do), you will be viewed as an important resource to help the congressional office deal with complex science or health related issues.

The most effective lobbyists constantly remind the staffs of their willingness to help on any issues unrelated to their own personal agendas. Individuals seeking to influence legislation should follow a similar course of action. The dividends may be great.

SUMMARY

By necessity this is a brief overview of the process of federal legislation. Detailed descriptions of how the federal government functions related to the important issues of health care and science is not possible using this format. If you are interested in learning more about federal legislation and the role of congressional staff, I would recommend the book Dance of Legislation by Eric Redman (published by Simon & Schuster November 1, 1988), at a cost of $11.00 U.S. If you are interested in a health or science policy fellowship appropriate information can be obtained directly from the Robert Wood Johnson Foundation and the American Association for the Advancement of Science. Both organizations offer an orientation as well as fellowship opportunities in congressional offices that provide you with a first hand experience of being a "staffer". Finally, if you would like to be more involved in the legislative process, you must take the initiative yourself. Remember the admonition from Plato, "The penalty that good men pay for not being interested in politics is to be governed by men worse than themselves".

INDEX